# Electronic Devices

| Device | Symbol | Model |
|---|---|---|
| NPN Bipolar Junction Transistor | B o—, C, E | **DC**: C, $\beta I_B$, $I_B$, B o—, .7 V, E<br>**AC**: $i_b$, B o—, C, $r_x$, $\beta i_b$, $r_o$, E |
| N-Channel Junction Field Effect Transistor | G o→, D, S | **AC**: G o—, D, $+$, $v_{gs}$, $-$, $g_m v_{gs}$, $r_d$, S |
| Operational Amplifier | $v_1$, $-$, $v_o$, $+$, $v_2$ | $v_1$, $-$, $v_i$, $r_i$, $+$, $v_2$, $r_o$, $v_o$, $+$, $A v_i$, $-$ |

# Introduction to Circuit Analysis

**Paul O. Neudorfer**
Seattle University

**Michael Hassul**
California State University, Long Beach

**Allyn and Bacon**
Boston    London    Sydney    Toronto

*This book is part of the* ALLYN AND BACON SERIES IN ENGINEERING
*Consulting Editor:* Frank Kreith, University of Colorado

*This book is dedicated*
 To my family, Cristine, Peter, and Anna—P. Neudorfer
 To my wife, Laurie, and in memory of my brother Steve—M. Hassul

Copyright © 1990 by Allyn and Bacon
A Division of Simon & Schuster, Inc.
160 Gould Street
Needham Heights, Massachusetts 02194

**Library of Congress Cataloging-in-Publication Data**

Neudorfer, Paul O.
 Introduction to circuit analysis / Paul O. Neudorfer, Michael
Hassul.
  p. cm.
 Includes index.
 ISBN 0-205-11373-7
 1. Electric circuit analysis.  I. Hassul, Michael.  II. Title.
TK454.N48 1989
621.381′32--dc19             89-30847
                       CIP

Printed in the United States of America
1  2  3  4  5  6  7  8  9  10—94  93  92  91  90  89

# Contents

* Optional section.

_____

* Optional section.

# Preface

The intent of this book is to introduce engineering students and other interested readers to the topic of electrical circuit analysis and to the discipline of electrical engineering in general. It is expected that most readers will be electrical engineering students in their sophomore or junior years. Some students may have had previous courses in electrical engineering, but for most, this text will be their first exposure to the profession. Students should have already studied differential and integral calculus as well as college physics. A background in additional topics such as differential equations will be helpful but not necessary. Whenever possible, the underlying mathematics and physics used in this book are described in enough detail that students who lack the particular background expected should be able to follow along without too much difficulty.

In most engineering schools, electrical circuit analysis is covered in a year-long sequence of courses. The sequence can consist of either two semester-length courses or two or three quarter-length courses, depending on how the school's academic calendar is organized. In semester schools, we expect Chapters 1 through 8 of the book to be covered in the first term and Chapters 9 through 14 in the second. For schools on the quarter system, a reasonable break-down would be Chapters 1 through 7, followed by 8 through 10, and then 11 through 14.

The order of presentation of topics in this book reflects our own preferred way of teaching circuit analysis. We do not claim that ours is the uniquely "best" approach. In fact, there are several very reasonable and equally valid ways of organizing the various topics of circuit analysis into a course of study. Debating their relative merits almost always generates a lively discussion among those who teach the subject. Accordingly, our goal has been to develop the book such that after Chapter 7, the order of presentation of the chapters can be changed to fit the particular needs of a wide variety of curricula and instructors. For further flexibility, several sections of the book have been marked with asterisks. These sections cover topics that we feel are of secondary importance and can be passed over without a loss of continuity to allow more time for other topics.

Electrical engineering education is under constant pressure to include the topics of a rapidly expanding technology within the framework of its traditional four-year degree program. There has been much discussion within the profession as to whether, in fact, the study of electrical engineering should be increased to a five-year program. At present, however, it seems certain that the four-year model will remain with us for the foreseeable future. Electrical engineering educators, then, are challenged to be creative in packaging an ever-increasing range of topics into a curriculum of fixed size. One response to this challenge has been to combine the studies of circuit analysis and electronics into one unified sequence. The hoped-for savings in overall time can then be used for other topics. We see much merit in this approach, but think that even if it becomes widely accepted, its implementation is several years away. We have chosen, instead, an intermediate approach. This book does not formally cover electronics in any unified way, but it does introduce electronics concepts where appropriate. For instance, operational amplifiers and various circuit models for transistors are used in examples of circuit analysis. In addition, concepts such as amplifier feedback and circuit loading, which are useful in electronics, are introduced.

In our experience, many topics of electrical engineering are fully appreciated only after several exposures. In this book, then, some topics are returned to from time-to-time rather than being treated in full detail in just one place. With each revisit to one of these topics, the student's general understanding of circuit analysis will have improved, and the student will be in a position to more completely understand the topic at hand. Operational amplifiers are an example of this type of treatment. Rather than restricting them to one chapter, their treatment is distributed throughout the book. They are first introduced in Chapter 2, where the student will gain the minimum background needed to understand their basic characteristics. Op amps are returned to in later chapters as students' analytical tools increase to the point that they can appreciate more fully the behavior and uses of these devices.

Each chapter of this book has a number of worked-out examples that demonstrate how the theory of circuit analysis is applied to specific problems. In addition, the end of each chapter contains problems of varying levels of difficulty that you can use to test your understanding of the material. The answers to many of these problems can be found at the end of the book. Also at the end of the book are several appendices that cover expected mathematical background and useful formulas.

**Acknowledgements:** A project of the magnitude of a textbook can only in the shallowest sense be considered to be a product of the efforts of its authors alone. Certainly, we are greatly indebted to our editors at Allyn and Bacon, Ray Short and Paul Solaqua. We also wish to thank many of our colleagues for their generous and inciteful comments and suggestions. These include Francis P. Wood, S.J., Hassan Babaie, Gordon Carpenter, Gene Hostetter, Ronald A. Rohrer, M. E. Van Valkenburg, Dalia Arbel, Narottam Shrestha, and Kevin DeAngelis. We are appreciative of the following individuals who reviewed and commented on parts or all of the manuscript: Wayne A. Anderson, University of Buffalo; Bennett L. Bassore, Oklahoma State University; William Russell Callen, Georgia Institute of Technology; Charles M. Close, Rensselaer Polytechnic Institute; James F. De-lansky, The Pennsylvania State University; William J. Eccles, University of South Carolina; Richard S. Gallagher, Kansas State University; R. W. Gilchrist, Clemson University; J. M. Googe, University of Tennessee–Knoxville; Donald W. Howe, Worcester Polytechnic Institute; Chin S. Hsu, Washington State University; Ralph A. Kinney, Louisiana State University; Frank Kreith, Solar Energy

Research Institute; Edward F. Kuester, University of Colorado–Boulder; T. N. Lee, George Washington University; Robert N. Martin, Northeastern University; Stephen Riter, University of Texas at El Paso; Charles E. Smith, University of Mississippi; John P. Stahl, Ohio Northern University; Hannis W. Thompson, Jr., Purdue University; and Leonard W. Weber, Oregon State University. These colleagues have individually and collectively steered us clear of many a pitfall. We are also indebted to the staff of Technical Texts, Inc., who guided us through the involved process of transforming a manuscript into a book. Finally, we are most thankful to our students who over the years have motivated our teaching.

Paul Neudorfer
Michael Hassul

# Electrical Circuit Fundamentals

1

## 1.1   INTRODUCTION

This textbook is probably your introduction to the field of electrical engineering. Electrical engineering is a diverse field with many areas of specialization. Regardless of specialty, however, the work of any engineer is to design devices and processes to perform specific tasks. For example, electronic engineers design and build transistorized devices such as stereo amplifiers; power engineers build large systems for generating and delivering electrical power; computer engineers design hardware to perform logical operations; communication engineers build radio transponders and also develop the techniques for processing transmitted information; and control engineers design and build devices that automatically regulate the operation of systems. The list could go on and on. These projects can be very complicated, requiring the cooperation of specialists from the many fields of electrical engineering as well as from other areas of science and engineering.

Common to most specialties of electrical engineering is the use of electrical circuits. Electrical circuits are interconnections of the electrical components that are used to build the various electronic, communication, computer, and control devices. It follows that understanding electrical circuits is an essential prerequisite for all who would be electrical engineers. While design is the end goal of most engineering, you must first learn the fundamentals. The design of electrical circuits is a sophisticated undertaking that requires both a thorough knowledge of circuit behavior and an intuitive "feel" for how circuit elements can be combined to perform specific tasks. The understanding of circuit behavior comes from learning how to mathematically model the circuit elements and their interconnections. This process is known as analysis. The intuitive feel, or insight, required of a designer comes from analyzing the behavior of many different circuits.

While the analysis of circuits is a prerequisite to their design, analysis should not be thought of as a mere stepping stone to that end. Analysis is important in its own right; the analytical techniques you will learn from this textbook are applicable to many systems and are not limited to electrical circuits.

The goal here is to introduce you to most of the commonly used techniques of circuit analysis. As you might suspect from the length of the book, this is a big task but one that is not beyond your capability. The subject is presented in a step-by-step manner, starting with the basic concepts and working toward more difficult ones. As you progress, you will be pleasantly surprised with the depth of your understanding of a subject that now might seem rather challenging.

Mastering a subject such as electrical circuit analysis is similar to learning a foreign language. We first begin with the basic vocabulary and then learn how to form sentences. Final mastery comes when we are able to express sophisticated ideas. The vocabulary of electrical circuits consists of the basic elements used to construct circuits and the variables that exist within them. The primary physical variables that we use in circuit analysis are voltage and current, but charge, power, and energy are also considered.

The circuit diagram shown in Figure 1.1A represents a lighting system. Analysis of this circuit requires the use of a simple algebraic equation, which you

FIGURE 1.1 *Example Circuits*

FIGURE 1.1  *Example Circuits*

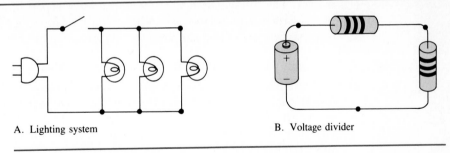

A.  Lighting system

B.  Voltage divider

FIGURE 1.2  *741 Operational Amplifier (Courtesy of National Semiconductor Corp., Santa Clara, California)*

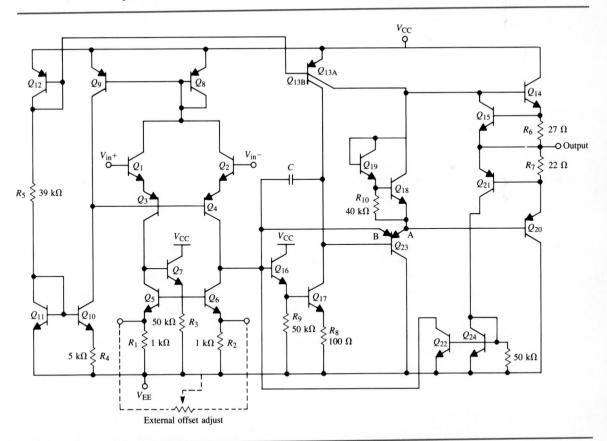

will learn later in this chapter. The circuit in Figure 1.1B is known as a voltage divider. The voltage divider frequently appears as part of a larger circuit. Analysis of this circuit requires the use of the circuit laws introduced in Chapter 2. The use of these laws leads to a simple set of algebraic equations that are easily solved by hand.

The circuit shown in Figure 1.2 represents an operational amplifier, one of the most useful modern electronic circuits. This circuit is very complicated and contains many transistors. However, by using the analytical techniques of Chapter 3 and the device models of this chapter, you will be able to write a set of equations that describe the basic behavior of this circuit. Because of the size of this circuit, a large set of algebraic equations results, and a computer or calculator program is necessary to obtain the desired answers. Chapter 4 will introduce additional techniques that help to simplify circuit analysis.

The response of the circuits shown in Figure 1.1 and the basic behavior of the operational amplifier shown in Figure 1.2 can be described with algebraic equations. Chapters 5 through 7 introduce circuit elements and circuits whose analyses require the use of differential equations. Later chapters will teach you some very powerful tools for analyzing and designing circuits. These mathematical tools will form the bases for most of the engineering analyses you will encounter in later courses and in engineering practice.

## 1.2 SYSTEM OF UNITS

All observed quantities in any system, such as the signals described in this textbook, are measured in some agreed-upon units. The units used in electrical engineering today are based on the SI (System Internationale) units. The SI units are similar to those in the metric system. For example, length is measured in meters. To save space, we often use symbols to represent quantities and their units. For instance, the symbols for length and meters are $l$ and m, respectively. Table 1.1 shows some common physical quantities, their units of measurement, and their symbols.

Some of the units shown in Table 1.1 have magnitudes that are typically very small (e.g., $1 \times 10^{-12}$) or very large (e.g., $1 \times 10^{12}$). To avoid writing such small or large numbers, we attach commonly accepted prefixes to the units. For example, 1,000 m is abbreviated as 1 km, where the prefix *kilo-* is symbolized by k and represents one thousand. Table 1.2 lists the standard prefixes, the multiplying power of ten they represent, and their symbols.

You should be careful when you interpret symbols. You should be aware of the context in which the letter is used.

## 1.3 ELECTRICAL VARIABLES

In mechanical systems, the variables that we usually use are force, displacement, velocity, and acceleration. Velocity and acceleration are, of course, the first and

**TABLE 1.1**   *SI Quantities, Units, and Symbols*

| Quantity | Symbol | Units | Symbol |
|----------|--------|-------|--------|
| Length | $l$ | Meter | m |
| Mass | $M$ | Kilogram | kg |
| Time | $t$ | Second | s |
| Electric current* | $I$ | Ampere | A |
| Temperature | $T$ | Kelvin | K |
| Force | $F$ | Newton | N |
| Energy* | $W$ | Joule | J |
| Power* | $P$ | Watt | W |
| Electric charge* | $Q$ | Coulomb | C |
| Potential difference* | $V$ | Volt | V |
| Resistance | $R$ | Ohm | $\Omega$ |
| Capacitance | $C$ | Farad | F |
| Inductance | $L$ | Henry | H |
| Magnetic flux | $\Phi$ | Weber | Wb |

* Electrical engineers often use lowercase symbols if these quantities are time varying.

second derivatives of displacement. The variables that we use to describe the behavior of electrical circuits are charge, current, and voltage. Since you should already be familiar with these variables, only a brief description of them and some related variables is given next.

**TABLE 1.2**   *Standard Prefixes of Units*

| Prefix | Power of Ten | Symbol |
|--------|-------------|--------|
| Exa | $10^{18}$ | E |
| Peta | $10^{15}$ | P |
| Tera | $10^{12}$ | T |
| Giga | $10^{9}$ | G |
| Mega | $10^{6}$ | M |
| Kilo | $10^{3}$ | k |
| Milli | $10^{-3}$ | m |
| Micro | $10^{-6}$ | $\mu$ |
| Nano | $10^{-9}$ | n |
| Pico | $10^{-12}$ | p |
| Femto | $10^{-15}$ | f |
| Atto | $10^{-18}$ | a |

## Charge

The basic building block of our physical universe is the atom. As you know, the atom itself is composed of protons and neutrons bound together at the center of the atom and electrons that circle this nucleus. Two protons, or two electrons, will repel each other. An electron and a proton will attract each other. Two neutrons will neither repel nor attract each other. Also, neutrons have no effect (other than gravitational) on, and are not affected by, electrons or protons.

The forces that exist between electrons and protons and that cause repulsion and attraction are electrical forces. Electrical force is one of the four recognized fundamental forces of the universe. Particles that are affected by electrical forces are described by the quantity **electrical charge,** or simply charge. Protons have one type of charge, which, by convention, is considered to be positive; electrons have a negative charge. Neutrons have no charge and are electrically neutral.

The symbol used to indicate charge is $Q$. Electrical engineers use the lowercase $q(t)$ for time-varying charge. This follows the general convention that uppercase symbols represent constant quantities, while lowercase symbols are used for quantities that may vary with time. The unit of charge is the coulomb, abbreviated with the letter C and named after the French physicist Charles A. de Coulomb (1736–1806), who studied the relationships between force and charge. It takes $6.242 \times 10^{18}$ electrons to make $-1$ C of charge!

## Current

While static electricity (i.e., stationary charge) is useful for parlor games such as getting balloons to stick to the ceiling and in some industrial applications such as photocopying, most electrical devices require moving charges for their operation. Moving charges are measured as **electrical currents.** The symbols for current are $I$ and $i(t)$, and its unit is the ampere, named after the French physicist André M. Ampère (1775–1836) and abbreviated as A. The relationship between current and charge in motion is given by

$$i(t) = \frac{dq(t)}{dt}$$

$(1.1)$

We see that current is the derivative or time rate of change of charge.

For a current to exist in a material, it is necessary that the material allows charge to move through it. For example, charge moves easily through metal but not through glass and wood. Also, for charge to flow at the energy levels considered in this text, a *closed path*, as shown in Figure 1.3A, is required. Figure 1.3B shows a conductor in an incomplete path. If charges are moving from left to right in this figure, then where do they go when they reach the end? At the energy levels considered in this textbook, they cannot go anywhere. Therefore, charge

FIGURE 1.3 *Electrical Current*

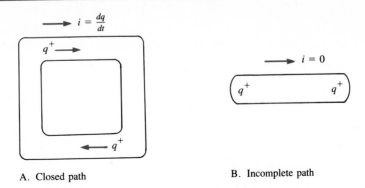

A. Closed path                    B. Incomplete path

builds up at this end. As this charge builds up, a force is created that prevents additional charges from moving toward this end. An equilibrium state is very quickly reached in which there is no net flow of charge.

We have used the term electrical circuit several times already in this textbook but have not yet formally defined it. An **electrical circuit** is any collection of electrical elements and devices that can support a current. This definition is somewhat limiting since it does not include as a circuit a collection of electrical elements that do not form a closed path. A more general definition of a circuit is simply *an interconnection of electrical elements*. This is the definition we will use. With this latter definition, the term **network** is synonymous with *circuit*, and we will use the terms interchangeably.

Example 1.1     *Finding Current*

a.   If $q(t) = 3t^2 + 2t - 1$ C, then $i(t) = 6t + 2$ A.
b.   See Figure 1.4 for a graphical example.

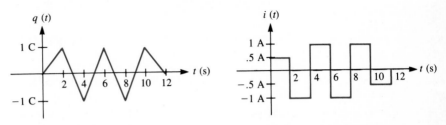

FIGURE 1.4

It is very common to talk of a current flowing or passing through some device. This wording is not technically correct. It is charge that moves or flows, not current. Current is a measure of charge flow and does not itself flow.

If current is the time derivative of charge, then charge can be found by integrating current with respect to time:

$$q(t) = \int_{-\infty}^{t} i(t) \, dt \qquad (1.2)$$

To be mathematically correct, $i(t) \, dt$ should be written with a dummy variable, for example, $i(x) \, dx$. However, when there can be no confusion, we will use the simpler notation as shown.

To evaluate the integral in Equation 1.2 requires knowledge of the current at $t = -\infty$. Since none of us were around at the beginning of the universe, it is unlikely that the current at that time is known. Rather, we know the current and charge at some time in the recent past, say, at the time $t_0$. We can rewrite the integral in Equation 1.2 as

$$q(t) = \int_{t_0}^{t} i(t) \, dt + q(t_0) \qquad (1.3)$$

The term $q(t_0)$ is the **initial value** of $q(t)$.

---

**Example 1.2**    *Finding Charge from Current*

a.   Given $i(t) = 3t - 10$ A and $q(0) = 3$ C, find $q(5)$. First find $q(t)$ and then evaluate at $t = 5$. Since the charge is known at $t = 0$, integrate the current from this time and use the initial value:

$$q(t) = \int_{0}^{t} i(t) \, dt + q(0) = \int_{0}^{t} (3t - 10) \, dt + 3 \text{ C}$$

$$= \left( \frac{3}{2} t^2 - 10t \right) \Big|_{0}^{t} + 3 = \frac{3}{2} t^2 - 10t + 3 \text{ C}$$

Thus,

$$q(5) = -9.5 \text{ C}$$

b.   See Figure 1.5 for a graphical example where $q(0) = 0$ C.

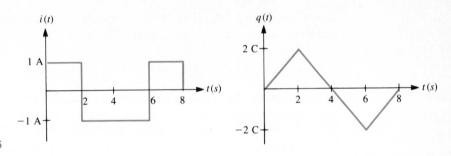

FIGURE 1.5

We have now defined charge in motion as an electrical current. However, the derivative relationship given in Equation 1.1 for current does not tell us in what direction the charges are moving. It also does not tell us whether positive or negative charges are in motion. In fact, current has direction as well as magnitude. Current direction is indicated by an arrow, as in Figure 1.3. By convention, the arrow indicates the direction of flow that positive charges, if present, would take.

It may be of some interest to you to know that this convention is due to the American statesman and scientist Benjamin Franklin (1706–1790). He held to the "one fluid theory" prevalent at the time that described electricity in terms of only one type of charge. Bodies that exhibited electrical properties were thought to have either an excess (plus) or deficit (minus) of this fluid. Fluid would naturally flow from excess to deficit—hence, the convention.

Figure 1.6 shows some of the possible charge flows that can exist in a section of a closed circuit. Let's assume the magnitude of the current in each case is the same and is indicated by $I$. The commonly used convention of assigning the current arrow in the direction of positive charge flow results in the direction

FIGURE 1.6  *Charge Movement and Current Direction*

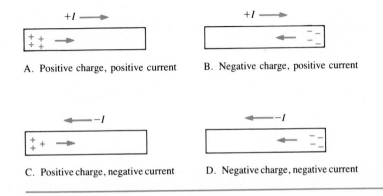

A. Positive charge, positive current

B. Negative charge, positive current

C. Positive charge, negative current

D. Negative charge, negative current

shown in Figure 1.6A. How do you explain the arrow direction assigned to the negative charge case in Figure 1.6B? The answer is that positive charges moving to the right or negative charges moving to the left create the same net effect of making the right side more positive than the left.

It is not necessary to know the actual direction of charge flow before you assign a current direction. For instance, in Figure 1.6C, a current direction from right to left has been assigned. Even though positive charges are moving left to right, the assigned current direction does *not* lead to a wrong answer. We simply get a value of $-I$ for the current in this case. Likewise, the assigned current direction in Figure 1.6D leads to an answer of $-I$ for the current.

Mathematically, a current of $+1$ A to the right is exactly the same as $-1$ A to the left. It is very important that you understand that there is no wrong way to assign current direction. The assigned current direction along with the algebraic sign of the current will result in a correct answer.

## Voltage

We know that forces exist between charges at rest and between charges in motion (current). For circuit analysis, it is rather cumbersome to write equations relating current and force. Instead, we define a new variable, voltage, which is related to both electrostatic and magnetic energies. Since the exact relationship between voltage and energy is rather complicated, only a simplified explanation will be given here.

In Figure 1.7, the charge $Q_a$ has been moved from infinity to point a, and the charge $Q_b$ has been moved from infinity to point b. To move these charges, we had to supply enough energy to overcome the repelling forces due to the charge at the reference point, $Q_{ref}$. We are assuming that the force between $Q_a$ and $Q_b$ is negligible. The potential energies (*PE*) gained in moving $Q_a$ and $Q_b$ are given by

$$PE_a = \frac{KQ_aQ_{ref}}{R_a} \quad \text{and} \quad PE_b = \frac{KQ_bQ_{ref}}{R_b}$$

FIGURE 1.7   *Potential Energy and Voltage*

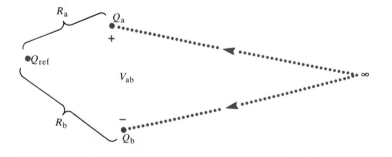

where $R_a$ and $R_b$ are, respectively, the distances from points a and b to the reference and $K$ is a constant related to the permittivity of free space. Energy has units of joules (J).

We define the **potential** at a point as the potential energy gained in moving a charge from infinity to the point, divided by the magnitude of the charge. Potential is also known as *voltage*, which has the symbol $V$. The potentials, or voltages, at points a and b are, therefore, given by

$$V_a = \frac{KQ_{ref}}{R_a} \quad \text{and} \quad V_b = \frac{KQ_{ref}}{R_b}$$

Potential has units of volts (V), where 1 volt = 1 joule/coulomb. The Italian physicist Alessandro C. Volta (1745–1827) constructed the first battery.

We are most often interested in the potential difference between the two points in a circuit, for example, the points a and b in Figure 1.7. The potential difference, or **voltage difference,** between the two points is denoted as $V_{ab}$ and is given by

$$V_{ab} = V_a - V_b \tag{1.4}$$

The plus and minus signs in Figure 1.7 indicate that we assume the potential at point a is more positive than the potential at point b. If this assumption turns out not to be the case, then $V_{ab}$ will be negative. With this assigned *polarity*, the voltage $V_{ab}$ is the voltage *rise* from point b to point a. Alternatively, $V_{ab}$ can be considered the voltage *drop* from point a to point b. Note that a voltage rise of +5 V is equal to a voltage drop of −5 V.

Energy is also imparted to the charges moving in a conductor by time-varying magnetic fields. If this magnetic energy is divided by the charge it is acting on, we again get a voltage, or an *electromotive force* (emf).

In analyzing circuit behavior, it does not matter whether voltage is electrically or magnetically produced. All voltages are treated alike in circuit equations.

## Energy and Power

We already briefly mentioned *electrical work*, or *energy*, and defined voltage based on it. The symbol used for energy is $w(t)$, and we measure energy in units of joules, abbreviated as J and named after the English physicist James P. Joule (1818–1889). **Power** is the rate at which work is done and is defined as

$$p(t) = \frac{dw(t)}{dt}$$

To define power in terms of electrical variables, we rewrite this equation as

$$p(t) = \frac{dw(t)}{dq} \cdot \frac{dq(t)}{dt}$$

We recognize $dw/dq$ as an incremental voltage and $dq/dt$ as a current and write electrical power as

$$p(t) = v(t) \cdot i(t) \qquad (1.5)$$

Power has units of watts (W), where 1 watt = 1 volt-ampere. The unit is named after the Scottish inventor James Watt (1736–1819).

Given power, we can find energy by integration:

$$w(t) = \int_{-\infty}^{t} p(t) \, dt = \int_{t_0}^{t} p(t) \, dt + w(t_0) \qquad (1.6)$$

where once again a dummy variable should properly be used in the integrand. It can be seen that the unit of electrical energy can also be written as the watt-second, where 1 watt-second = 1 joule. When you pay your electric bill, you are paying for kilowatt-hours; customers of a power company pay for energy consumed, not power.

More will be said about energy and power as the various circuit elements are introduced in this chapter. Chapter 10 discusses power and energy in even greater detail.

## 1.4   CIRCUIT ELEMENTS AND MODELS

Now that the variables used to describe electrical circuit behavior have been introduced, some of the basic elements that are used to construct circuits can be described.

The box in Figure 1.8 represents a general circuit element with two external points of connection known as *terminals*. We assume that the element shown is part of a closed circuit so that current may exist in it. The two terminals form what is known as a *port*. The behavior of any two-terminal element is completely described by the relationship between the current and voltage at the device port. This relationship, called the *V–I characteristic*, can be modeled mathematically or graphically.

You can see that we have assigned the current direction so that positive charges are assumed to move from the positive terminal to the negative terminal within the device. For the assigned current direction and voltage polarity, the energy dissipated or stored in this device is given by Equation 1.6 and is repeated

**FIGURE 1.8** *Passive Sign Convention*

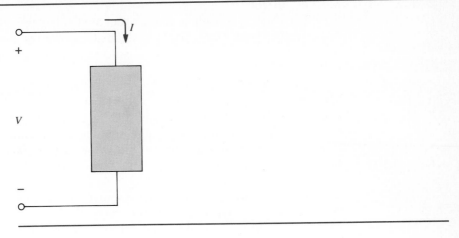

here in terms of electrical variables:

$$w(t) = \int_{-\infty}^{t} p(t)\, dt = \int_{-\infty}^{t} v(t) \cdot i(t)\, dt \qquad (1.7)$$

If the energy given by Equation 1.7 is always positive, then the device is *passive* and either dissipates or stores the energy. If the device is *active*, the energy will be negative. Negative energy dissipated means that energy is actually being delivered to some other circuit element or elements.

Since the convention of assigning a current entering at the positive terminal results in positive energy for passive devices, it is known as the **passive sign convention.** We will follow this convention for all devices in this textbook.

## Ideal Independent Energy Sources

For passive elements to be useful, they must be supplied with energy. The devices that supply energy to an electrical circuit are **independent sources.** The importance of the term *independent* will become clear later in this chapter. There are two types of ideal sources that circuit analysts recognize: a voltage source and a current source.

Voltage and current sources are obtained by converting various forms of energy to electrical energy. Batteries convert chemical energy to electrical energy. Mechanical energy is converted to electrical energy by use of generators or dynamos (alternators in automobiles). Batteries and generators provide an independent source of voltage. Current sources are constructed with electronic circuitry that converts the output of voltage sources into the necessary current. In addition, heat, light, and pressure can be applied to semiconductor devices to produce current.

The ideal independent voltage source ($V_s$) has the circuit symbol shown in Figure 1.9A. Note that current is defined as entering the plus terminal of the source, which is consistent with the passive sign convention that was introduced earlier.

The $V–I$ characteristic of the ideal voltage source is shown in Figure 1.9B. This characteristic shows that the voltage across an ideal voltage source does not depend on the current in it. An ideal voltage source can deliver any amount of current to the circuit it is supplying. Another way to explain this characteristic is that an ideal voltage source constrains the voltage between its two terminals to the value of the voltage source, no matter what else is connected to these terminals.

The ideal independent current source ($I_s$) and its $V–I$ characteristic are shown in Figures 1.10A and 1.10B, respectively. Again, note the use of the passive sign convention. An ideal current source will supply the circuit with a current that is not dependent on the voltage across the current source or the circuit elements to which it is connected.

There are, of course, no ideal sources. All practical voltage sources are limited in the current they can supply, and all practical current sources are limited in the voltage they can supply. In other words, all practical sources can supply only a finite amount of power. In addition, practical sources have $V–I$ characteristics that are not purely vertical or horizontal (Figure 1.11). Thus, voltage across and current in a practical source are not independent of each other. However, in this text, the symbols shown in Figures 1.9 and 1.10 will always indicate an ideal voltage or current source. If we want to accurately model a practical source, we will add additional components to the ideal source.

Energy sources have classically been divided into two types depending on their behavior with respect to time. Many sources, such as batteries, produce a

**FIGURE 1.9**   *Ideal Voltage Source*

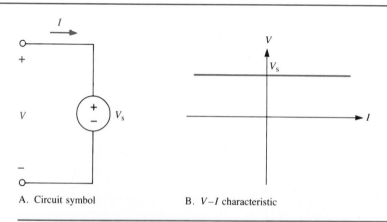

A. Circuit symbol                    B. $V–I$ characteristic

FIGURE 1.10 *Ideal Current Source*

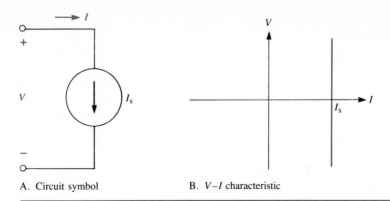

A. Circuit symbol  B. *V–I* characteristic

voltage or a current that is constant—that is, that does not vary with time. These sources are known as direct current, or DC, sources. A typical ideal battery voltage as a function of time is shown in Figure 1.12A. Note that Figure 1.12A and Figure 1.9B are not the same. In Figure 1.9B, voltage is plotted as a function of current. In Figure 1.12, voltage is plotted as a function of time.

Special symbols are sometimes used to indicate DC, or constant, sources. These symbols are shown in Figure 1.12B for a voltage source (battery) and in Figure 1.12C for a current source. In Figure 1.12B, the long line always represents the positive terminal. Although there are special symbols for DC sources, they are often represented with the standard symbols of Figures 1.9A and 1.10A. This is especially true for the current source.

The voltage that is supplied to a home is generated by rotating coils of wires in magnetic fields, which produce a voltage that varies sinusoidally with time, as

FIGURE 1.11 *Practical Source V–I Characteristics*

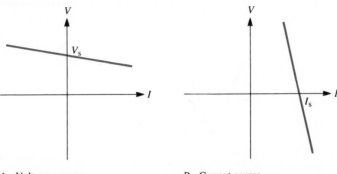

A. Voltage source  B. Current source

**FIGURE 1.12**    *Constant, or DC, Sources*

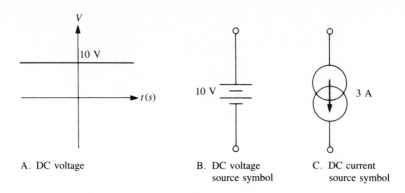

A. DC voltage          B. DC voltage            C. DC current
                          source symbol              source symbol

shown in Figure 1.13A. This type of source is known as an alternating current, or AC, source. The AC voltage source symbol is shown in Figure 1.13B. There is no special symbol for an AC current source.

Although the term *AC* technically implies that a sinusoidal signal is being produced, we generally refer to any source that produces a time-varying signal (as opposed to a constant signal) as an AC source. Some of the signals that we will study in this text are constants, exponentials, ramps, and sinusoids. Figures 1.14A, 1.14B, and 1.14C show examples of some of these functions. We will discuss each signal in more detail as required in our studies.

In electrical circuits, a variable that is a pure DC signal is commonly written as an uppercase letter with an uppercase subscript, for example, $I_E$, $I_B$, and $V_{BE}$. A variable that is a pure AC signal is written as a lowercase letter with a lowercase subscript, for example, $i_e$, $i_b$, and $v_{be}$.

**FIGURE 1.13**    *Sinusoidal, or AC, Sources*

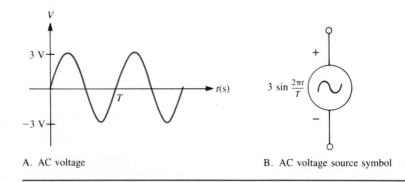

A. AC voltage                                    B. AC voltage source symbol

**FIGURE 1.14** *Time-Varying Signals*

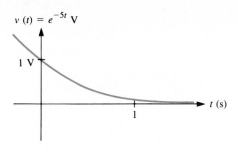

A. Exponential

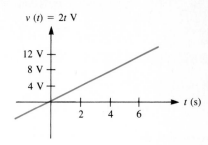

B. Ramp

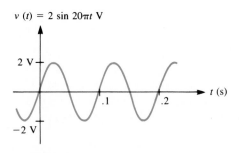

C. Sinusoid

Power absorbed by voltage and current sources is given simply by

$$p_s(t) = v_s(t) \cdot i_s(t)$$

For a voltage source, $v_s(t)$ is set by the source and $i_s(t)$ is determined by the circuit. For a current source, $i_s(t)$ is known and $v_s(t)$ is determined by the circuit.

Negative power absorbed by a device means that the device is actually supplying power. In circuits with more than one source, some sources may be delivering power ($p_s(t) < 0$), while others are absorbing power ($p_s(t) > 0$). For example, an automobile has two voltage sources: the battery and the alternator. When you first start a car, the battery is delivering power to the cranking system and the ignition circuit. Once the motor is running, the alternator produces a voltage that supplies power to the spark plugs and also charges the battery. At this time, the battery is absorbing power.

**Example 1.3** *Source Power and Energy*

Given the electrical circuit shown in Figure 1.15, first find the power

(*continues*)

**Example 1.3**   *Continued*

absorbed by the voltage and current sources. Then, find the energy absorbed by the voltage and current sources.

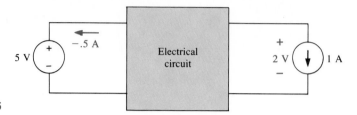

**FIGURE 1.15**

Power:   $p_s(t) = v_s(t) \cdot i_s(t)$

Voltage source:   $p_v(t) = 5 \cdot -.5 = -2.5 \text{ W}$

Current source:   $p_i(t) = 2 \cdot 1 = 2 \text{ W}$

The signs of the powers tell us that the voltage source is actually delivering power, while the current source is absorbing power.

Energy:   $w(t) = \int_0^t p(t)\, dt$     (assume $w(0) = 0$)

Voltage source:   $w_v(t) = -2.5t \text{ J}$

Current source:   $w_i(t) = 2t \text{ J}$

## Resistance and Resistors

In the 1800s, the German physicist Georg S. Ohm (1787–1854) performed a series of experiments to determine how the current in a metal wire was related to the voltage across it (Figure 1.16A). Figure 1.16B shows the results of one such experiment. Ohm found that voltage was directly proportional to the current it produced and could be described by the simple equation known as Ohm's law:

$$V = IR \qquad\qquad\qquad (1.8)$$

When voltage is plotted as a function of current and the passive sign convention is assumed, $R$ is the slope of the straight line.

Figure 1.16C shows the $V$–$I$ characteristics for several metals: silver (Ag), copper (Cu), and aluminum (Al). For a given voltage, the current is the least in aluminum and greatest in silver. This relationship corresponds to the fact that the slope of the $V$–$I$ characteristic is the smallest in silver and greatest in aluminum.

**FIGURE 1.16** *Ohm's Law*

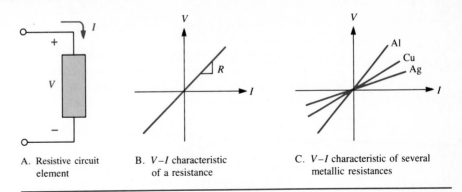

A. Resistive circuit
   element

B. *V–I* characteristic
   of a resistance

C. *V–I* characteristic of several
   metallic resistances

The slope $R$ represents the **resistance** of a particular metal to charge flow (current). The resistance of all passive elements is positive.

A device that is specifically constructed to take advantage of this straight line *V–I* characteristic is known as a **resistor.** Resistance and resistors are measured in units of ohms. The symbol for ohms is the uppercase Greek omega, $\Omega$. It is clear from Ohm's law that 1 ohm = 1 volt/ampere.

The symbol representing a resistor or resistance in a circuit is shown in Figure 1.17A. Figure 1.17B shows the circuit symbol for a type of variable resistor known as a *potentiometer*. This device is a three-terminal element in which the arrow represents an internal wiper. The wiper is a mechanical means of connecting one of the external terminals to various points along the internal resistance. The position of the wiper determines the resistances between the fixed terminals and the wiper terminal.

Although Ohm's law was experimentally derived only for metals and only for constant currents and voltages, it is an accurate model for many materials and for time-varying signals. Ohm's law can be written variously as

**FIGURE 1.17** *Resistance*

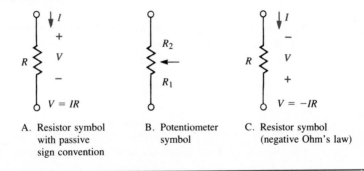

A. Resistor symbol
   with passive
   sign convention

B. Potentiometer
   symbol

C. Resistor symbol
   (negative Ohm's law)

*Ohm's Law*

$$V = IR \quad \text{or} \quad I = \frac{V}{R} \quad \text{or} \quad R = \frac{V}{I} \tag{1.9}$$

Keep in mind that these equations assume that current direction and voltage polarity have been assigned according to the passive sign convention. If the assigned current enters at the negative terminal, as shown in Figure 1.17C, then Ohm's law is

$$V = -IR$$

The resistance of any element is a function of many factors: cross-sectional area, length, temperature, and molecular structure. Materials that have very low resistance at room temperature are classified as **conductors.** The ideal conductor would have zero resistance. All pure metals and metal alloys are good conductors with very low but finite resistance. Supercooling certain metal alloys and ceramic compounds will reduce their resistance to approximately zero. **Insulators** are materials with very high resistance and do not support much charge flow. Wood, plastic, glass, and other ceramics are examples of insulators.

There are materials that fall between conductors and insulators. These conduct poorly unless sufficient energy is applied to them. They can then become good conductors. These materials are **semiconductors** and are the heart of the modern electronics industry. You will study them in great detail in electronics courses.

The resistance of any device can be found from the physical properties of the material and its physical dimensions and is given by

*Physical Description*

$$R = \rho\left(\frac{l}{A}\right) \tag{1.10}$$

where $l$ = length of the material

$A$ = cross-sectional area

$\rho$ = resistivity of the material

Resistors that can be bought in most hobby stores range in value from $10\,\Omega$ to $10\,M\Omega$. Resistors that you might see in commercial circuits range in value from a few tenths of an ohm to hundreds of megohms.

---

**Example 1.4**   *Resistance*

The resistivities of gold, aluminum, and carbon are $2.4 \times 10^{-6}$, $2.6 \times 10^{-6}$, and $3.5 \times 10^{-3}$ $\Omega$-cm (ohm-centimeters), respectively. If a resistor is formed from each of these conductors with a cross-sectional area of $10^{-6}\,\mathrm{cm}^2$, find the lengths required to achieve a resistance of $1\,\mathrm{k}\Omega$.

The required length is found from Equation 1.10: $l = AR/\rho$.

For gold:   $l = 417\,\mathrm{cm}$

For aluminum:   $l = 385\,\mathrm{cm}$

For carbon:   $l = .29\,\mathrm{cm}$

You can see that carbon is the most suitable element for making resistors of reasonable physical size.

---

There are two extreme cases of resistance that you should be aware of: $R = 0\,\Omega$, corresponding to a perfect, or ideal, conductor; and $R \to \infty\,\Omega$, corresponding to a perfect insulator. These cases are shown in Figures 1.18A and 1.18B. A perfect conductor has a horizontal $V\text{--}I$ characteristic (slope = 0). Therefore, regardless of the magnitude of the current in it, a perfect conductor has zero potential difference across it and is known as a **short circuit.** Figure 1.18A also shows the circuit model for an element with zero resistance. We assume in this book that the wires connecting various devices are perfect conductors.

An ideal voltage source that has a value of 0 V has the same $V\text{--}I$ characteristic as the short circuit. In fact, whenever we turn off a voltage source in circuit analysis, we can replace it with a short circuit.

The perfect insulator, as shown in Figure 1.18B, supports no current. This is an **open circuit,** and its circuit model is also shown in Figure 1.18B. A current source set to 0 A has the $V\text{--}I$ characteristic of Figure 1.18B and so must be an open circuit. Whenever we set a current source to zero in circuit analysis, we can replace it with an open circuit. Perfection is never achieved in the real world, and the resistance of all devices falls somewhere between 0 and $\infty$. Bear in mind that devices that are insulators under normal conditions can become conductors if

**FIGURE 1.18**   *Short and Open Circuits*

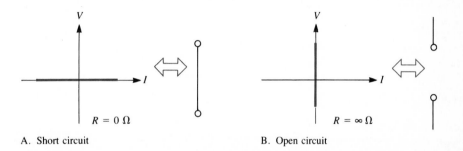

A. Short circuit                          B. Open circuit

enough voltage is applied to them. For example, if a lightning bolt hits a ceramic insulator, do not expect it to remain an insulator! Also, any real conductor has a limit to how much current it can support before it burns up and becomes an open circuit.

A final word, for now, about resistance. Resistance is not always linear. That is, many devices have $V$–$I$ characteristics that are not straight lines. Figure 1.19 shows the $V$–$I$ curve for a particular semiconductor device, the diode. Note that for semiconductors, we usually plot $I$ versus $V$ instead of $V$ versus $I$. This device exhibits nonlinear resistance. The equation that relates diode current to diode voltage is shown in the figure. Some methods for analyzing circuits containing these nonlinear devices will be discussed in Chapter 4.

The power absorbed by a resistor is given by

$$p_R(t) = v_R(t) \cdot i_R(t) \tag{1.11}$$

where voltage and current have been assigned according to the passive sign convention.

We can use Ohm's law to find other forms of the preceding relationship. Since $v_R(t) = i_R(t) \cdot R$,

$$p_R(t) = i_R^2(t) \cdot R = \frac{v_R^2(t)}{R} \tag{1.12}$$

The power absorbed by physical resistors is always positive and is dissipated as heat, which can be useful in a room heater or a hair dryer. However, in most circuits, heat is an undesirable side effect that may damage components.

**FIGURE 1.19**    *Diode V–I Characteristic*

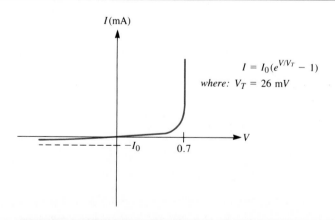

$$I = I_0(e^{V/V_T} - 1)$$
*where:* $V_T = 26$ mV

In addition to the resistance values of a resistor, you must also pay attention to its power rating. If you use a resistor rated at $\frac{1}{4}$ W in a circuit where it has to dissipate 1 W, the resistor will smoke, smell bad, and eventually burn up.

---

**Example 1.5**   *Practical Resistors*

Practical resistors that you buy in electronics stores have at least three parameters associated with them: nominal value, tolerance, and power rating. The nominal value is the presumed magnitude of the resistor, for example, $1\,k\Omega$. The tolerance recognizes the fact that it is impossible to manufacture a large quantity of $1\,k\Omega$ resistors whose values are exactly $1\,k\Omega$. The tolerance is the allowable error between the nominal and actual value. Typical tolerances available are $\pm 5\%$ and $\pm 10\%$, while precision resistors with $\pm 1\%$ tolerance can also be purchased. Resistors are also rated in watts. Carbon resistors are typically available in $\frac{1}{8}$, $\frac{1}{4}$, and $\frac{1}{2}$ W ratings. Wire-wound power resistors are readily available in ratings up to 10 W.

Assume the following situation. A 100 V DC source will supply current to a resistive load. The desired current in the resistor is 100 mA, but a $\pm 7\%$ variation in current is acceptable. Find the required resistor nominal value, tolerance, and power rating.

The nominal resistance is found from Ohm's law:

$$R = \frac{V}{I} = \frac{100}{.1} = 1\,k\Omega$$

Because a $\pm 7\%$ variation in current is allowed, the current can vary from a minimum of 93 mA to a maximum of 107 mA. Therefore, the allowable range for $R$ is

$$R_{max} = \frac{V}{I_{min}} = \frac{100}{.093} = 1.075\,k\Omega$$

and

$$R_{min} = \frac{V}{I_{max}} = \frac{100}{.107} = .935\,k\Omega$$

Thus, resistor tolerance is $+7.5\%$ to $-6.5\%$. Since we have only 5% and 10% resistors available, we must use a 5% resistor.

Nominal power rating:   $P = V \cdot I = 100 \cdot .1 = 10\,W$

Maximum power dissipated:   $P_{max} = V \cdot I_{max} = 100 \cdot .107 = 10.7\,W$

We would have to find a $1\,k\Omega$, 5%, 11 W power resistor for this requirement. In fact, we would need a resistor with a larger power rating to give us a margin of safety.

---

## Controlled (Dependent) Sources

The elements that we have described so far have two terminals. Many devices in circuits and electronics have several terminals. A "black box" representation of a four-terminal device is shown in Figure 1.20A. The box may contain anything from a relatively simple circuit element to a very complicated circuit such as an operational amplifier. For now, we want to examine the special cases shown in Figures 1.20B through 1.20E. The diamond-shaped symbols represent either current or voltage sources. They are different from independent sources in that the values of these sources are controlled by (or are dependent on) some other circuit variable.

You can see from Figure 1.20 that there are four types of **controlled sources:** a voltage source whose value is controlled by a voltage somewhere else in the circuit (VCVS), a voltage source whose value is controlled by a current somewhere else in the circuit (ICVS), a current source whose value is controlled by a current somewhere else in the circuit (ICIS), and a current source whose value is controlled by a voltage somewhere else in the circuit (VCIS). The terms in parentheses here contain shorthand notation where V stands for voltage, I for current, C for controlled, and S for source.

**FIGURE 1.20**  *Controlled Sources*

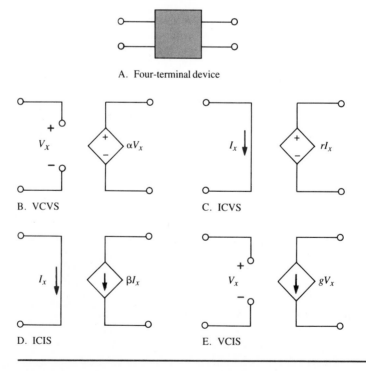

A. Four-terminal device

B. VCVS

C. ICVS

D. ICIS

E. VCIS

Controlled sources usually are not devices that are constructed for a specific purpose, as are independent sources and resistors. Rather, these elements are used to model either physical effects within a device or the input-output characteristics of a larger circuit.

Controlled sources are frequently found in models of semiconductor devices. Several important semiconductor devices and simple models for them are discussed in the next section. The scope of this text does not permit a complete description of the behavior of these devices. That is best left to an electronics course. They are mentioned here so that you will become used to analyzing these devices as circuit elements.

## 1.5  SEMICONDUCTOR DEVICES

Semiconductors are the backbone of the modern electronics industry. The basic semiconductor device is the transistor, from which radios, stereo amplifiers, and computers are constructed. There are two main types of transistors: the bipolar junction transistor (BJT) and the field effect transistor (FET). An extremely important device built from transistors is the operational amplifier (op amp).

### Bipolar Junction Transistor

The circuit symbol for a *bipolar junction transistor* (BJT) is shown in Figure 1.21A. The BJT has three terminals labeled the base (B), the collector (C), and the emitter (E). Two models used for the BJT are shown in Figure 1.21. The model in Figure 1.21B is for DC voltages and currents, while the model in Figure 1.21C is used for AC calculations. In both cases, the current-controlled current source models the physical property that the collector current $(I_C)$ is proportional to the base current $(I_B)$.

FIGURE 1.21  *Bipolar Junction Transistor (BJT)*

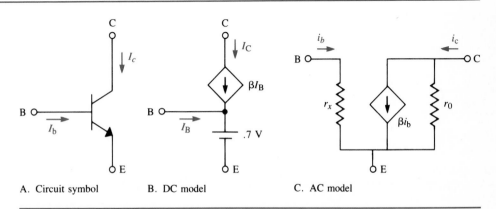

A.  Circuit symbol   B.  DC model   C.  AC model

## Field Effect Transistor

Figure 1.22A shows the circuit symbol for another type of transistor, the *field effect transistor* (FET). The three terminals of the FET are labeled the gate (G), the source (S), and the drain (D). There are several types of FETs; each type differs in its DC behavior. Since the DC behavior of these devices is very complicated, we will not model it here. However, for AC signals, we can derive a simple model for the FET. The output current of an FET ($i_d$) is directly dependent on the voltage ($v_{gs}$) that exists between the gate and the source of the device. Therefore, we model the FET with a voltage-controlled current source, as shown in Figure 1.22B. Because of the physics of this device, there is an approximate open circuit (infinite resistance) between the gate (G) and the source (S).

## Operational Amplifier

The *operational amplifier* is a device that is constructed by connecting many transistors together, as was shown earlier in Figure 1.2. An exact analysis of this device would require replacing each of the transistors with an appropriate model and analyzing the resultant very large circuit. This analysis can only be reasonably done with a computer, which, fortunately, for most applications, is unnecessary. It is enough to know how the output of the device depends on its input.

Figure 1.23A shows the circuit symbol for the operational amplifier (op amp) and the simplified model (Figure 1.23B) that is most often used to represent its behavior. In an op amp, the output voltage is dependent on the voltage between the plus and minus terminals, so we use a voltage-controlled voltage source in the model. The scale factor of the controlled source is the internal gain of the amplifier and is very large ($>10^5$). Since the input resistance ($r_{in}$) is very large (greater than 1 M$\Omega$) and the output resistance ($r_o$) is very small (less than 100 $\Omega$), the model in Figure 1.23B is often further simplified to the one in Figure

**FIGURE 1.22**   *Field Effect Transistor (FET)*

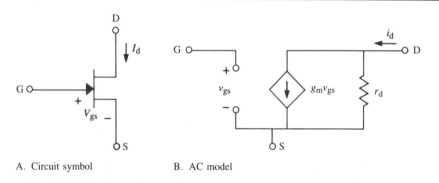

A.  Circuit symbol                    B.  AC model

**FIGURE 1.23** *Operational Amplifier*

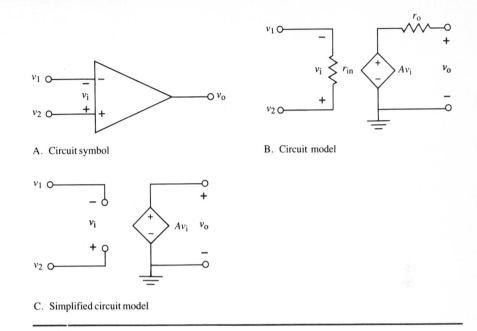

A. Circuit symbol

B. Circuit model

C. Simplified circuit model

1.23C, in which the output resistance is assumed to be zero and the input resistance is assumed to be infinite.

Since most circuits built these days include electronic devices such as transistors and operational amplifiers, it is very important that you learn to analyze circuits containing these devices as early as possible. Keep in mind that you do not need to know the physics of these devices just yet. It is enough to know their circuit models and to be able to analyze circuits containing controlled sources. To that end, this text presents many examples and problems that include controlled sources in them.

## Ground

There is a circuit symbol in Figure 1.23 that has not been seen before. What is the meaning of the three horizontal lines at the bottom of the circuit? This symbol indicates that this point in the circuit is the ground for the circuit. The term *ground* comes from the field of electrical power delivery.

The earth itself is a very large conductor and is used as the return conducting path for power systems. Does this mean that any time you see a ground symbol at a point in a circuit that the point is physically connected to the earth? This would prove rather cumbersome. The ground in a circuit has a related but slightly different meaning from the ground in a power system. The ground in a circuit is usually a conductor that forms a common junction to which many of the circuit elements are connected. For example, the metal chassis of a car serves as a conductor to which several electrical devices (e.g., battery, starter motor) are connected. The ground in a circuit is usually used as a common reference point for the measurement of voltage. The voltage at the ground is always assumed to be zero.

**Example 1.6**  *Ground*

The op amp circuit in Figure 1.24A has a voltage source that is connected between ground and the negative terminal. The positive terminal of the op amp is connected to ground. If we replace the op amp with its model, we get the circuit in Figure 1.24B. This circuit can be very confusing to the beginning student since the interconnections are not explicit. Remember, however, that a circuit ground is a common point. All of the ground symbols are electrically connected. If we actually draw these connections, we get the circuit in Figure 1.24C.

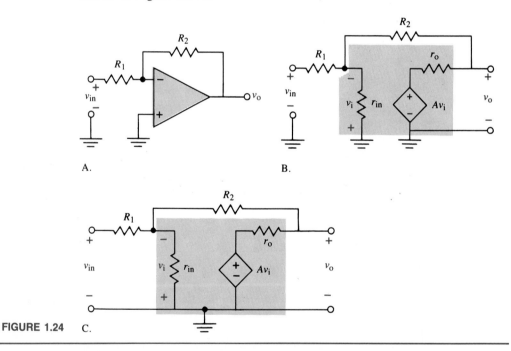

**FIGURE 1.24**   C.

## 1.6  SYSTEMS

An electrical circuit is a special type of a system. The mathematics of systems analysis are, therefore, very useful in the analysis of a circuit. A **system** is a

**FIGURE 1.25**  *System Representation*

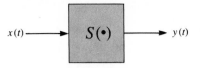

collection of component parts organized to produce a desired result. For example, an automobile is a system composed of mechanical (body, springs, shock absorbers, etc.), chemical (gasoline), electrical (the ignition system), and biological (the driver and passengers) components. The desired result of the use of an automobile is to provide transportation from one point to another.

A general representation of a system with a single input and a single output is shown in Figure 1.25. The symbol $S(\cdot)$ represents the mathematical model of the behavior of the system and indicates that the system acts on the input $x(t)$ to produce an output $y(t)$, where

$$y(t) = S[x(t)] \tag{1.13}$$

The system can be mechanical, electrical, chemical, hydraulic, biological, or any combination of these. The system symbol $S(\cdot)$ might represent a set of algebraic equations, differential equations, or perhaps a nonlinear operation such as the diode $V$–$I$ equation (Figure 1.19).

We begin our study of systems by discussing some concepts used for their classification—namely, linearity, time invariance, spatial distribution, and passivity. While all of these properties have mathematical descriptions, only linearity can be easily and usefully so described. The other properties will be discussed qualitatively.

## Linearity

A system is **linear** if it exhibits the two properties of homogeneity and superposition (also known as additivity). These are fancy names for simple but extremely important properties. Figure 1.26 shows the same system with two different inputs. In Figure 1.26A, the input to the system is $x(t)$ and the output is

$$y_1(t) = S[x(t)]$$

In Figure 1.26B, the input from Figure 1.26A has been multiplied by the scale factor a. The output is now

$$y_2(t) = S[ax(t)]$$

**FIGURE 1.26**   *Homogeneity*

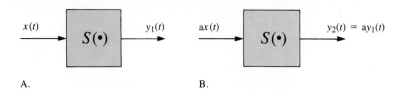

A.                                                    B.

If the system is linear, then the property of *homogeneity* states that this output is

$$y_2(t) = aS[x(t)] = ay_1(t)$$

That is, if a system is linear, then scaling the input scales the output by the same amount.

**Example 1.7**   *Homogeneity*

Determine which of the following systems satisfy the property of homogeneity:

System 1:   $S(x) = \dfrac{d(x)}{dt}$

$$y_1(t) = \dfrac{dx(t)}{dt}$$

$$y_2(t) = \dfrac{d[ax(t)]}{dt} = \dfrac{a\,dx(t)}{dt} = ay_1(t)$$

System 2:   $S(x) = (x)^2$

$$y_1(t) = x^2(t)$$

$$y_2(t) = [ax(t)]^2 = a^2x^2(t) \neq ay_1(t)$$

Ohm's law:   $V = RI$

$$V_1 = RI$$

$$V_2 = R(aI) = aRI = aV_1$$

Therefore, the first system (taking the derivative of the input) satisfies the homogeneity property and *may* be linear. The second system (squaring the input) does not satisfy this property and so is not linear. Ohm's law satisfies homogeneity.

Figure 1.27A shows a system with two different inputs and outputs. Figure 1.27B shows the same system with an input that is the sum of the two inputs in

FIGURE 1.27   *Superposition*

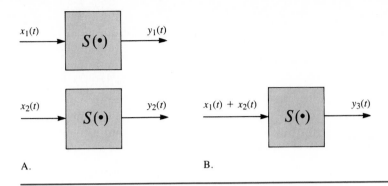

A.                                          B.

part A. The output of part B is given by

$$y_3(t) = S[x_1(t) + x_2(t)]$$

In linear systems, this output can also be written as

$$y_3(t) = S[x_1(t)] + S[x_2(t)] = y_1(t) + y_2(t)$$

where $y_1(t)$ and $y_2(t)$ are the outputs to $x_1(t)$ and $x_2(t)$, respectively. This is the property of **superposition.** If a system is linear, then its output to the sum of inputs is equal to the sum of the outputs to the individual inputs. We will make great use of superposition in circuit analysis. Note that superposition does *not* apply to power and energy calculations, since these are nonlinear.

**Example 1.8**   *Superposition*

In all of the examples that follow, $y_1(t)$ is the output to $x_1(t)$, $y_2(t)$ is the output to $x_2(t)$, and $y_3(t)$ is the output to $x_1(t) + x_2(t)$. Determine which of these systems satisfy superposition.

$$\text{System 1:} \quad S(x) = \frac{d(x)}{dt}$$

$$y_1(t) = \frac{dx_1(t)}{dt}$$

$$y_2(t) = \frac{dx_2(t)}{dt}$$

$$y_3(t) = \frac{d[x_1(t) + x_2(t)]}{dt} = \frac{dx_1(t)}{dt} + \frac{dx_2(t)}{dt} = y_1(t) + y_2(t)$$

*(continues)*

**Example 1.8**   *Continued*

System 2:   $S(x) = 5 \cdot (x)$

$y_1(t) = 5x_1(t)$

$y_2(t) = 5x_2(t)$

$y_3(t) = 5[x_1(t) + x_2(t)] = 5x_1(t) + 5x_2(t) = y_1(t) + y_2(t)$

System 3:   $S(x) = \cos(x)$

$y_1(t) = \cos x_1$

$y(t) = \cos x_2$

$y_3(t) = \cos(x_1 + x_2) = \cos x_1 \cos x_2 - \sin x_1 \sin x_2 \neq y_1(t) + y_2(t)$

System 4:   $S(x) = (x)^2$

$y_1(t) = x_1^2(t)$

$y_2(t) = x_2^2(t)$

$y_3(t) = [x_1(t) + x_2(t)]^2 = x_1^2(t) + 2x_1(t)x_2(t) + x_2^2(t) \neq y_1(t) + y_2(t)$

Ohm's law:   $V = RI$

$V_1 = RI_1$

$V_2 = RI_2$

$V_3 = R(I_1 + I_2) = RI_1 + RI_2 = V_1 + V_2$

Therefore, systems 3 and 4 do not satisfy the superposition property and are not linear. Since Ohm's law satisfies both homogeneity and superposition, it is a linear operation.

---

A system is linear if it exhibits *both* properties of homogeneity and superposition. Linear systems are reasonably easy to analyze, while the mathematics describing nonlinear systems can be quite involved. Fortunately, the devices we use in basic electrical circuits can be modeled as linear systems. A system that contains independent energy sources is technically not linear; however, we may still be able to use linear techniques to analyze it.

---

**Example 1.9**   *Linearity of System with Independent Sources*

**a.**   A linear system *without* independent sources will have the input-output characteristic shown in Figure 1.28A:

$y = mx$

If   $x = a_1x_1 + a_2x_2$

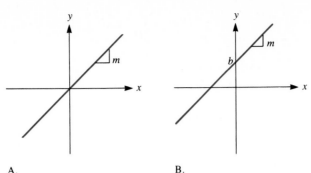

**FIGURE 1.28**    A.                                                B.

then

$$y = m(a_1x_1 + a_2x_2) = a_1mx_1 + a_2mx_2 = a_1y_1 + a_2y_2$$

Therefore, this system is linear.

b.   A system containing internal independent sources will have an output even when there is no input to the system, as shown in Figure 1.28B:

$$y = mx + b$$

If      $x = a_1x_1 + a_2x_2$

then

$$y = m(a_1x_1 + a_2x_2) + b = a_1mx_1 + a_2mx_2 + b \neq y_1 + y_2$$

Therefore, this system is not linear, even though it is described with a straight line. However, if we redefine the input variable as

$$\hat{x} = x + \frac{b}{m}$$

then

$$y = m\hat{x}$$

The equation for $\hat{x}$ is linear. This procedure allows us to apply linear techniques to this system.

## Time Invariance

If the response of a system is not dependent on the time at which the input is applied, the system is **time invariant.** That is, if the input is shifted by $T$ seconds, the output is shifted by the same $T$ seconds, but the shape of the output does not change. A time invariant relationship is expressed mathematically as follows:

If

$$y(t) = S[x(t)]$$

then

$$y_1(t) = S[x(t-T)] = y(t-T)$$

A system whose components change their behavior with time is known as a time varying system. You are an example of a time varying system. All biological systems have internal clocks that control their behavior. You do not respond the same way to the same stimulus at different times of the day. Jet lag occurs because your internal clock does not correspond to the time of day at your destination.

---

**Example 1.10**   *Time Invariance*

    **a.**  Let $y(t) = dx(t)/dt$, where $x(t) = 3e^{-5t}$:

$$y(t) = -15e^{-5t}$$

Now, let $x_1(t) = 3e^{-5(t-T)}$:

$$y_1(t) = -15e^{-5(t-T)} = y(t-T)$$

    **b.**  Let $y(t) = t \cdot dx(t)/dt$, where $x(t) = 3e^{-5t}$:

$$y(t) = -15t \cdot e^{-5t}$$

Now, let $x_1(t) = 3e^{-5(t-T)}$:

$$y_1(t) = -15t \cdot e^{-5(t-T)} \neq y(t-T)$$

The system represented in part a is time invariant, while the system represented in part b is time varying.

---

All systems exhibit some time variations of their parameters, particularly as the systems age. However, for the sake of simplicity, we will assume that the electrical circuits that we deal with in this book are time invariant.

## Spatial Distribution

If the speed of moving charges were infinite, then it would take no time for the charges to move around a circuit. However, the speed is not infinite, and it takes a finite time for charges to move from one place in a circuit to another. If this transit time is significant, then we would have to include the spatial parameters of the circuit, and we would be working with a **distributed** system. Transmission lines are very long, and the transit time of the charges that they carry must be accounted for. Although nerve cells are very short, the chemical processes involved in propagating neural impulses are slow enough that a nerve cell should be treated as a distributed system. Electrical systems that are physically small but are excited by very rapidly changing signals may also have to be modeled as distributed systems.

Fortunately, for the electrical components and the frequencies used in this textbook, we can assume that all electrical effects happen instantaneously throughout the circuit. This type of circuit is known as a **lumped** circuit to distinguish it from a distributed circuit. The $V$–$I$ characteristics of lumped elements are described by simple scalar quantities, such as $R$ for a resistor. The physical dimensions of the resistor are not important as long as $R$ is known.

### Passivity

Systems are defined as active or passive depending on their energy usage. If the energy absorbed by a system is always nonnegative, the system is **passive.** Otherwise, it is an **active** system. A more useful definition for us is that an electrical element, device, or network is passive if it contains no sources of energy. Resistors are passive devices, while transistor circuits and operational amplifiers are active devices since they require energy sources for their operation.

## 1.7  SOME SIMPLE CIRCUITS

Let's consider a simple electrical circuit. An ideal voltage source with terminals a and b and a resistor with terminals a′ and b′ are shown in Figure 1.29A. The voltage source has a constant value of 5 V. Remember, this means that the voltage between terminals a and b is 5 V no matter what else is connected to these terminals. In Figure 1.29B, the resistor is joined to the voltage source by connecting a′ to a and b′ to b. Note that in Figure 1.29B we have assigned the resistor voltage ($V_R$) and current ($I_R$) and the source current ($I_s$) according to the passive sign convention.

Figure 1.29C shows the more conventional way to represent this circuit in a schematic diagram. Unlike drawings in mechanical or civil engineering, circuit diagrams are not literal pictures of what the circuits actually look like. They simply represent the electrical connections present and are drawn for easy visualization. From either representation, Figure 1.29B or 1.29C, you should be able to see that there are only two terminals in this circuit: one at the top and one at the bottom, indicated by the dashed ellipses in Figure 1.29C.

Even without the knowledge of the circuit laws presented in Chapter 2, you can analyze the circuit in Figure 1.29C based on what you already know and by

FIGURE 1.29    *Voltage Source and Resistor Circuit*

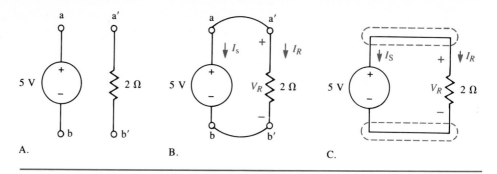

A.                              B.                              C.

using your intuition. The voltage between the top and bottom terminals must be 5 V because of the voltage source. The unknown voltage across the resistor, $V_R$, also exists between the same two points. Since only one voltage can exist between any two points, we have

$$V_R = 5\,\text{V}$$

(Note: The polarity of $V_R$ and the voltage source are the same.)
   Ohm's law now tells us that

$$I_R = \frac{V_R}{R} = \frac{5}{2} = 2.5\,\text{A}$$

What is $I_s$? This is where you must use your intuition. Since there is only a single closed path in this circuit, there can be only one current. We simply have two labels, $I_R$ and $I_s$, for the same current. Since their arrows have been assigned in opposite directions, we conclude that

$$I_s = -I_R = -2.5\,\text{A}$$

To finish the analysis, we calculate the power absorbed by each device:

$$P_R = V_R \times I_R = 5 \times 2.5 = 12.5\,\text{W}$$

and

$$P_s = V_s \times I_s = 5 \times -2.5 = -12.5\,\text{W}$$

Note that resistor power is positive and that source power is negative. The resistor is dissipating energy, and the source is supplying energy. Note also that the total power absorbed by elements in the circuit is zero. In a closed system, the total power supplied always equals the total power dissipated or stored.

**Example 1.11**    *Circuit of Figure 1.29 with Polarities Reversed*

Figure 1.30 shows the same circuit as in Figure 1.29, but with the assigned polarities reversed.

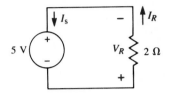

**FIGURE 1.30**

$$V_R = -5 \text{ V}$$

$$I_R = -2.5 \text{ A}$$

$$I_s = I_R = -2.5 \text{ A}$$

$$P_R = 12.5 \text{ W}$$

$$P_s = -12.5 \text{ W}$$

This example demonstrates that the polarity you assign to a resistor voltage does not matter as long as you follow the passive sign convention for the current. If you choose the "wrong" polarity, you will get a negative answer. However, the negative answer, along with the assigned polarity, is still a correct answer.

**Example 1.12**    *Current Source and Resistor*

Find the indicated voltages and currents for the circuit in Figure 1.31. Find the power absorbed by the resistor and current source.

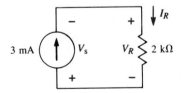

**FIGURE 1.31**

$$I_R = I_s = 3 \text{ mA} \quad \text{(only one current in circuit)}$$

$$V_R = I_R \cdot R = 6 \text{ V}$$

$$V_s = -V_R = -6 \text{ V} \quad \text{(note the polarity)}$$

$$P_R = 18 \text{ mW}$$

$$P_s = -18 \text{ mW}$$

**Example 1.13**   *Circuit of Figure 1.31 with Polarities Reversed*

Figure 1.32 shows the same circuit as in Figure 1.31 with the assigned polarities reversed.

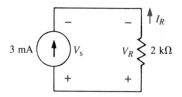

**FIGURE 1.32**

$$I_R = -I_s = -3\,\text{mA}$$
$$V_R = -6\,\text{V}$$
$$V_s = -6\,\text{V}$$
$$P_R = 18\,\text{mW}$$
$$P_s = -18\,\text{mW}$$

**Example 1.14**   *Controlled Source*

Figure 1.33 shows a typical AC small-signal transistor (BJT) circuit. Although it may look complicated, it really is not. This is not one circuit with two sources, but, rather, two circuits with one source each. Note that there is no closed path between the two parts of the circuit. To analyze, examine each part separately and then put the pieces together.

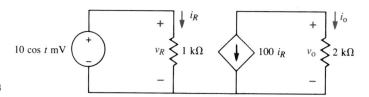

**FIGURE 1.33**

On the left, we have

$$v_R = 10\cos t\,\text{mV}$$
$$i_R = 10\cos t\,\mu\text{A}$$

On the right, we have

$$i_o = -100 i_R$$
$$v_o = i_o \times (2 \times 10^3) = -2 \times 10^5 (i_R)$$

Since $i_R = 10 \cos t \, \mu A$, we get

$$i_o = -1 \cos t \, \text{mA}$$

$$v_o = -2 \cos t \, \text{V}$$

## 1.8 SUMMARY

In this introductory chapter, we have described the two primary circuit variables of interest, voltage and current, and the power and energy relationships for these variables. Remember to observe the passive sign convention when you assign voltage and current polarities.

We introduced you to several electrical circuit devices: the ideal independent voltage and current sources, the resistor, and controlled sources. Independent sources supply the necessary energy to circuits. The resistor is a device whose voltage is proportional to the current in it, as described by Ohm's law. Controlled sources are used to model physical effects within a device or input-output characteristics of a larger circuit.

We discussed several systems concepts, including superposition. Superposition states that, for linear systems, the output for the sum of several inputs is equal to the sum of the outputs given the inputs one at a time.

We concluded this chapter by showing you how to analyze some simple circuits. In Chapter 2, we will show you the laws that govern general circuit behavior. With these laws and Ohm's law, you will be able to analyze any resistor circuit, no matter how complicated. In Chapter 3, we will show you some systematic ways to efficiently analyze larger circuits.

## ▪ PROBLEMS

**1.1** Find the current in a conductor for the charge profile given in Figure P1.1.

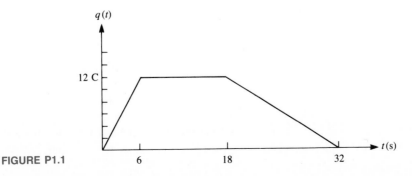

**FIGURE P1.1**

**1.2** Find the current in a conductor for the charge flow given by:

   **a.**  $q(t) = 10 \cos 50t \ \mu C$

   **b.**  $q(t) = e^{-5t} \sin 1000t \ mC$

   **c.**  $q(t) = te^{-10t} \cos 500t \ nC$

**1.3** Find the total charge delivered at $t = 5$ s for the current profile given in Figure P1.3. Assume that $q(0) = 0$.

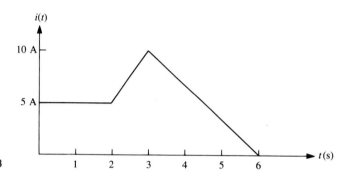

**FIGURE P1.3**

**1.4** Find $q(t)$ for the following currents and initial conditions:

   **a.**  $i(t) = 5 \ mA, \ q(0) = 0$

   **b.**  $i(t) = 3 + 2t + t^2 \ \mu A, \ q(0) = 5 \ \mu C$

   **c.**  $i(t) = e^{-10t} \sin 10t \ mA, \ q(0) = 0$

   **d.**  $i(t) = \cos 100t \sin 100t \ A, \ q(0) = 3 \ C$

   (*Hint:* See Appendix C for useful identities.)

**1.5** For Figure P1.5, find the power absorbed by $N_1$ and $N_2$ for the following voltages and currents. In each case, determine whether $N_1$ and $N_2$ are actually absorbing or delivering power.

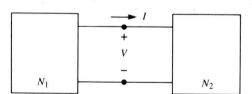

**FIGURE P1.5**

   **a.**  $V = 10 \ V, \ I = 5 \ mA$

   **b.**  $V = -\cos t \ V, \ I = 10 \ mA$

   **c.**  $V = -\cos t \ V, \ I = -\cos t \ A$

**1.6** Find the energy absorbed by $N_2$ for the three cases in Problem 1.5. (*Hint:* See Appendix C for useful identities for part c.)

**1.7** For the circuit shown in Figure P1.7, $v_1 = 6 - .2e^{-2t} \ V$, $v_2 = 4 - .8e^{-2t} \ V$, and $v_3 = 4 + .2e^{-2t} \ V$.

   **a.**  Label and find all branch currents and the power absorbed by each element.

   **b.**  What is the total power absorbed by all elements in this circuit?

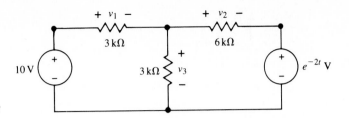

**FIGURE P1.7**

**1.8** For the circuit shown in Figure P1.8, $i_1 = 5 + .2 \cos 10t$ A, $i_2 = 5 - .2 \cos 10t$ A, $i_3 = 5 + .8 \cos 10t$ A.

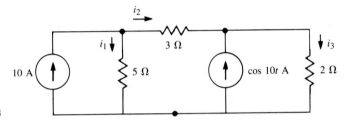

**FIGURE P1.8**

**a.**   Label and find all branch voltages and the power absorbed by each element.

**b.**   What is the total power absorbed by all elements in this circuit?

**1.9** A flashlight bulb is lit by a 1.5 V battery. The bulb is rated at 6 W.

**a.**   Find the current in the bulb.

**b.**   Find the resistance of the bulb.

**c.**   Draw a circuit model for the flashlight.

**1.10** Practical sources cannot supply infinite energy. For example, the 12 V storage battery in a car may be rated at 100 ampere-hours.

**a.**   What is the total energy capacity of this battery? Express your answer in joules.

**b.**   If the starter motor of the car has a resistance of .1 Ω, how long can you turn over an engine that won't start before you drain the battery?

**c.**   You have ignored the calculation in part b and you now have a dead battery. Your friend has a battery charger that can supply a constant 2 A at 12 V. How long will it take to fully recharge the battery?

**1.11** Figure P1.11 shows a practical carbon resistor and its color bands (or color codes). The first three color bands indicate the nominal resistance of the resistor by the following formula:

$$R_{\text{nom}} = AB \times 10^C$$

where        black = 0          green = 5
             brown = 1          blue = 6
             red = 2            violet = 7
             orange = 3         gray = 8
             yellow = 4         white = 9

**FIGURE P1.11**
A B C D

The fourth color band gives the tolerance of the resistor. A silver band indicates that the actual value of the resistor can vary $\pm 10\%$ from the nominal value. A gold band indicates a $\pm 5\%$ variation. Find the nominal, minimum, and maximum values for the resistors with the following color codes:

**a.** brown, black, brown, silver

**b.** yellow, violet, orange, silver

**c.** red, black, yellow, gold

**d.** green, green, black, gold

**1.12** A 10 V source is applied to the resistors given in Problem 1.11a–d.

**a.** Find the nominal current, minimum current, and maximum current for each resistor.

**b.** Find the nominal power, minimum power, and maximum power for each resistor.

**1.13** A 1,000 W heater is to be constructed from copper wiring with a cross-sectional area of $2 \times 10^{-5}$ cm$^2$ and is connected to a 120 V source. Copper has a resistivity of $1.72 \times 10^{-6}$ $\Omega$-cm.

**a.** Determine the resistance of the heater.

**b.** Determine the length of copper wire needed.

**1.14** All of the independent sources in the circuit in Figure P1.14 are turned off. Redraw the circuit using open and short circuits.

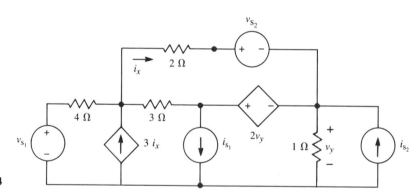

**FIGURE P1.14**

**1.15** Using the appropriate model, draw an equivalent circuit diagram for the BJT circuit shown in Figure P1.15.

**1.16** Using the appropriate model, draw an equivalent circuit diagram for the BJT circuit shown in Figure P1.16.

**1.17** Draw an equivalent circuit diagram for the FET circuit shown in Figure P1.17.

**1.18** For the FET circuit shown in Figure P1.17, $R_G = 10$ k$\Omega$, $R_D = 1$ k$\Omega$, $R_s = 0\,\Omega$ (short circuit), $r_d = \infty\,\Omega$ and $g_m = .001$ S. Find $v_o$.

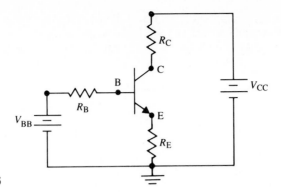

**FIGURE P1.15**

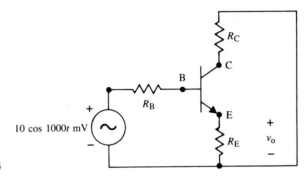

**FIGURE P1.16**

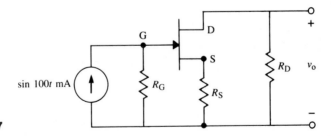

**FIGURE P1.17**

**1.19** Determine the linearity of the systems described by the following equations:

a. $y(t) = \dfrac{dx(t)}{dt} + 3x(t)$

b. $y(t) = \dfrac{dx(t)}{dt} + 3x^2(t)$

c. $y(t) = \displaystyle\int_0^t x(t)\, dt + y_0$

d. $y(t) = \ln x(t)$

e. $y(t) = \text{INT}[x(t)]$, where INT truncates the value of $x(t)$ to nearest integer (e.g., $\text{INT}[3.8] = 3$)

**1.20** The linear system represented in Figure P1.20 contains no independent sources and is driven by an external current and a voltage source. When $V_s = 5$ V and $I_s = 0$ A, $I = 10$ A. When $V_s = 0$ V and $I_s = -10$ A, $I = 5$ A. Find $I$ if $V_s = 20$ V and $I_s = 10$ A.

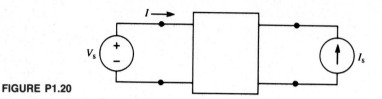

**FIGURE P1.20**

# Circuit Laws and Simple Equivalent Circuits

2

## 2.1 INTRODUCTION

In Chapter 1, you were introduced to independent voltage and current sources, resistors, and controlled sources. With the aid of Ohm's law and some intuition, several very simple resistive circuits were analyzed. If we try to use the same intuition to analyze the simple circuit in Figure 2.1, we quickly run into trouble.

In this circuit, two resistors are connected to the ideal voltage source. The location of the source tells us that the voltage between points a and c is 10 V. What, then, are the values of $V_{R_1}$ and $V_{R_2}$, the resistor voltages? We do not know—yet. A general circuit cannot be analyzed using Ohm's law alone.

In this chapter, we will consider the two basic laws that govern the behavior of voltage and current in electrical circuits. These laws were discovered experimentally by the German physicist Gustav R. Kirchhoff (1824–1887). With Kirchhoff's laws and Ohm's law, you will be able to analyze all resistor circuits.

When we analyze certain small circuits, such as the ones you will see in electronics courses, the straightforward application of Kirchhoff's and Ohm's laws leads to unnecessarily complicated sets of equations. It is often easier to represent a circuit with an equivalent circuit that can be analyzed by inspection. We will discuss the meaning of equivalent circuits and examine several common ones in this chapter.

First, however, we should review some notation used in analyzing circuits. Figure 2.2 shows a general circuit in which the rectangles represent arbitrary two-terminal elements. They may be sources, resistors, or elements yet to be described; it does not matter. The rectangles are often referred to as **black boxes.**

The points at which terminals are connected together are called **nodes.** The circuit in Figure 2.2 has nodes lettered from a to f. The lines drawn from the nodes to the boxes are always assumed to be perfect conductors; that is, they have zero resistance. Thus, node b, for example, includes everything within the shaded area. Elements (and their terminals) connected between nodes are known as **branches.** The branches of this circuit are labeled from 1 to 7. The voltages and currents for the elements are known as **branch variables.**

As you may recall from Chapter 1, a closed conducting path is required for charges to flow continuously. A closed path is a **loop.** This circuit has three loops.

**FIGURE 2.1**   *Simple Circuit*

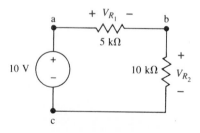

**FIGURE 2.2** *General Circuit*

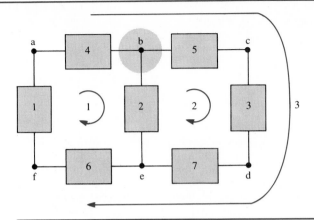

One loop contains branches 1, 4, 2, and 6; the second loop contains branches 2, 5, 3, and 7; and the third loop contains branches 1, 4, 5, 3, 7, and 6.

## 2.2 KIRCHHOFF'S CURRENT LAW (KCL)

Kirchhoff examined empirically the relationships among the currents that enter and leave nodes in an electrical circuit. He found that these relationships could be described by the following statement, now known as Kirchhoff's current law (KCL).

*Kirchhoff's Current Law:* The sum of the currents entering a node equals the sum of the currents leaving the node.

Figure 2.3 demonstrates Kirchhoff's current law. This particular circuit has six nodes. The currents shown in the figure are the **branch currents.** With the

**FIGURE 2.3** *KCL Example Circuit*

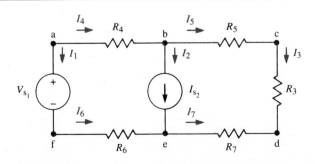

exception of current sources, branch current directions may be arbitrarily assigned. For a current source, the current direction is given. The KCL equations at each node are as follows:

$$\text{Entering} = \text{Leaving}$$

Node a: $\qquad 0 = I_1 + I_4$

Node b: $\qquad I_4 = I_2 + I_5$

Node c: $\qquad I_5 = I_3$

Node d: $\quad I_3 + I_7 = 0$

Node e: $\quad I_2 + I_6 = I_7$

Node f: $\qquad I_1 = I_6$

Kirchhoff's current law confirms what you might suspect if you remember that current is a measure of charge flow. If KCL is not true, then it would be possible to take away more charge from a node than is delivered. This would mean that charge is somehow being created at the node. Since charge is conserved, this situation is an impossibility. Likewise, in conducting circuits, charge cannot accumulate at any point. It is not possible, therefore, to deliver more charge than is removed.

---

**Example 2.1**    *KCL at One Node*

Find the current $I$ shown in Figure 2.4.

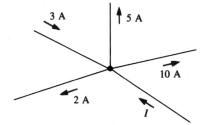

**FIGURE 2.4**

Applying KCL, we sum the currents entering and equate to the sum of the currents leaving:

$$3\,\text{A} + I = 5\,\text{A} + 10\,\text{A} + 2\,\text{A}$$

$$I = 14\,\text{A}$$

---

**Example 2.2**    *KCL at Several Nodes*

Find the unknown currents $I_1$, $I_2$, and $I_3$ shown in Figure 2.5.

When you use KCL at several nodes, concentrate on one node at a time. For instance, in Figure 2.5A, when considering node a, do not worry

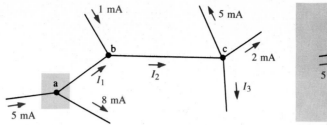
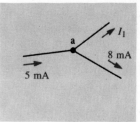

FIGURE 2.5     A. Three-node circuit                     B. Isolated node a

about what is happening at nodes b or c. This point is emphasized here (in this example only) by highlighting node a and showing this node in isolation in Figure 2.5B. Applying KCL at node a, we get

$$5\,\text{mA} = 8\,\text{mA} + I_1$$
$$I_1 = -3\,\text{mA}$$

Then, focusing on node b, we get

$$I_1 + 1\,\text{mA} = I_2$$
$$-3\,\text{mA} + 1\,\text{mA} = I_2$$
$$I_2 = -2\,\text{mA}$$

And finally, at node c, we get

$$I_2 = 5\,\text{mA} + 2\,\text{mA} + I_3$$
$$-2\,\text{mA} = 7\,\text{mA} + I_3$$
$$I_3 = -9\,\text{mA}$$

This example demonstrates a very important point. When you solve for circuit variables, you may get negative answers. This simply means that the currents solved for have directions opposite to the ones assigned in the example. Remember, negative answers are not *wrong*. The sign of the current and the arrow direction in the circuit provide a complete and correct answer.

There are two alternate expressions for Kirchhoff's current law.

*KCL Alternate #1:*   The sum of all currents entering a node is zero.

*KCL Alternate #2:*   The sum of all currents leaving a node is zero.

In Alternate #1, currents entering a node are considered to be positive and currents leaving a node to be negative. In Alternate #2, the reverse is true. If we look back to Figure 2.3, we get the following two sets of equations:

KCL alternate #1

Node a:  $-I_1 - I_4 = 0$

Node b:  $-I_2 + I_4 - I_5 = 0$

Node c:  $-I_3 + I_5 = 0$

Node d:  $I_3 + I_7 = 0$

Node e:  $I_2 + I_6 - I_7 = 0$

Node f:  $I_1 - I_6 = 0$

KCL alternate #2

Node a:  $I_1 + I_4 = 0$

Node b:  $I_2 - I_4 + I_5 = 0$

Node c:  $I_3 - I_5 = 0$

Node d:  $-I_3 - I_7 = 0$

Node e:  $-I_2 - I_6 + I_7 = 0$

Node f:  $-I_1 + I_6 = 0$

You can see that the three ways of expressing Kirchhoff's current law are mathematically identical. The method used is a matter of personal preference. It is also possible to use one formulation of KCL at one node and a different formulation at another. We will be discussing some systematic and algorithmic approaches to analyzing circuits in later chapters. In these approaches, we usually adopt one formulation and use it throughout.

Kirchhoff's current law is not limited in use just to nodes. It is also used to describe the relationships among the currents that are entering or leaving any part of a circuit that can be enclosed within an imaginary surface. For example, Figure 2.6 shows the same nodes and currents as those in Example 2.2. However, in Figure 2.6, all three nodes are enclosed within the highlighted surface, and therefore it is not necessary to show the currents $I_1$ and $I_2$ (they are within the colored surface). The current $I_3$ is leaving and is the variable we will solve for. Equating the currents entering the surface to the currents leaving the surface yields

$$5\,\text{mA} + 1\,\text{mA} = 8\,\text{mA} + 5\,\text{mA} + 2\,\text{mA} + I_3$$

$$I_3 = -9\,\text{mA}$$

This answer is the same one obtained in Example 2.2. However, we have solved one equation instead of three. The set of branches that are cut by the highlighted surface is a **cut-set.** Cut-set analysis is a very powerful tool in advanced circuit analysis. We will make limited use of it in this book.

**FIGURE 2.6**   *Cut-Set*

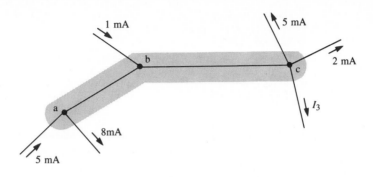

Now, let's use KCL and Ohm's law to analyze a simple yet important circuit. Figure 2.7 shows a current source connected to two resistors. Note that in this circuit arrangement there are only two nodes.

We start the analysis by writing the KCL equation for the top node to get

$$I_{in} = I_1 + I_2$$

Again, you must use your insight to recognize that there is only one voltage in this circuit. (Shortly, we will examine a formal method for proving this.) The voltage $V$ here appears across both resistors. From Ohm's law, we have

$$I_1 = \frac{V}{R_1} \qquad \text{and} \qquad I_2 = \frac{V}{R_2}$$

Substituting these expressions into the KCL equation gives

$$I_{in} = \frac{V}{R_1} + \frac{V}{R_2} = V\left(\frac{1}{R_1} + \frac{1}{R_2}\right) = V\left(\frac{R_1 + R_2}{R_1 R_2}\right)$$

**FIGURE 2.7**   *Two-Node Circuit*

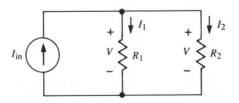

We can now find the unknown variable $V$:

$$V = I_{in}\left(\frac{R_1 R_2}{R_1 + R_2}\right)$$

The two currents are found from Ohm's law to be

$$I_1 = I_{in}\left(\frac{R_2}{R_1 + R_2}\right) \quad \text{and} \quad I_2 = I_{in}\left(\frac{R_1}{R_1 + R_2}\right)$$

We will reconsider this particular circuit later in the chapter.

This example demonstrates a very common procedure in analyzing circuits. We started with an equation relating currents; we used Ohm's law to replace the branch currents with a branch voltage; we solved for the branch voltage; and finally, we again used Ohm's law to solve for the branch currents.

## 2.3   KIRCHHOFF'S VOLTAGE LAW (KVL)

When Kirchhoff examined the voltages that exist in a circuit, he found the following relationship, now known as Kirchhoff's voltage law (KVL).

*Kirchhoff's Voltage Law:*   The sum of the voltage rises equals the sum of the voltage drops around any closed path.

Figure 2.8 shows the circuit of Figure 2.3 but with *branch voltages* assigned. Here, the polarities of the branch voltages are arbitrary, except for the voltage source. Again, for a voltage source, the polarity is given.

The circuit in Figure 2.8 has three complete paths, or loops. The arrows indicate the loops and the direction we will take around each loop. Applying KVL

FIGURE 2.8   *KVL Example Circuit*

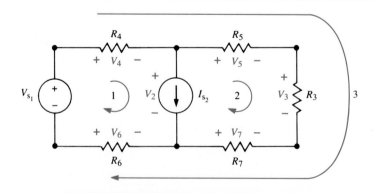

to each loop results in

$$\text{Loop 1:} \quad V_{s_1} + V_6 = V_2 + V_4$$
$$\text{Loop 2:} \quad V_2 + V_7 = V_3 + V_5$$
$$\text{Loop 3:} \quad V_{s_1} + V_6 + V_7 = V_3 + V_4 + V_5$$

---

**Example 2.3**  *KVL Around One Loop*

Find the voltage $V_1$ shown in Figure 2.9.

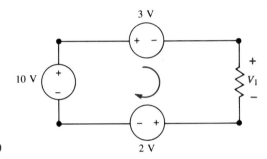

**FIGURE 2.9**

$$10\text{ V} = V_1 + 2\text{ V} + 3\text{ V}$$
$$V_1 = 5\text{ V}$$

---

The expression of KVL stated at the beginning of this section does not lead to a systematic approach to writing loop equations. Two alternate expressions for Kirchhoff's voltage law provide systematic approaches to loop analysis.

***KVL Alternate #1:***  The sum of all voltage rises around a loop is zero. (A voltage rise in the direction of travel is positive, while a voltage drop is negative.)

***KVL Alternate #2:***  The sum of all voltage drops around a loop is zero. (A voltage drop in the direction of travel is positive, while a voltage rise is negative.)

---

**Example 2.4**  *KVL Alternate #1 (Counterclockwise Direction)*

Find the voltage $V_1$ shown in Figure 2.10.
Since voltage rises are considered positive, the KVL equation is

*(continues)*

**Example 2.4**   *Continued*

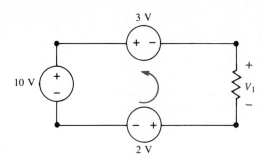

**FIGURE 2.10**

$$V_1 + 3\,V - 10\,V + 2\,V = 0$$
$$V_1 = 5\,V$$

**Example 2.5**   *KVL Alternate #1 (Clockwise Direction)*

Find the voltage $V_1$ shown in Figure 2.11.

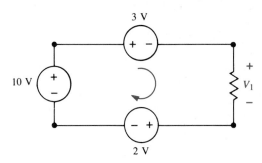

**FIGURE 2.11**

$$-V_1 - 2\,V + 10\,V - 3\,V = 0$$
$$V_1 = 5\,V$$

**Example 2.6**   *KVL Alternate #2 (Clockwise Direction)*

Find the voltage $V_1$ shown in Figure 2.12.

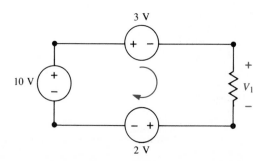

**FIGURE 2.12**

Since voltage drops are considered positive, the KVL equation is

$$V_1 + 2\,V - 10\,V + 3\,V = 0$$
$$V_1 = 5\,V$$

Examples 2.4, 2.5, and 2.6 demonstrate that the direction taken around a loop and the definition of a positive voltage are arbitrary. You must, however, be consistent within each loop.

**Example 2.7**   *Incomplete Circuit*

The circuit in Figure 2.13 has an open circuit between terminals a and b. Because of this open circuit, a current cannot exist. However, a voltage does exist between a and b! This voltage is found using KVL around the loop (which includes the open circuit):

$$-5\,mV - 3.2\,mV + V_{ab} + 4\,mV = 0$$
$$V_{ab} = 4.2\,mV$$

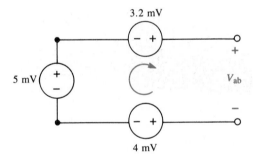

**FIGURE 2.13**

**Example 2.8**   *Incomplete Circuit with Resistor*

Find the voltage $V_o$ shown in Figure 2.14.

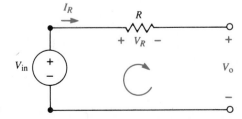

**FIGURE 2.14**

*(continues)*

**Example 2.8** *Continued*

Because there is an open circuit,

$$I_R = 0$$

$$V_R = I_R R = 0$$

From KVL,

$$V_o = V_{in} - V_R = V_{in}$$

---

Now, let's analyze a simple one-loop source–resistor circuit by using KVL, Ohm's law, and KCL. Figure 2.15 shows such a single-loop circuit. The application of KCL implies that since there is only one closed path, there is only one current in this circuit. We call this current $i(t)$ and label each branch current accordingly.

If we write KVL around this loop, we get

$$-\cos 3t + v_1 + v_2 + v_3 = 0$$

We use Ohm's law to write each branch voltage in terms of the single branch current $i$:

$$-\cos 3t + 2000i + 3000i + 5000i = 0$$

$$10{,}000i = \cos 3t$$

$$i(t) = .1 \cos 3t \text{ mA}$$

We again use Ohm's law to find the branch voltages:

$$v_1(t) = .2 \cos 3t \text{ V}$$

$$v_2(t) = .3 \cos 3t \text{ V}$$

$$v_3(t) = .5 \cos 3t \text{ V}$$

**FIGURE 2.15** *Single-Loop Circuit*

---

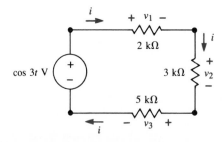

## 2.4  BRANCH VARIABLE METHOD

To analyze general resistive circuits, we must use KCL, KVL, and Ohm's law. Consider the circuit in Figure 2.16A. This circuit has five branches. Since each branch has two variables (a voltage and a current) associated with it, there are ten branch variables in this circuit. Thus, we must find ten *independent* equations relating the branch voltages and currents.

For each branch, we have an equation relating its voltage to its current (its *V–I* characteristics). If the branch is a resistor, we use Ohm's law. If the branch is a voltage source, the branch voltage equals the value of the source. If the branch is a current source, the branch current equals the value of the source. We have, therefore, *five independent branch equations* for this circuit.

At this point, let's write the KCL equations:

Node a:   $I_1 + I_2 = 0$

Node b:   $-I_2 + I_3 + I_4 = 0$

Node c:   $-I_4 + I_5 = 0$

Node d:   $-I_1 - I_3 - I_5 = 0$

There are four KCL node equations for this circuit. Are they independent equations? The answer is no. Any one of the node equations can be found by adding together the other three. Try it yourself. We state, without proof, the following formula:

> Number of independent KCL node equations = $N - 1$,
> where $N$ = total number of nodes

FIGURE 2.16   *Example Circuit for Branch Variable Method*

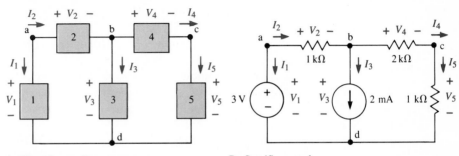

A. General example                                 B. Specific example

Fortunately, any $N-1$ set of node equations will be independent. We simply ignore one of the nodes of the circuit. The node at which we do not write a KCL equation is known as the **datum node.** If a node in a particular circuit is connected to ground, we usually use this node as the datum. However, any node in the circuit can be used as the datum node. In this circuit, there are *three independent node equations.* Combining these equations with the five branch *V–I* characteristics raises the total number of equations derived thus far to eight.

We need two more independent equations (remember, there are ten variables). We have not yet used KVL. This circuit has three loops. The KVL equations for these loops are

Loop 1:   $-V_1 + V_2 + V_3 = 0$
Loop 2:   $-V_3 + V_4 + V_5 = 0$
Loop 3:   $-V_1 + V_2 + V_4 + V_5 = 0$

Again, the total set of KVL equations for a circuit is, in general, not independent. The combination of any two of the preceding equations will yield the third. Thus, there are only *two independent loop equations* in this example. Do not let this fool you! The number of independent KVL equations is *not* equal to the total number of loops minus 1.

We derive the formula for the number of independent KVL loops with the following reasoning. Any circuit with $B$ branches has $2B$ unknowns and requires $2B$ independent equations for solution. The branch voltage–current relationships provide $B$ equations, while the KCL node equations provide $N-1$. The KVL loop equations must supply the difference $2B - B - (N-1)$ so that

$$\text{Number of independent KVL loop equations} = B - N + 1$$

Since the circuit in Figure 2.16A has five branches and four nodes, according to this formula, there are two independent KVL loop equations. This is consistent with what we have already determined.

Finding the independent set of KVL equations is not so simple in general, as this particular example implies. To find the independent sets of loop equations for a large circuit requires the knowledge of circuit concepts that are beyond the scope of this book.

There is a simple technique that will always yield an independent set of KVL equations for planar networks. A **planar** network can be drawn on a flat sheet of paper without any lines crossing. The concept of planar networks is examined in more detail later in the problems section. Most of the circuits that we will deal with in this text are planar networks. For these circuits, we define a special type of loop known as a mesh. A **mesh** is a loop that does not enclose any other loops. In Figure 2.16, for example, the loops composed of branches 1, 2, and 3 and 3, 4, and 5 are meshes. The loop of branches 1, 2, 4, and 5 is not a mesh because it encloses the other two loops.

We can see that in this circuit there are two meshes. We have also determined that there are two independent KVL equations for this circuit. This is no coincidence. If we confine our KVL equations in planar networks to meshes, we will always have a complete set of independent loop equations. Thus, we have the alternate formula:

> Number of independent KVL loop equations = number of meshes

Applying the branch variable method to the circuit in Figure 2.16B, using node d as the datum node, and writing KVL for the meshes, we get the following set of equations:

Branch $V$–$I$ relationships

$V_1 = 3\,V$     (voltage source)

$I_3 = 2\,mA$     (current source)

$V_2 = 1000I_2$

$V_4 = 2000I_4$

$V_5 = 1000I_5$

KCL node equations

Node a:   $I_1 + I_2 = 0$

Node b:   $-I_2 + I_3 + I_4 = 0$

Node c:   $-I_4 + I_5 = 0$

KVL mesh equations

Mesh 1:   $-V_1 + V_2 + V_3 = 0$

Mesh 2:   $-V_3 + V_4 + V_5 = 0$

We thus have ten equations in ten variables. Usually, when we are using this technique to find circuit variables, we can greatly simplify the analysis. For example, in this circuit, since we know that $V_1 = -3\,V$ and $I_3 = 2\,mA$, we can use these values immediately. Also, since we see by inspection of this circuit that $I_1 = -I_2$ and $I_4 = I_5$, we do not have to write KCL equations at nodes a and c. In addition, we can avoid explicitly writing the Ohm's law equations by using the resistor $V$–$I$ relationships directly in the KCL or the KVL equations. In this example, we can write the KVL equations using branch currents instead of branch voltages for all resistors. The actual set of equations we would then have to solve is as follows:

$$-I_2 + I_4 = -0.002$$

$$1000I_2 + V_3 = 3$$

$$-V_3 + 2000I_4 + 1000I_4 = 0 \qquad \text{(remember, } I_5 = I_4\text{)}$$

From these equations, we find $I_2$, $I_4$ and $V_3$. By using the original equations for this circuit, the branch voltages and currents are found to be

$$V_1 = 3 \text{ V} \qquad I_1 = -2.25 \text{ mA}$$
$$V_2 = 2.25 \text{ V} \qquad I_2 = 2.25 \text{ mA}$$
$$V_3 = .75 \text{ V} \qquad I_3 = 2 \text{ mA}$$
$$V_4 = .5 \text{ V} \qquad I_4 = I_5 = .25 \text{ mA}$$
$$V_5 = .25 \text{ V}$$

---

**Example 2.9** *Application of Branch Variable Method*

Find the branch voltages and currents indicated in Figure 2.17.

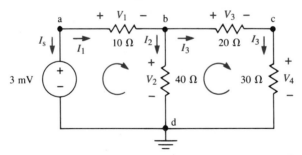

**FIGURE 2.17**

Since node d is grounded, we will use it as the datum node. Also, we immediately recognize that the same current exists in the 20 Ω and 30 Ω resistors and label it accordingly. It is also obvious that $I_s = -I_1$. We therefore need a KCL equation only at node b. We write the KVL mesh equations in terms of branch currents by using Ohm's law.

$$\text{KCL at node b:} \quad -I_1 + I_2 + I_3 = 0$$
$$\text{KVL for mesh 1:} \quad -3 \cdot 10^{-3} + V_1 + V_2 = 0$$
$$\text{By Ohm's law:} \quad -3 \cdot 10^{-3} + 10I_1 + 40I_2 = 0$$
$$\text{KVL for mesh 2:} \quad -V_2 + V_3 + V_4 = 0$$
$$\text{By Ohm's law:} \quad -40I_2 + 20I_3 + 30I_3 = 0$$

We now have three equations in three unknowns:

$$-I_1 + I_2 + I_3 = 0$$

$$10I_1 + 40I_2 = 3 \cdot 10^{-3}$$

$$-40I_2 + 50I_3 = 0$$

These equations can be solved to find

$$I_1 = \frac{27}{290} \text{ mA}$$

$$I_2 = \frac{15}{290} \text{ mA}$$

$$I_3 = \frac{12}{290} \text{ mA}$$

Application of Ohm's law then leads to

$$V_1 = \frac{27}{29} \text{ mV}$$

$$V_2 = \frac{60}{29} \text{ mV}$$

$$V_3 = \frac{24}{290} \text{ mV}$$

$$V_4 = \frac{36}{29} \text{ mV}$$

---

**Example 2.10**   *Another Application of Branch Variable Method*

Find the branch voltages and currents indicated in Figure 2.18.

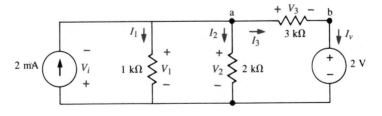

**FIGURE 2.18**

By inspection:   $I_v = I_3$

$$V_1 = V_2 = -V_i$$

(*continues*)

**Example 2.10** *Continued*

KCL at node a:    $-2 \cdot 10^{-3} + I_1 + I_2 + I_3 = 0$

By Ohm's law:    $\dfrac{V_1}{10^3} + \dfrac{V_1}{2 \cdot 10^3} + \dfrac{V_3}{3 \cdot 10^3} = 2 \cdot 10^{-3}$

KVL on right side:    $V_1 - V_3 = 2\,\text{V}$

The answers are as follows:

$$V_1 = V_2 = \frac{16}{11}\,\text{V}$$

$$V_3 = -\frac{6}{11}\,\text{V}$$

$$I_1 = \frac{16}{11}\,\text{mA}$$

$$I_2 = \frac{8}{11}\,\text{mA}$$

$$I_3 = -\frac{2}{11}\,\text{mA}$$

---

**Example 2.11** *Transistor Amplifier*

Find $v_o$ for the AC transistor circuit modeled in Figure 2.19.

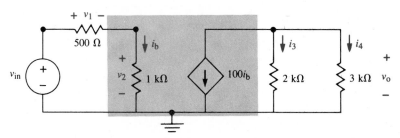

**FIGURE 2.19**

By inspection, the left side has only one current, $i_b$, and the right side has only one voltage, $v_o$.

KVL on left side:    $-v_{in} + v_1 + v_2 = 0$

$$(500 + 1000)i_b = v_{in}$$

Thus,

$$i_b = \frac{v_{in}}{1.5 \cdot 10^3}\,\text{A}$$

KCL on right side:   $100i_b + i_3 + i_4 = 0$

$$\left(\frac{1}{2000} + \frac{1}{3000}\right)v_o = -100i_b$$

Answer:   $v_o = -1.2 \cdot 10^5 i_b = -80v_{in}$

---

**Example 2.12**   *Op Amp Circuit*

Find $v_o$ for the op amp circuit in Figure 2.20A. Assume $r_{in} = 1\,M\Omega$, $r_o = 0\,\Omega$, and $A = 10^5$.

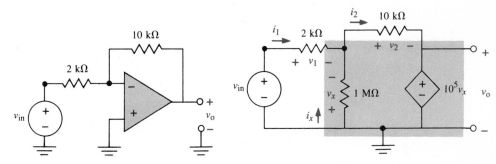

**FIGURE 2.20**

From the modeled circuit shown in Figure 2.20B, we note that due to the voltage source at the output, $v_o = 10^5 v_x$. Therefore, once we find $v_x$, we know $v_o$.

KCL:   $-i_1 - i_x + i_2 = 0$      (1)

By Ohm's law:   $-\dfrac{v_1}{2 \cdot 10^3} - \dfrac{v_x}{10^6} + \dfrac{v_2}{10^4} = 0$   (2)

KVL for right mesh:   $v_x + v_2 + 10^5 v_x = 0$

$\qquad\qquad\qquad\qquad v_2 = -(10^5 + 1)v_x \cong -10^5 v_x$

KVL for left mesh:   $-v_{in} + v_1 - v_x = 0$

$\qquad\qquad\qquad\qquad v_1 = v_{in} + v_x$

Substituting for $v_1$ and $v_2$ in equation 2, we get

$$-\frac{(v_{in} + v_x)}{2 \cdot 10^3} - \frac{v_x}{10^6} - \frac{10^5 v_x}{10^4} = 0$$

So,

$$v_x = -\frac{500}{10^7} v_{in}$$

$$v_o = 10^5 v_x = -5v_{in}$$

**Example 2.13**   *Design Example 1*

In design problems, desired circuit variable values are known. The engineer must choose circuit element values to obtain the given output values. The circuit equations are the same as in analysis; the unknown variables are different.

For the circuit in Figure 2.21, find $V_{in}$ so that $I_1 = 2\,mA$.

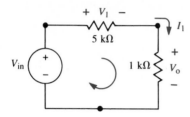

**FIGURE 2.21**

Given $I_1 = 2\,mA$:   $V_o = 2\,V$
$$V_1 = 10\,V$$

KVL:   $-V_{in} + V_1 + V_o = 0$

Answer:   $V_{in} = 12\,V$

**Example 2.14**   *Design Example 2*

For the circuit in Figure 2.22, find $R$ so that $V_o = 3\,V$.

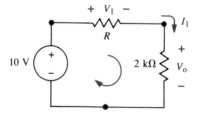

**FIGURE 2.22**

Given $V_o = 3\,V$:   $I_1 = 1.5\,mA$

KVL:   $-10 + V_1 + 3 = 0$
$$V_1 = 7\,V$$

Ohm's law:   $R = \dfrac{V_1}{I_1} = \dfrac{7}{1.5 \cdot 10^{-3}} = 4.67\,k\Omega$

The power dissipated in the resistor is $I_1^2 R = 10.5\,mW$. Therefore, we can safely use a 1/8 W resistor.

**Example 2.15**   *Design Example 3*

Find $V_{in}$ in Figure 2.23 so that $V_4$ will be 5 V.

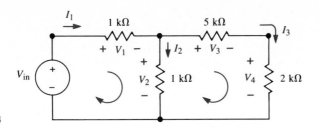

**FIGURE 2.23**

By Ohm's law: $I_3 = \dfrac{V_4}{2000} = \dfrac{5}{2000} = 2.5 \text{ mA}$

If $I_3 = 2.5 \text{ mA}$: $V_3 = (5000)(2.5 \cdot 10^{-3}) = 12.5 \text{ V}$

KVL: $V_2 = V_3 + V_4 = 17.5 \text{ V}$

Ohm's law: $I_2 = \dfrac{V_2}{1000} = 17.5 \text{ mA}$

KCL: $I_1 = I_2 + I_3 = 20 \text{ mA}$

Ohm's law: $V_1 = I_1 \cdot 1000 = 20 \text{ V}$

KVL: $V_{\text{in}} = V_1 + V_2 = 37.5 \text{ V}$

## 2.5  EQUIVALENT CIRCUITS

The two terminal boxes in Figure 2.24A represent two different circuits, or networks. Network 1 may contain a few resistors and sources, while network 2

**FIGURE 2.24**   *Equivalent Circuits*

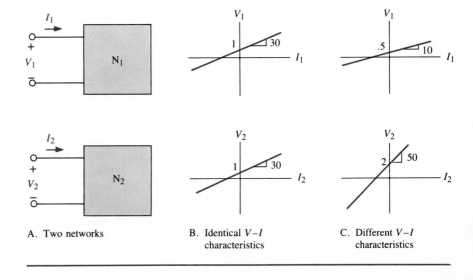

A. Two networks

B. Identical V–I
   characteristics

C. Different V–I
   characteristics

may contain thousands. If the two networks have identical $V$–$I$ relationships at their terminals, they are said to be **equivalent.** Examples of their $V$–$I$ relationships are shown graphically in Figures 2.24B and 2.24C. In Figure 2.24B, the two network relationships are the same. Under these conditions, we say that network 1 and network 2 are equivalent and either can be used to produce the desired result at the terminals. In Figure 2.24C, on the other hand, the two networks have different terminal $V$–$I$ characteristics. In this case, the two networks are not equivalent to each other.

Note that if two networks are equivalent at one pair of their terminals, it does not imply that they are equivalent at any other pair. If the first network contains few elements and the second network contains many, it is not reasonable to assume that the voltages and currents inside the two boxes have any relationship to each other. All we are saying is that if we are "outside the boxes looking in," then we cannot tell the difference between the two.

Certain equivalent circuits are commonly used in analysis or design. These equivalents are series and parallel combinations of elements and source transformations and are discussed next.

## Series Combinations of Elements

Many circuits have nodes to which only two branches are connected. See, for example, Figure 2.25A. The two branches are said to be in **series** with each other. An important result of two branches in series is that the current in each branch is the same. This result is easily verified using KCL.

In the circuit shown in Figure 2.25B, the two voltage sources are in series, $R_2$ and $R_3$ are in series, and $R_5$ and $R_6$ are in series. At nodes b, d, and f, more than two elements are connected. Therefore, there are no series combinations at these nodes.

**Resistors in Series.**  Let's examine two resistors in series, as shown in Figure 2.26A. We claim that the single resistor $R_{eq}$ in Figure 2.26B is equivalent to the

**FIGURE 2.25**   *Elements in Series*

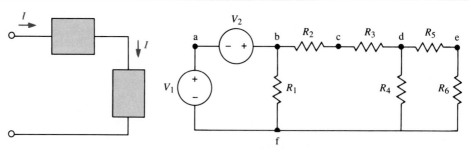

A. Two elements in series          B. Circuit containing series connections

**FIGURE 2.26**   *Resistors in Series*

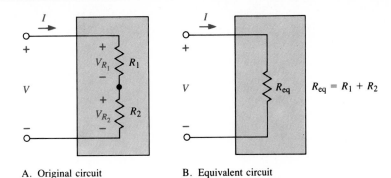

A. Original circuit          B. Equivalent circuit

two resistors in Figure 2.26A. For this equivalence to be true, the *V–I* characteristic must be the same in both cases. We will use circuit laws to verify the equivalence. In Figure 2.26A, we have

$$V = V_{R_1} + V_{R_2} = IR_1 + IR_2 = I(R_1 + R_2)$$

From Figure 2.26B,

$$V = IR_{eq}$$

The two network functions will be identical if and only if

$$R_{eq} = R_1 + R_2$$

Thus, we have our first important equivalence: *Two resistors in series can be replaced by a single resistor whose value is the sum of the individual resistors.* This relationship can be easily extended to any number of series resistors to get

*Series Resistors*

$$R_{eq} = R_1 + R_2 + R_3 + \cdots$$

(2.1)

It is clear from Equation 2.1 that the largest resistor in a series combination dominates.

**Example 2.16**   *Resistors in Series*

Given the series resistors in the original circuits shown in Figure 2.27A, find the equivalent resistor for each.

*(continues)*

**Example 2.16**   *Continued*

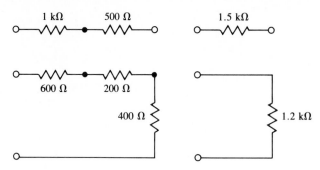

**FIGURE 2.27**   A. Original circuits                   B. Equivalent circuits

As shown in Figure 2.27B,

$$1 \text{ k}\Omega + 500 \ \Omega = 1.5 \text{ k}\Omega$$

$$100 \ \Omega + 200 \ \Omega + 400 \ \Omega + 600 \ \Omega = 1.2 \text{ k}\Omega$$

---

**Example 2.17**   *Constructing Required Resistor*

Given four resistors whose values are $100 \ \Omega$, $300 \ \Omega$, $500 \ \Omega$, and $1 \text{ k}\Omega$, combine them in series to produce the following required resistor values:

**a.**   $400 \ \Omega$     **b.**   $1.3 \text{ k}\Omega$     **c.**   $1.9 \text{ k}\Omega$

Figure 2.28 shows how the resistors are combined.

       100 Ω  300 Ω
**a.** 400 Ω = o—⋁⋁—•—⋁⋁—o

       1 kΩ   300 Ω
**b.** 1.3 kΩ = o—⋁⋁—•—⋁⋁—o

       1 kΩ   500 Ω   300 Ω   100 Ω
**FIGURE 2.28**   **c.** 1.9 kΩ = o—⋁⋁—•—⋁⋁—•—⋁⋁—•—⋁⋁—o

---

**Example 2.18**   *Analyzing Series Resistor Circuit*

Solve for $v_o(t)$ in the original circuit shown in Figure 2.29A. This can be done fairly simply with the branch variable method. However, it is quicker to use equivalent circuits.

  First recognize that since all of the resistors are in series, they may be added together to get the single-resistor equivalent circuit shown in Figure 2.29B. Two points to consider are: (1) the output voltage $v_o$ does not appear in the equivalent circuit, and (2) the current $i(t)$ is the same in both circuits. From the equivalent circuit, $i(t)$ is found by inspection to be

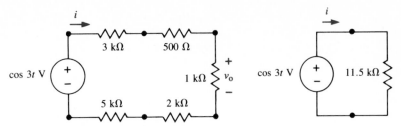

**FIGURE 2.29**    A. Original circuit                                B. Equivalent circuit

$$i(t) = \frac{\cos 3t}{11{,}500} = 87 \cos 3t \; \mu A$$

Now return to the original circuit to find

$$v_o(t) = i(t) \cdot 10^3 = 87 \cos 3t \; mV$$

This example demonstrates a common occurrence when equivalent circuits are used. Often, the variable of interest is lost when the circuit is simplified, and you must return to the original circuit to find the desired variable. However, this approach is often faster than applying the branch variable method or the formal techniques of Chapter 3.

**Voltage Sources in Series.**    Figure 2.30A shows a series connection of two voltage sources; the equivalent circuit is shown in Figure 2.30B. If the terminal voltage $V_{ab}$ is to be the same for the two circuits, then by using KVL we find

$$V_s = V_1 + V_2$$

**FIGURE 2.30**    *Voltage Sources in Series*

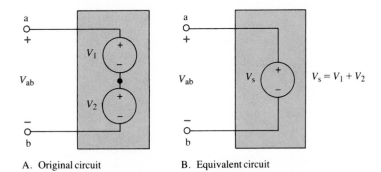

A. Original circuit                    B. Equivalent circuit

That is, *voltage sources in series add*. This statement is true for any number of voltage sources in series. You must be careful of source polarities, however, as the following example demonstrates.

---

**Example 2.19**    *Voltage Sources in Series*

Refer to Figure 2.31 and note the following:

a.  If the voltage sources are connected together at both positive terminals, or both negative terminals, their magnitudes subtract.

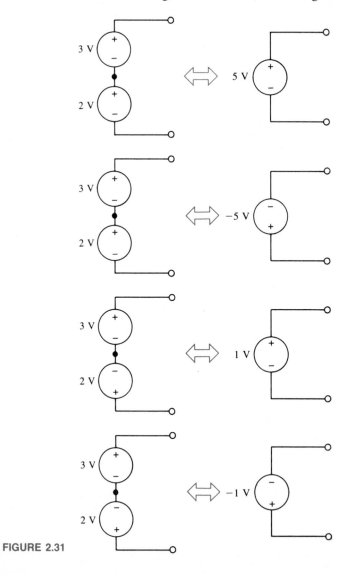

**FIGURE 2.31**

b. If a positive terminal is connected to a negative terminal, then the magnitudes add.
c. The final sign of the equivalent voltage source depends on the polarity assigned to it.

**Example 2.20**   *Series Voltage–Resistor Circuit*

Use series combinations to find $I$ and $V_o$ in Figure 2.32A.

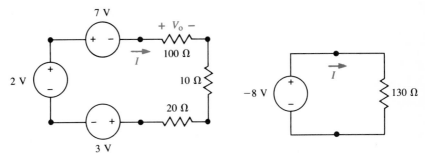

**FIGURE 2.32**   A. Original circuit          B. Equivalent circuit

Since the equivalent source has its plus side up, its value is found from

$$V_s = -3\,\text{V} + 2\,\text{V} - 7\,\text{V} = -8\,\text{V}$$

The equivalent resistor is 130 $\Omega$, so we find from the equivalent circuit in Figure 2.32B,

$$I = -\frac{8}{130}\,\text{A}$$

$$V_o = 100I = -\frac{80}{13}\,\text{V}$$

**Example 2.21**   *Another Series Voltage–Resistor Circuit*

Find $V_o$ in the circuit shown in Figure 2.33.

This circuit is a bit trickier than the one in Figure 2.32 because the resistors do not appear to be in series, nor do the voltage sources. However, it is clear that the top resistor and voltage source are in series with each other. We do not change the behavior of the circuit by redrawing the top part of the circuit as shown on the right. The same is true of the bottom part of the circuit.

The equivalent circuit has a single voltage source with a value of 20 mV (plus on top) and a single resistor of 600 $\Omega$. Thus,

*(continues)*

**Example 2.21**   *Continued*

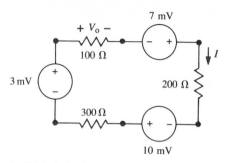 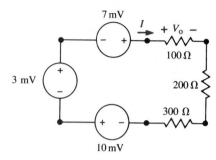

**FIGURE 2.33**   A. Original circuit

B. Rearranged circuit

$$I = \frac{0.02}{600} = \frac{1}{30} \text{ mA}$$

$$V_o = \frac{10}{3} \text{ mV}$$

**Current Sources in Series.**   The circuit in Figure 2.34A represents a possible contradiction. A current source in a branch means that the branch current must equal the value of the source. Therefore, the current between node b and node a must equal $I_1$, and the current between node c and node b must equal $I_2$. However, since the two sources are in series, their branch currents must be the same. This is not possible unless the two current sources have the same magnitude and direction. If this is true, then the equivalent current source, shown in Figure 2.34B, is just equal to the value of the common current.

**FIGURE 2.34**   *Current Sources in Series*

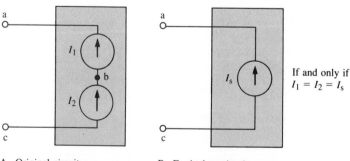

A. Original circuit

B. Equivalent circuit

What is gained by having current sources in series? In a practical sense, the use of two identical current sources instead of one will lead to more available power to drive the circuit. While ideal current sources can supply limitless power, real sources cannot. Increasing the number of identical current sources in series increases the available power.

**Current Source in Series with Resistor.** Figure 2.35A shows a circuit with a current source in series with a resistor. Does the current that enters node a depend on the series resistor $R$? The answer is no. The current source and the resistor are in series, so they must have the same current in them. The value of this current is set by the current source and is not dependent on the value of the resistor. Since the current entering node a is not dependent on $R$, this resistor can be removed from the circuit—that is, set to 0—as shown in Figure 2.35B.

Keep in mind that the voltage across the current source in the original circuit does depend on $R$. Analyzing Figure 2.35B yields

$$I_1 = \frac{10}{3} \text{ mA} \quad \text{and} \quad I_2 = \frac{20}{3} \text{ mA}$$

To find $V_I$, we use this information in the original circuit. From KVL, we get

$$-V_I = V_R + V_1 = (0.01 \cdot R) + (I_1 \cdot 20) = 0.01 \cdot R + 0.067 \text{ V}$$

It is obvious that $V_I$ is a function of $R$.

## Parallel Combinations of Elements

A parallel connection of two elements is shown in Figure 2.36A. For two elements to be in parallel, they must be connected together at *both* terminals. This concept can be extended to several elements in parallel as shown in Figure 2.36B.

**FIGURE 2.35** *Current Source in Series with Resistor*

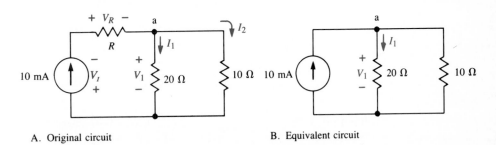

A. Original circuit          B. Equivalent circuit

**FIGURE 2.36**  *Elements in Parallel*

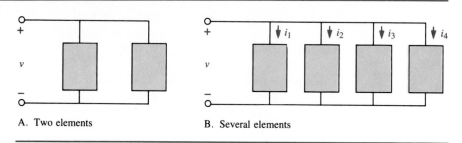

A. Two elements          B. Several elements

The important circuit consideration in parallel elements is that the voltage across them is the same, while the current in each is different.

**Example 2.22**  *Series–Parallel Combinations*

Consider the circuit shown in Figure 2.37.

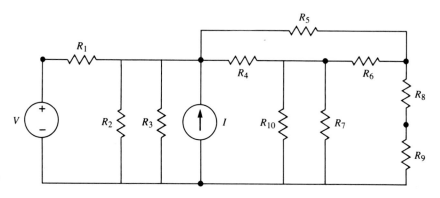

**FIGURE 2.37**

a.  Which elements are in series?
   Answer:  $V$ and $R_1$
   $R_8$ and $R_9$
b.  Which elements are in parallel?
   Answer:  $R_2$, $R_3$, and $I$
   $R_{10}$ and $R_7$

**Resistors in Parallel.**  Figure 2.38A shows two resistors in parallel; their equivalent resistor is shown in Figure 2.38B. The voltage across the two parallel resistors is the same as the voltage across the equivalent resistor. In Figure 2.38A, the current at the terminals, $I$, can be found using KCL to equal

**FIGURE 2.38**    *Resistors in Parallel*

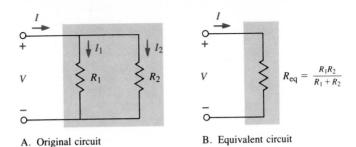

A. Original circuit                    B. Equivalent circuit

$$I = I_1 + I_2 = \frac{V}{R_1} + \frac{V}{R_2} = V\left(\frac{1}{R_1} + \frac{1}{R_2}\right)$$

For the equivalent circuit shown in Figure 2.38B,

$$I = \frac{V}{R_{eq}}$$

The two *V–I* relationships will be the same if and only if

*Two Parallel Resistors*

$$R_{eq} = \left(\frac{1}{R_1} + \frac{1}{R_2}\right)^{-1} \tag{2.2}$$

or

$$R_{eq} = \frac{R_1 R_2}{R_1 + R_2} \tag{2.3}$$

Thus, we have another important equivalence: *The equivalent resistance of two resistors in parallel is equal to the product of the individual resistances divided by the sum.* It is not obvious, but a little thought should convince you that the value of the parallel equivalent resistor is less than the smaller of the two resistors. The smallest resistor in a parallel combination dominates.

---

**Example 2.23**    *Resistors in Parallel*

a. To find the equivalent resistance of two resistors in parallel, take the product of the individual resistances and divide by their sum. See Figure 2.39A.

b. To find the equivalent resistance of more than two resistors in parallel, first replace two resistors with their equivalent and then proceed. The order does not matter. See the two alternatives shown in Figure 2.39B.

*(continues)*

**Example 2.23**   *Continued*

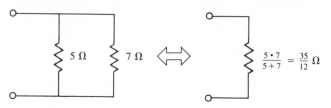

A.  Two resistors in parallel

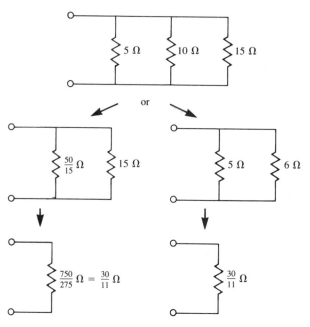

**FIGURE 2.39**     B.  Three resistors in parallel

Part b of Example 2.23 demonstrates that combining many resistors in parallel is more difficult than combining many resistors in series. However, Equation 2.2 can be extended to the general case as follows:

*Parallel Resistors*

$$R_{eq} = \left( \frac{1}{R_1} + \frac{1}{R_2} + \frac{1}{R_3} + \cdots \right)^{-1}$$

(2.4)

This formula is easily handled with calculators. For example, the parallel resistance for the circuit in Figure 2.39B is quickly found as follows:

$$R_{eq} = \left( \frac{1}{5} + \frac{1}{10} + \frac{1}{15} \right)^{-1} = \frac{30}{11} \, \Omega$$

The inverse of resistance occurs frequently in analyzing circuits. It is useful, therefore, to define the term **conductance** $(G)$:

*Conductance*

$$G = \frac{1}{R} \tag{2.5}$$

Conductance is the inverse of resistance. The units of conductance have been classically termed **mhos,** which is *ohms* spelled backward, and symbolized by an upside-down omega $(\mho)$. Recently, the unit of conductance has been changed to **siemens** (S) to honor the British engineer William Siemens (1823–1883). Although we will use the modern terminology, you should become familiar with both notations.

We can use conductance to rewrite Ohm's law and the power relationship for a resistor as follows:

$$I = GV \quad \text{and} \quad P_R = GV^2 \tag{2.6}$$

If you look at the reciprocal formula for equivalent parallel resistance (Equation 2.4), you will see that it is easier to find equivalent parallel conductance:

$$G_{eq} = G_1 + G_2 + G_3 + \cdots \tag{2.7}$$

Simply put, *conductances in parallel add.*

For the sake of clarity, all resistors in this book are labeled with the value of their resistance. If you want to use conductance, you will have to invert the given resistance.

---

**Example 2.24**   *Parallel Equivalents Using Conductance*

Find the equivalent resistance for the parallel combination shown in Figure 2.40.

$G_1 = .01 \, \text{S}$

$G_2 = .05 \, \text{S}$

*(continues)*

**Example 2.24**   *Continued*

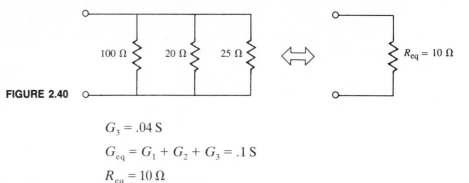

FIGURE 2.40

$$G_3 = .04 \, \text{S}$$
$$G_{eq} = G_1 + G_2 + G_3 = .1 \, \text{S}$$
$$R_{eq} = 10 \, \Omega$$

---

**Example 2.25**   *Analyzing Parallel Circuit*

Analyze the parallel circuit shown in Figure 2.41A. The equivalent circuit is shown in Figure 2.41B. Note that the voltage in both circuits is the same, but the branch currents in the original circuit do not appear in the parallel equivalent.

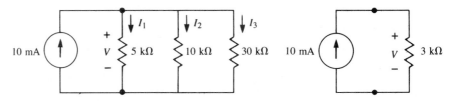

**FIGURE 2.41**   A. Original circuit                                        B. Equivalent circuit

From the equivalent circuit,

$$V = 0.01 \cdot R_{eq} = 0.01 \cdot 3000 = 30 \, \text{V}$$

Now go back to the original circuit to find the branch currents:

$$I_1 = G_1 V = 6 \, \text{mA}$$
$$I_2 = G_2 V = 3 \, \text{mA}$$
$$I_3 = G_3 V = 1 \, \text{mA}$$

---

**Example 2.26**   *Parallel Combinations of Identical Resistors*

In the circuit of Figure 2.42, there are six 47 Ω resistors in parallel. The equivalent resistance is

$$R_{eq} = \left( \frac{1}{47} + \frac{1}{47} + \frac{1}{47} + \frac{1}{47} + \frac{1}{47} + \frac{1}{47} \right)^{-1} = 7.83 \, \Omega$$

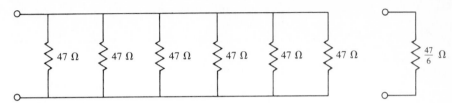

**FIGURE 2.42**

This result can be found easily by factoring out the common term in the equation above to get

$$R_{eq} = \frac{47}{6} = 7.83 \ \Omega$$

Example 2.26 demonstrates the following general statement: *The equivalent resistance of n identical resistors in parallel is equal to the individual resistance divided by the number of resistors.*

***n Identical Parallel Resistors***

$$R_{eq} = \frac{R}{n} \qquad\qquad (2.8)$$

**Current Sources in Parallel.** Figure 2.43A shows a circuit with several current sources connected in parallel. These sources can be represented with a single current source ($I_p$), as shown in Figure 2.43B.

The equivalent source can be found by applying KCL to the original circuit. The terminal current for this circuit is

$$I_{ab} = I_1 + I_2 + I_3$$

Since the terminal current for the equivalent circuit is $I_p$, we get the relationship

**FIGURE 2.43** *Current Sources in Parallel*

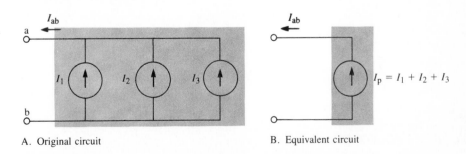

A. Original circuit

B. Equivalent circuit

$$I_P = I_1 + I_2 + I_3 \qquad (2.9)$$

That is, *current sources in parallel can be added together*. If the current source arrow is in the same direction as the arrow assigned to the equivalent current source, the current magnitude is added. If the arrow directions are opposite, the current magnitude is subtracted.

**Example 2.27**    *Parallel Circuit Analysis*

Find $V_o$ in Figure 2.44A.

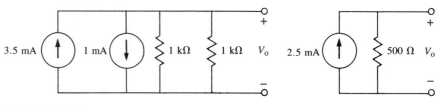

**FIGURE 2.44**    A. Original circuit                          B. Equivalent circuit

From the equivalent circuit in Figure 2.44B,

$$V_o = R_{eq} \cdot I_P = (500)(2.5 \cdot 10^{-3}) = 1.25 \text{ V}$$

**Example 2.28**    *Another Parallel Circuit Analysis*

Find $v_o(t)$ in Figure 2.45A.

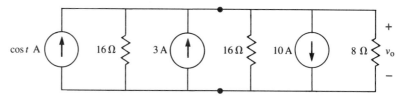

A. Original circuit

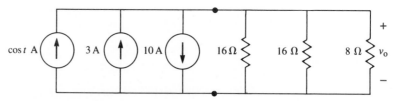

**FIGURE 2.45**    B. Rearranged circuit

This example demonstrates that even though circuit elements may not be physically next to one another, they may still be in parallel. The original circuit has been redrawn in Figure 2.45B to emphasize the parallel connections of the sources and resistors.

$$i_\mathrm{P}(t) = \cos t - 7\,\mathrm{A}$$

$$R_\mathrm{eq} = 4\,\Omega$$

$$v_\mathrm{o}(t) = 4\cos t - 28\,\mathrm{V}$$

**Voltage Sources in Parallel.** Figure 2.46A shows two ideal voltage sources in parallel. In general, this is an impossible situation. The voltage across the two terminals of the parallel combination cannot be, for example, 6 V and 12 V at the same time. This would mean that there is a potential difference across the connecting wire of 6 V. Since the connecting wire has very little resistance (ideally, zero), an extremely large current would result. If you have ever tried to connect a 6 V car battery to a 12 V battery, you know what is meant here. The resulting sparks and melting metal make quite a light show. Ideal voltage sources only can be put in parallel if they have the same value (Figure 2.46B). The equivalent source is then equal to that value. Again, as with current sources in series, voltage sources are put in parallel to increase available supply power.

**Voltage Source in Parallel with Resistor.** Figure 2.47A shows a case analogous to a current-source–series-resistor combination. The voltage across node a and node c is equal to 120 V, no matter what $R$ equals. Therefore, the circuit in Figure 2.47B is equivalent to the original. We find

$$V_1 = 40\,\mathrm{V} \qquad \text{and} \qquad V_2 = 80\,\mathrm{V}$$

To find the source current $I_\mathrm{s}$, we return to the original circuit and use KCL to get

$$I_\mathrm{s} = -I_R - I_1 = -\frac{120}{R} - 16\,\mathrm{mA}$$

**FIGURE 2.46** *Voltage Sources in Parallel*

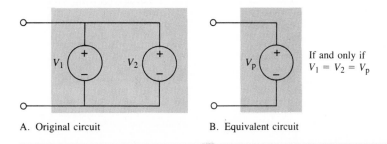

A. Original circuit    B. Equivalent circuit

FIGURE 2.47    *Voltage Source in Parallel with Resistor*

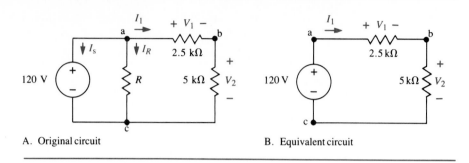

A. Original circuit                              B. Equivalent circuit

Note that the voltage source current is a function of the parallel resistor $R$.

## Series–Parallel Circuits

The circuit shown in Figure 2.48 is typical of a class of circuits known as **ladder** networks. The term *ladder* comes from the physical appearance of the circuit, which looks like a ladder turned on its side. Although this type of circuit can be analyzed with the branch variable method, it is often simpler to use equivalent circuits.

We start the analysis by recognizing that the last two resistors on the right are in series and then adding them. This equivalent resistance is now in parallel with the next resistor on the left. We continue from right to left, combining resistors in series and parallel until we reduce the circuit to a single equivalent resistance. The five steps taken to perform this reduction are shown in Figure 2.49.

We see from the end result that $I_1$ is 5 mA. Note that this current and the source voltage are the only branch variables that exist in both the final equivalent and the original circuit. However, we now can work backward through the equivalent circuits to find all of the other branch variables. Since the current $I_1$ is

FIGURE 2.48    *Ladder Network*

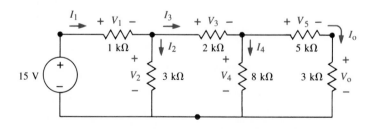

**FIGURE 2.49**  *Series–Parallel Reduction of Circuit in Figure 2.48*

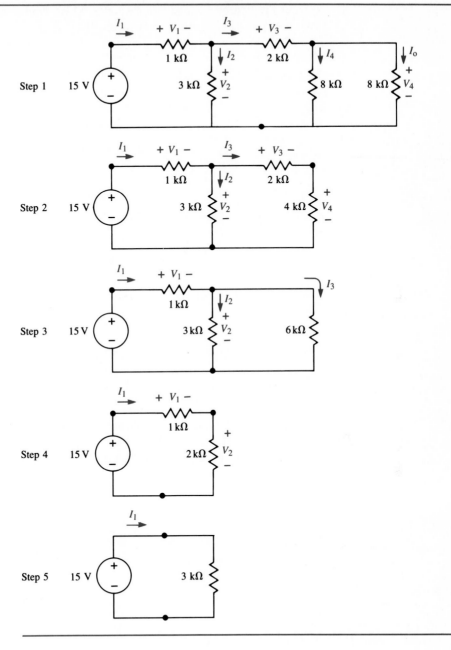

the same in the last step and the fourth step, $V_1$ is 5 V and $V_2$ is 10 V. Note that $V_1$, $V_2$, and the source voltage sum to zero, which confirms KVL.

We then go back to the circuit in the third step and use Ohm's law to find that $I_2 = V_2/3000$ or 10/3 mA and that $I_3 = V_2/6000$ or 5/3 mA. We next find that $V_3 = 2000I_3$ or 10/3 V and that $V_4 = 4000I_3$ or 20/3 V. We find that both $I_4$ and $I_o$ are 20/24 mA. Finally, from the original circuit, we find that $V_5 = 100/24$ mA and $V_o = 60/24$ V.

---

**Example 2.29**    *Ladder Network*

Find $v_o(t)$ in Figure 2.50A.

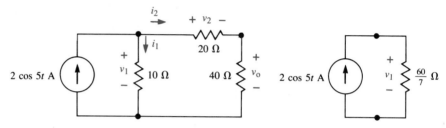

**FIGURE 2.50**    A. Original circuit                                                  B. Equivalent circuit

The current source "sees" an equivalent resistance of

$$R_{eq} = (20 + 40)\|10 = \frac{60}{7} \ \Omega$$

where the double bars ($\|$) indicate the parallel combination of the 60 $\Omega$ series equivalent and the 10 $\Omega$ resistor. The solution follows from analysis of the equivalent circuit in Figure 2.50B.

$$v_1 = \frac{120}{7} \cos 5t \text{ V}$$

$$i_1 = \frac{12}{7} \cos 5t \text{ A}$$

$$i_2 = \frac{2}{7} \cos 5t \text{ A}$$

$$v_2 = \frac{40}{7} \cos 5t \text{ V}$$

$$v_o = \frac{80}{7} \cos 5t \text{ V}$$

---

**Example 2.30**    *Transistor Circuit*

An AC transistor circuit is modeled in Figure 2.51A. Find $i_b$ and $v_o$.

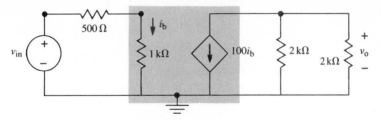

A. Original circuit

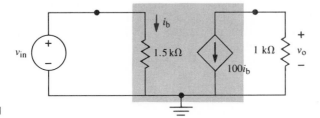

**FIGURE 2.51**

From the equivalent circuit in Figure 2.51B,

$$i_b = \frac{v_{in}}{1500}$$

and

$$v_o = -100i_b \cdot 1000 \qquad \text{(note negative sign)}$$

So,

$$v_o = -66.67 v_{in}$$

## 2.6  SOURCE TRANSFORMATIONS

You may have noticed that the ladder circuits analyzed in Section 2.5 had only a single source in them. Ladder circuits with more than one source usually cannot be analyzed with the techniques described thus far. We now will consider a circuit equivalence that is an extremely powerful tool in circuit analysis.

We make the rather startling assertion that for the proper choice of element values, the two circuits in Figure 2.52 are equivalent at the terminal pair a–b. To verify this equivalence, let us find the relationship between $V_{ab}$ and $I_{ab}$ for both circuits. We are assuming that these source–resistor combinations are part of a larger circuit so that $I_{ab}$ does not have to be zero.

For the voltage source circuit, we use KVL:

$$V_{ab} + I_{ab}R_1 = V_s$$

For the current source circuit, we use KCL:

**FIGURE 2.52**     *Source Transformations*

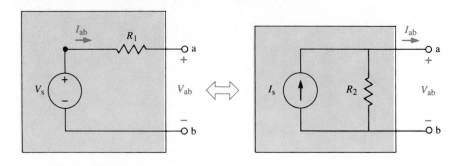

$$\frac{V_{ab}}{R_2} + I_{ab} = I_s \quad \text{and} \quad V_{ab} + I_{ab}R_2 = I_sR_2$$

If we want the *V–I* characteristics at terminal pair a–b to be the same for both circuits, then the following must be true:

$$R_1 = R_2$$

Because the two resistors are equal, we give them the one symbol of simply *R*. For equivalence, it must also be true that

$$V_s = I_sR$$

or

$$I_s = \frac{V_s}{R}$$

(2.10)

    If we are given a voltage source in series with a resistor, we can replace the combination with a current source in parallel with the same resistor. The value of the current source is equal to the value of the voltage source divided by the resistance. The arrowhead describing the direction of the current source is on the same side as the plus sign of the voltage source.

    If we are given a current source in parallel with a resistor, we can replace them with a voltage source in series with the resistor. The value of the voltage source is equal to the value of the current source times the resistance. The plus sign of the voltage source is on the same side as the arrowhead of the current source.

**Example 2.31**     *Voltage-to-Current Source Transformations*

Figure 2.53 shows two examples of voltage source to current source transformation.

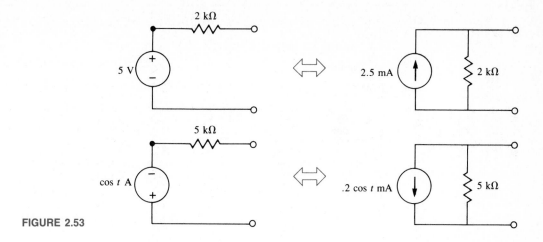

**FIGURE 2.53**

---

**Example 2.32**    *Current-to-Voltage Source Transformations*

Figure 2.54 shows two examples of current source to voltage source transformation.

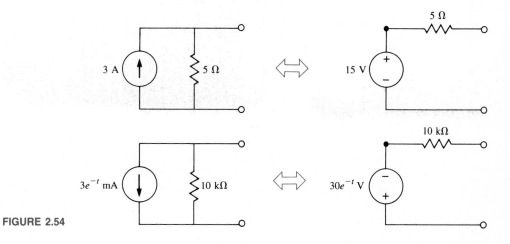

**FIGURE 2.54**

---

Circuit analysis of ladder networks is very easily accomplished by using source transformations along with parallel and series combinations. The technique is to make a source transformation, combine resistors, make another source transformation, combine resistors, and so on. The next set of examples demonstrates this technique.

**Example 2.33**   *Use of Source Transformations in Circuit Analysis*

Find $V_o$ in the circuit in Figure 2.55A.

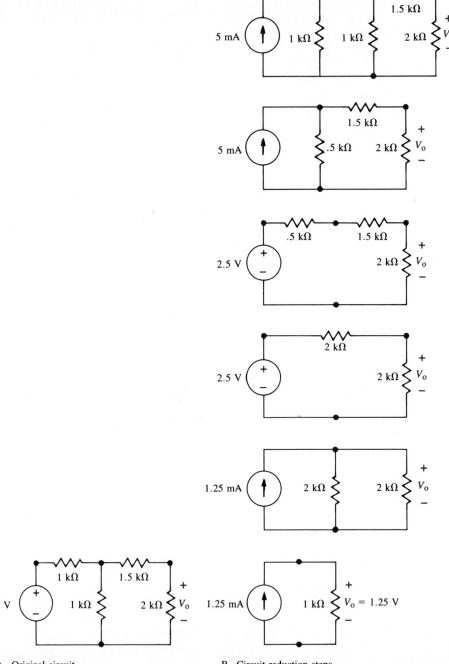

**FIGURE 2.55**   A. Original circuit                     B. Circuit reduction steps

Figure 2.55B shows successive source transformations and series and parallel combinations that lead to the answer.

Note that in Example 2.33, even though the 1.5 kΩ and the 2 kΩ resistors are in series, we did not add them together. We would have lost the desired output variable, $V_o$, if we had done so. A good rule to follow when you use source transformations is to start as far away in the circuit as you can from the branch of interest; do not involve the branch of interest in any transformation or resistor combination.

**Example 2.34**    *Another Source Transformation Analysis*

Find $V_o$ in the circuit in Figure 2.56A.

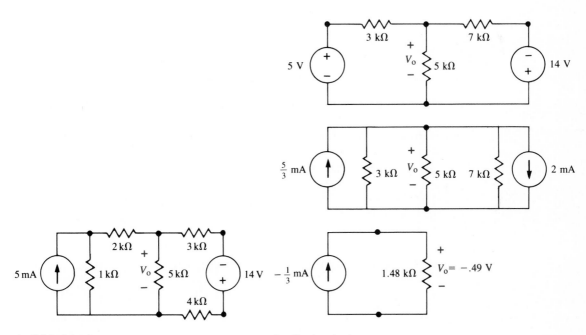

A. Original circuit                                 B. Circuit reduction steps

**FIGURE 2.56**

The desired output is in the middle of the circuit. We, therefore, work from both ends toward the middle as shown in Figure 2.56B. Note that the 3 kΩ and the 4 kΩ resistors on the right of the circuit are in series. They are added before the source transformation is performed on the voltage source.

## 2.7 VOLTAGE AND CURRENT DIVIDERS

In this section, we focus on two additional circuit analysis techniques: *voltage division* and *current division*. The voltage and current divider circuits are not equivalent circuits, but because their use does simplify analysis, they will be discussed here. The voltage divider is described first.

### Voltage Divider

We have already analyzed the simple circuit shown in Figure 2.57. The total resistance "seen" by the source is the sum of the two series resistors $R_1$ and $R_2$. The current $I$ in this single-loop circuit is, therefore, equal to $V_{in}/(R_1 + R_2)$. The two branch voltages are as follows:

$$V_1 = \frac{R_1 V_{in}}{R_1 + R_2} \tag{2.11}$$

and

$$V_2 = \frac{R_2 V_{in}}{R_1 + R_2} \tag{2.12}$$

The supply voltage in this single-loop circuit divides between the two resistors according to the simple formulas just given. The circuit is known as a **voltage divider** circuit. It can be used to produce a desired voltage from a larger supply voltage.

We can make a few intuitive points:

1. The ratio of resistors is always less than or equal to 1.
2. If $R_1$ is 0, then $V_1$ is 0 and $V_2$ equals the supply voltage.
3. If $R_2$ is 0, then $V_2$ is 0 and $V_1$ equals the supply voltage.
4. If $R_1 >> R_2$, then $V_1$ approximately equals the supply voltage.
5. If $R_2 >> R_1$, then $V_2$ approximately equals the supply voltage.

**FIGURE 2.57**   *Voltage Divider*

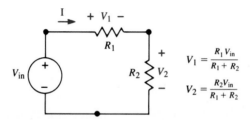

**Example 2.35**   *Voltage Divider Circuits*

Find $V_1$ and $V_2$ in Figure 2.58A.

$$V_1 = \frac{5 \cdot 10}{10 + 5} = \frac{50}{15} \text{ V}$$

$$V_2 = \frac{10 \cdot 10}{10 + 5} = \frac{100}{15} \text{ V}$$

Find $V_3$ and $V_4$ in Figure 2.58B.

$$V_3 = \frac{3000 \cdot 8}{10,000} = 2.4 \text{ V}$$

$$V_4 = -\frac{7000 \cdot 8}{10,000} = -5.6 \text{ V}$$

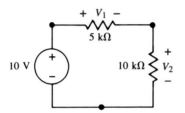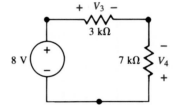

**FIGURE 2.58**   A.                                          B.

Note the negative sign for $V_4$, which occurs because the plus sign for $V_4$ has been assigned at the bottom, while the source is positive at the top. A simple technique for finding the proper sign of the answer is to trace the plus sign of the desired voltage back to the source. If you come to the positive terminal of the source, then the answer is positive. If you arrive at the negative terminal of the source, the answer is negative.

**Example 2.36**   *Voltage Divider with More Than Two Resistors*

Find the branch voltages shown in Figure 2.59.

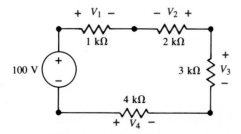

**FIGURE 2.59**

(*continues*)

**Example 2.36** *Continued*

The circuit in Figure 2.59 has more than two resistors. However, all of the resistors are still in series since this is a single-loop circuit. The total resistance seen by the source $(R_s)$ is their sum, or 10 k$\Omega$. The current in this circuit is equal to $V_{in}/R_s$. Each branch voltage is its branch resistance times this current. We have, therefore, the following:

$$V_1 = \frac{V_{in} \cdot R_1}{R_s} = \frac{100 \cdot 1}{10} = 10 \text{ V}$$

$$V_2 = -\frac{V_{in} \cdot R_2}{R_s} = -\frac{100 \cdot 2}{10} = -20 \text{ V}$$

$$V_3 = \frac{V_{in} \cdot R_3}{R_s} = \frac{100 \cdot 3}{10} = 30 \text{ V}$$

$$V_4 = -\frac{V_{in} \cdot R_4}{R_s} = -\frac{100 \cdot 4}{10} = -40 \text{ V}$$

Again, note the polarities of $V_2$ and $V_4$.

---

**Example 2.37** *The Voltage Divider in Circuit Analysis*

Find $v_o(t)$ in Figure 2.60A.

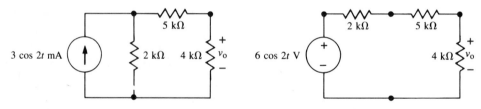

**FIGURE 2.60**    A. Original circuit                              B. Transformed circuit

A source transformation is performed to obtain the circuit shown in Figure 2.60B. Since this equivalent circuit is a voltage divider circuit, we simply write the answer by inspection:

$$v_o = 6 \cos 2t \cdot \frac{4}{11} = \frac{24}{11} \cos 2t \text{ V}$$

---

## Current Divider

The *current divider*, as shown in Figure 2.61, is a parallel circuit driven by a current source. The voltage across each resistor equals the total parallel resistance times the supply current:

**FIGURE 2.61** *Current Divider*

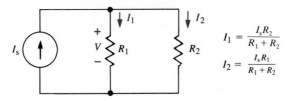

$$I_1 = \frac{I_s R_2}{R_1 + R_2}$$

$$I_2 = \frac{I_s R_1}{R_1 + R_2}$$

$$V = I_s \frac{R_1 R_2}{R_1 + R_2}$$

The current in each resistor is found by Ohm's law to be

$$I_1 = I_s \frac{R_2}{R_1 + R_2} \qquad \text{(2.13)}$$

and

$$I_2 = I_s \frac{R_1}{R_1 + R_2} \qquad \text{(2.14)}$$

In the current divider, the current in a resistor is equal to the opposite resistor divided by the sum of the resistances times the supply current. This means that the smaller resistor will carry the larger current. In the extreme, if one resistor equals zero—that is, a short circuit—all of the supply current is in that resistor. The other resistor has been "shorted out" and carries no current.

**Example 2.38** *Current Divider*

Find $I_1$ and $I_2$ in Figure 2.62

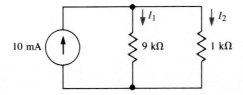

**FIGURE 2.62**

(*continues*)

**Example 2.38**  *Continued*

$$I_1 = \frac{(0.01)(1000)}{1000 + 9000} = 1 \, \text{mA}$$

$$I_2 = \frac{(0.01)(9000)}{1000 + 9000} = 9 \, \text{mA}$$

**Example 2.39**  *Design Example*

Given the circuit in Figure 2.63, find $R_1$ so that $I_1 = 10 \, \text{mA}$.

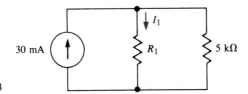

**FIGURE 2.63**

The current divider equation is still used to get

$$I_1 = \frac{(0.030)(5000)}{5000 + R_1} = 10 \, \text{mA}$$

This equation is solved for $R_1 = 10 \, \text{k}\Omega$.

The current divider rule given in Equations 2.13 and 2.14 is difficult to extend to a circuit with more than two resistors in parallel. For a multiple-resistor current divider, you would proceed as in the following example.

**Example 2.40**  *Multiple-Resistor Current Divider*

Find $I_{10}$ and $I_{20}$ in Figure 2.64A.

If we want the current in the $10 \, \Omega$ resistor, we must first combine all of the other resistances into a single equivalent and then apply the current divider rule. This has been done in Figure 2.64B, and the answer is $I_{10} = 2 \, \text{A}$.

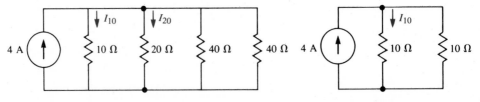

**FIGURE 2.64**    A. Original circuit                                    B. Equivalent circuit

If we also want the current in the 20 Ω resistor, we combine the set of resistors, 10 Ω, 40 Ω, and 40 Ω, and then use the current divider rule. The answer in this case is $I_{20} = 1\,\text{A}$.

---

As you can see, this is a fairly clumsy way to proceed. A more efficient current divider rule can be derived if we use conductances instead of resistances. We know that several conductors in parallel add to form an equivalent conductance. The common voltage across the conductances is equal to the supply current divided by the equivalent conductance, so

$$V = \frac{I_s}{G_{eq}} = \frac{I_s}{G_1 + G_2 + G_3 + \cdots}$$

Since each resistor current is equal to its voltage times its conductance, we get the following:

$$I_1 = \frac{I_s G_1}{G_1 + G_2 + G_3 + \cdots}$$

and

$$I_2 = \frac{I_s G_2}{G_1 + G_2 + G_3 + \cdots}$$

and so on

(2.15)

The current divider is simple if you use conductances. The current in one resistor of a parallel combination is equal to the source current times the conductance of that resistor divided by the total conductance. The smallest resistor has the largest conductance and so carries the largest amount of current.

---

**Example 2.41**   *Current Divider Using Conductances*

Find $I_1$ and $I_2$ in Figure 2.65A.

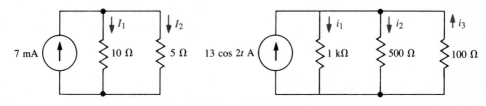

**FIGURE 2.65**   A.

B.

(*continues*)

**Example 2.41**   *Continued*

$$I_1 = \frac{(.007)(.1)}{.1 + .2} = \frac{7}{3} \text{ mA}$$

$$I_2 = \frac{(.007)(.2)}{.1 + .2} = \frac{14}{3} \text{ mA}$$

Find $i_1(t)$, $i_2(t)$ and $i_3(t)$ in Figure 2.65B.

$$i_1(t) = \frac{13 \cos 2t \text{ A} \cdot .001}{.001 + .002 + .01} = \cos 2t \text{ A}$$

$$i_2(t) = \frac{13 \cos 2t \text{ A} \cdot .002}{.013} = 2 \cos 2t \text{ A}$$

$$i_3(t) = -\frac{13 \cos 2t \text{ A} \cdot .01}{.013} = -10 \cos 2t \text{ A}$$

Note the minus sign for $i_3(t)$, which occurs because its assigned direction is opposite that of the source current.

---

**Example 2.42**   *Transistor Circuit*

Find $i_o$ in the AC transistor circuit modeled in Figure 2.66A.

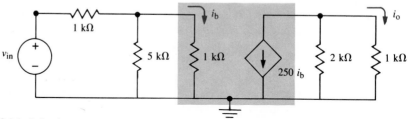

A. Original circuit

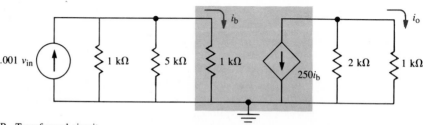

**FIGURE 2.66**   B. Transformed circuit

    First, transform the voltage source as shown in Figure 2.66B. Applying current dividers to both sides of the equivalent circuit, we get

$$i_b = .001 v_{in} \cdot .454 = (.454 \cdot 10^{-3}) v_{in}$$

$$i_o = -250 i_b \cdot .667 = -.075 v_{in}$$

## 2.8 SUPERPOSITION

Chapter 1 introduced you to the concept of superposition. Figure 2.67 shows how superposition may be used to find the output of a *linear* system that has many inputs.

We use superposition here as follows:

1. Turn off (set to zero) all inputs except $x_1$, and solve for the output $y_1$.
2. Set $x_1$ to zero, turn $x_2$ on, and solve for $y_2$.
3. Turn $x_2$ off, turn $x_3$ on, and solve for $y_3$, and so on.
4. Having found all of the individual output values, take their sum to find the answer:

$$y = y_1 + y_2 + y_3 + \cdots$$

We can apply this technique to electrical circuits if we remember a few facts. The inputs to electrical circuits are independent voltage and current sources. We

**FIGURE 2.67**  *Superposition*

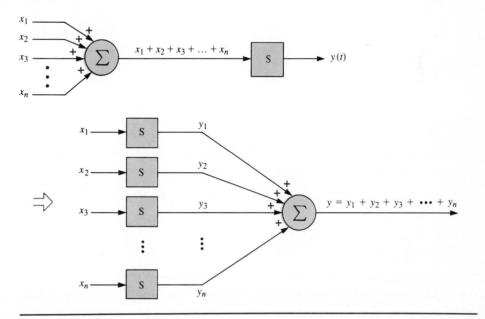

emphasize that *only independent sources are inputs*. Dependent sources are models for physical behavior of the circuit and so cannot be turned off. A source is turned off by setting its value to zero. As a reminder, some circuits from Chapter 1 are repeated here. Figure 2.68 shows that a zero-valued voltage source is a short circuit, while a zero-valued current source is an open circuit.

We can use superposition to very quickly analyze the circuit in Figure 2.69A. First, we turn off the current source to get the circuit in Figure 2.69B. Since we now have a single-loop circuit, we can use the voltage divider rule to find that

$$V_{o_1} = (10)\left(\frac{3}{9}\right) = \frac{10}{3} \text{ V}$$

Next, we turn off the voltage source and turn the current source back on. The circuit in Figure 2.69C is a two-node circuit, so we can use the current divider rule to find the current in the 3 kΩ resistor and then Ohm's law to find the voltage:

$$I_{3 \text{ k}\Omega} = (0.006)\left(\frac{6}{9}\right) = 4 \text{ mA}$$

$$V_{o_2} = (0.004)(3000) = 12 \text{ V}$$

**FIGURE 2.68**   *Zero-Valued Sources*

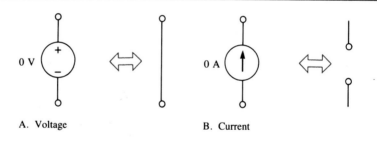

A.  Voltage                                        B.  Current

**FIGURE 2.69**   *Circuit Analysis Using Superposition*

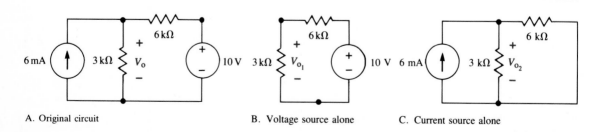

A. Original circuit                B. Voltage source alone          C. Current source alone

The answer is

$$V_o = V_{o_1} + V_{o_2} = \frac{46}{3} \text{ V}$$

---

**Example 2.43** *Superposition*

Find $I_o$ in Figure 2.70A.

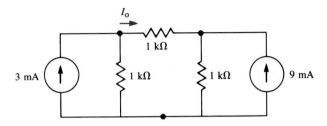

A. Original circuit

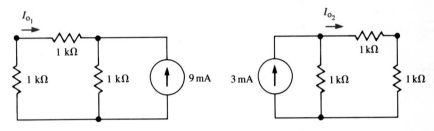

**FIGURE 2.70** B. 9 mA source alone      C. 3 mA source alone

Turn off the 3 mA source as shown in Figure 2.70B and find

$$I_{o_1} = -\frac{(1000)(0.009)}{1000 + 2000} = -3 \text{ mA}$$

Next, turn off the 9 mA source, turn on the 3 mA source, and find

$$I_{o_2} = \frac{(1000)(0.003)}{1000 + 2000} = 1 \text{ mA}$$

The total answer is the sum of $I_{o_1}$ and $I_{o_2}$:

$$I_o = -3 \text{ mA} + 1 \text{ mA} = -2 \text{ mA}$$

---

**Example 2.44** *Superposition with a Controlled Source*

Find $V_o$ in Figure 2.71A.

(*continues*)

**Example 2.44** *Continued*

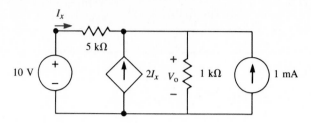

A. Original circuit

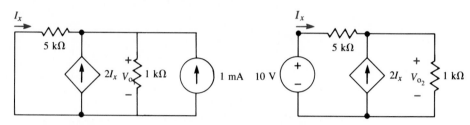

**FIGURE 2.71**    B. Current source alone                    C. Voltage source alone

We find $V_{o_1}$ from Figure 2.71B and $V_{o_2}$ from Figure 2.71C. *Note:* The controlled source is never turned off.

The node equation for $V_{o_1}$ is

$$\frac{V_{o_1}}{1000} + \frac{V_{o_1}}{5000} - 2I_x = 1\,\text{mA}$$

and

$$I_x = -\frac{V_{o_1}}{5000}$$

So

$$V_{o_1}\left(\frac{1}{1000} + \frac{1}{5000} + \frac{2}{5000}\right) = 1\,\text{mA}$$

$$V_{o_1} = \frac{5}{8}\,\text{V}$$

The node equation for $V_{o_2}$ is

$$\frac{V_{o_2}}{1000} + \frac{V_{o_2}}{5000} - 2I_x = \frac{10}{5000}$$

and

$$I_x = \frac{10 - V_{o_2}}{5000} = 2\,\text{mA} - \frac{V_{o_2}}{5000}$$

$$V_{o_2}\left(\frac{1}{1000} + \frac{1}{5000} + \frac{2}{5000}\right) = 6 \, \text{mA}$$

$$V_{o_2} = \frac{30}{8} \, \text{V}$$

The answer is

$$V_o = \frac{5}{8} + \frac{30}{8} = 4.375 \, \text{V}$$

## 2.9 SUMMARY

This chapter introduced several circuit terms: branch, node, loop, cut-set, and mesh. Kirchhoff's laws for current and voltage behavior in a circuit were described. With these two laws and Ohm's law, resistive circuits of any complexity can be analyzed. One analysis technique, the branch variable method, was shown here. Other circuit analysis techniques will be examined in the next chapter.

The systematic techniques for analyzing circuits can lead to a large set of simultaneous equations that need to be solved. With the current availability of calculator and computer programs, this presents little problem. However, for many small circuits, analysis can be simplified using equivalent circuits. Equivalent circuits can yield additional insight into circuit behavior. We discussed series and parallel combinations of elements and also considered the very powerful tool of source transformations. In Chapter 4, we will discuss equivalent circuits in greater detail.

Two special circuits, the voltage divider and the current divider, occur so frequently that it is worthwhile to memorize their responses. When you see a single-loop circuit or a two-node circuit, you should be able to find the circuit variables by inspection. The chapter concluded with a section on how to use the concept of superposition to simplify analysis of linear circuits.

## ▪ PROBLEMS

**2.1** Apply KCL to the circuit in Figure P2.1 to find all branch currents.

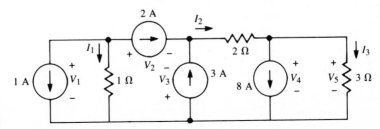

**FIGURE P2.1**

**2.2 a.** Apply Ohm's law and KVL to the circuit in Figure P2.1 to find $V_1$, $V_2$, $V_3$, $V_4$, and $V_5$.

  **b.** Find the power absorbed by each element and the total absorbed power.

  **c.** Which sources are delivering power?

**2.3** Apply KVL to the circuit in Figure P2.3 to find all branch voltages.

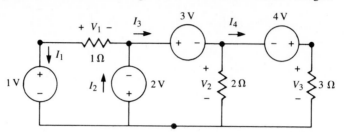

**FIGURE P2.3**

**2.4 a.** Apply Ohm's law and KCL to the circuit in Figure P2.3 to find $I_1$, $I_2$, $I_3$, and $I_4$.

  **b.** Find the power absorbed by each element and the total absorbed power.

  **c.** Which sources are delivering power?

**2.5** Use KCL applied to cut-sets to find $I_1$ and $I_2$ in Figure P2.5.

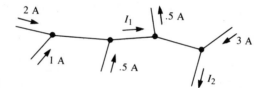

**FIGURE P2.5**

**2.6 a.** Determine which circuits in Figure P2.6 are planar.

  **b.** Redraw the planar circuits on a flat plane with no crossing lines.

  **c.** Find the meshes in the planar circuits.

**2.7** For the cube shown in Figure P2.6D, find the current in each resistor. (*Hint:* Consider the symmetry of the cube.)

**2.8** Use the branch variable method to find $V_1$, $V_2$, and $V_3$ in the circuit in Figure P2.8.

**2.9** Use the branch variable method to find $I_1$, $I_2$, and $I_3$ in the circuit in Figure P2.9.

**2.10** Given the operational amplifier circuit in Figure P2.10 and the device parameters $r_{in} = 1\,\mathrm{M\Omega}$, $r_o = 0$, and $A = 10^5$.

  **a.** Draw the equivalent circuit using the op amp model introduced in Chapter 1.

  **b.** Use the branch variable method to find $V_o$.

**2.11** Use the branch variable method to find $v_o$ in the transistor circuit of Figure P1.16 ($\beta = 100$, $r_x = 1\,\mathrm{k\Omega}$, $r_o = \infty$).

**2.12** Prove Equation 2.1.

**2.13** Prove Equation 2.4.

**2.14** A voltmeter is a device that measures the voltage across a circuit element. Its symbol is a circle with a "V" inside. Show how you would connect a voltmeter to the circuit in Figure P2.3 to measure $V_3$. Should the resistance of a voltmeter be large or small? Why?

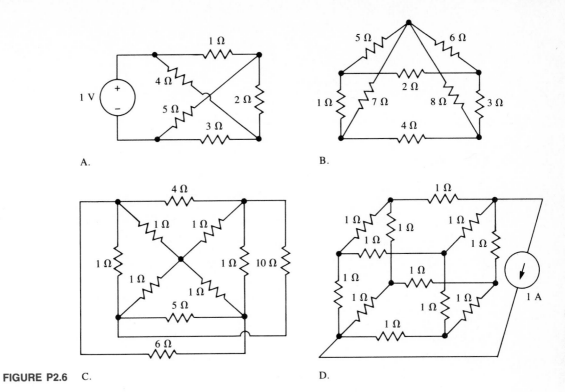

**FIGURE P2.6**   A.   B.   C.   D.

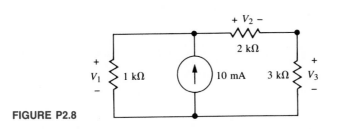

**FIGURE P2.8**

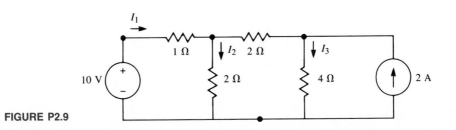

**FIGURE P2.9**

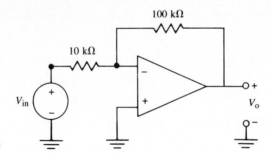

**FIGURE P2.10**

**2.15** An ammeter is a device that measures current in a circuit element. Its symbol is a circle with an "A" inside. Show how you would connect an ammeter to the circuit in Figure P2.3 to measure $I_3$. Should the resistance of an ammeter be large or small? Why?

**2.16** Find the equivalent resistance for the circuit in Figure P2.16.

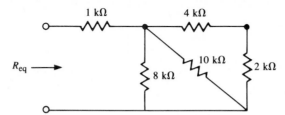

**FIGURE P2.16**

**2.17** Find the equivalent conductance for the circuit in Figure P2.17.

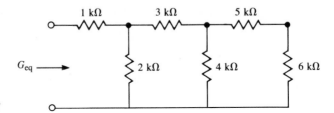

**FIGURE P2.17**

**2.18** Find the equivalent resistance of the circuit in Figure P2.18.

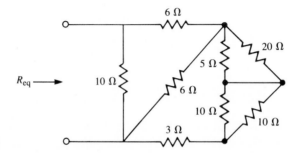

**FIGURE P2.18**

**2.19** Find the equivalent resistance between node a and node b in Figure P2.19.

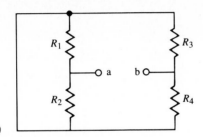

**FIGURE P2.19**

**2.20** Several resistors are connected in series. The equivalent series resistance is $R_{eq}$. If the value of each resistor is multiplied (scaled) by the same factor, say $a$, what is the new series resistance?

**2.21** Several resistors are connected in parallel. The equivalent parallel resistance is $R_{eq}$. If the value of each resistor is multiplied (scaled) by the same factor, say $a$, what is the new parallel resistance?

**2.22** A 120 V source is connected in parallel with four resistors. You are told that the source delivers 288 W to the parallel combination and that the resistors have the ratios 1:2:3:6 (for example, 5 Ω, 10 Ω, 15 Ω, and 30 Ω.) Find the values of the four resistors.

**2.23** The equivalent resistance of $n$ equal resistors in parallel is given by $R/n$. This fact can simplify some parallel resistance calculations. For example, consider the parallel combination of 10 Ω and 20 Ω resistors. The 10 Ω resistor can be treated as two 20 Ω resistors in parallel. This combination is, therefore, the same as three 20 Ω resistors in parallel. The parallel resistance is then equal to 20/3 Ω. Use this technique to find the parallel combination of 100 Ω, 300 Ω, 600 Ω, and 1200 Ω resistors.

**2.24** You have a large number of the following practical resistors available: 4.7 Ω, 100 Ω, 220 Ω, 1 kΩ, 2.2 kΩ, 4.7 kΩ, and 10 kΩ. Construct the following resistances using series and parallel combinations of the available resistors:

   **a.**   50 Ω

   **b.**   114.7 Ω

   **c.**   5.5 kΩ

   **d.**   .47 Ω

   **e.**   24.55 kΩ

**2.25** If each resistor you used in Problem 2.24 has ±10% tolerance, determine the minimum and maximum values your resistor combinations in parts a–e can have.

**2.26** Find $V_s$ in the circuit of Figure P2.26.

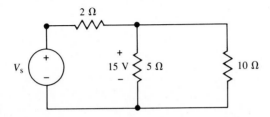

**FIGURE P2.26**

**2.27** Find $I_s$ in the circuit of Figure P2.27.

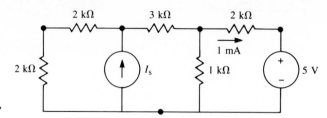

**FIGURE P2.27**

**2.28** For the circuit in Figure P2.28, $I_o = 5\,\text{mA}$. Find $V_{in}$.

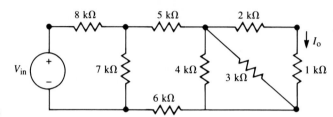

**FIGURE P2.28**

**2.29** For the circuit in Figure P2.28, find the resistance seen by the source.

**2.30** For the circuit in Figure P2.30:

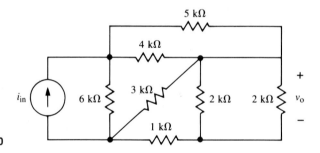

**FIGURE P2.30**

   **a.** Find $i_{in}$ if $v_o = 1\,\text{V}$

   **b.** Find $i_{in}$ if $v_o = 10t\,\text{V}$

   **c.** Compare the answers to parts a and b.

**2.31** For the circuit in Figure P2.30, find the conductance seen by the source.

**2.32** Use source transformations to find $V_o$ in the circuit of Figure P2.32.

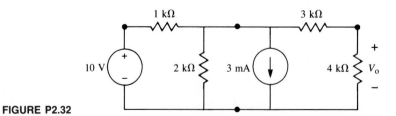

**FIGURE P2.32**

**2.33** Use source transformations to find $V_o$ in the circuit of Figure P2.33.

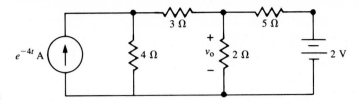

**FIGURE P2.33**

**2.34** For the circuit in Figure P2.28:

    **a.** Find $I_o$ if $V_{in} = 5$ V.

    **b.** Find $I_o$ if $V_{in} = 5e^{-5t}$ V.

    **c.** Compare the answers to parts a and b.

**2.35** For the circuit in Figure P2.30, $i_{in} = \cos 10t$ mA. Find $v_o$.

**2.36** The circuit in Figure P2.36 is known as a bridge circuit.

    **a.** Use the voltage divider relationship to find $V_o$.

    **b.** Find $R_4$ in terms of the other resistors so that $V_o = 0$.

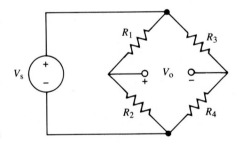

**FIGURE P2.36**

**2.37** For the BJT circuit in Figure P2.37, $\beta = 200$ and $r_x = 2$ k$\Omega$. Find $v_o$ and $i_o$ (using source transformations and current dividers) if

    **a.** $r_o = \infty \, \Omega$

    **b.** $r_o = 100$ k$\Omega$

    **c.** $r_o = 2$ k$\Omega$

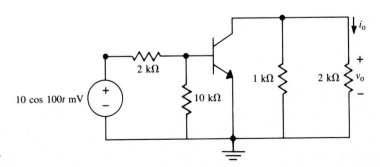

**FIGURE P2.37**

**2.38** Use superposition and voltage and current dividers to find $v_o$ in the circuit in Figure P2.38.

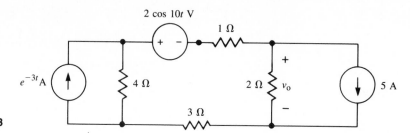

**FIGURE P2.38**

**2.39** The transistor circuit in Figure P2.39 has both DC and AC sources. We can use superposition to reduce this circuit to a pure DC circuit and a pure AC circuit. The total output is equal to the sum of the DC and the AC outputs.

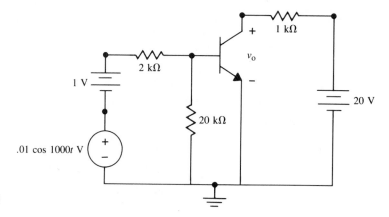

**FIGURE P2.39**

a. Turn off the AC source, replace the transistor with its DC model ($\beta = 100$), and find the output voltage.

b. Turn off the DC sources, replace the transistor with its AC model ($\beta = 100$, $r_x = 236\ \Omega$, $r_o = \infty$), and find the output voltage.

c. Find and sketch the total output voltage.

# Node and Mesh Analysis

3

## 3.1 INTRODUCTION

In this chapter, you will learn how to analyze circuits by using either of two systematic approaches. As we have seen, any circuit can be analyzed by writing a sufficient set of Kirchhoff's current and voltage equations and the branch $V$–$I$ relationships. This branch variable method usually results in a large number of equations that need to be solved. The techniques of node and mesh analysis shown in this chapter will allow us to solve reduced sets of circuit equations.

## 3.2 NODE ANALYSIS

In the *node analysis* technique, we explicitly write only Kirchhoff's current law (KCL) equations. The branch $V$–$I$ and Kirchhoff's voltage law (KVL) equations are also used but only implicitly; we never actually write them down. We will develop this technique by analyzing the circuit shown in Figure 3.1.

This circuit has three nodes, five branches, and three meshes. We assign node c as the datum node and write the KCL equations at nodes a and b:

$$\text{Node a:} \quad -i_{s_1} + i_2 + i_3 = 0$$
$$\text{Node b:} \quad -i_3 + i_4 + i_5 = 0$$

We use Ohm's law to replace branch currents with branch voltages:

$$\text{Node a:} \quad -i_{s_1} + \frac{v_2}{R_2} + \frac{v_3}{R_3} = 0$$

$$\text{Node b:} \quad -\frac{v_3}{R_3} + \frac{v_4}{R_4} + \frac{v_5}{R_5} = 0$$

An algebraically more compact form of the preceding equations can result from using conductances rather than resistances. Remember, conductance is the inverse of resistance—for example, $G_3 = 1/R_3$. Rewriting these equations using

**FIGURE 3.1**   *Node Analysis*

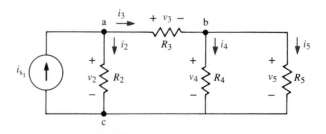

conductances, we get

$$\text{Node a:} \quad -i_{s_1} + G_2 v_2 + G_3 v_3 = 0$$
$$\text{Node b:} \quad -G_3 v_3 + G_4 v_4 + G_5 v_5 = 0$$

We now have two independent equations for four unknowns. At this point, we could use the branch variable technique of Chapter 2. However, there is another method that, once understood, is easier to use. With this method, we do not introduce more equations but, rather, reduce the number of variables.

Thus far, we have analyzed circuits in terms of branch voltages. A voltage difference can also be measured between any node in the circuit and a reference point. This voltage is known as a **node voltage.** The voltage at the reference point is always assumed to be zero. Generally, the datum node is used as the reference node, although this is not a necessary or unique choice.

It is common practice to talk about voltages "at" a node. This terminology is not strictly correct since voltages are measured between two points. A voltage at a node is understood to exist between the node and the reference point. Also, the polarity for a node voltage is not explicitly assigned on the circuit diagram. We always assume that the polarity of the node of interest is positive with respect to the reference node. However, the value of a node voltage will often be negative.

How does all of this information help us analyze the circuit? The key is to replace the branch voltages in the KCL equations with node voltages. Note that the circuit that we are analyzing has three nodes. Since node c is used as the datum node, we have in this circuit only two unknown node voltages, one at node a and one at node b. We also have two equations. If we can replace the four branch voltages with the two node voltages, then we have a solvable set of equations. The relationship between branch voltages and node voltages is fairly simple, but it is very important.

To demonstrate this relationship, we will isolate the branch $R_3$ together with the nodes it is connected to and the datum node, as shown in Figure 3.2. We have defined two new voltages, $v_{ac}$ and $v_{bc}$, which are the node voltages from nodes a and b relative to the datum node. You can see that the polarity of both node voltages is positive at the node of interest. We can apply KVL to this subcircuit to

**FIGURE 3.2** *Node Voltages*

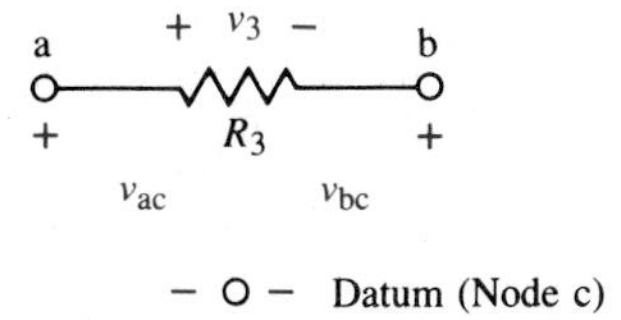

get

$$v_3 + v_{bc} - v_{ac} = 0$$

which leads to the branch voltage–node voltage relationship:

$$v_3 = v_{ac} - v_{bc}$$

Since all node voltages in any circuit are referenced to the same datum node, the second subscript is common to them all. We can, for the sake of simplicity, drop this subscript and represent all node voltages with a single subscript. Therefore,

$$v_3 = v_a - v_b \tag{3.1}$$

In general, we have the following relationship:

$$v_{ab} = v_a - v_b$$

where   $v_{ab}$ = branch voltage
$v_a$ = node voltage at positive side
$v_b$ = node voltage at negative side

It is important to remember that the choice of the datum node is arbitrary. Therefore, node voltages in a circuit are not unique. If two engineers analyze the same circuit but choose different datum nodes, they will not get the same node voltages. However, the branch voltages and currents for a given circuit are unique. For example, the first engineer may find that the node voltage on one side of a resistor is 10 V and on the other side is 6 V. The second engineer may find these same voltages to be 12 V and 8 V. In both cases, however, the resistor branch voltage will be 4 V.

The node voltage at one side of a branch is equal to the node voltage at the other side plus or minus the branch voltage. So, again referring to Figure 3.2, we get

$$v_a = v_b + v_3 \quad \text{and} \quad v_b = v_a - v_3 \tag{3.2}$$

You add branch voltage to node voltage if you are going from the negative side to the positive side. You subtract the branch voltage if you are going from the positive side to the negative side.

Applying this branch voltage–node voltage relationship to the circuit in Figure 3.1, we get

$$v_2 = v_a \quad \text{(remember, datum node voltage is zero)}$$

$$v_3 = v_a - v_b$$

$$v_4 = v_5 = v_b \quad \text{(again, datum node voltage is zero)}$$

If we replace the branch voltages with the node voltages in our equations for this circuit, we get

$$G_2 v_a + G_3(v_a - v_b) = i_{s_1}$$

$$-G_3(v_a - v_b) + G_4 v_b + G_5 v_b = 0$$

The known current has been moved to the right side. We are left with two equations in two unknowns, the node voltages. We rearrange the equations to get them in a final form:

$$(G_2 + G_3)v_a - G_3 v_b = i_{s_1}$$

$$-G_3 v_a + (G_3 + G_4 + G_5)v_b = 0$$

It is common practice in engineering to represent simultaneous equations with matrices and vectors. So represented, the preceding equations become

$$\begin{bmatrix} G_2 + G_3 & -G_3 \\ -G_3 & G_3 + G_4 + G_5 \end{bmatrix} \begin{bmatrix} v_a \\ v_b \end{bmatrix} = \begin{bmatrix} i_{s_1} \\ 0 \end{bmatrix}$$

This matrix representation is known as the *node conductance matrix* equation and is written as follows:

---

**GV = I**                                                                **(3.3)**

where   **G** = node conductance matrix

          **V** = output vector

          **I** = input vector

---

Matrix techniques are reviewed in Appendix A.

---

**Example 3.1**   *Node Analysis I*

Find the node voltages $V_a$ and $V_b$ for the circuit in Figure 3.3.

*(continues)*

**Example 3.1**   *Continued*

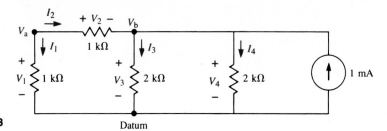

**FIGURE 3.3**

Datum

Node a:   $\dfrac{V_1}{1000} + \dfrac{V_2}{1000} = 0$

Node b:   $-\dfrac{V_2}{1000} + \dfrac{V_3}{2000} + \dfrac{V_4}{2000} - 0.001 = 0$

Now, the branch voltages can be written in terms of node voltages:

$V_1 = V_a$

$V_2 = V_a - V_b$

$V_3 = V_b$

$V_4 = V_b$

Replacing the branch voltages with node voltages leads to

$$\dfrac{V_a}{1000} + \dfrac{V_a - V_b}{1000} = 0$$

$$-\dfrac{V_a - V_b}{1000} + \dfrac{V_b}{2000} + \dfrac{V_b}{2000} = 0.001$$

Rearranging, we get

$$\left(\dfrac{1}{1000} + \dfrac{1}{1000}\right)V_a - \dfrac{V_b}{1000} = 0$$

$$-\dfrac{V_a}{1000} + \left(\dfrac{1}{1000} + \dfrac{1}{2000} + \dfrac{1}{2000}\right)V_b = 0.001$$

In matrix form, these last two equations become

$$\begin{bmatrix} .002 & -.001 \\ -.001 & .002 \end{bmatrix} \begin{bmatrix} V_a \\ V_b \end{bmatrix} = \begin{bmatrix} 0 \\ .001 \end{bmatrix}$$

The solution is

$$V_a = .33 \text{ V} \qquad \text{and} \qquad V_b = .67 \text{ V}$$

The node analysis technique provides a solvable set of linear independent equations in terms of the $N-1$ node voltages of the circuit. Frequently, however, you are really trying to find a branch voltage or a branch current. This presents no problem if you are patient, as we see in the next example.

**Example 3.2** *Finding Branch Variables Given Node Voltages*

Find all resistor branch voltages and currents in the circuit in Example 3.1.
We know that the node voltages are as follows: $V_a = .33$ and $V_b = .67$. Therefore, for the branch voltages and currents, we get

$$V_1 = V_a = .33 \text{ V} \qquad \text{and} \qquad I_1 = .33 \text{ mA}$$
$$V_2 = V_a - V_b = .34 \text{ V} \qquad \text{and} \qquad I_2 = -.34 \text{ mA}$$
$$V_3 = V_b = .67 \text{ V} \qquad \text{and} \qquad I_3 = .33 \text{ mA}$$
$$V_4 = V_b = .67 \text{ V} \qquad \text{and} \qquad I_4 = .33 \text{ mA}$$

**Example 3.3** *Node Analysis II*

Find all branch variables shown in the circuit in Figure 3.4.

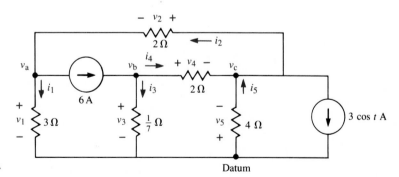

**FIGURE 3.4**                                                          Datum

We first write the KCL node equations in terms of branch voltages:

Node a:  $\dfrac{v_1}{3} - \dfrac{v_2}{2} + 6 = 0$

(*continues*)

**Example 3.3**   *Continued*

$$\text{Node b:}\quad -6 + 7v_3 + \frac{v_4}{2} = 0$$

$$\text{Node c:}\quad \frac{v_2}{2} - \frac{v_4}{2} - \frac{v_5}{4} + 3\cos t = 0$$

We now find the relationships between the branch and node voltages:

$$v_1 = v_a$$

$$v_2 = v_c - v_a$$

$$v_3 = v_b$$

$$v_4 = v_b - v_c$$

$$v_5 = -v_c$$

Substituting, we get

$$\frac{v_a}{3} - \frac{v_c - v_a}{2} + 6 = 0$$

$$-6 + 7v_b + \frac{v_b - v_c}{2} = 0$$

$$\frac{v_c - v_a}{2} - \frac{v_b - v_c}{2} - \frac{-v_c}{4} + 3\cos t = 0$$

Rearranging, we have

$$\left(\frac{1}{3} + \frac{1}{2}\right)v_a - \frac{1}{2}v_c = -6$$

$$\left(7 + \frac{1}{2}\right)v_b - \frac{1}{2}v_c = 6$$

$$-\frac{1}{2}v_a - \frac{1}{2}v_b + \left(\frac{1}{2} + \frac{1}{2} + \frac{1}{4}\right)v_c = -3\cos t$$

Writing these equations in matrix form yields

$$\begin{bmatrix} .833 & 0 & -.5 \\ 0 & 7.5 & -.5 \\ -.5 & -.5 & 1.25 \end{bmatrix}\begin{bmatrix} v_a \\ v_b \\ v_c \end{bmatrix} = \begin{bmatrix} -6 \\ 6 \\ -3\cos t \end{bmatrix}$$

Solving the matrix equation yields

$$v_a = -9.29 - 1.96\cos t \text{ V}$$

$$v_b = .57 - .22\cos t \text{ V}$$

$$v_c = -3.49 - 3.27\cos t \text{ V}$$

Therefore, for the branch voltages and currents, we get

$$v_1 = -9.29 - 1.96 \cos t \text{ V} \qquad i_1 = -3.1 - .65 \cos t \text{ A}$$

$$v_2 = 5.8 - 1.31 \cos t \text{ V} \qquad i_2 = 2.9 - .65 \cos t \text{ A}$$

$$v_3 = .57 - .22 \cos t \text{ V} \qquad i_3 = 3.99 - 1.54 \cos t \text{ A}$$

$$v_4 = 4.06 + 3.05 \cos t \text{ V} \qquad i_4 = 2.03 + 1.52 \cos t \text{ A}$$

$$v_5 = 3.49 + 3.27 \cos t \text{ V} \qquad i_5 = .87 + .82 \cos t \text{ A}$$

The node analysis procedure is simpler than these examples imply. Rather than starting with branch voltages, we can immediately write the KCL equations in terms of node voltages. For instance, in the circuit in Example 3.3, we see by inspection that

$$i_1 = \frac{v_a}{3}$$

$$i_2 = \frac{v_c - v_a}{2}$$

$$i_3 = \frac{v_b}{1/7}$$

$$i_4 = \frac{v_b - v_c}{2}$$

$$i_5 = -\frac{v_c}{4}$$

We could use these relationships to directly write a set of KCL equations for the circuit of Example 3.3 in terms of its node voltages without going through any preliminary steps.

It is not necessary to assign branch current directions to find the node voltages. Consider Figure 3.5, in which the branches attached to node c from Example 3.3 are shown highlighted within the original circuit. Note that the current directions have been reversed from the original analysis.

The branch currents now have the following relationships to the node voltages:

$$i_2 = \frac{v_a - v_c}{2}$$

$$i_4 = \frac{v_c - v_b}{2}$$

$$i_5 = \frac{v_c}{4}$$

Thus, by reversing the branch polarities, we have multiplied the previous branch

**FIGURE 3.5**    *Effect of Branch Current Direction on Node Equation*

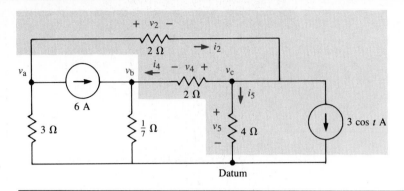

current–node voltage relationship by $-1$. At node c, KCL gives

$$-i_2 + i_4 + i_5 = -3\cos t$$

$$-\frac{v_a - v_c}{2} + \frac{v_c - v_b}{2} + \frac{v_c}{4} = -3\cos t$$

$$-\frac{v_a}{2} - \frac{v_b}{2} + v_c\left(\frac{1}{2} + \frac{1}{2} + \frac{1}{4}\right) = -3\cos t$$

This equation at node c is the same as the one we arrived at in Example 3.3! By changing the assigned direction of branch currents, we changed the sign of both the KCL equation and the branch variable–node voltage relationship. The two changes in sign cancel each other.

Since the final node voltage equations do not depend on the branch polarities, we do not really need to assign them to the circuit. How can we write the KCL equations without knowing the branch current directions? At a node for which we are writing a KCL equation, we assume that all currents are leaving the node. We can then write the node equation in terms of node voltages without ever actually drawing the current arrows. When we move to another node, we again assume that all currents are leaving. Remember, KCL applied to one node is independent of KCL applied to another node. Of course, if you need to solve for a particular branch voltage or current, then you do have to assign the appropriate polarity or direction.

**Example 3.4**    *Quick Node Analysis*

Find the node voltages for the circuit in Figure 3.6.

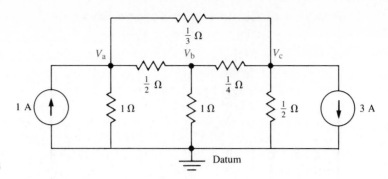

**FIGURE 3.6**

Node a:  $\dfrac{V_a}{1} + (V_a - V_b)2 + (V_a - V_c)3 - 1 = 0$

Node b:  $(V_b - V_a)2 + \dfrac{V_b}{1} + (V_b - V_c)4 = 0$

Node c:  $(V_c - V_a)3 + (V_c - V_b)4 + 2V_c + 3 = 0$

Rearranging, we get

Node a:  $6V_a - 2V_b - 3V_c = 1$
Node b:  $-2V_a + 7V_b - 4V_c = 0$
Node c:  $-3V_a - 4V_b + 9V_c = -3$

So,

$V_a = -.3 \text{ V}$

$V_b = -.44 \text{ V}$

$V_c = -.63 \text{ V}$

---

The simplified node analysis procedure can be summarized as follows:

*Step 1:* Label nondatum node voltages.
*Step 2:* Write KCL at each node in terms of node voltages, assuming at each node that current is leaving.
*Step 3:* Reformat equations to separate variables.
*Step 4:* Solve for node voltages.
*Step 5:* If necessary, solve for branch variables.

As you analyze circuits and work practice problems, look for patterns that develop during the analysis procedure. Later in this chapter (Section 3.6), we will take advantage of these patterns to write the final set of equations (step 3) by inspection of the circuit.

## Analysis of Circuits with Voltage Sources

Node analysis is most easily applied to circuits that contain only current sources. Let's try to apply node analysis to the circuit in Figure 3.7 and see what happens. This circuit contains a voltage source as well as a current source. We say that such circuits are driven by mixed sources.

If we try to write the KCL equation at node c in terms of the node voltages, we discover that it can't be done. The current in a voltage source is not dependent on the voltage across it! Several methods are available for dealing with this situation. Each method has advantages and disadvantages. We will consider three methods and a special case here.

**Source Transformation Method.** This method, where directly applicable, is probably the easiest to use. It is based on the source transformation equivalence described in Chapter 1. We learned there that a voltage source in series with a resistor is equivalent to a current source in parallel with the same resistor. The value of the current source is equal to the value of the voltage source divided by the resistance.

Now, re-examine the circuit in Figure 3.7. You can see that the troublesome voltage source is in series with the 4 Ω resistor. We can use a source transformation to convert this series combination of resistor and voltage source into a parallel combination of resistor and current source. The parallel combination is connected between node b and the datum node. We now have the circuit of Figure 3.8.

A few points should be made about this method. A common error in its application is to misplace the transformed parallel combination of resistor and current source. It must be connected between the same nodes to which the series combination of resistor and voltage source was connected. Also, you must remember the proper orientation of the current source.

One advantage of the source transformation method is that the number of nodes is always reduced. The circuit in Figure 3.8 has only two nodes of interest, while the original circuit had three. Thus, once the transformation has been performed, fewer node equations are required. We analyze the transformed

**FIGURE 3.7**   *Node Analysis with Voltage Source*

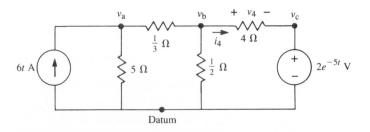

**FIGURE 3.8**    *Source Transformation*

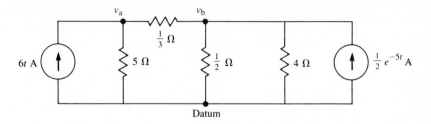

circuit as follows:

$$\text{Node a:} \quad -6t + \frac{v_a}{5} + 3(v_a - v_b) = 0$$

$$\text{Node b:} \quad 3(v_b - v_a) + 2v_b + \frac{v_b}{4} - \frac{e^{-5t}}{2} = 0$$

In matrix form, we get

$$\begin{bmatrix} 3.2 & -3 \\ -3 & 5.25 \end{bmatrix} \begin{bmatrix} v_a \\ v_b \end{bmatrix} = \begin{bmatrix} 6t \\ .5e^{-5t} \end{bmatrix}$$

Solving the matrix equation yields

$$v_a = 4.04t + .19e^{-5t} \text{ V}$$

$$v_b = 2.3t + .21e^{-5t} \text{ V}$$

Remember, in any transformation some information that is internal to the transformation is lost. For example, the branch variables for the $4\,\Omega$ resistor and the voltage source do not appear in Figure 3.8. They have been lost as explicit circuit quantities. If you are interested in these quantities, then you must write some equations for them before you do the transformation and come back to these equations after you have analyzed the transformed circuit.

In the original circuit in Figure 3.7, we note that the node voltage $v_c$ is the voltage across the voltage source and must, therefore, be

$$v_c = 2e^{-5t} \text{ V}$$

The branch voltage across the $4\,\Omega$ resistor is

$$v_4 = v_b - v_c = v_b - 2e^{-5t} = 2.3t - 1.79e^{-5t} \text{ V}$$

and the current in the $4\,\Omega$ resistor and voltage source is

$$i_4 = \frac{v_4}{4} = \frac{v_b - 2e^{-5t}}{4} = .57t - .45e^{-5t} \text{ A}$$

In addition to the possible loss of information, there are two other disadvantages to this method. If there is no resistor in series with the voltage source, the transformation cannot be easily done. If there is a series resistor but the resistor's branch voltage or current is used to control a dependent source, then you must be very careful about using the resistor in a transformation.

**Voltage Source Current Method.** In this second method for analyzing circuits with mixed sources, we do not make any transformations in the original circuit. Instead, if we have a voltage source, we assign a current to the voltage source. We include this current as a variable in the node equations, thus increasing the number of variables.

The circuit in Figure 3.7 has been redrawn in Figure 3.9, and we have assigned a current to the voltage source. We write the node equations for this circuit in terms of the node voltages and the *voltage source current:*

$$\text{Node a:} \quad -6t + \frac{v_a}{5} + 3(v_a - v_b) = 0$$

$$\text{Node b:} \quad 3(v_b - v_a) + 2v_b + \frac{v_b - v_c}{4} = 0$$

$$\text{Node c:} \quad \frac{v_c - v_b}{4} + i_v = 0$$

Note that we have not made any substitution for the current $i_v$. Rearranging the node equations, we get

$$\text{Node a:} \quad \left(\frac{1}{5} + 3\right)v_a - 3v_b = 6t$$

$$\text{Node b:} \quad -3v_a + \left(3 + 2 + \frac{1}{4}\right)v_b - \frac{v_c}{4} = 0$$

$$\text{Node c:} \quad -\frac{v_b}{4} + \frac{v_c}{4} + i_v = 0$$

**FIGURE 3.9**    *Voltage Source Current*

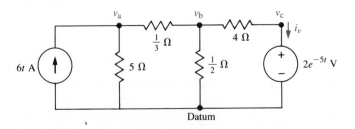

At this point, we must stop and consider what we have. There are four unknowns, the three node voltages and $i_v$, but only three equations. We need another equation. What information have we not used yet? The value of the voltage source has not appeared anywhere in our equations. So, we go back to the circuit in Figure 3.9 and look for a relationship between the voltage source and the node voltages. Examination of the circuit shows that the voltage at node c is the voltage across the voltage source. Therefore,

$$v_c = 2e^{-5t} \text{ V}$$

This equation is the fourth equation that we need. It is commonly called a *constraint* equation because a voltage source in a circuit constrains the values of some of the node voltages.

The final set of equations can be written in matrix form as follows:

$$\begin{bmatrix} 3.2 & -3 & 0 & 0 \\ -3 & 5.25 & -.25 & 0 \\ 0 & -.25 & .25 & 1 \\ 0 & 0 & 1 & 0 \end{bmatrix} \begin{bmatrix} v_a \\ v_b \\ v_c \\ i_v \end{bmatrix} = \begin{bmatrix} 6t \\ 0 \\ 0 \\ 2e^{-5t} \end{bmatrix}$$

The first three rows represent the node equations, and the last row is the constraint equation.

We solve the matrix equations and get

$$v_a = 4.04t + .19e^{-5t} \text{ V}$$

$$v_b = 2.3t + .21e^{-5t} \text{ V}$$

These answers are the same ones we found using the source transformation method. Continuing the solution leads to

$$i_v = .57t - .45e^{-5t} \text{ A}$$

The voltage source current method has an obvious disadvantage. It increases the number of simultaneous equations that need to be solved. It is, however, very straightforward and can be applied to any circuit, particularly if a computer or calculator program is available. The source transformation technique, on the other hand, cannot be applied, without modification, to many circuits. Consider the circuit shown in Figure 3.10.

As you can see, the voltage source is not in series with any single resistor. It cannot, therefore, be directly transformed into a current source (there are techniques that do allow the transformation). The voltage source current method, however, applies, and we write

$$\text{Node a:} \quad -7 + \frac{V_a}{2} + (V_a - V_b)5 + \frac{V_a - V_c}{3} = 0$$

**FIGURE 3.10**    *Voltage Source Current Example Circuit*

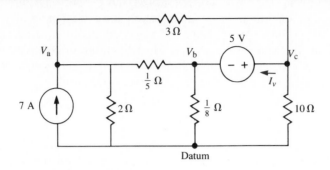

Datum

Node b:    $(V_b - V_a)5 + 8V_b - I_v = 0$

Node c:    $\dfrac{V_c - V_a}{3} + I_v + \dfrac{V_c}{10} = 0$

Constraint:   $V_c - V_b = 5$      (note polarity)

Rearranging yields

Node a:    $\left(\dfrac{1}{2} + \dfrac{1}{3} + 5\right)V_a - 5V_b - \dfrac{V_c}{3} = 7$

Node b:    $-5V_a + (5 + 8)V_b - I_v = 0$

Node c:    $-\dfrac{V_a}{3} + \left(\dfrac{1}{3} + \dfrac{1}{10}\right)V_c + I_v = 0$

Constraint:   $-V_b + V_c = 5$

From these equations, the solution proceeds directly.

**Cut-Set Method.** If you examine the equations just given at nodes b and c, you may notice that by adding these two equations, the voltage source current $I_v$ disappears. You then have only the following three equations to solve:

Node a:    $\left(\dfrac{1}{2} + \dfrac{1}{3} + 5\right)V_a - 5V_b - \dfrac{V_c}{3} = 7$

Nodes b + c:    $-\left(5 + \dfrac{1}{3}\right)V_a + (5 + 8)V_b + \left(\dfrac{1}{3} + \dfrac{1}{10}\right)V_c = 0$

Constraint:   $-V_b + V_c = 5$

This set of equations can be arrived at without ever assigning the voltage source current. When KCL was described in Chapter 2, we saw that it applied to any closed surface. Remember, the set of branches whose terminals are crossed

**FIGURE 3.11**  *Cut-Set*

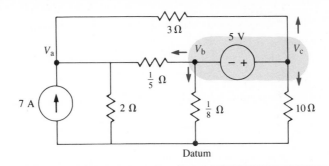

by the closed surface is a cut-set. Sometimes the term *super-node* is used instead of cut-set. Let's isolate the voltage source and the two nodes to which it is connected, as shown in Figure 3.11.

If we sum the cut-set currents leaving the surface, we get

$$\text{Cut-set:} \quad (V_b - V_a)5 + 8V_b + \frac{V_c - V_a}{3} + \frac{V_c}{10} = 0$$

Reformatting yields

$$\text{Cut-set:} \quad -\left(5 + \frac{1}{3}\right)V_a + (5 + 8)V_b + \left(\frac{1}{3} + \frac{1}{10}\right)V_c = 0$$

which is the same equation that we obtained by adding the KCL equations at nodes b and c. Note that the voltage source–node voltage constraint equation is still needed.

**Example 3.5**  *Cut-Set Analysis*

Find the node voltages for the circuit in Figure 3.12.

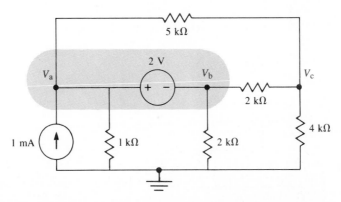

**FIGURE 3.12**

(*continues*)

**Example 3.5**  *Continued*

We write a cut-set equation for nodes a and b, a node equation for node c, and a constraint equation.

Cut-set:   $.001V_a + .0002(V_a - V_c) + .0005V_b + .0005(V_b - V_c) = .001$

Node c:   $.0002(V_c - V_a) + .0005(V_c - V_b) + .00025V_c = 0$

Constraint:   $V_a - V_b = 2$

Rearranging and solving, we get

$V_a = 1.34$ V

$V_b = -.66$ V

$V_c = -.063$ V

The obvious advantage of the cut-set approach is that you do not have to worry about the voltage source current. It may seem to you that there is no reason to ever use the voltage source current method. However, the cut-set method requires some care when more than one voltage source is connected to the same nondatum node or when a voltage source is directly connected to the datum node (see Problem 3.9). The voltage source current technique is always directly applicable.

**Special Case Technique.**   The circuit in Figure 3.13 has a voltage source connected to the datum node. Although the cut-set approach can be used in this circuit, it is not obvious how to do it. There is a much simpler method for dealing with the voltage source here. We note immediately that since one end of the

**FIGURE 3.13**   *A Special Case*

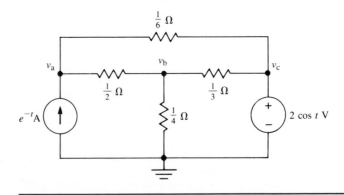

voltage source is connected to the datum node, $v_c$ is simply the voltage across the source:

$$v_c = 2 \cos t \text{ V}$$

We use this value for $v_c$ in the KCL equations. Since $v_c$ is known, there is no reason to write a KCL equation at node c. So, we need only the equations at nodes a and b, as follows:

Node a: $2(v_a - v_b) + 6(v_a - 2 \cos t) - e^{-t} = 0$
Node b: $2(v_b - v_a) + 4v_b + 3(v_b - 2 \cos t) = 0$

Note that the known value of $v_c$ is used in the node equations. Reformatting, we get

Node a: $(2 + 6)v_a - 2v_b = e^{-t} + 12 \cos t$
Node b: $-2v_a + (2 + 4 + 3)v_b = 6 \cos t$

The node voltages are

$$v_a = \frac{9}{68} e^{-t} + \frac{120}{68} \cos t \text{ V}$$

$$v_b = \frac{1}{34} e^{-t} + \frac{18}{17} \cos t \text{ V}$$

$$v_c = 2 \cos t \text{ V}$$

---

**Example 3.6**    *Special Case Technique*

Find the node voltages for the circuit in Figure 3.14.

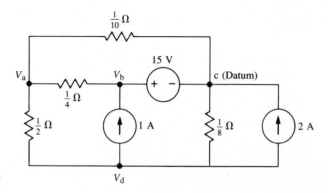

**FIGURE 3.14**

*(continues)*

**Example 3.6**    *Continued*

Since we are free to choose, we will use node c as the datum. This choice allows us to apply the special case technique. Here $V_b = 15$ V, and we write KCL equations at nodes a and d:

Node a:    $4(V_a - 15) + 10V_a + 2(V_a - V_d) = 0$
Node d:    $2(V_d - V_a) + 8V_d = -1 - 2$

Reformatting, we get

Node a:    $16V_a - 2V_d = 60$
Node d:    $-2V_a + 10V_d = -3$

The node voltages are

$$V_a = \frac{297}{78} \text{ V}$$

$$V_b = 15 \text{ V}$$

$$V_d = \frac{36}{78} \text{ V}$$

---

The choice of which mixed-source technique to use is a personal one. However, some guidelines can be suggested. When you are working by hand with a nonprogrammable calculator, you will want to minimize the number of simultaneous equations that you have to solve. With this in mind, you should use the techniques with the following priority: (1) special case, (2) source transformation, and (3) cut-set approach. If a calculator or computer program that solves matrix equations is available, then the voltage source current method is preferable, because it is applied in a straightforward manner.

## Analysis of Circuits with Controlled Sources

Now, we consider how the presence of controlled sources in a circuit affects the node analysis procedure. Keep in mind that a controlled source has the same branch characteristics as an independent source. The only difference is that the magnitude of a controlled source is an algebraic quantity. To analyze these circuits, we simply write the controlling variable in terms of node voltages and proceed with the node analysis. The next four examples illustrate this procedure.

---

**Example 3.7**    *Voltage-Controlled Current Source (VCIS)*

Find the node voltages for the circuit in Figure 3.15.

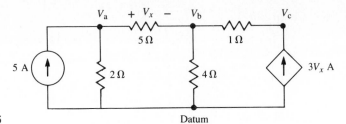

**FIGURE 3.15**

First note that since $V_x = V_a - V_b$, $3V_x = 3(V_a - V_b)$.

Node a: $\quad \dfrac{V_a}{2} + \dfrac{V_a - V_b}{5} - 5 = 0$

Node b: $\quad \dfrac{V_b - V_a}{5} + \dfrac{V_b}{4} + \dfrac{V_b - V_c}{1} = 0$

Node c: $\quad \dfrac{V_c - V_b}{1} - \underbrace{3(V_a - V_b)}_{\text{VCIS}} = 0$

Reformatting, we get

$$\frac{7}{10}V_a - \frac{1}{5}V_b = 5$$

$$-\frac{1}{5}V_a + \frac{29}{20}V_b - V_c = 0$$

$$-3V_a + 2V_b + V_c = 0$$

Thus,

$$V_a = 9.7 \text{ V}$$

$$V_b = 9.0 \text{ V}$$

$$V_c = 11.11 \text{ V}$$

---

**Example 3.8**  *Current-Controlled Current Source (ICIS)*

Find the node voltages for the circuit in Figure 3.16.

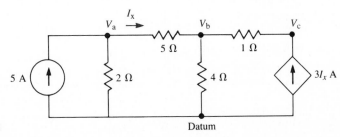

**FIGURE 3.16**

*(continues)*

**Example 3.8**   *Continued*

First note that since $I_x = (V_a - V_b)/5$, $3I_x = [3(V_a - V_b)]/5$.

Node a:   $\dfrac{V_a}{2} + \dfrac{V_a - V_b}{5} - 5 = 0$

Node b:   $\dfrac{V_b - V_a}{5} + \dfrac{V_b}{4} + \dfrac{V_b - V_c}{1} = 0$

Node c:   $\dfrac{V_c - V_b}{1} - \underbrace{\dfrac{3(V_a - V_b)}{5}}_{\text{ICIS}} = 0$

Rearranging, we get

$$\dfrac{7}{10}\, V_a - \dfrac{V_b}{5} = 5$$

$$-\dfrac{V_a}{5} + \dfrac{29}{20}\, V_b - V_c = 0$$

$$-\dfrac{3}{5}\, V_a - \dfrac{2}{5}\, V_b + V_c = 0$$

Thus,

$$V_a = 9.13 \text{ V}$$

$$V_b = 6.96 \text{ V}$$

$$V_c = 8.26 \text{ V}$$

---

**Example 3.9**   *Voltage-Controlled Voltage Source (VCVS)*

Find the node voltages for the circuit in Figure 3.17.

Since this circuit contains a voltage source, we must first decide which mixed-source method to use. The simplest technique to use here is to recognize that $V_c = 5V_x$ and use the special case technique. In terms of node voltages, $V_x = V_a - V_b$ so that $V_c = 5(V_a - V_b)$.

Node a:   $\dfrac{V_a}{2} + \dfrac{V_a - V_b}{5} - 5 = 0$

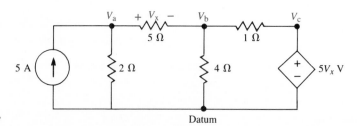

**FIGURE 3.17**

$$\text{Node b:} \quad \frac{V_b - V_a}{5} + \frac{V_b}{4} + \frac{\overbrace{V_b - 5(V_a - V_b)}^{\text{VCVS}}}{1} = 0$$

Reformatting, we get

$$\frac{7}{10} V_a - \frac{V_b}{5} = 5$$

$$-\frac{26}{5} V_a + \frac{129}{20} V_b = 0$$

Thus,

$$V_a = 9.28 \text{ V}$$

$$V_b = 7.48 \text{ V}$$

$$V_x = 1.8 \text{ V}$$

$$V_c = 9.0 \text{ V}$$

**Example 3.10** *Current-Controlled Voltage Source (ICVS)*

Find the node voltages for the circuit in Figure 3.18.

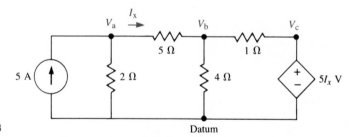

**FIGURE 3.18**

Since the voltage source is between node c and the datum, we proceed as in Example 3.9, except now $V_c = 5I_x = [5(V_a - V_b)]/5 = V_a - V_b$.

$$\text{Node a:} \quad \frac{V_a}{2} + \frac{V_a - V_b}{5} - 5 = 0$$

$$\text{Node b:} \quad \frac{V_b - V_a}{5} + \frac{V_b}{4} + \frac{\overbrace{V_b - (V_a - V_b)}^{\text{ICVS}}}{1} = 0$$

(*continues*)

**Example 3.10** *Continued*

Rearranging, we get

$$\frac{7}{10} V_a - \frac{V_b}{5} = 5$$

$$-\frac{6}{5} V_a + \frac{49}{20} V_b = 0$$

Thus,

$$V_a = 8.3 \text{ V}$$

$$V_b = 4.1 \text{ V}$$

$$I_x = .84 \text{ A}$$

$$V_c = 4.2 \text{ V}$$

---

Finally, let's consider the circuit in Figure 3.19. It contains one independent current source and all four types of controlled sources! Since the voltage-controlled voltage source, $5v_4$, is connected to the datum node, we will use the special case technique for this source. We will use a cut-set for the current-controlled voltage source.

In this circuit, the controlling variables are $i_1$, $i_5$, $v_4$, and $v_5$. In terms of

**FIGURE 3.19** *Controlled-Source Circuit*

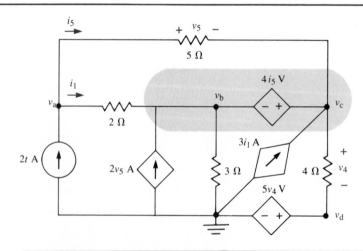

node voltages, they can be written as follows:

$$i_1 = \frac{v_a - v_b}{2}$$

$$i_5 = \frac{v_a - v_c}{5}$$

$$v_4 = v_c - v_d$$

$$v_5 = v_a - v_c$$

and since $v_d = 5v_4$

$$v_4 = v_c - 5v_4 = \frac{v_c}{6}$$

For clarity, we will first write the equations using the controlling branch variables and then make the node voltage substitutions. It would be quicker to use the relationships immediately. Three equations are needed to analyze this circuit using the cut-set method—the KCL equation at node a, the cut-set equation at nodes b and c, and the constraint equation for the current-controlled voltage source (we have already accounted for the voltage-controlled voltage source):

$$\text{Node a:} \quad \frac{v_a - v_b}{2} + \frac{v_a - v_c}{5} - 2t = 0$$

$$\text{Cut-set:} \quad \frac{v_b - v_a}{2} + \frac{v_b}{3} - 2v_5 + \frac{v_c - v_a}{5} + \frac{v_c - 5v_4}{4} - 3i_1 = 0$$

$$\text{Constraint:} \quad v_c - v_b = 4i_5$$

Substituting for branch variables yields

$$\text{Node a:} \quad \frac{v_a - v_b}{2} + \frac{v_a - v_c}{5} - 2t = 0$$

$$\text{Cut-set:} \quad \frac{v_b - v_a}{2} + \frac{v_b}{3} - 2(v_a - v_c) + \frac{v_c - v_a}{5} + \frac{v_c - 5v_c/6}{4} - \frac{3(v_a - v_b)}{2} = 0$$

$$\text{Constraint:} \quad v_c - v_b = \frac{4(v_a - v_c)}{5}$$

Reformatting, we get

$$\begin{bmatrix} .7 & -.5 & -.2 \\ -4.2 & 2.33 & 2.24 \\ -.8 & -1 & 1.8 \end{bmatrix} \begin{bmatrix} v_a \\ v_b \\ v_c \end{bmatrix} = \begin{bmatrix} 2t \\ 0 \\ 0 \end{bmatrix}$$

Solving the matrix equation gives

$$v_a = 31.62t \text{ V}$$

$$v_b = 28.34t \text{ V}$$

$$v_c = 29.79t \text{ V}$$

## 3.3  MESH ANALYSIS

In Section 3.2, we analyzed circuits by starting with KCL equations and then making appropriate substitutions. An alternative approach would be to first write the KVL equations for the circuit. If we limit our discussion to planar networks, we can find the circuit variables by writing KVL equations for the meshes of the circuit. This method of solution is known as *mesh analysis*. The development of the mesh analysis procedure is directly analogous to node analysis.

Let's examine the circuit in Figure 3.20A. Branch voltages and currents have been assigned as shown. Since this circuit has two meshes, we expect to find two independent KVL equations for it. We label each mesh with a lettered (or numbered, if you wish) arrow. The arrow indicates the direction that we will take around the mesh. Although the direction of travel around each mesh is arbitrary, we will always move in a clockwise manner. This consistent approach will enable us to develop the algorithmic techniques presented later in the chapter.

Mesh a is composed of $v_{s_1}$, $R_1$, and $R_3$. Mesh b is composed of $v_{s_2}$, $R_2$, and $R_3$. If we sum voltage drops in the clockwise direction, the KVL equations for the two meshes are

Mesh a:    $-v_{s_1} + v_1 + v_3 = 0$
Mesh b:    $-v_{s_2} - v_2 - v_3 = 0$

Just as we did in node analysis, we use the branch $V\text{--}I$ relationships to

**FIGURE 3.20    *Mesh Current***

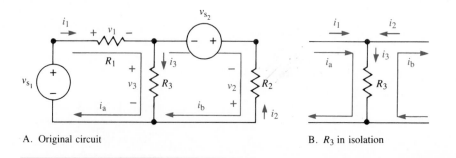

A.  Original circuit                                          B.  $R_3$ in isolation

replace, in this case, the resistor branch voltages with branch currents:

$$\text{Mesh a:} \quad -v_{s_1} + R_1 i_1 + R_3 i_3 = 0$$
$$\text{Mesh b:} \quad -v_{s_2} - R_2 i_2 - R_3 i_3 = 0$$

At this point, we are left with two equations in three unknowns. Our intention, again, is to reduce the number of unknowns. In node analysis, we introduced the node voltage and found relationships between this new type of voltage and the branch voltages. Here, we will proceed in a similar manner.

If you look back at the circuit in Figure 3.20A, you will see that an arrow was used to represent the pathway around each mesh. This symbol looks very much like the symbol used for branch currents. In fact, we can assume that each curved arrow denotes a current. This current is known as the **mesh current.** The mesh current is analogous to the node voltage. Unlike the node voltage, however, the mesh current does not necessarily represent any measurable current in the circuit, which will be demonstrated shortly. It does, however, provide us with a mathematical framework for solving the mesh equations.

What is the relationship between mesh currents and branch currents? If you examine the resistor $R_1$ in the circuit of Figure 3.20, you can see that two arrows are passing through it. One arrow represents the branch current, while the other arrow represents the mesh current. Since there is only one current in $R_1$, we are, in fact, using two different symbols for the same current. Therefore, the magnitudes of $i_1$ and $i_a$ must be the same. Since both arrows have the same direction, these two currents must have the same sign. Thus, we have the following relationship for $R_1$:

$$i_1 = i_a$$

Resistor $R_2$ also has one branch current and one mesh current symbol representing its current. Again, the magnitude of $i_2$ must equal the magnitude of $i_b$. However, in this resistor, the branch and mesh currents have opposite directions so that

$$i_2 = -i_b$$

The last resistor, $R_3$, has two mesh current arrows as well as the branch current arrow passing through it. To see what happens in a resistor like $R_3$, which borders two meshes, let's isolate $R_3$ and the nodes to which it is connected (see Figure 3.20B). If we apply KCL to the top node, we get

$$-i_1 - i_2 + i_3 = 0$$

or

$$-i_a + i_b + i_3 = 0$$

This gives us the branch current–mesh current relationship for $R_3$:

---

$$i_3 = i_a - i_b \tag{3.4}$$

where   $i_3$ = branch current

$i_a$ = mesh current in branch in same direction as assigned branch current

$i_b$ = mesh current in branch in opposite direction as assigned branch current

---

This relationship also applies to those branches that do not border two meshes and thus have only a single mesh current in them. For example, in $R_1$, since there is no mesh current of opposite direction to the branch current, the second term in Equation 3.4 is zero. In $R_2$, since there is no mesh current in the same direction as the branch current, the first term is zero.

We complete the analysis of the circuit in Figure 3.20A by substituting the branch current–mesh current relationships into the mesh equations:

$$\text{Mesh a:} \quad -v_{s_1} + i_a R_1 + (i_a - i_b)R_3 = 0$$
$$\text{Mesh b:} \quad -v_{s_2} - (-i_b)R_2 - (i_a - i_b)R_3 = 0$$

Rearranging gives

$$\text{Mesh a:} \quad (R_1 + R_3)i_a - R_3 i_b = v_{s_1}$$
$$\text{Mesh b:} \quad -R_3 i_a + (R_2 + R_3)i_b = v_{s_2}$$

In matrix form,

$$\begin{bmatrix} R_1 + R_3 & -R_3 \\ -R_3 & R_2 + R_3 \end{bmatrix} \begin{bmatrix} i_a \\ i_b \end{bmatrix} = \begin{bmatrix} v_{s_1} \\ v_{s_2} \end{bmatrix}$$

We now have two equations in two unknowns and can solve for the mesh currents. The matrix form is known as the *mesh resistance matrix* equation and has the general form

---

$$\mathbf{R}\mathbf{I} = \mathbf{V} \tag{3.5}$$

where   $\mathbf{R}$ = mesh resistance matrix

$\mathbf{I}$ = output vector

$\mathbf{V}$ = input vector

---

As in node analysis, the variables we find with the mesh analysis technique may not be the circuit variables of interest. This presents no problem if you go back to the branch current–mesh current relationships to find the desired branch current and then, if necessary, the branch voltage.

It is important to repeat that the mesh current does not, in general, represent a physically measurable current. In the circuit just analyzed, the mesh currents can, indeed, be measured by placing an ammeter in series with $R_1$ or $R_2$. The mesh currents in these branches are the same (at least in magnitude) as the actual branch currents. However, in the circuit of Figure 3.21, it is not possible to measure directly the mesh current $i_b$. Mesh current $i_b$ is strictly a mathematical variable that we have introduced to solve the mesh equations.

Before we consider how a more compact mesh analysis is done, let's analyze the circuit of Figure 3.22. This circuit has three meshes, so we will get three KVL equations for it.

Branch variables and the mesh currents have been assigned as shown in the figure. The solution begins with the mesh equations:

Mesh a:   $-2t + v_1 + v_2 = 0$

Mesh b:   $-v_2 - v_3 - \cos t = 0$

Mesh c:   $-v_1 + e^{-3t} - v_4 + v_3 = 0$

We use Ohm's law to replace branch voltages with branch currents:

Mesh a:   $-2t + 100i_1 + 10i_2 = 0$

Mesh b:   $-10i_2 - 10i_3 - \cos t = 0$

Mesh c:   $-100i_1 + e^{-3t} - 1000i_4 + 10i_3 = 0$

FIGURE 3.21   *Unmeasurable Mesh Current*

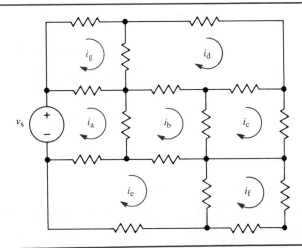

**FIGURE 3.22**   *Mesh Analysis*

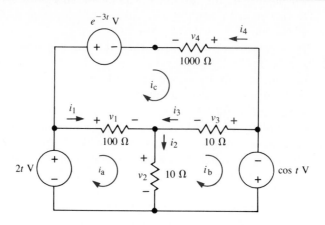

We then find the branch currents in terms of the mesh currents:

$$i_1 = i_a - i_c$$
$$i_2 = i_a - i_b$$
$$i_3 = i_c - i_b$$
$$i_4 = -i_c$$

Substituting for the branch currents, we get

Mesh a:   $-2t + 100(i_a - i_c) + 10(i_a - i_b) = 0$
Mesh b:   $-10(i_a - i_b) - 10(i_c - i_b) - \cos t = 0$
Mesh c:   $-100(i_a - i_c) + e^{-3t} - 1000(-i_c) + 10(i_c - i_b) = 0$

The final set of mesh equations is as follows:

Mesh a:   $110i_a - 10i_b - 100i_c = 2t$
Mesh b:   $-10i_a + 20i_b - 10i_c = \cos t$
Mesh c:   $-100i_a - 10i_b + 1110i_c = -e^{-3t}$

In matrix form, this set of equations is

$$\begin{bmatrix} 110 & -10 & -100 \\ -10 & 20 & -10 \\ -100 & -10 & 1110 \end{bmatrix} \begin{bmatrix} i_1 \\ i_2 \\ i_3 \end{bmatrix} = \begin{bmatrix} 2t \\ \cos t \\ -e^{-3t} \end{bmatrix}$$

The mesh currents are

$$i_a = .021t + .0058 \cos t - .001e^{-3t} \text{ A}$$

$$i_b = .0115t + .0534 \cos t - .001e^{-3t} \text{ A}$$

$$i_c = .002t + .001 \cos t - .001e^{-3t} \text{ A}$$

Now, if needed, we can go back to the original equations to find the branch currents. Ohm's law is then used to find the branch voltages. For example, if we solve for the first two branch voltages, we get

$$v_1 = 100i_1 = 100(i_a - i_c) = 1.9t + .48 \cos t \text{ V}$$

$$v_2 = 10i_2 = 10(i_a - i_b) = .095t - .476 \cos t \text{ V}$$

The final set of mesh current equations does not depend on the assigned polarity of the branch voltages or directions of the branch currents. Altering a branch voltage polarity will change its sign in the mesh equation. It will also change the sign of the branch voltage—mesh current relationship. These two sign changes will cancel each other.

In this text, we will adopt the strategy that each mesh will be circled in a clockwise direction. We will also assume that when we write the KVL equation for the $n$th mesh, all resistor voltage polarities are assumed to be positive on the side that $i_n$ enters. Thus, for mesh a in Figure 3.22, for example, the mesh current $i_a$ is positive and all other mesh currents are negative. In this manner, we can skip most of the steps shown in the mesh analysis of the circuit in Figure 3.22.

The simplified mesh analysis procedure, demonstrated in Example 3.11, can be summarized as follows:

*Step 1:* Label mesh currents in the clockwise direction.
*Step 2:* For the $n$th mesh, write KVL equations, assuming resistor voltages are positive on the side that $i_n$ enters.
*Step 3:* Reformat equations to separate variables.
*Step 4:* Solve for mesh currents.
*Step 5:* If necessary, solve for branch variables.

---

**Example 3.11**   *Quick Mesh Analysis*

Find the mesh currents for the circuit in Figure 3.23.

*(continues)*

**Example 3.11**  *Continued*

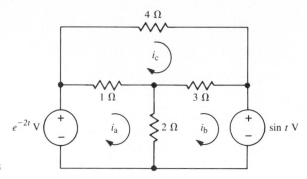

**FIGURE 3.23**

In mesh a, assume that the voltage across the $1\,\Omega$ resistor is positive on the left, while the voltage across the $2\,\Omega$ resistor is positive on the top.

Mesh a:  $1(i_a - i_c) + 2(i_a - i_b) - e^{-2t} = 0$

In mesh b, the voltage across the $2\,\Omega$ resistor is now assumed to be positive on the bottom, and that across the $3\,\Omega$ resistor is positive on the left.

Mesh b:  $2(i_b - i_a) + 3(i_b - i_c) + \sin t = 0$

In mesh c, assume the voltages across both the $1\,\Omega$ and $2\,\Omega$ resistors to be positive on the right, while that of the $4\,\Omega$ resistor is positive on the left.

Mesh c:  $1(i_c - i_a) + 3(i_c - i_b) + 4i_c = 0$

The final set of mesh equations is as follows:

Mesh a:  $(1+2)i_a - 2i_b - i_c = e^{-2t}$
Mesh b:  $-2i_a + (2+3)i_b - 3i_c = -\sin t$
Mesh c:  $-i_a - 3i_b + (1+3+4)i_c = 0$

The mesh currents are

$$i_a = \frac{31}{44} e^{-2t} - \frac{19}{44} \sin t \text{ A}$$

$$i_b = \frac{19}{44} e^{-2t} - \frac{23}{44} \sin t \text{ A}$$

$$i_c = \frac{11}{44} e^{-2t} - \frac{11}{44} \sin t \text{ A}$$

## Analysis of Circuits with Current Sources

The mesh analysis procedure just described is straightforward if there are only voltage sources in the network. How do we deal with current sources? Consider, for example, the circuit in Figure 3.24. If we follow the steps of the mesh analysis procedure as listed earlier, we would have a problem. We cannot find the branch voltage of the current source in terms of mesh currents. As in node analysis, several techniques can be used to deal with the mixed-source problem. Let's examine how the source transformation and the constraint equation approaches are used in mesh analysis.

**Source Transformation Method.**   You can see that in the circuit of Figure 3.24 there is a resistor in parallel with the current source. We can, therefore, make the transformation shown in Figure 3.25.

Be careful that you maintain the proper polarity of the voltage source and connect the series combination between the correct nodes. The same advantages and disadvantages of this technique that were discussed in the previous section on mixed sources apply here. The number of equations that need to be written is

**FIGURE 3.24**   *Mesh Analysis with Current Source*

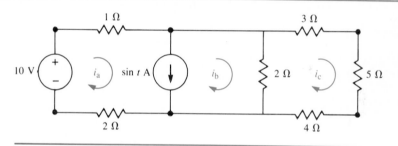

**FIGURE 3.25**   *Source Transformation*

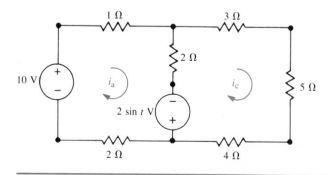

reduced (we have eliminated one mesh), but the branch variables of the current source–resistor combination have been lost. The set of mesh equations for the transformed circuit and the mesh current solutions are as follows:

Mesh a:   $1i_a + 2(i_a - i_c) - 2\sin t + 2i_a - 10 = 0$

Mesh c:   $3i_c + 5i_c + 4i_c + 2\sin t + 2(i_c - i_a) = 0$

So,

Mesh a:   $5i_a - 2i_c = 2\sin t + 10$

Mesh c:   $-2i_a + 14i_c = -2\sin t$

Thus,

$i_a(t) = .36\sin t + 2.12$ A

$i_c(t) = -.09\sin t + .3$ A

**Current Source Voltage Method.**   There are, of course, many circuits in which the troublesome current source is not in parallel with a resistor. In this case, we can assign a branch voltage to the current source and write equations for the original, untransformed circuit. For each current source, a constraint equation is written that relates the value of the current source to its mesh currents. Let's try this approach on the circuit in Figure 3.24, which is redrawn in Figure 3.26:

Mesh a:   $1i_a + v_I + 2i_a - 10 = 0$

Mesh b:   $-v_I + 2(i_b - i_c) = 0$

Mesh c:   $2(i_c - i_b) + 3i_c + 5i_c + 4i_c = 0$

Constraint:   $i_a - i_b = \sin t$

So,

Mesh a:   $3i_a + v_I = 10$

Mesh b:   $-v_I + 2i_b - 2i_c = 0$

Mesh c:   $-2i_b + 14i_c = 0$

Constraint:   $i_a - i_b = \sin t$

**FIGURE 3.26**   *Current Source Voltage*

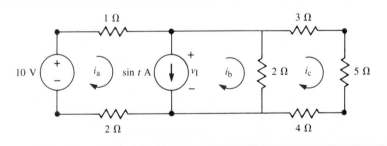

These equations are solved to find

$$i_a(t) = .36 \sin t + 2.12 \text{ A}$$

$$i_b(t) = -.64 \sin t + 2.12 \text{ A}$$

$$i_c(t) = -.09 \sin t + .3 \text{ A}$$

**Loop Analysis Method.** You can see that by adding the first and second mesh equations in the current source voltage example, the current source voltage disappears so that

Meshes a + b:    $3i_a + 2i_b - 2i_c = 10$

Mesh c:    $-2i_b + 14i_c = 0$

Constraint:    $i_a - i_b = \sin t$

We have, here, a situation analogous to our development of the cut-set approach. In fact, the equation just given for meshes a and b can be arrived at directly by writing a loop equation around the path that includes the 1 Ω, 2 Ω, and 2 Ω resistors and the 10 V source. In general, you can find the proper loop to use by incorporating all of the resistors and voltage sources of the two meshes that contain the current source. Sometimes the term *super-mesh* is used instead of loop.

**Example 3.12**    *Loop Analysis*

Find the mesh currents for the circuit in Figure 3.27.

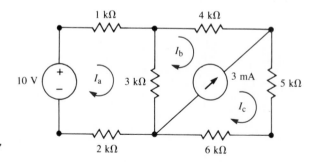

**FIGURE 3.27**

Mesh a:    $1000I_a + 3000(I_a - I_b) + 2000I_a - 10 = 0$

Loop:    $3000(I_b - I_a) + 4000I_b + 5000I_c + 6000I_c = 0$

Constraint:    $I_c - I_b = .003$

*(continues)*

**Example 3.12**   *Continued*

Rearranging gives

$$6000I_a - 3000I_b = 10$$

$$-3000I_a + 7000I_b + 11{,}000I_c = 0$$

$$-I_b + I_c = .003$$

The mesh currents are

$$I_a = .82 \text{ mA}$$

$$I_b = -1.69 \text{ mA}$$

$$I_c = 1.3 \text{ mA}$$

---

**Special Case Technique.**   The loop analysis method is difficult to use if the current source is in only one mesh. Figure 3.28 shows such a case. However, if you examine the mesh that contains the current source, you will see by inspection that

$$i_c = \sin t \text{ mA}$$

Since $i_c$ is known, there is no need to write an equation for mesh c. You simply use the value of the current source in the equations for the other meshes:

Mesh a:   $1000i_a + 2000(i_a - i_b) - 5 = 0$

Mesh b:   $2000(i_b - i_a) + 2000i_b + 4000(i_b - .001 \sin t) = 0$

So,

$$3000i_a - 2000i_b = 5$$

$$-2000i_a + 8000i_b = 4 \sin t$$

**FIGURE 3.28**   *A Special Case*

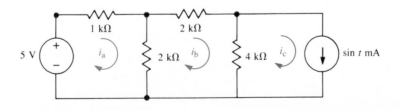

Thus,

$$i_a(t) = 2 + .4 \sin t \text{ mA}$$

$$i_b(t) = .5 + .6 \sin t \text{ mA}$$

---

**Example 3.13**   *Special Case Technique*

Find the mesh currents for the circuit in Figure 3.29.

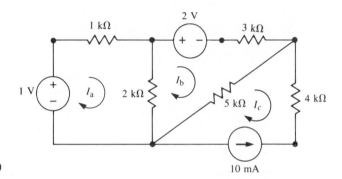

**FIGURE 3.29**

Since $I_c$ is in the opposite direction of the current source, $I_c = -10\,\text{mA}$. (Note that the $4\,\text{k}\Omega$ resistor is in series with the current source.)

Mesh a:   $1000I_a + 2000(I_a - I_b) - 1 = 0$

Mesh b:   $2000(I_b - I_a) + 2 + 3000I_b + 5000(I_b + .01) = 0$

So,

$$3000I_a - 2000I_b = 1$$

$$-2000I_a + 10{,}000I_b = -52$$

Thus,

$$I_a = -3.62\,\text{mA} \qquad \text{and} \qquad I_b = -5.93\,\text{mA}$$

---

## Analysis of Circuit with Controlled Sources

Here, we consider how the presence of controlled voltage and current sources in a circuit affects the mesh analysis procedure. There is really nothing new to learn. You already know how to do mesh analysis in circuits containing voltage and current sources. To analyze circuits with controlled sources, we simply write the controlling variable in terms of mesh currents and proceed with the mesh analysis. The next four examples illustrate this procedure.

**Example 3.14**    *Current-Controlled Voltage Source (ICVS)*

Find the mesh currents for the circuit in Figure 3.30.

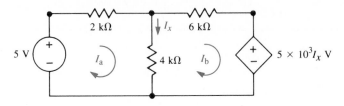

**FIGURE 3.30**

First note that since $I_x = I_a - I_b$, $5000I_x = 5000(I_a - I_b)$.

Mesh 1:    $2000I_a + 4000(I_a - I_b) - 5 = 0$

Mesh 2:    $4000(I_b - I_a) + 6000I_b + \underbrace{5000(I_a - I_b)}_{\text{ICVS}} = 0$

Reformatting, we get

$$6000I_a - 4000I_b = 5$$

$$1000I_a + 5000I_b = 0$$

Thus,

$$I_a = .75\,\text{mA} \qquad \text{and} \qquad I_b = -.15\,\text{mA}$$

**Example 3.15**    *Voltage-Controlled Voltage Source (VCVS)*

Find the mesh current for the circuit in Figure 3.31.

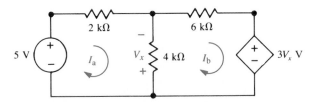

**FIGURE 3.31**

First note that since $V_x = 4000(I_b - I_a)$, $3V_x = 12,000(I_b - I_a)$.

Mesh 1:    $2000I_a + 4000(I_a - I_b) - 5 = 0$

Mesh 2:    $4000(I_b - I_a) + 6000I_b + \underbrace{12,000(I_b - I_a)}_{\text{VCVS}} = 0$

Rearranging, we get

$$6000I_a - 4000I_b = 5$$

$$-16,000I_a + 22,000I_b = 0$$

Thus,

$$I_a = 1.62 \text{ mA} \quad \text{and} \quad I_b = 1.18 \text{ mA}$$

**Example 3.16**   *Current-Controlled Current Source (ICIS)*

Find the mesh current for the circuit in Figure 3.32.

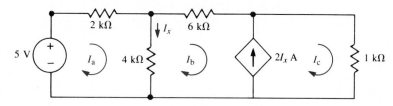

**FIGURE 3.32**

We take two initial steps before we start. We note that $I_x = (I_a - I_b)$, and we decide to use loop analysis for meshes b and c.

Mesh 1:   $2000I_a + 4000(I_a - I_b) - 5 = 0$

Loop:   $4000(I_b - I_a) + 6000I_b + 1000I_c = 0$

Constraint:   $I_c - I_b = 2I_x = \underbrace{2(I_a - I_b)}_{\text{ICIS}}$

Reformatting, we get

$$6000I_a - 4000I_b = 5$$

$$-4000I_a + 10{,}000I_b + 1000I_c = 0$$

$$-2I_a + I_b + I_c = 0$$

Thus,

$$I_a = .98 \text{ mA}$$

$$I_b = .22 \text{ mA}$$

$$I_c = 1.74 \text{ mA}$$

**Example 3.17**   *Voltage-Controlled Current Source (VCIS)*

Find the mesh currents for the circuit in Figure 3.33.

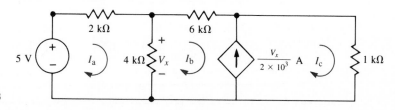

**FIGURE 3.33**

*(continues)*

**Example 3.17**   *Continued*

As in Example 3.16, we will use loop analysis for meshes b and c. However, in this example, the current source is controlled by $V_x$, where $V_x = 4000(I_a - I_b)$. The magnitude of the current source is, therefore, $2(I_a - I_b)$.

Mesh 1:   $2000I_a + 4000(I_a - I_b) - 5 = 0$

Loop:   $4000(I_b - I_a) + 6000I_b + 1000I_c = 0$

Constraint:   $I_c - I_b = \underbrace{2(I_a - I_b)}_{\text{VCIS}}$

Rearranging, we get

$$6000I_a - 4000I_b = 5$$

$$-4000I_a + 10{,}000I_b + 1000I_c = 0$$

$$-2I_a + I_b + I_c = 0$$

Thus,

$$I_a = .98\,\text{mA}$$

$$I_b = .22\,\text{mA}$$

$$I_c = 1.74\,\text{mA}$$

Finally, let's consider the circuit in Figure 3.34. It contains one independent voltage source and all four types of controlled sources. Since this circuit contains

**FIGURE 3.34**   *Controlled-Source Circuit*

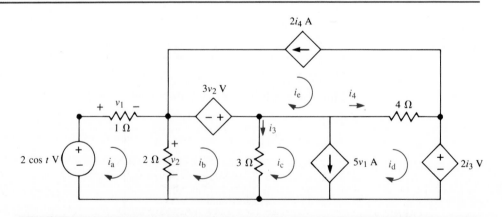

two current sources, we must decide which of the approaches just discussed will be used to deal with this mixed-source circuit.

The branch current through the $3\,\Omega$ resistor, $i_3$, is used to control the voltage source at the far right of the circuit. We cannot, therefore, involve this resistor in a source transformation. We will use a loop equation for the two meshes that contain the voltage-controlled current source $(5v_1)$. The $2i_4$ controlled current source is present only in mesh e, so we will treat this source as a special case and use $-2i_4$ in place of $i_e$ in the analysis (note the direction of the $2i_4$ current source and the mesh current).

Remember, with patience and care, you can solve circuits with controlled sources without difficulty. Treat the controlled sources as you would independent sources, and replace the controlling variables with their mesh current representations. For clarity, we will first write the mesh equations in terms of the controlling branch variables and then make the appropriate substitutions:

$$\text{Mesh a:} \quad 1i_a + 2(i_a - i_b) - 2\cos t = 0$$
$$\text{Mesh b:} \quad 2(i_b - i_a) - 3v_2 + 3(i_b - i_c) = 0$$
$$\text{Loop:} \quad 3(i_c - i_b) + 4[i_d - (-2i_4)] + 2i_3 = 0$$
$$\text{Constraint:} \quad i_c - i_d = 5v_1$$

Replacing $v_1$, $v_2$, $i_3$, and $i_4$ with their mesh current relationships gives $v_1 = i_a$, $v_2 = 2(i_a - i_b)$, $i_3 = i_b - i_c$, and $i_4 = i_d - i_e = i_d + 2i_4$. So, $i_4 = -i_d$. Substituting these relationships into the KVL equations yields

$$\text{Mesh a:} \quad 1i_a + 2(i_a - i_b) - 2\cos t = 0$$
$$\text{Mesh b:} \quad 2(i_b - i_a) - 3 \cdot 2(i_a - i_b) + 3(i_b - i_c) = 0$$
$$\text{Loop:} \quad 3(i_c - i_b) + 4(i_d - 2i_d) + 2(i_b - i_c) = 0$$
$$\text{Constraint:} \quad i_c - i_d = 5i_a$$

Reformatting, we get

$$\text{Mesh a:} \quad 3i_a - 2i_b = 2\cos t$$
$$\text{Mesh b:} \quad -8i_a + 11i_b - 3i_c = 0$$
$$\text{Loop:} \quad -i_b + i_c - 4i_d = 0$$
$$\text{Constraint:} \quad -5i_a + i_c - i_d = 0$$

We are left with four equations in four variables that can be solved to find

$$i_a = -1.2\cos t \text{ A}$$
$$i_b = -2.8\cos t \text{ A}$$
$$i_c = -7.06\cos t \text{ A}$$
$$i_d = -1.065\cos t \text{ A}$$

In summary, to deal with controlled sources in mesh analysis, treat the controlled sources exactly as you treat independent sources. Simply find the relationship between the branch variable that controls each source and the mesh currents in the circuit. Once you make these substitutions, proceed in the usual manner.

## 3.4  COMPARISON OF CIRCUIT ANALYSIS METHODS

Three different methods for analyzing circuits have been shown thus far: (1) the branch variable method, (2) node analysis, and (3) mesh analysis. Which of these methods is best? There is no clear-cut answer. Most computer programs use node analysis; when you work by hand, you will have to use your own judgment based on your experience. As an example, we will use all three techniques to analyze the DC transistor circuit shown in Figure 3.35 (the DC model for the transistor is enclosed by the colored box).

Let's assume that we want to solve for the branch current $I_6$. We note that in this circuit, we need to write only a KCL equation at node b. All of the other nodes (except, of course, the datum node) have just two branches connected to them. Therefore, the KCL equations at these nodes lead to the trivial conclusion that the same current exists in both of the branches connected to the node. We get the following relationships from inspection of the circuit:

$$I_1 = -I_2$$

$$I_3 = I_4$$

$$I_5 = I_6 = -I_7$$

**FIGURE 3.35**   *DC Transistor Circuit*

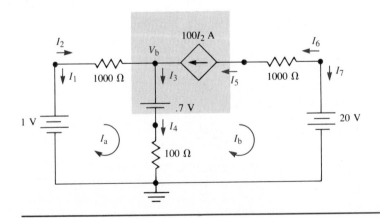

There are, then, only three unknown branch currents. We will use $I_2$, $I_4$, and $I_6$ in our solution.

We will use the branch variable method first. Since $I_6 = I_5$, we get, by inspection,

$$I_6 = I_5 = 100 I_2$$

Applying KCL at node b results in

$$-I_2 + I_4 - 100 I_2 = 0$$

so

$$I_4 = 101 I_2$$

Applying KVL around the left-hand mesh gives

$$-1 + 1000 I_2 + .7 + 100 I_4 = 0$$

Substituting for $I_4$ yields

$$1000 I_2 + 100(101 I_2) = .3$$

which leads to

$$I_2 = .027 \, \text{mA}$$

and we get the following answer:

$$I_6 = 100 I_2 = 2.7 \, \text{mA}$$

Does it surprise you that we did not have to write a KVL equation for the right-hand mesh? If it does, re-examine this mesh. The 1 kΩ resistor and the 20 V source are in series with a current source. Remember, elements in series with a current source cannot affect circuit variables in other parts of the circuit.

Now let's try node analysis. In a formal node analysis, we would write KCL equations at nodes a, b, and c. However, because of the series current source, we do not need to include the right-hand 1 kΩ resistor and 20 V source. The simplest approach, here, is to transform the .7 V source and 100 Ω resistor combination and use the special case technique at node a. The transformed circuit is shown in Figure 3.36.

The KCL equation at node b is

$$\frac{V_b - 1}{1000} + \frac{V_b}{100} - .007 - 100 I_2 = 0$$

FIGURE 3.36 *Transformed Transistor Circuit*

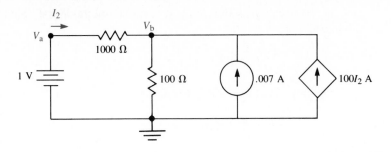

which leads to

$$\frac{V_b - 1}{1000} + \frac{V_b}{100} - .007 - \frac{100(1 - V_b)}{1000} = 0$$

where the ICIS current is $[100(1 - V_b)]/1000$. This equation is easily solved to give

$$V_b = .973 \text{ V}$$

Finally, we find $I_6$ with the following calculations:

$$I_6 = 100I_2 = \frac{100(1 - V_b)}{1000} = .1(1 - .973)$$

so

$$I_6 = 2.7 \text{ mA}$$

For mesh analysis, we will return to Figure 3.35 and write a KVL equation for the left-hand mesh:

$$-1 + 1000I_a + .7 + 100(I_a - I_b) = 0$$

which leads to

$$1100I_a - 100I_b = .3$$

From the current source in the right-hand mesh, we get the following constraint equation:

$$I_b = -100I_2 = -100I_a \qquad (\text{since } I_2 = I_a)$$

Substituting the constraint equation into the mesh equation gives

$$(1100 + 10,000)I_a = .3$$

so

$$I_a = \frac{.3}{11,100} = .027 \, \text{mA}$$

The desired branch current is found from

$$I_6 = 100I_2 = 100I_a = 2.7 \, \text{mA}$$

Notice that any one of the three methods can be used to analyze a given circuit. In this particular circuit, node analysis required slightly more involved calculations than either of the other two methods. There was no significant difference between mesh analysis and the branch variable approach. However, the branch variable approach does not require the introduction of any new circuit variables. It is especially useful in solving the simple electronic circuits that you may encounter in future studies.

## 3.5 NODE ANALYSIS BY INSPECTION

Node analysis of circuits that contain only independent current sources is very easy. For these circuits, the node conductance matrix equation can be written by inspection. Consider the circuit shown in Figure 3.37.

The node equations for this circuit are

Node a:  $V_a G_1 + (V_a - V_b)G_2 + (V_a - V_c)G_5 = I_1$

Node b:  $(V_b - V_a)G_2 + V_b G_3 = -I_2$

Node c:  $(V_c - V_a)G_5 + V_c G_4 = I_2$

**FIGURE 3.37**   *Node Analysis by Inspection*

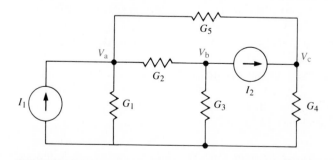

Rearranging, we get

Node a:   $V_a(G_1 + G_2 + G_5) - V_bG_2 - V_cG_5 = I_1$

Node b:   $-V_aG_2 + V_b(G_2 + G_3) = -I_2$

Node c:   $-V_aG_5 + V_c(G_4 + G_5) = I_2$

In matrix form,

$$\begin{bmatrix} G_1 + G_2 + G_5 & -G_2 & -G_5 \\ -G_2 & G_2 + G_3 & 0 \\ -G_5 & 0 & G_4 + G_5 \end{bmatrix} \begin{bmatrix} V_a \\ V_b \\ V_c \end{bmatrix} = \begin{bmatrix} I_1 \\ -I_2 \\ I_2 \end{bmatrix}$$

which has the form $\mathbf{GV} = \mathbf{I}$.

If we compare the entries in $\mathbf{G}$ with the circuit in Figure 3.37, a simple pattern becomes clear. Row 1 of $\mathbf{G}$ corresponds to the KCL equation at node a. The diagonal element $G(1, 1)$ is simply the sum of all conductances connected to the first node, node a. The off-diagonal element $G(1, 2)$ is equal to the negative of the conductance between node a and node b. Element $G(1, 3)$ is equal to the negative of the conductance between node a and node c.

The second row of $\mathbf{G}$ corresponds to KCL at node b and is filled in the same way. The diagonal element $G(2, 2)$ is the sum of all conductances connected to the second node, node b. Element $G(2, 1)$ is the negative of the conductance between the second and first nodes, nodes b and a. Element $G(2, 3)$ is the negative of the conductance between the second and third nodes, nodes b and c.

These observations lead to the following general rules for the entries of $\mathbf{G}$:

*Diagonal G Entries:*   $G(k, k) =$ sum of all conductances connected to node $k$
*Off-Diagonal G Entries:*   $G(k, l) =$ negative of total conductance connected between nodes $k$ and $l$

If you follow these rules, and the circuit does not contain controlled sources, then the node conductance matrix must be symmetric. It is important to keep in mind that the node number corresponds to the order of the node voltages in $\mathbf{V}$. In our example, $\mathbf{V}$ was ordered as $V_a$, $V_b$, and $V_c$ so that node a is the first node, node b is the second node, and node c is the third node. In Example 3.19, we will change this order.

We see that the first entry in $\mathbf{I}$ is the independent current source current, $I_1$, entering node a. At node b, the current source current $I_2$ is leaving, and the second entry in $\mathbf{I}$ is $-I_2$. The last entry in $\mathbf{I}$ is $I_2$, which is entering node c. The general rule for $\mathbf{I}$ is:

*Input Current Vector:*   $I(k) =$ sum of all independent current source currents *entering* node $k$.

**Example 3.18**   *Node Equation by Inspection*

Write the node conductance matrix equation for the circuit in Figure 3.38.

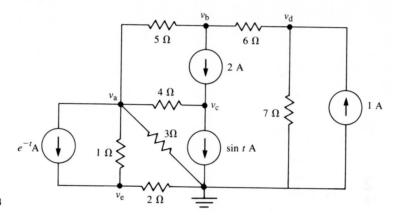

**FIGURE 3.38**

Let $\mathbf{V} = (v_a, v_b, v_c, v_d, v_e)^T$, so that

$$
\begin{bmatrix}
1.78 & -.2 & -.25 & 0 & -1 \\
-.2 & .36 & 0 & -.17 & 0 \\
-.25 & 0 & .25 & 0 & 0 \\
0 & -.17 & 0 & .31 & 0 \\
-1 & 0 & 0 & 0 & 1.5
\end{bmatrix}
\begin{bmatrix}
v_a \\ v_b \\ v_c \\ v_d \\ v_e
\end{bmatrix}
=
\begin{bmatrix}
-e^{-t} \\
-2 \\
2 - \sin t \\
1 \\
e^{-t}
\end{bmatrix}
$$

where the diagonal values of $\mathbf{G}$ are

$$G(1, 1) = \frac{1}{1} + \frac{1}{3} + \frac{1}{4} + \frac{1}{5} = 1.78$$

$$G(2, 2) = \frac{1}{5} + \frac{1}{6} = .36$$

$$G(3, 3) = \frac{1}{4} = .25$$

$$G(4, 4) = \frac{1}{6} + \frac{1}{7} = .31$$

$$G(5, 5) = \frac{1}{1} + \frac{1}{2} = 1.5$$

Verify the off-diagonal elements.

**Example 3.19**   *Reordering Voltage Vector*

For the circuit in Example 3.18, reorder $\mathbf{V}$ and reanalyze with the first node as node c, followed by nodes a, b, d, and e.

*(continues)*

**Example 3.19**   *Continued*

In matrix form,

$$
\begin{bmatrix}
.25 & -.25 & 0 & 0 & 0 \\
-.25 & 1.78 & -.2 & 0 & -1 \\
0 & -.2 & .36 & -.17 & 0 \\
0 & 0 & -.17 & .31 & 0 \\
0 & -1 & 0 & 0 & 1.5
\end{bmatrix}
\begin{bmatrix}
v_c \\ v_a \\ v_b \\ v_d \\ v_e
\end{bmatrix}
=
\begin{bmatrix}
2 - \sin t \\ -e^{-t} \\ -2 \\ 1 \\ e^{-t}
\end{bmatrix}
$$

Note that the matrix is symmetric, regardless of the ordering of the voltage vector.

---

Node analysis by inspection can be extended to include circuits that contain independent voltage sources. For these circuits, we use the voltage source current approach. An unknown current is assigned to each voltage source, and these currents become additional variables in the KCL equations. Node voltage constraint equations are then written for each voltage source.

The final set of KCL equations for the mixed-source circuit in Figure 3.39 is as follows:

Node a:    $1.1v_a - 1v_b - .1v_c + i_{v_1} = 3$
Node b:    $-v_a + 2.5v_b - .5v_d - i_{v_2} = 0$
Node c:    $-.1v_a + .55v_c - .25v_d + i_{v_2} = 0$
Node d:    $-.5v_b - .25v_c + .75v_d - i_{v_1} = 0$
Constraint 1:   $1v_a - 1v_d = 10$
Constraint 2:   $-1v_b + 1v_c = \sin 10t$

**FIGURE 3.39**   *Mixed-Source Circuit*

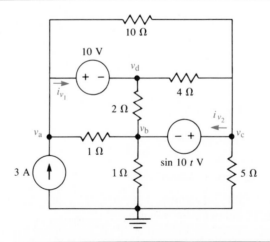

which leads to the following matrix equation:

$$
\left[
\begin{array}{cccc|cc}
1.1 & -1 & -.1 & 0 & 1 & 0 \\
-1 & 2.5 & 0 & -.5 & 0 & -1 \\
-.1 & 0 & .55 & -.25 & 0 & 1 \\
0 & -.5 & -.25 & .75 & -1 & 0 \\
\hline
1 & 0 & 0 & -1 & 0 & 0 \\
0 & -1 & 1 & 0 & 0 & 0
\end{array}
\right]
\left[
\begin{array}{c}
v_a \\
v_b \\
v_c \\
v_d \\
i_{v_1} \\
i_{v_2}
\end{array}
\right]
=
\left[
\begin{array}{c}
3 \\
0 \\
0 \\
0 \\
10 \\
\sin 10t
\end{array}
\right]
$$

The entries in the upper left-hand area are found as before. We have, however, added two new variables $i_{v_1}$ and $i_{v_2}$. Since $i_{v_1}$ is leaving node a, +1 is entered in the appropriate column of the first row. Since $i_{v_2}$ is entering node $b$, $-1$ is entered in the appropriate column of the second row, and so on.

The last two rows of the matrix contain the node voltage constraint equations. This matrix is a hybrid since it has both current and voltage equations. If it is set up properly, the variables will have their correct units.

Node analysis by inspection is not an efficient way to analyze circuits by hand because a large matrix results from small circuits. However, with a computer or calculator package that can solve matrix equations, this technique is excellent.

---

**Example 3.20**   *Mixed-Source Circuit Analysis*

Write the node conductance matrix equation for the circuit in Figure 3.40.

$$
\left[
\begin{array}{ccccccc}
.003 & -.001 & -.001 & 0 & 0 & 0 & 0 \\
-.001 & .002 & 0 & -.001 & 0 & -1 & 0 \\
-.001 & 0 & .002 & 0 & -.001 & 1 & 0 \\
0 & -.001 & 0 & .001 & 0 & 0 & 1 \\
0 & 0 & -.001 & 0 & .002 & 0 & -1 \\
0 & -1 & 1 & 0 & 0 & 0 & 0 \\
0 & 0 & 0 & 1 & -1 & 0 & 0
\end{array}
\right]
$$

$$
\times
\left[
\begin{array}{c}
V_a \\
V_b \\
V_c \\
V_d \\
V_e \\
I_{v_1} \\
I_{v_2}
\end{array}
\right]
=
\left[
\begin{array}{c}
-.002 \\
0 \\
0 \\
0 \\
.003 \\
1 \\
2
\end{array}
\right]
$$

(*continues*)

**Example 3.20**   *Continued*

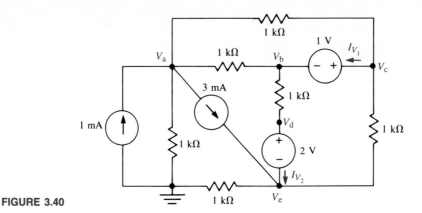

**FIGURE 3.40**

## 3.6   MESH ANALYSIS BY INSPECTION

Like node equations, mesh equations can often be written by inspection. The development of the following rules is similar to the development of the node analysis rules. Therefore, the derivation of the rules is omitted here. We will start with circuits that contain only independent voltage sources and then consider mixed-source circuits. The general mesh matrix equation for circuits with only independent current sources has the form $\mathbf{RI} = \mathbf{V}$.

Again, the order of the output vector, in this case $\mathbf{I}$, determines the elements of the mesh resistance matrix. The first row of $\mathbf{R}$ corresponds to the KVL equation for the mesh of the first entry in $\mathbf{I}$. With this understanding, the rules for $\mathbf{R}$ are as follows:

> ***Diagonal R Entries:***   $R(k, k) =$ sum of all resistances contained in mesh $k$
> ***Off-Diagonal R Entries:***   $R(k, l) =$ negative of total resistance that borders
> meshes $k$ and $l$

The mesh resistance matrix is always symmetric for circuits without controlled sources.

The input voltage vector $\mathbf{V}$ is found as follows:

> ***Input Voltage Vector:***   $V(k) =$ sum of all voltage source voltages around
> mesh $k$. A voltage is positive if mesh current
> $i_k$ leaves the positive terminal of the source.

**Example 3.21**   *Mesh Equation by Inspection I*

Write the mesh resistance matrix equation for the circuit in Figure 3.41.

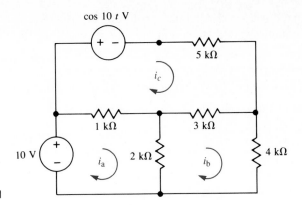

**FIGURE 3.41**

$$\begin{bmatrix} 3000 & -2000 & -1000 \\ -2000 & 9000 & -3000 \\ -1000 & -3000 & 9000 \end{bmatrix} \begin{bmatrix} i_a \\ i_b \\ i_c \end{bmatrix} = \begin{bmatrix} 10 \\ 0 \\ -\cos 10t \end{bmatrix}$$

where the diagonal values of $R$ are

$$R(1, 1) = 1000 + 2000 = 3000$$

$$R(2, 2) = 2000 + 3000 + 4000 = 9000$$

$$R(3, 3) = 1000 + 5000 + 3000 = 9000$$

Verify the off-diagonal elements.

---

**Example 3.22** *Mesh Equation by Inspection II*

Write the mesh resistance matrix equation for the circuit in Figure 3.42.

$$\begin{bmatrix} 3000 & -1000 & 0 \\ -1000 & 3000 & -1000 \\ 0 & -1000 & 2000 \end{bmatrix} \begin{bmatrix} i_a \\ i_b \\ i_c \end{bmatrix} = \begin{bmatrix} -\cos t \\ \cos t - \sin t \\ \sin t \end{bmatrix}$$

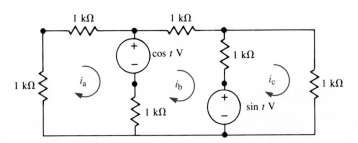

**FIGURE 3.42**

Mixed-source circuits are analyzed using current source voltages and constraint equations. This technique is demonstrated in the following example.

---

**Example 3.23**    *Mixed-Source Circuit Analysis*

Write the mesh resistance matrix equation for the circuit in Figure 3.43.

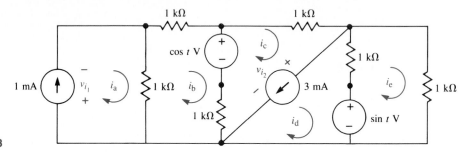

**FIGURE 3.43**

The current source voltage $v_{i_1}$ appears in mesh a, while the current
source voltage $v_{i_2}$ appears between meshes c and d. The constraint equations
are $i_a = .001$ and $i_c - i_d = .003$. When these constraint equations are combined with the KVL equations, the following matrix equation results:

| | | | | | | | |
|---|---|---|---|---|---|---|---|
| Mesh a | 1000 | −1000 | 0 | 0 | 0 | 1 | 0 |
| Mesh b | −1000 | 3000 | −1000 | 0 | 0 | 0 | 0 |
| Mesh c | 0 | −1000 | 2000 | 0 | 0 | 0 | 1 |
| Mesh d | 0 | 0 | 0 | 1000 | −1000 | 0 | −1 |
| Mesh e | 0 | 0 | 0 | −1000 | 2000 | 0 | 0 |
| Constraint 1 | 1 | 0 | 0 | 0 | 0 | 0 | 0 |
| Constraint 2 | 0 | 0 | 1 | −1 | 0 | 0 | 0 |

$$
\times
\begin{bmatrix} i_a \\ i_b \\ i_c \\ i_d \\ i_e \\ v_{i_1} \\ v_{i_2} \end{bmatrix}
=
\begin{bmatrix} 0 \\ -\cos t \\ \cos t \\ -\sin t \\ \sin t \\ .001 \\ .003 \end{bmatrix}
$$

---

## 3.7 SUMMARY

This chapter introduced two very powerful circuit analysis techniques, node
analysis and mesh analysis. Either one of these techniques can be used to analyze

any resistor circuit of any size. Node analysis is very straightforward if the circuit contains only current sources (independent or dependent). Mesh analysis is easily applied to circuits containing only voltage sources.

Analyzing circuits with both voltage and current sources requires some thought. Several techniques for dealing with this situation were presented. Practice them all so you can make the most efficient choice for a given circuit.

While the systematic techniques described in this chapter will enable you to analyze any circuit, it is often easier to analyze small networks with equivalent circuit methods. Some simple equivalences were shown in Chapter 2. Chapter 4 will describe more advanced circuit analysis techniques and equivalences.

## ■ PROBLEMS

3.1 For the circuit in Figure P3.1, use node c as the datum.

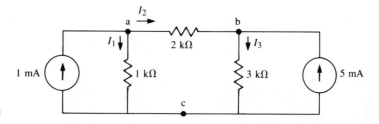

**FIGURE P3.1**

   a.   Find the node voltages at node a and node b.
   b.   Find the branch currents $I_1$, $I_2$, and $I_3$.
   c.   Find the power absorbed by each element.

3.2 Repeat Problem 3.1 with node b as the datum.
   a.   Find the node voltages at node a and node c.
   b.   Find the branch currents $I_1$, $I_2$, and $I_3$.
   c.   Compare your results for Problems 3.1 and 3.2.

3.3 As shown in the circuit in Figure P3.3, a 10 V source is connected from node c in the previous circuit (Figure P3.1) to ground.
   a.   Use node c as the datum node and write node equations at nodes a and b. (Do not forget that node c is *not* at zero potential now.)
   b.   Find the node voltages at node a and node b.
   c.   Find $I_1$, $I_2$, and $I_3$.
   d.   Compare your results for Problems 3.1 and 3.3.

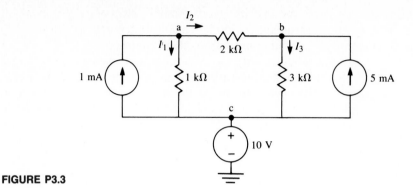

**FIGURE P3.3**

**3.4** Apply node analysis to the circuit in Figure P3.4 and find all branch variables and branch powers. What is the total power absorbed by all elements in the circuit?

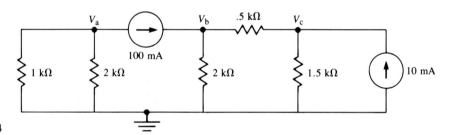

**FIGURE P3.4**

**3.5** Find the node voltages of the circuit in Figure P3.5.

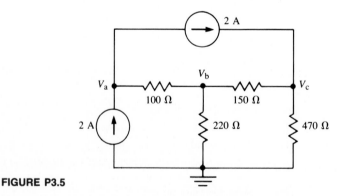

**FIGURE P3.5**

**3.6** Find $V_a$, $V_b$, and $V_c$ for the circuit in Figure P3.6.

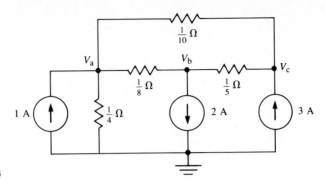

**FIGURE P3.6**

**3.7** The circuit in Figure P3.7 has two voltage sources. Use each of the following methods to write a set of solvable simultaneous equations in matrix form:

**a.** Source transformation and cut-set

**b.** Only cut-sets

**c.** Voltage source currents

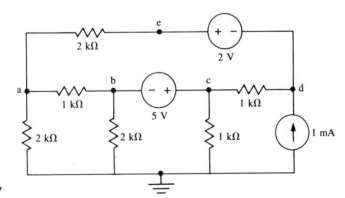

**FIGURE P3.7**

**3.8** Solve the matrix equation(s) derived in Problem 3.7. Use a calculator or computer, if available.

**3.9** The circuit in Figure P3.9 has six nodes. However, a judicious choice of the datum node can greatly simplify the analysis.

**a.** Use node e as the datum node and use the special case approach to find the node voltages. (*Hint:* You need to write only two node equations.)

**b.** Use node f as the datum node and use the cut-set approach to find the node voltages. (*Hint:* You will need five equations.)

**c.** Use node f as the datum node and use the voltage source current approach to write the node equations. Do not find node voltages unless a calculator or computer program is available to you.

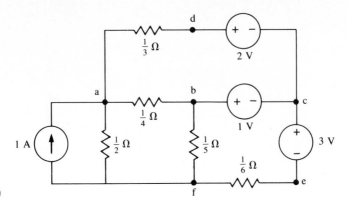

**FIGURE P3.9**

**3.10** It is possible to use the source transformation technique in circuits that do not have a resistor directly in series with the voltage source. Figure P3.10A shows such a circuit. The voltage source can be shifted into each of the branches that are connected to it, as shown in Figure P3.10B. The technique is known as *voltage source shifting*, or *splitting*.

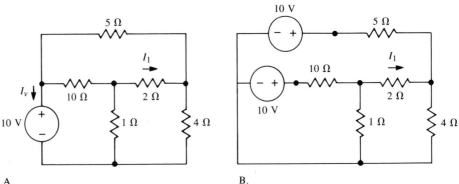

**FIGURE P3.10**    A.                                                              B.

    **a.** Use KVL to prove that the two circuits of Figure P3.10 are equivalent.

    **b.** Where is $I_v$ in Figure P3.10B?

    **c.** Using source transformations, redraw Figure P3.10B so that it contains only current sources.

    **d.** Find $I_1$ and $I_v$ by using node analysis.

**3.11** Given the operational amplifier circuit in Figure P3.10, replace the op amp with its model $(r_i = 1\,M\Omega, r_o = 100\,\Omega,$ and $A = 10^5)$.

    **a.** Write the node equations for this circuit.

    **b.** Find $V_o$.

**3.12** Given the following node conductance matrix equation

$$\begin{bmatrix} 1.6 & -.5 & -.1 \\ -.5 & .75 & -.25 \\ -.1 & -.25 & .55 \end{bmatrix} \begin{bmatrix} v_a \\ v_b \\ v_c \end{bmatrix} = \begin{bmatrix} 1 \\ 2 \\ -3 \end{bmatrix}$$

    **a.** Find $v_a$, $v_b$, and $v_c$.

    **b.** Draw a circuit that would yield this equation. (*Hint:* Start with the off-diagonal terms.)

**3.13** Repeat Problem 3.12 for the following node conductance matrix equation:

$$\begin{bmatrix} 5.5 & -4.5 & -1 \\ -.5 & 1.75 & -1.25 \\ -1 & -.25 & 1.45 \end{bmatrix} \begin{bmatrix} v_a \\ v_b \\ v_c \end{bmatrix} = \begin{bmatrix} 0 \\ 0 \\ 2 \end{bmatrix}$$

(*Hint:* You will need controlled sources.)

**3.14** For the BJT circuit in Figure P3.14, $r_x = 2\,\text{k}\Omega$, $r_o = 100\,\text{k}\Omega$, and $\beta = 200$. Replace the transistor with its AC model and use node analysis to find $v_o$.

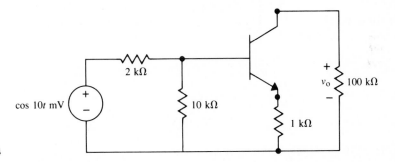

**FIGURE P3.14**

**3.15** For the FET circuit in Figure P3.15, $r_d = 5\,\text{k}\Omega$ and $g_m = .01\,\text{S}$. Replace the FET with its model and use node analysis to find $v_0$.

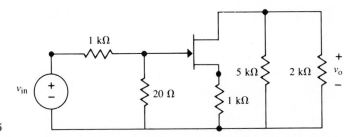

**FIGURE P3.15**

**3.16** Use node analysis to find $v_{R_1}$ and $v_{R_2}$ in the circuit in Figure P3.16.

**3.17** Use node analysis to find $V_{R_1}$ and $I_{R_2}$ in the circuit in Figure P3.17.

**3.18** Write the node equations for the circuit in Figure P3.18.

**3.19** Use mesh analysis to find the branch currents in the circuit in Figure P3.19. Find the power absorbed by each element in the circuit.

**3.20** The circuit shown in Figure P3.20 is a bridge circuit.

    **a.** Write the mesh equations for this circuit.

    **b.** The bridge circuit is in balance when $i_L = 0$. Make this substitution to simplify the mesh equations.

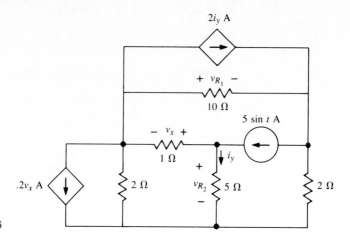

**FIGURE P3.16**

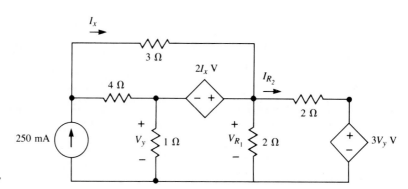

**FIGURE P3.17**

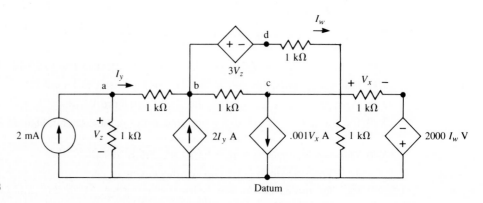

**FIGURE P3.18**

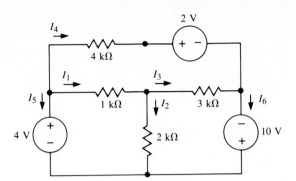

**FIGURE P3.19**

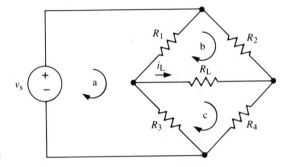

**FIGURE P3.20**

    **c.**    Use the simplified mesh equations to show that $R_4/R_3 = R_2/R_1$ when the bridge circuit is in balance.

**3.21** Write a set of mesh equations for the circuit in Figure P3.7. Find the power absorbed by the sources in the circuit.

**3.22** Write a set of mesh equations for the circuit in Figure P3.9. Find the power absorbed by the sources in the circuit.

**3.23** For the circuit in Figure P3.23, use mesh analysis to find all resistor currents.

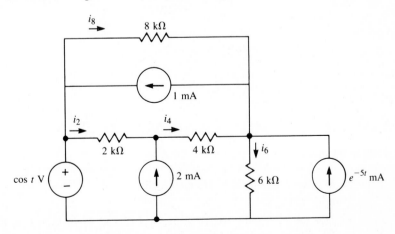

**FIGURE P3.23**

**3.24** It is possible to use the source transformation technique in circuits that do not have a resistor directly in parallel with the current source. Figure P3.24A shows such a circuit. The current source can be replaced with two identical current sources in series, as shown in Figure P3.24B. A conductor is connected between the current sources and any desired node in the circuit.

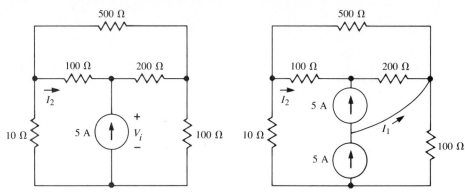

**FIGURE P3.24**    A.                                                                    B.

   **a.** Use KCL to prove that the current ($I_1$) in the additional conductor is zero. This is our justification for adding the conductor to the circuit. If it carries no current, then it can have no effect on the rest of the circuit.

   **b.** Using source transformations, redraw Figure P3.24B so that it contains only voltage sources.

   **c.** Where is $V_I$ in the transformed circuit?

   **d.** Find $I_2$ and $V_I$ by using mesh analysis.

**3.25** The operational amplifier circuit in Figure P3.25 is a noninverting amplifier. Replace the op amp with its model ($r_i = 1\ M\Omega$, $r_o = 0$, and $A = 10^5$).

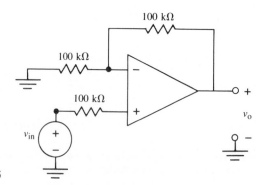

**FIGURE P3.25**

   **a.** Write the mesh equations for this circuit.

   **b.** Find $v_o$.

**3.26** Use the model for the FET ($g_M = .015$ and $r_d = 5\ k\Omega$) to redraw the circuit in Figure P3.26. Use mesh analysis to find $v_o$.

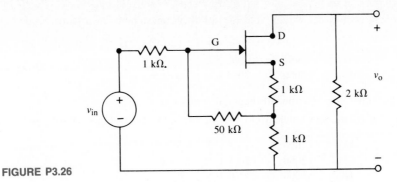

**FIGURE P3.26**

**3.27** Use mesh equations to find $i_{R_1}$ and $v_{R_2}$ in Figure P3.27.

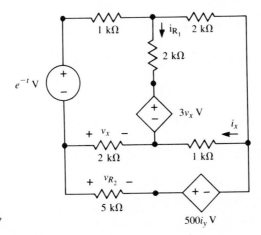

**FIGURE P3.27**

**3.28** Use mesh equations to find $I_{R_1}$, $I_y$, and $V_x$ in Figure P3.28.

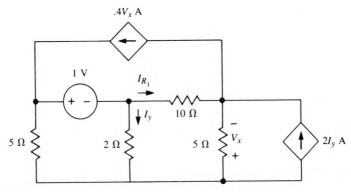

**FIGURE P3.28**

**3.29** Use mesh analysis to find $I_y$, $I_w$, $V_x$, and $V_z$ in Figure P3.18. (*Hint:* You can write two mesh, one loop, and two constraint equations.)

**3.30** Draw a circuit that would yield the following mesh resistance matrix equation:

$$\begin{bmatrix} 5000 & -1000 & -2000 \\ -1000 & 10000 & -3000 \\ -2000 & -3000 & 7000 \end{bmatrix} \begin{bmatrix} i_a \\ i_b \\ i_c \end{bmatrix} = \begin{bmatrix} 10 \\ 0 \\ -\cos 10t \end{bmatrix}$$

(*Hint:* First consider the off-diagonal terms.)

**3.31** Draw a circuit that would yield the following mesh resistance matrix equation:

$$\begin{bmatrix} 5000 & -1000 & -1000 \\ -2000 & 3000 & -1000 \\ -1000 & -1000 & 1000 \end{bmatrix} \begin{bmatrix} i_a \\ i_b \\ i_c \end{bmatrix} = \begin{bmatrix} 5 \\ e^{-7t} \\ -e^{-7t} \end{bmatrix}$$

(*Hint:* You will need controlled sources in the circuit.)

# Advanced Circuit Analysis Techniques

## 4.1  INTRODUCTION

In Chapter 3, we analyzed circuits using the formal procedures of node analysis and mesh analysis. These procedures are necessary for the analysis of large circuits and can be easily implemented in computer-aided analysis programs. For smaller circuits, however, the equivalent circuit techniques discussed in Chapter 2 often lead to quicker analysis and provide more insight into circuit behavior. These equivalent circuit techniques are particularly well suited to ladder networks without controlled sources. In this chapter, we will extend the equivalent circuit concept to nonladder networks and to controlled-source networks.

We will begin by discussing two very important equivalent circuits, the Thévenin and Norton equivalents. These circuits allow us to replace a very large resistive circuit with a single resistor and a single independent source. This replacement technique is a very powerful tool in circuit analysis and design. We will use these equivalent circuits to derive voltage, current, and power transfer relationships.

The circuit model for an operational amplifier (op amp) was shown in Chapter 1. By now, you have probably solved several op amp problems by using this model. Because the input resistance and the gain of an op amp are so large, we can derive a new model, the ideal op amp model. Its use greatly simplifies the analysis of op amp circuits. We will discuss this model toward the end of this chapter. We will conclude the chapter with a brief section on the analysis of nonlinear resistive circuits.

## 4.2  THÉVENIN EQUIVALENT CIRCUITS

Figure 4.1 shows two different circuits, each composed of two subcircuits. The subcircuit $N_{1_1}$ in Figure 4.1A has several sources and resistors, while $N_{1_2}$ in Figure 4.1B contains one source and one resistor. The subcircuits $N_{2_1}$ and $N_{2_2}$ are the same single load resistor $R_L$. It is obvious that the $V$–$I$ relationship at terminals a–a′ looking right toward the load resistor is the same for both circuits—that is, $V = IR_L$. Let's find the $V$–$I$ relationships looking left toward $N_{1_1}$ and $N_{1_2}$.

$$N_{1_1} \text{ by KCL at node b:} \qquad \frac{3}{1000} V_b - \frac{V}{1000} = .001 + \frac{10}{1000}$$

$$V_b = \frac{11}{3} + \frac{V}{3}$$

$$\text{By KVL:} \qquad V = V_b - 1000I = \frac{11}{3} + \frac{V}{3} - 1000I$$

$$V = 5.5 - 1500I \tag{4.1}$$

$$N_{1_2} \text{ by inspection:} \qquad V = 5.5 - 1500I \tag{4.2}$$

Notice that the two $V$–$I$ relationships are identical. This means that the subcircuits $N_{1_1}$ and $N_{1_2}$ are equivalent at the terminals a–a′. If we are interested in

**FIGURE 4.1**   *Equivalent Circuits*

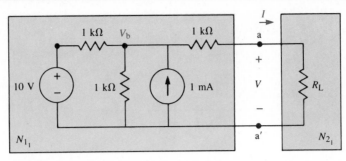

A. Original circuit

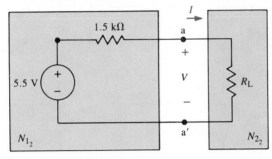

B. Equivalent circuit

only the voltage and current of the load resistor $R_L$, we can replace the large subcircuit $N_{1_1}$ in Figure 4.1A with the smaller subcircuit $N_{1_2}$ in Figure 4.1B.

Although we just developed the equivalency for a specific circuit, it can be demonstrated that all resistive circuits can be modeled in the same way. The circuit $N_1$ in Figure 4.2A contains linear resistors, controlled sources that are controlled from within $N_1$, and independent sources. We have applied an external current source $I_a$. To be consistent with Figure 4.1, the current source arrow points down. Node analysis of this circuit results in the node conductance matrix equation

$$\mathbf{GV} = \mathbf{I}$$

which leads to

$$
\begin{bmatrix}
G_{11} & \cdots & G_{1k} \\
G_{21} & \cdots & G_{2k} \\
\vdots & \ddots & \vdots \\
G_{k1} & \cdots & G_{kk}
\end{bmatrix}
\begin{bmatrix}
V_a \\
V_2 \\
\vdots \\
V_k
\end{bmatrix}
=
\begin{bmatrix}
-I_a + I_{s_1} \\
I_{s_2} \\
\vdots \\
I_{s_k}
\end{bmatrix}
\tag{4.3}
$$

**FIGURE 4.2**    *Thévenin Equivalent Circuit*

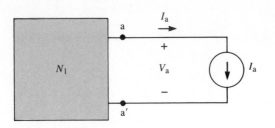

A. Original circuit

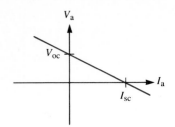

B. *V–I* characteristic

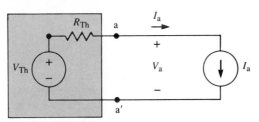

C. Equivalent circuit

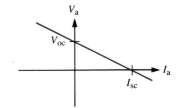

D. *V–I* characteristic

where     $V_a$ = voltage at the external node

$V_2 \ldots V_k$ = voltages at the internal nodes of $N_1$

$I_a$ = externally applied current

$I_{s_1} \ldots I_{s_k}$ = internal current sources

We use Cramer's rule to solve for $V_a$:

$$V_a = \frac{\begin{vmatrix} (-I_a + I_{s_1}) & \cdots & G_{1k} \\ I_{s_2} & \cdots & G_{2k} \\ \cdot & \cdot & \cdot \\ \cdot & \cdots & \cdot \\ \cdot & \cdot & \cdot \\ I_{s_k} & \cdots & G_{kk} \end{vmatrix}}{\det \mathbf{G}}$$

We can find the determinant of the numerator by multiplying the current source entries in the first column by their cofactors (see Appendix A):

$$V_a = \frac{(-I_a + I_{s_1})C_{11} + I_{s_2}C_{21} + \cdots + I_{s_k}C_{k1}}{\det \mathbf{G}}$$

$$= \frac{-I_a C_{11}}{\det \mathbf{G}} + \frac{I_{s_1}C_{11} + \cdots + I_{s_k}C_{k1}}{\det \mathbf{G}} \tag{4.4}$$

**FIGURE 4.3** *Ladder Network Equivalent Circuit*

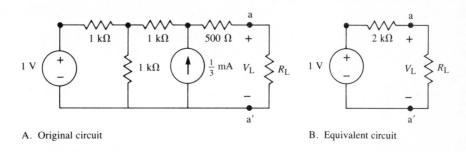

A. Original circuit          B. Equivalent circuit

where $C_{j1}$ is the cofactor of the $j$th entry in the first column of the numerator determinant. The cofactors and det **G** are simple constants.

The final equation is a straight line, as shown in Figure 4.2B. The slope of the straight line is $-C_{11}/(\det \mathbf{G})$, and the $V$ axis intercept is $(I_{s_1}C_{11} + \cdots + I_{s_k}C_{k1})/(\det \mathbf{G})$. In fact, the $V$ axis intercept is found by setting $I_a = 0$. Since a zero-valued current source can be replaced by an open circuit, the $V$ axis intercept is known as the *open circuit voltage* ($V_{oc}$). The $I$ axis intercept is found by setting the terminal voltage ($V_a$) to zero. In other words, if we replace the current source with a short circuit, the current in that short circuit will be the $I$ axis intercept. This current is known as the *short circuit current* ($I_{sc}$).

The simple circuit shown in Figure 4.2C has the straight-line characteristic shown in Figure 4.2D. By proper choice of voltage source and resistance values, the straight-line $V$–$I$ characteristic of Figure 4.2D can be made identical to the one of Figure 4.2B. For these values, this simple circuit is equivalent at terminals a–a′ to the larger circuit of Figure 4.2A. Therefore, it can be used in place of $N_1$. This equivalent circuit is known as a **Thévenin equivalent circuit,** developed by the French engineer M.L. Thévenin (1857–1926).

For ladder networks without controlled sources, it is very easy to obtain the Thévenin equivalent circuit. Consider, for example, the circuit in Figure 4.3A. We can start on the left side of the circuit with a source transformation. We then proceed with series–parallel combinations and additional source transformations to get the circuit in Figure 4.3B.

The Thévenin equivalent for the circuit to the left of the terminals a–a′ is, therefore, a 1 V source in series with a 2 kΩ resistor. Note that the final circuit is a simple voltage divider, and $V_L$ can be found by inspection. One of the advantages of the Thévenin approach is that it can reduce many circuits to simple voltage dividers.

---

**Example 4.1**   *Circuit Analysis by Thévenin Equivalent*

Find $V_L$ for the circuit in Figure 4.4.

(*continues*)

**Example 4.1** *Continued*

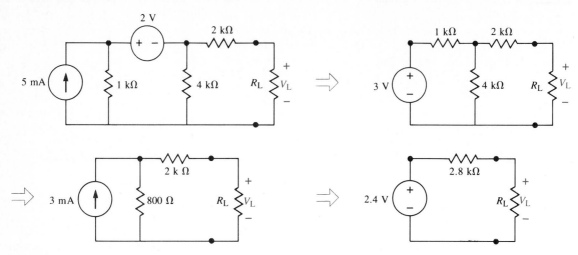

**FIGURE 4.4**

The circuit is reduced as shown. We find that

$$V_{Th} = 2.4 \text{ V} \quad \text{and} \quad R_{Th} = 2.8 \text{ k}\Omega$$

So,

$$V_L = \frac{R_L}{2800 + R_L} \cdot 2.4$$

We can find $V_L$ for any load resistor.

---

If a circuit contains controlled sources or is not a ladder network, then we need a more formal approach to find the Thévenin equivalent circuit voltage source and resistance. We can derive this approach if we keep in mind that the external terminals (e.g., a–a′ in Figures 4.1A and 4.1B) are available to us for manipulation. That is, we can place anything we like between these two terminals in both the original and equivalent circuits.

In Figure 4.5, we have replaced the load resistors with an open circuit $(R_L \to \infty)$. The voltage between the external terminals is the open circuit voltage, $V_{oc}$. It is clear from the equivalent circuit on the right that

$$V_{oc} = V_{Th}$$

Note that in Figure 4.5B $V_{oc}$ is the value of $V_{aa'}$ when $I = 0$. The Thévenin voltage between two terminals is found, therefore, by opening the circuit at those terminals and solving for the open circuit voltage. You may use any of the techniques of Chapters 2 or 3 for this purpose.

**FIGURE 4.5**   *Thévenin Voltage*

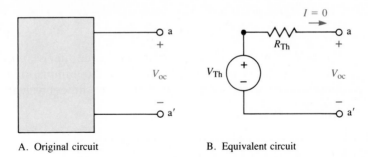

A. Original circuit                    B. Equivalent circuit

**Example 4.2**   *Thévenin Voltage by Node Analysis*

Find the Thévenin voltage between nodes b and c for the circuit in Figure 4.6.

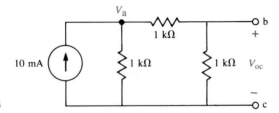

**FIGURE 4.6**

From node analysis, we find the matrix equation

$$\begin{bmatrix} .002 & -.001 \\ -.001 & .002 \end{bmatrix} \begin{bmatrix} V_a \\ V_{oc} \end{bmatrix} = \begin{bmatrix} .01 \\ 0 \end{bmatrix}$$

Solution of this matrix equation leads to

$$V_{Th} = V_{oc} = 3.33 \text{ V}$$

Finding the Thévenin equivalent voltage is very straightforward: We simply remove the load at the terminals of interest and solve for the open circuit voltage. This can be done by hand for simple circuits or with a computer for large and complicated circuits. For real circuits, we can find the open circuit voltage in the laboratory. We simply remove the load and measure the open circuit voltage with a voltmeter.

To find the Thévenin equivalent resistance, we must find the slope of the straight line in Figures 4.2B or 4.2D. This task is not always easy. Several techniques are available. The simplest technique can be applied to actual circuits.

If you refer back to Figure 4.1B, you find $V_L$ by inspection to be

$$V_L = V_{Th} \cdot \frac{R_L}{R_L + R_{Th}} \tag{4.5}$$

Remember, we found $V_{Th}$ by opening the terminals and measuring the voltage there. Now, measure the voltage between the terminals for a known load resistor. The only unknown left in Equation 4.5 is the equivalent resistance, which can now be calculated.

---

**Example 4.3**    *Laboratory Measurements*

We wish to determine in the laboratory the Thévenin equivalent circuit of a two-terminal device. We connect various known load resistors to the device and measure the resultant load voltages. These data are tabulated as follows:

| $R_L$ ($\Omega$) | $V_L$ (V) |
|---|---|
| 100 | .196 |
| 500 | .91 |
| 1,000 | 1.67 |
| 5,000 | 5 |
| 10,000 | 6.67 |
| open | 10 |

From the measured data, we find that $V_{Th} = 10$ V. We can use any other data point to find $R_{Th}$. For example, we can use the values $V = 1.67$ V and $R_L = 1$ k$\Omega$ in Equation 4.5 to find $R_{Th}$:

$$1.67 = 10 \cdot \frac{1000}{1000 + R_{Th}}$$

$$R_{Th} = 5 \text{ k}\Omega$$

In fact, the easiest way to find $R_{Th}$ is to adjust the load resistor $R_L$ until $V_L$ is exactly half of $V_{Th}$. At this point, $R_{Th}$ must equal $R_L$. For this example, half of the Thévenin voltage is 5 V, which occurs when $R_L = 5$ k$\Omega$.

---

To use circuit analysis techniques to find the Thévenin equivalent resistance requires several steps. Figure 4.7A shows the Thévenin equivalent circuit again. If we turn off the voltage source, we get the circuit in Figure 4.7B. Remember, turning off a source means that we set its value to zero. A zero-valued voltage source is a short circuit. It is obvious from Figure 4.7B that the resistance seen between the two terminals is the Thévenin equivalent resistance we are looking for.

**FIGURE 4.7**   *Thévenin Resistance*

**FIGURE 4.7**   *Thévenin Resistance*

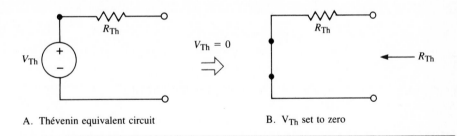

A. Thévenin equivalent circuit          B. $V_{Th}$ set to zero

It should not surprise you that the only way to set the Thévenin equivalent voltage to zero is to turn off all of the *independent* sources in the original network. Remember, we *never* turn off controlled sources. With all the independent sources turned off, the Thévenin resistance is simply the resistance between the two terminals of interest, as shown in Figure 4.7B.

**Example 4.4**   *Thévenin Equivalent I*

Find $V_{Th}$ and $R_{Th}$ for the circuit in Figure 4.8A.

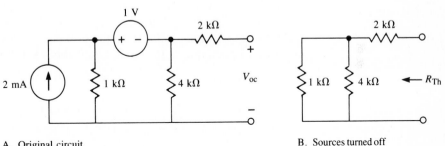

**FIGURE 4.8**   A. Original circuit                     B. Sources turned off

Analysis of Figure 4.8A results in

$$V_{Th} = V_{oc} = .8 \text{ V}$$

To find $R_{Th}$, we short the voltage source and open the current source to get the circuit in Figure 4.8B. The total resistance between the two terminals of interest is found by combining the 1 kΩ and 4 kΩ resistors in parallel and adding that combination to the series 2 kΩ resistor. The equivalent resistance is

$$R_{Th} = 2000 + (1000 \,\|\, 4000)$$

(*continues*)

**Example 4.4**   *Continued*

So,

$$R_{Th} = 2.8 \text{ k}\Omega$$

---

**Example 4.5**   *Thévenin Equivalent II*

Find $V_{Th}$ and $R_{Th}$ for the circuit in Figure 4.9A.

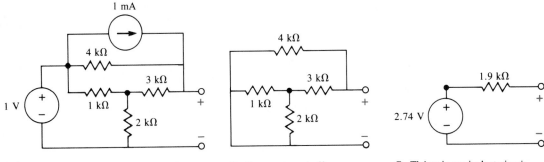

A. Original circuit                     B. Sources turned off          C. Thévenin equivalent circuit

**FIGURE 4.9**

The Thévenin voltage can be found by node analysis to be $V_{Th} = 2.74$ V.
The circuit in Figure 4.9B is obtained by turning off the sources. It may
be a little tricky to see, but after turning off the sources, the 1 kΩ and 2 kΩ
resistors are in parallel; that combination is in series with the 3 kΩ resistor;
and this combination is in parallel with the 4 kΩ resistor. Therefore,

$$R_{Th} = [(1000 \| 2000) + 3000] \| 4000 = 1.9 \text{ k}\Omega$$

The Thévenin equivalent circuit is shown in Figure 4.9C.

---

Many circuits do not reduce to simple series–parallel combinations of
resistors when the independent sources are turned off. Figure 4.10 shows two such
circuits. Finding the equivalent resistances for these types of circuits requires an
additional step. We could actually build the circuits on the right in Figure 4.10 and
connect an ohmmeter between the terminals of interest, but we construct the
ohmmeter with paper and pencil (or computer) instead. An ohmmeter works by
applying a known voltage across the terminals and measuring the resultant current
(Figure 4.11A). The resistance is then found by Ohm's law. Note that we have
reversed the direction of the terminal current so that we keep the definition of
$R_{Th}$ positive.

Using either mesh or node analysis, we find the current $I_m$:

$$I_m = \frac{V_m}{500}$$

**FIGURE 4.10**   *Controlled-Source and Nonladder Networks*

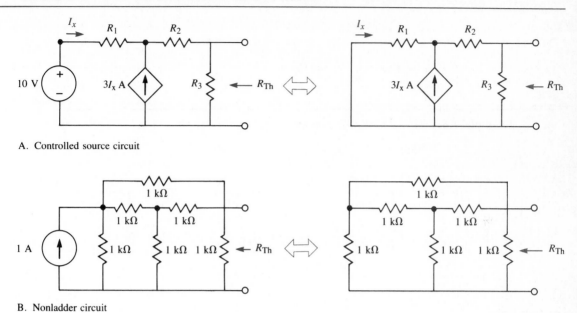

A. Controlled source circuit

B. Nonladder circuit

**FIGURE 4.11**   *Finding $R_{Th}$ with Ohmmeter*

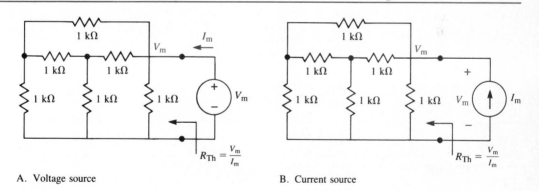

A. Voltage source          B. Current source

The Thévenin resistance is, therefore,

$$R_{Th} = \frac{V_m}{I_m} = 500 \ \Omega$$

We can also find the Thévenin resistance by applying a current source at the terminals of interest and solving for the voltage across the terminals (Figure

4.11B). Analysis of the circuit in Figure 4.11B results in

$$V_m = 500 I_m$$

and

$$R_{Th} = \frac{V_m}{I_m} = 500 \ \Omega$$

---

**Example 4.6**    *Thévenin Resistance by Mesh Analysis*

Find $R_{Th}$ for the circuit in Figure 4.12A.

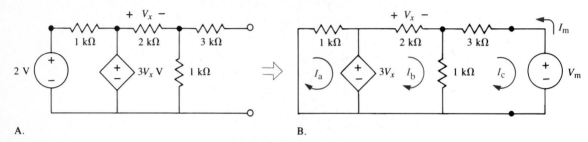

A.

B.

**FIGURE 4.12**

We add a voltage source as shown in Figure 4.12B and solve for $I_m$. We first note two facts: (1) $V_x = 2000 I_b$ and (2) $I_m$ is the mesh current in the third mesh and circulates counterclockwise so that $I_c = -I_m$.

$$\begin{bmatrix} 1000 & 6000 & 0 \\ 0 & -3000 & -1000 \\ 0 & -1000 & 4000 \end{bmatrix} \begin{bmatrix} I_1 \\ I_2 \\ -I_m \end{bmatrix} = \begin{bmatrix} 0 \\ 0 \\ -V_m \end{bmatrix}$$

$$I_m = \frac{3}{13,000} V_m$$

So,

$$R_{Th} = 4.33 \ \text{k}\Omega$$

---

**Example 4.7**    *Finding $R_{Th}$ Using a Current Source*

Find $R_{Th}$ for the circuit in Figure 4.13.

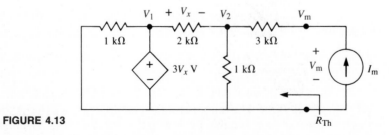

**FIGURE 4.13**

The circuit in Figure 4.13 is the same as the one in Example 4.6, except that here we apply a current source at the appropriate terminals and solve for $V_m$. A natural choice for solution in this case is node analysis. We note that since $V_1 = 3V_x$ and $V_x = V_1 - V_2$, $V_1 = 1.5V_2$. The node equation is

$$\begin{bmatrix} 6.5/6000 & -1/3000 \\ -1/3000 & 1/3000 \end{bmatrix} \begin{bmatrix} V_2 \\ V_m \end{bmatrix} = \begin{bmatrix} 0 \\ I_m \end{bmatrix}$$

Once again, we can solve for $V_m$ in terms of $I_m$ or set $I_m$ equal to a convenient number and solve for $V_m$. In either case, $R_{Th}$ is found from

$$V_m = 4333 \cdot I_m$$

So,

$$R_{Th} = \frac{V_m}{I_m} = 4.33 \text{ k}\Omega$$

In summary, to find the complete Thévenin equivalent circuit, we have essentially analyzed two different circuits:

1. The Thévenin voltage is found by opening up the terminals of interest and solving for the open circuit voltage.
2. The Thévenin resistance is found by turning off all independent sources and solving for the resulting equivalent resistance.

The Thévenin equivalent circuit can also be found with a single circuit analysis. Figure 4.14 shows the general linear resistive network and the expected terminal $V$–$I$ straight-line characteristic. Here, we have returned to the conven-

**FIGURE 4.14**  *Thévenin Straight-Line Equation*

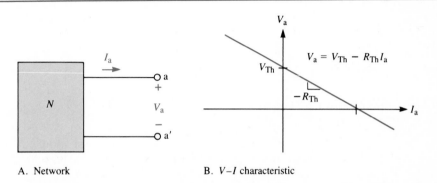

A. Network                                      B. *V–I* characteristic

tion that the terminal current leaves at the top terminal. We now know that $V_a = V_{Th}$ for $I_a = 0$, and the slope of the line is $-R_{Th}$. Therefore,

$$V_a = V_{Th} - R_{Th}I_a \qquad (4.6)$$

Equation 4.6 is true for all $I$, so we can connect a current source between the terminals of interest and solve for the resulting voltage. Node analysis is recommended in this procedure. Note that to be consistent with the equation, the current source must go from terminal a to terminal a'. The next example demonstrates this technique.

---

**Example 4.8**   *Single-Step Thévenin Analysis*

For the circuit of Figure 4.15A, find the Thévenin equivalent between terminals a and a', looking to the right.

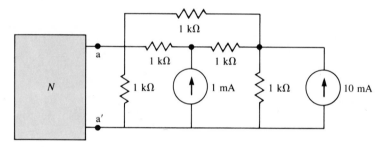

A. Original circuit

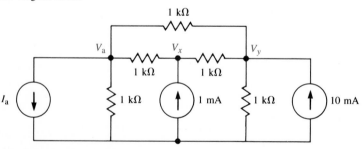

**FIGURE 4.15**   B. Applied current source $I_a$

First remove $N$ and connect a current source as shown in Figure 4.15B. Note the direction of the current source. The node equations for this circuit are as follows:

$$.003V_a - .001V_x - .001V_y = -I_a$$

$$-.001V_a + .002V_x - .001V_y = .001$$

$$-.001V_a - .001V_x + .003V_y = .01$$

Solve for $V_a$ to get

$$V_a = 4.25 - 625I_a$$

Compare this result to Equation 4.6 and conclude that

$$V_{Th} = 4.25 \text{ V} \quad \text{and} \quad R_{Th} = 625 \ \Omega$$

We will make extensive use of Thévenin equivalent circuits in some of the following chapters. An immediate example of the practical use of this technique is the analysis of the op amp circuit shown in Figure 4.16.

You may have already analyzed the noninverting op amp circuit shown in Figure 4.16A. Assuming the gain and input resistance of the op amp are very large, the output for a single input is given by the following equation:

$$v_o = \left(1 + \frac{R_F}{R_A}\right)v_{in} \quad \text{(single input)}$$

To find the output for the multi-input case, we can analyze the entire op amp circuit given in Figure 4.16B. More simply, we can replace the sources and resistors inside the box with their Thévenin equivalent, as is done in Figure 4.16C.

If we compare the Thévenin equivalent circuit in Figure 4.16C to the single-input circuit in Figure 4.16A, it is clear that the response of the noninverting op amp is, in general,

$$v_o = \left(1 + \frac{R_F}{R_A}\right)v_{Th} \quad \text{(noninverting)}$$

For this particular circuit,

$$v_o = 3v_{Th}$$

$$v_{Th} = .55v_1 + .27v_2 + .18v_3$$

$$R_{Th} = .55 \text{ k}\Omega$$

So, $\quad v_o = 1.65v_1 + .81v_2 + .54v_3$

**FIGURE 4.16**    *Op Amp Example Circuit*

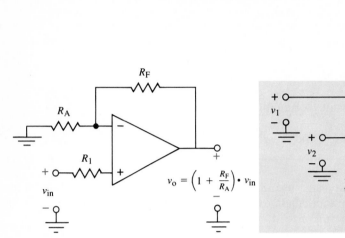

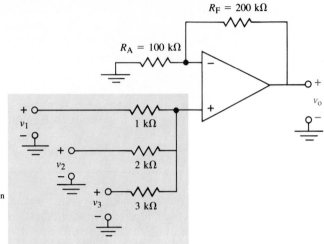

A.  Noninverting amplifier, single input

B.  Noninverting amplifier, multiple inputs

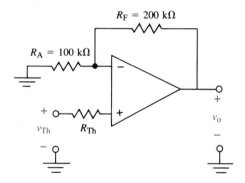

C.  Thévenin equivalent circuit

## 4.3  NORTON EQUIVALENT CIRCUITS

The network $N$ shown in Figure 4.17A can be represented by the Thévenin equivalent circuit shown in Figure 4.17B. Furthermore, since we know from Chapter 2 that a voltage source in series with a resistor can be modeled as a current source in parallel with the same resistor, the network $N$ has the current source–resistor equivalent shown in Figure 4.17C. This circuit is known as the *Norton equivalent circuit* for $N$. In fact, the source transformation is also known as the Thévenin–Norton transformation.

**FIGURE 4.17**   *Norton Equivalent Circuit*

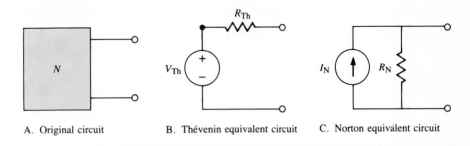

A. Original circuit          B. Thévenin equivalent circuit          C. Norton equivalent circuit

It is clear from what we know about source transformations that the Thévenin and Norton resistances are the same. That is,

$$R_{Th} = R_N \overset{\triangle}{=} R_{eq} \tag{4.7}$$

The relationships between the Thévenin voltage and the Norton current are as follows:

$$V_{Th} = I_N R_{eq} \quad \text{and} \quad I_N = \frac{V_{Th}}{R_{eq}} \tag{4.8}$$

where $R_{eq}$ is the equivalent resistance for either the Thévenin or Norton circuits.

One way of finding the Norton equivalent is to first find the Thévenin equivalent and then transform it to the current source form. While this method will usually work, it does not lend itself to gaining any insight into the current behavior of the equivalent circuit. (To demonstrate that it does not always work, try transforming a Thévenin equivalent with $R_{Th} = 0$ into a Norton equivalent.) To find some methods for directly determining $I_N$, let's take a look at Figure 4.18.

Figure 4.18 shows the original network and its Norton equivalent with the output terminals short-circuited. You can see from the Norton equivalent that the short circuit current $(I_{sc})$ is

$$I_{sc} = I_N$$

Thus, the Norton current can be found by shorting the terminals of interest and finding the current in the short. Be sure you understand the proper current

**FIGURE 4.18**  *Finding $I_N$*

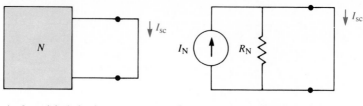

A. $I_{sc}$, original circuit          B. $I_{sc}$, Norton equivalent circuit

directions. If the Norton current source is defined with the arrow up, then the short circuit current is defined with the arrow down.

**Example 4.9**  *Norton Equivalent I*

Find the Norton equivalent for the circuit in Figure 4.19A.

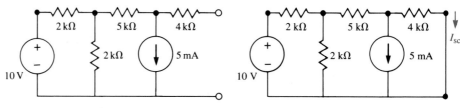

**FIGURE 4.19**   A. Original circuit                          B. Short circuit current, $I_{sc}$

The equivalent resistance is found to be

$$R_N = 10\,k\Omega$$

To find $I_N$, we short the output terminals, as shown in Figure 4.19B. We can find $I_{sc}$ by first transforming the voltage source and then using superposition:

$$I_{sc} = \frac{1/9000}{1/9000 + 1/2000 + 1/2000} \cdot \frac{10}{2000} - \frac{1/4000}{1/4000 + 1/6000}(.005)$$

Thus,

$$I_N = I_{sc} = -2.5\,\text{mA}$$

**Example 4.10**  *Norton Equivalent II*

Find $I_N$ for the circuit in Figure 4.20.

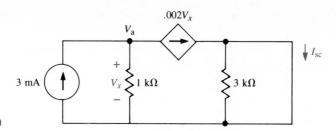

**FIGURE 4.20**

Note that the 3 kΩ resistor is shorted out so that $I_{3\,k\Omega} = 0$. Since we will use node analysis, we first find that

$$V_x = V_a$$

From KCL at node a,

$$.001V_a + .002V_a = .003$$

so that

$$V_a = 1\,\text{V}$$

and

$$I_{sc} = .002V_a = .002\,\text{A} = 2\,\text{mA} = I_N$$

The Norton equivalent circuit can also be derived by examining a single set of equations. To see how this is done, consider Figure 4.21, where the Norton equivalent is redrawn along with its *V–I* characteristic. Here, we recognize the

**FIGURE 4.21**    *Norton Straight-Line Equation*

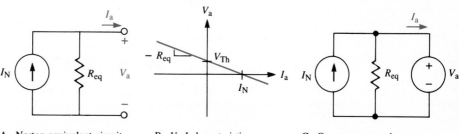

A. Norton equivalent circuit        B. *V–I* characteristic        C. One-step approach

voltage axis intercept as $V_{\text{Th}}$ ($V_{\text{oc}}$) and the current axis intercept as $I_{\text{N}}$ ($I_{\text{sc}}$). The equation for this straight line is, again,

$$V_{\text{a}} = V_{\text{Th}} - R_{\text{eq}}I_{\text{a}} \tag{4.9}$$

We can rewrite this equation as

$$I_{\text{a}} = \frac{V_{\text{Th}}}{R_{\text{eq}}} - \frac{V_{\text{a}}}{R_{\text{eq}}} = I_{\text{N}} - \frac{V_{\text{a}}}{R_{\text{eq}}} \tag{4.10}$$

In the single-step Norton equivalence technique, we connect a voltage source to the terminals (Figure 4.21C) and use mesh analysis to find the current in the voltage source.

---

**Example 4.11**   *Single-Step Norton Analysis*

Find the Norton equivalent for the circuit in Figure 4.22.

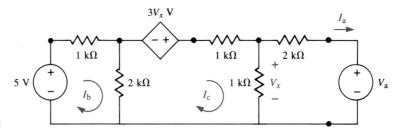

**FIGURE 4.22**

Note that $V_x = 1000(I_{\text{c}} - I_{\text{a}})$.

$$\begin{bmatrix} 3000 & -2000 & 0 \\ -2000 & 1000 & 2000 \\ 0 & -1000 & 3000 \end{bmatrix} \begin{bmatrix} I_{\text{b}} \\ I_{\text{c}} \\ I_{\text{a}} \end{bmatrix} = \begin{bmatrix} 5 \\ 0 \\ -V_{\text{a}} \end{bmatrix}$$

From this equation, we find $I_{\text{a}}$:

$$I_{\text{a}} = .00333 + \frac{V_{\text{a}}}{3000}$$

Comparing to Equation 4.10, we get

$$I_N = 3.33 \text{ mA} \quad \text{and} \quad R_{eq} = -3 \text{ k}\Omega$$

Note that with controlled sources, $R_{eq}$ can be negative.

---

## 4.4 VOLTAGE, CURRENT, AND POWER TRANSFER

Very often, an electrical engineer is confronted with the situation depicted in Figure 4.23A. The network $N$ may be an audio amplifier, and the load may be a speaker. The network could also be a physical system, for example, a patient with electrocardiogram electrodes attached. The load, in this case, would be the recording device used. The question that arises is: Given the network $N$, how do you choose the load resistance to maximize the voltage, current, or power delivered to this passive load?

Since $N$ can be a very complicated network, the answer to the preceding question may require extensive analysis. However, if $N$ is a linear resistive network, then we can replace it with either its Thévenin (Figure 4.23B) or its Norton (Figure 4.23C) equivalent. From these simpler equivalent circuits, we can quickly derive the voltage, current, and power transfer relationships. We can then derive some very useful design rules for the load resistance.

### Voltage Transfer

Let's look at voltage transfer first. We find the load voltage very easily from the Thévenin equivalent circuit to be

$$V_L = V_{Th} \cdot \frac{R_L}{R_L + R_{eq}} \tag{4.11}$$

**FIGURE 4.23**  *Voltage, Current, and Power Transfer*

---

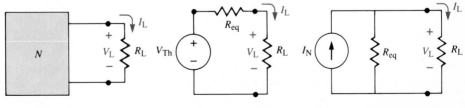

A. Original circuit    B. Thévenin equivalent circuit    C. Norton equivalent circuit

If the Thévenin equivalent resistance, $R_{eq}$, is given, how do we maximize the load voltage? An examination of the voltage divider formula in Equation 4.11 shows that as $R_L \to \infty$, the load voltage approaches the Thévenin voltage. Therefore, to maximize load voltage, we should set the load resistance to infinity—that is, an open circuit. An open is usually not practical since all real loads have finite resistance. However, we can approximate the maximum response by choosing the load resistor $R_L$ to be much greater than $R_{eq}$ to get

$$R_L \gg R_{eq} \qquad \text{so that} \qquad V_L \simeq V_{eq}$$

If the load resistor is much greater than the network's equivalent resistance, then most of the network's equivalent voltage is transferred to the load. Much greater usually means a factor of 10.

If the load resistor is not much greater than the equivalent resistance, then only a fraction of the network's voltage will appear across the load. In this case, we say that the network has been "loaded down".

---

**Example 4.12**    *Voltage Transfer*

Suppose that in the lab, you have inserted a microelectrode into a nerve cell to measure its voltage output. The microelectrode has a resistance of 1 MΩ. If your recording device has an input resistance of only 100 kΩ, what fraction of the nerve voltage will it see? What input resistance should the recording device have?

At 100 kΩ, the recording device will see only $10^5/(10^5 + 10^6)$ or 9% of the actual nerve voltage. The recording device should have an input resistance of at least 10 MΩ, which would result in a 91% transfer of the source voltage.

---

## Current Transfer

The rule for maximum current transfer can be found from the Norton equivalent of Figure 4.13C. This current divider yields

$$I_L = I_N \cdot \frac{R_{eq}}{R_L + R_{eq}} \tag{4.12}$$

From Equation 4.12, we find that maximum current transfer takes place when $R_L = 0$, a short circuit. Again, a short circuit is seldom used as a practical load, so we choose a load resistor as small as possible, preferably at least 10 times smaller than the equivalent resistance.

From the preceding discussion on voltage and current transfer, it is clear that you must first determine whether you are trying to measure voltage or current. If you are trying to measure, or transfer, *voltage*, then the load resistance must be *large*. If you are trying to measure, or transfer, *current*, then the load resistance must be *small*.

## Power Transfer

With many loads, such as speakers, motors, and heaters, we are interested in maximizing the power delivered. The choice of the load resistor in this case is quite different from the voltage and current transfer cases. Remember, power is voltage delivered times current delivered.

If we use an infinite load, we will maximize voltage delivered but load current will be zero, resulting in zero power delivered. If, at the other extreme, we use a short circuit load, current delivered will be maximized but voltage delivered will be zero, again resulting in zero power delivered. The correct load for maximum power delivery must lie somewhere between these two extremes.

We will now derive the maximum power relationship by using the Thévenin equivalent. (In an end-of-chapter problem, you will repeat this derivation by using the Norton equivalent.) From Figure 4.23B, we know that

$$V_L = V_{Th} \cdot \frac{R_L}{R_L + R_{eq}}$$

The power delivered to the load resistor is

$$P_L = \frac{V_L^2}{R_L} = V_{Th}^2 \cdot \frac{R_L}{(R_L + R_{eq})^2} \qquad \text{(4.13)}$$

We assume that $R_{eq}$ is fixed and maximize the power with respect to $R_L$ by taking its derivative and setting it to zero:

$$\frac{dP_L}{dR_L} = V_{Th}^2 \cdot \frac{(R_L + R_{eq})^2 - 2R_L(R_L + R_{eq})}{(R_L + R_{eq})^3} = 0$$

Solving this equation leads to

$$-R_L + R_{eq} = 0 \qquad \text{so that} \qquad R_L = R_{eq} \qquad \text{(4.14)}$$

for maximum power transfer. We get the simple relationship that if $R_{eq}$ is given, then maximum power is transferred to the load if the load resistance equals the equivalent resistance.

**Example 4.13**    *Maximum Power Transfer*

Find $R_L$ to maximize power transfer in the circuit in Figure 4.24.

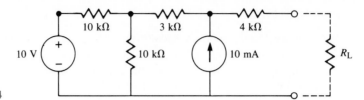

FIGURE 4.24

To find $R_L$ for maximum power transfer, we first find the equivalent resistance of the network:

$$R_{eq} = 5000 + 3000 + 4000 = 12 \text{ k}\Omega$$

Therefore, for maximum power transfer,

$$R_L = 12 \text{ k}\Omega$$

Note that for maximum voltage transfer,

$$R_L > 120 \text{ k}\Omega$$

and for maximum current transfer,

$$R_L < 1.2 \text{ k}\Omega$$

## Load Resistor Given

A note of caution is in order about all of the transfer relationships we have just discussed. In all of these cases, we assumed that the equivalent resistance of the network was given and that we had to choose the load. In some circumstances, the load resistance is known, and we must design a network to have the required equivalent resistance for maximum transfer. In this case, we would ideally like to connect the equivalent source directly to the load. This means that an equivalent voltage source should have a very small equivalent resistance ($R_{eq} \ll R_L$). And, conversely, an equivalent current source should have a very large equivalent resistance ($R_{eq} \gg R_L$). (End-of-chapter problems call for proving these relationships.)

## 4.5 TEE–PI EQUIVALENCE

Figure 4.25 shows a circuit equivalence that can greatly simplify the analysis of nonladder networks. The circuit in Figure 4.25A is termed a *tee* circuit, for obvious reasons. The circuit in Figure 4.25B is known as a *pi* circuit because it resembles the Greek letter $\pi$.

The circuits can also be drawn as shown in Figure 4.26. For these forms we use the terminology of *wye* (Y) and *delta* ($\Delta$) circuits. You will find both tee–pi and wye–delta used in the engineering literature.

The pi equivalent circuit for the tee network is not as simple to derive as the equivalent circuit of series or parallel resistors. The difference is that this circuit has three external terminals instead of two.

Circuits with two external terminals are equivalent if the *V–I* characteristics at those terminals are the same for both circuits. We are still interested here in the *V–I* characteristics at corresponding pairs of terminals. The tee and pi circuits

**FIGURE 4.25**  *Tee–Pi Equivalence*

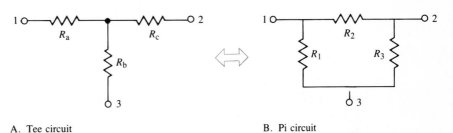

A.  Tee circuit

B.  Pi circuit

**FIGURE 4.26**  *Wye–Delta Equivalence*

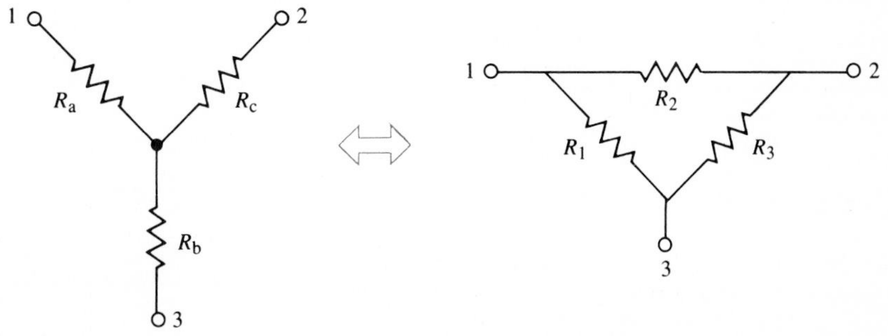

A.  Wye circuit

B.  Delta circuit

have three pairs of terminals: 1,2; 1,3; and 2,3. The two circuits in Figure 4.25 (or in Figure 4.26) are equivalent if at each terminal pair the $V$–$I$ characteristic is the same.

In Figure 4.27A, we have assigned voltages and currents to the tee and pi circuits. To find the relationships between the two circuits, we will apply a voltage source between the same two terminals of each circuit and solve for the resulting current. In all cases, we will leave the other terminal open.

This procedure has been explicitly shown in Figure 4.27B for one set of terminals. For the tee and pi circuits in Figure 4.27B, we get the following:

$$V_{13} = I_1 \cdot (R_a + R_b) \qquad \text{(tee)}$$
$$V_{13} = I_1 \cdot [R_1 \| (R_2 + R_3)] \qquad \text{(pi)}$$

For the two circuits to be equivalent,

$$R_a + R_b = \frac{R_1 R_2 + R_1 R_3}{R_1 + R_2 + R_3}$$

We get similar results at the other two pairs of terminals.

**FIGURE 4.27**  *Finding Tee–Pi Equivalence*

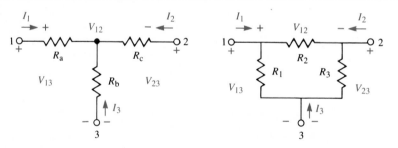

A. Original circuits

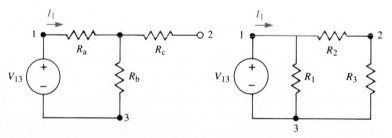

B. Resistance between terminals 1 and 3

The total set of equations that result from this analysis is

$$R_a + R_b = \frac{R_1 R_2 + R_1 R_3}{R_1 + R_2 + R_3}$$

$$R_a + R_c = \frac{R_1 R_2 + R_2 R_3}{R_1 + R_2 + R_3}$$

$$R_b + R_c = \frac{R_2 R_3 + R_1 R_3}{R_1 + R_2 + R_3}$$

If we are given the pi network and want to find its equivalent tee network, then we get

*Pi to Equivalent Tee*

$$R_a = \frac{R_1 R_2}{R_1 + R_2 + R_3}$$

$$R_b = \frac{R_1 R_3}{R_1 + R_2 + R_3}$$

$$R_c = \frac{R_2 R_3}{R_1 + R_2 + R_3}$$

(4.15)

If we are given the tee network and want to find its equivalent pi, then since $R_a$, $R_b$, and $R_c$ are known, we solve for $R_1$, $R_2$, and $R_3$ to get

*Tee to Equivalent Pi*

$$R_1 = \frac{R_a R_b + R_a R_c + R_b R_c}{R_c}$$

$$R_2 = \frac{R_a R_b + R_a R_c + R_b R_c}{R_b}$$

$$R_3 = \frac{R_a R_b + R_a R_c + R_b R_c}{R_a}$$

(4.16)

Clearly, this equivalence is not easy to remember or to derive. Nevertheless, it can prove very useful in circuit analysis. Consider the circuit shown in Figure 4.28.

To find $V_o$ from the original circuit in Figure 4.28A, we would have to use node analysis and solve a $4 \times 4$ matrix equation. The solution is easily done with a computer or a calculator, but it is rather involved when done by hand. We can, however, transform the highlighted tee network to a pi network to get the circuit in Figure 4.28B.

Although it may not be obvious at first glance, the top 3 kΩ resistor of the pi is in parallel with the 5 kΩ resistor. Also, the 3 kΩ resistor on the right leg of the

FIGURE 4.28    *Tee–Pi Equivalence in Circuit Analysis*

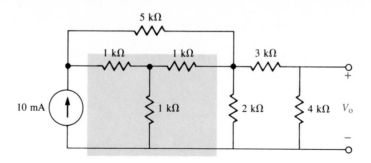

A. Original circuit

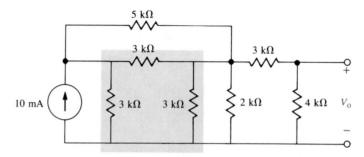

B. Tee-to-pi transformation

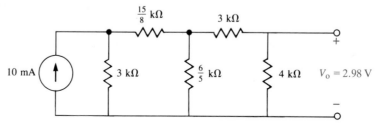

C. Final circuit

pi is in parallel with the 2 kΩ resistor. After combining these resistors, we get the final circuit shown in Figure 4.28C. The final equivalent circuit is a ladder network and can be easily analyzed to get the voltage shown ($V_o = 2.98$ V).

## *4.6   CONTROLLED SOURCES AS RESISTORS

In this section, we will examine how to model some controlled sources as resistors and how to use this model to simplify the analysis of some controlled-source circuits. The technique can be quite tricky to master. However, it is well worth the effort, particularly for op amp circuits.

Figure 4.29 shows two of the controlled sources that you are familiar with by now. In Figure 4.29A, the voltage between terminals a–a' is

$$V_{aa'} = rI_x$$

where $I_x$ is a current that exists somewhere in the circuit. In Figure 4.29B, the current is

$$I_{aa'} = gV_y$$

where $V_y$ is a voltage that exists somewhere in the circuit. The letters r and g indicate scale factors for the controlled sources. It is interesting to note that r has units of ohms and that g has units of siemens.

Let's examine the very special case of the current-controlled voltage source shown in Figure 4.30A. Here, the controlling current is also the voltage source current. The controlling current does not always have to exist in some distant branch.

Is there any difference in the V–I characteristic of the controlled source or the resistor shown in Figure 4.30A? The answer is no. In both cases, the voltage across the terminals is r times the current entering at node a. These two elements are equivalent! Thus, we can replace all resistors with current-controlled voltage sources.

The voltage-controlled current source shown in Figure 4.30B can also be replaced by a resistor. This figure shows the special case of a VCIS in which the controlling voltage appears across the source itself. We get the same V–I relationship from the resistor whose value is 1/g. Therefore, we can also replace resistors with voltage-controlled current sources.

**FIGURE 4.29**   *Controlled Sources*

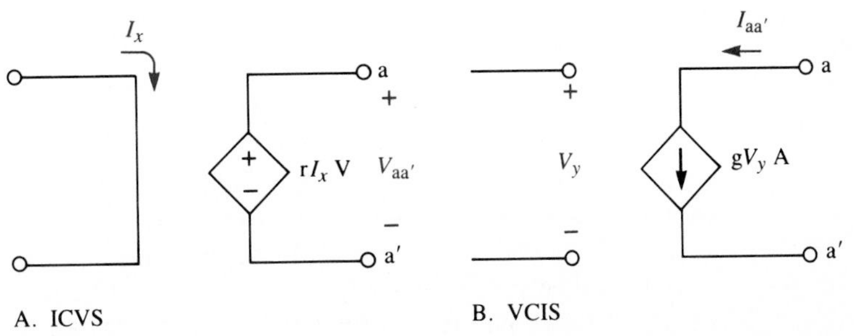

A. ICVS

B. VCIS

FIGURE 4.30    *Controlled Source as Resistor*

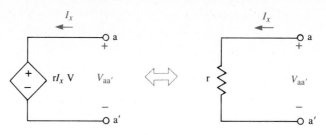

A. Current-controlled voltage source as resistor

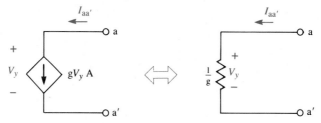

B. Voltage-controlled current source as resistor

**Example 4.14**    *Replacing Resistors with Controlled Sources*

Replace all resistors in Figure 4.31A with controlled sources.

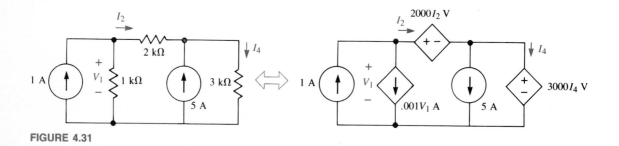

FIGURE 4.31

Figure 4.31B shows the equivalent circuit. Note that the passive sign convention is followed for the controlled sources.

What do we gain by replacing a resistor with a controlled source? Not much. The importance of this equivalence is in replacing a controlled source with a resistor. You will rarely, if ever, run across a circuit that, in its original state, has a current-controlled voltage source where the controlling current exists in the source. However, a circuit can often be manipulated to achieve this. Consider the circuit shown in Figure 4.32.

We wish to find the current $I_x$ in Figure 4.32A. The first step is to transform the current-controlled current source into a current-controlled voltage source, as shown in Figure 4.32B. The tricky part comes next. We recognize that in this transformed circuit, the controlling current $I_x$ is also the current-controlled voltage source current. We can, therefore, replace the controlled source with a resistor of 10 kΩ. The current $I_x$ is found by inspection of Figure 4.32C to be .77 mA.

This technique is used quite often to help analyze electronic circuits. Practical op amp circuits are particularly well suited to this type of analysis. Figure 4.33 shows a noninverting op amp circuit in part A and its practical model in part B. The output resistance of the op amp has been omitted to simplify the analysis.

We are interested in finding both the output voltage, $v_o$, and the input resistance, $R_{in}$, for this circuit. The controlled voltage source is dependent on $v_i$ rather than on a current. We will have to manipulate the circuit. First note that $v_o = Av_i$ so that we need to solve for $v_i$. We can transform the controlled source into a current source to get the circuit in Figure 4.33C. The resistors $R_A$ and $R_F$ are now in parallel and can be combined (they are connected to each other at one node, and both are connected to ground). We transform back to a voltage source and obtain the circuit in Figure 4.33D. We have a single-loop circuit, but we still have a voltage-controlled voltage source. However, we can use Ohm's law to replace $v_i$ with $r_i i_i$ (Figure 4.33E). At this point, we see that the controlling current enters the controlled source and replace it with a resistor to get the final circuit shown in Figure 4.33F. From the final circuit, which is a voltage divider, we get

$$R_{in} = r_i + R_A \| R_F + \frac{AR_A r_i}{R_A + R_F}$$

**FIGURE 4.32** *Controlled Source as Resistor in Circuit Analysis*

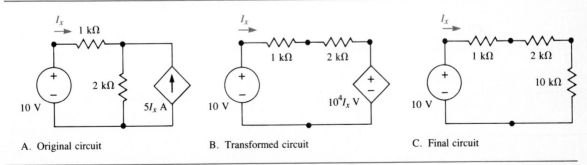

A. Original circuit        B. Transformed circuit        C. Final circuit

**FIGURE 4.33**   *Noninverting Op Amp Circuit*

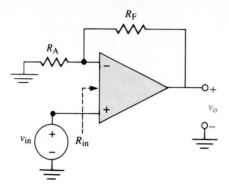

A. Original circuit

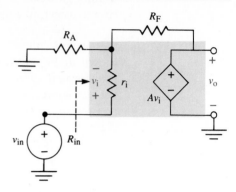

B. Practical model

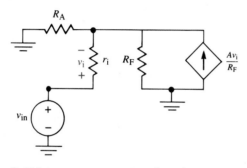

C. Voltage-to-current source transformation

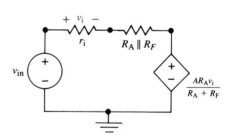

D. Current-to-voltage source transformation

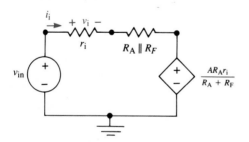

E. Replace $v_i$ with $i_i$

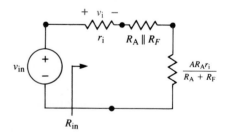

F. Final circuit

and

$$v_i = v_{in} \cdot \frac{r_i}{R_{in}}$$

so that

$$v_o = v_{in} \cdot \frac{Ar_i}{R_{in}}$$

A voltage-controlled current source can also be replaced with a resistor under the right circumstances. If a circuit can be manipulated so that the controlling voltage of a VCIS is across the VCIS itself, then the controlled source is replaced by a resistor.

Figure 4.34 shows an application of this technique to the inverting op amp circuit. We proceed as we did for the noninverting op amp. First note that $v_0 = Av_i$. Now transform the controlled source to a current source. At this point, we see that the controlling voltage exists across the controlled source. Therefore, we replace the controlled source with a resistor, as shown in part D. Note that the resistance is the inverse of the VCIS scale factor. We find by inspection that

$$R_{in} = R_A + r_i \parallel R_F \parallel \frac{R_F}{A}$$

and

$$v_i = -v_{in} \cdot \frac{r_i \parallel R_F \parallel (R_F/A)}{R_{in}}$$

so that

$$v_o = -v_{in} \cdot \frac{A[r_i \parallel R_F \parallel (R_F/A)]}{R_{in}}$$

**FIGURE 4.34** *Inverting Op Amp Circuit*

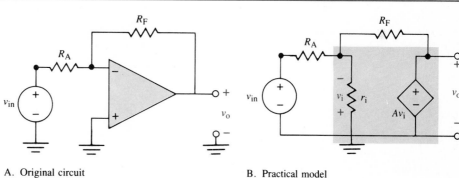

A. Original circuit  B. Practical model

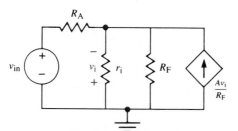

C. Source transformation  D. Final circuit

**Example 4.15**    *Transistor Circuit*

Find $R_{in}$ for the transistor circuit modeled in Figure 4.35.

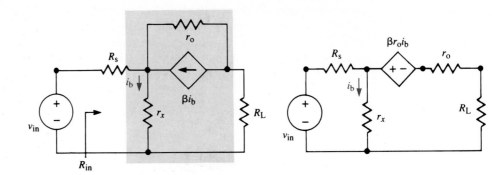

A.  Transistor circuit                                              B.  Current-to-voltage source transformation

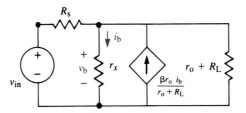

C.  Voltage-to-current source transformation          D. $i_b = \dfrac{v_b}{r_x}$

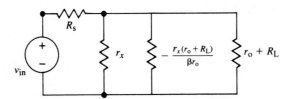

**FIGURE 4.35**    E.  Final circuit

Note that the output resistance, $r_o$, of the transistor has been included. The steps to solution are shown in Figures 4.35B through 4.35E. In Figure 4.35D, we have used the fact that $i_b = v_b/r_x$. The current source is replaced with a negative resistor (Figure 4.35E). The total input resistance is, therefore,

$$R_{in} = R_s + r_x \| (r_o + R_L) \| \left[ -\frac{r_x(r_o + R_L)}{\beta r_o} \right]$$

## *4.7   THE IDEAL OP AMP

We already know that the op amp (Figure 4.36A) can be modeled as a four-terminal device with an input resistance $r_i$, an output resistance $r_o$, and a voltage-controlled voltage source (Figure 4.36B). We have analyzed several op amp circuits by using this model. Since the output resistance of an op amp is very small, we usually ignore $r_o$. We can take advantage of some other op amp characteristics to simplify its analysis.

The input resistance $r_i$ and the voltage gain $A$ of op amps are very large. In fact, it is common practice to assume that they both approach infinity. To see how these properties affect the output of op amp circuits, we will re-examine the inverting and noninverting op amp circuits of Section 4.6.

From the inverting op amp circuit in Figure 4.34, we found that

$$R_{in} = R_A + r_i \| R_F \| \frac{R_F}{A}$$

and

$$v_o = - \frac{v_{in} \cdot A[r_i \| R_F \| (R_F/A)]}{R_{in}}$$

If we let $r_i \rightarrow \infty$ and $A \rightarrow \infty$, then the total input resistance becomes

$$R_{in} \simeq R_A \qquad \text{(inverting)} \tag{4.17}$$

The output becomes

$$v_o \simeq -v_{in} \cdot \frac{A(R_F/A)}{R_A} = -v_{in} \cdot \frac{R_F}{R_A} \qquad \text{(inverting)} \tag{4.18}$$

These approximations are so good that we usually accept these answers as exact.

FIGURE 4.36   *Practical Op Amp Model*

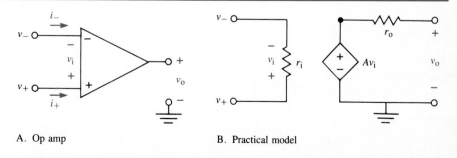

A. Op amp          B. Practical model

A similar analysis of the noninverting op amp equations derived from the circuit in Figure 4.33 results in the following approximations:

$$R_{in} \rightarrow \infty \quad \text{(noninverting)}$$

and                                                                                             **(4.19)**

$$v_o \simeq v_{in} \cdot \left(1 + \frac{R_F}{R_A}\right) \quad \text{(noninverting)}$$

Again, these approximations are very accurate.

The ideal approximations lead to very simple relationships for op amp circuits. We can always find these relationships by using the practical model for the op amp (Figure 4.36B) and then setting $r_o$ to zero and $r_i$ and $A$ to infinity. However, this procedure is quite involved. We can simplify op amp analysis if we take advantage of the ideal characteristics at the start.

What are the consequences of an infinite input resistance? An infinite resistance between the plus and minus terminals in Figure 4.36 implies that an open circuit exists there; that is, current cannot enter or leave the op amp inputs so that

$$i_- = i_+ = 0$$                                                                   **(4.20)**

The effect of an infinite voltage gain is a bit more difficult to understand. If we assume that $r_o$ is zero, then the output voltage of an op amp is always equal to

$$v_o = A \cdot v_i = A \cdot (v_+ - v_-)$$

Now, the output voltage of any real device is limited; that is, an op amp with $\pm 15$ V supplies cannot produce an output that is greater than 15 V in magnitude. All electronic devices require a DC supply voltage or voltages. If the maximum output is 15 V, what is the maximum allowable input voltage for linear operation? We can find the answer by dividing the output voltage by the voltage gain:

$$v_{i\,max} = \frac{v_{o\,max}}{A} = \frac{15}{A}$$

For a typical op amp gain of $10^5$, the maximum allowable input is $15 \times 10^{-5}$ V, or .15 mV. As the gain increases, the maximum allowable input voltage decreases. As the gain goes to infinity, the input voltage goes to zero. Thus, our ideal approximation is that $v_i$ is zero, which implies that

$$v_+ = v_-$$                                                                       **(4.21)**

The important result for an ideal op amp analysis is that the voltage at the plus terminal equals the voltage at the minus terminal. Most op amp circuits can be analyzed with these ideal approximations. You will learn in an electronics course when you must use the more practical model.

Figure 4.37 shows the ideal op amp model. Note that we do not use resistors and controlled sources to model the ideal op amp. Let's use the ideal op amp analysis to find the input resistance and output voltage for the inverting circuit shown in Figure 4.38.

Note that the positive terminal of the op amp is grounded so that $v_+ = 0$. For the ideal op amp, $v_- = v_+$. So,

$$v_- = 0$$

If we write a KCL equation at the negative terminal, we get

$$\frac{0 - v_{in}}{R_A} + \frac{0 - v_o}{R_F} + i_- = 0$$

**FIGURE 4.37**  *Ideal Op Amp Model*

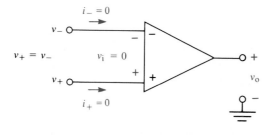

**FIGURE 4.38**  *Inverting Op Amp Analysis with Ideal Model*

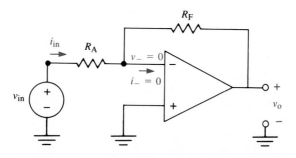

For the ideal op amp, $i_- = 0$. So, this equation reduces to

$$-\frac{v_{in}}{R_A} + \frac{-v_o}{R_F} = 0$$

which gives

$$v_o = -v_{in} \cdot \frac{R_F}{R_A}$$

The input resistance is

$$R_{in} = \frac{v_{in}}{i_{in}} = \frac{v_{in}}{(v_{in} - 0)/R_A} = R_A$$

These answers are the same as the ones we obtained by analyzing the practical circuit and then setting $r_i$ and $A$ to infinity.

In the noninverting op amp circuit shown in Figure 4.39, the voltage at the positive terminal is $v_+ = v_{in}$. Therefore,

$$v_- = v_+ = v_{in}$$

Again, if we write the KCL equation at the negative terminal, we get

$$\frac{v_{in}}{R_A} + \frac{v_{in} - v_o}{R_F} + i_- = 0$$

Since $i_- = 0$,

$$\frac{v_{in}}{R_A} + \frac{v_{in} - v_o}{R_F} = 0$$

**FIGURE 4.39**   *Noninverting Op Amp Analysis with Ideal Model*

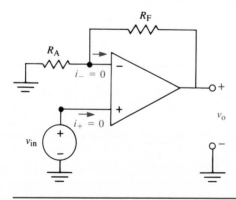

We get, therefore,

$$v_o = v_{in} \cdot \left(1 + \frac{R_F}{R_A}\right)$$

and

$$R_{in} = \frac{v_{in}}{i_{in}} = \frac{v_{in}}{i_+} = \frac{v_{in}}{0} \rightarrow \infty$$

---

**Example 4.16**  *Inverting Op Amp with Multiple Inputs*

Find $v_o$ for the circuit in Figure 4.40.

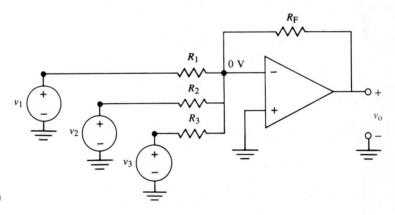

**FIGURE 4.40**

Since the + terminal is grounded,

$$v_- = v_+ = 0$$

KCL at the − terminal yields

$$-\frac{v_1}{R_1} - \frac{v_2}{R_2} - \frac{v_3}{R_3} - \frac{v_o}{R_F} = 0$$

So,

$$v_o = -v_1 \cdot \frac{R_F}{R_1} - v_2 \cdot \frac{R_F}{R_2} - v_3 \cdot \frac{R_F}{R_3}$$

---

**Example 4.17**  *Op Amp Differencer*

Find $v_o$ for the circuit in Figure 4.41.

(*continues*)

**Example 4.17**   *Continued*

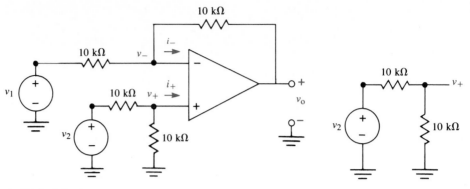

**FIGURE 4.41**   A. Original circuit                                   B. Finding $v_+$

Since $i_+ = 0$, we find $v_+$ and $v_-$ by analyzing the subcircuit in Figure 4.41B to get

$$v_- = v_+ = v_2 \cdot \frac{10^4}{10^4 + 10^4} = .5v_2$$

KCL at the $-$ terminal yields

$$\frac{.5v_2 - v_1}{10^4} + \frac{.5v_2 - v_o}{10^4} = 0$$

So,

$$v_o = v_2 - v_1$$

## *4.8   NONLINEAR RESISTANCE CIRCUITS

Most electronic devices have *V–I* characteristics that are nonlinear. Very often, these devices are analyzed under special operating conditions that allow us to use linear models for them. We have already considered some examples of these devices in this text. However, you should be aware that these nonlinear devices can still be analyzed with some of the circuit techniques you have learned. Although superposition *cannot* be used in nonlinear circuits, KCL and KVL techniques still apply. An example of nonlinear circuit analysis follows.

Figure 4.42 shows a simple one-loop circuit that contains a voltage source, a linear resistor, and a diode. A diode is a semiconductor device that is a rectifier. That is, it allows current to pass in one direction but not the other. In Figure 4.42A, current passes easily only in a clockwise direction. One simple model for the diode is a short circuit for current passing in one direction and an open circuit

**FIGURE 4.42** *Diode Circuit*

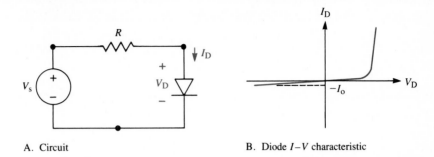

A. Circuit

B. Diode $I-V$ characteristic

for current in the other direction. Alternate models are used if more accuracy is desired. The actual $V–I$ characteristic is shown in Figure 4.42B. The equation for the diode $V–I$ characteristic is

$$I_D = I_o(e^{-V_D/V_T} - 1) \tag{4.22}$$

where $I_o$ and $V_T$ are device parameters with typical values of $I_o = 10^{-14}$ A and $V_T = 26$ mV at room temperature.

The two circuit laws, KVL and KCL, apply to nonlinear as well as to linear circuits. KCL tells us the obvious fact that a single current exists in this circuit. From KVL, we find

$$V_s = V_D + I_D R$$

Substituting for the diode current, we get a single equation with the unknown $V_D$:

$$V_s = V_D + I_o R(e^{V_D/V_T} - 1)$$

This equation cannot be directly solved for $V_D$. Instead we can solve for $V_D$ by iteration. We will demonstrate this procedure for a specific example. Let $V_s = .7$ V and $R = 1$ kΩ with $I_o$ and $V_T$ given their typical values. Then,

$$.7 = V_D + 10^{-11}(e^{V_D/.026} - 1)$$

We first try $V_D = .6$ V. For this value, the right side of the preceding equation is .705. Since this value is too large, we try a smaller value. With $V_D = .58$, we get .629, which is too small. We quickly find the correct value, to three places, to be $V_D = .598$ V.

Some computer programs use numerical techniques to find the answers to these types of nonlinear equations. For a circuit containing more than a single nonlinear device, these computer programs are a necessity. However, for a circuit

**FIGURE 4.43**   *Load-Line Analysis*

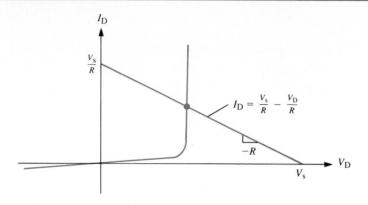

with a single nonlinear device, we can obtain solutions as just shown, or we can use a graphical technique known as a *load-line analysis*.

The load line for the circuit in Figure 4.42 is derived from the KVL equation

$$V_s = V_D + I_D R$$

We reformat to obtain

$$I_D = \frac{V_s}{R} - \frac{V_D}{R}$$

We also have the device *V–I* characteristic:

$$I_D = I_o(e^{V_D/V_T} - 1)$$

We plot the two equations for $I_D$ on the same graph, as in Figure 4.43. The diode current must satisfy both of these equations, so the answer must be the intersection of the two curves in Figure 4.43. Load-line analysis is used extensively in electronics.

## 4.9   SUMMARY

This chapter described how any linear resistive circuit can be represented with a single source and resistor. The Thévenin equivalent circuit is composed of a voltage source in series with a resistor. The Norton equivalent contains a current source in parallel with the same resistor.

The Thévenin equivalent voltage between two terminals is found by leaving the two terminals open and solving for the open circuit voltage at the terminals.

The equivalent resistance can be found in several ways. The most general method involves placing a source (a current source is recommended) between the terminals and solving for the resistance as $V/I$.

The Norton equivalent current is found by placing a short between the two terminals and solving for the short circuit current. When necessary, a source (a voltage source is recommended) is placed at the terminals to find the equivalent resistance.

We used the Thévenin and Norton equivalent circuits to discuss transfer of voltage, current, and power from the network sources to the load, assuming that the source resistance is given. To transfer voltage, the load resistance should be large compared to the equivalent resistance. To transfer current, the load resistance should be small compared to the equivalent resistance. To transfer power, the load resistance should equal the equivalent resistance.

A useful equivalence, but one that is difficult to remember, is the tee-to-pi transformation. It can help reduce nonladder networks to ladder networks. Since ladder networks are so easy to analyze, learning this transformation is worthwhile.

We examined a technique in which certain controlled-source circuits can be manipulated so that the controlled source can be replaced with a resistor. This technique is particularly applicable to electronic circuits.

We discussed the ideal op amp. Two important properties of the ideal op amp are as follows:

1. If the input resistance is infinite, then the input current is zero, so $i_- = i_+ = 0$.
2. If the voltage gain is infinite, then the input voltage is zero, so $v_+ = v_-$.

An ideal op amp circuit is analyzed by first solving for $v_+$, which is equal to $v_-$. A KCL equation is then written at the negative terminal to find the output voltage.

Finally, you were introduced to the concept of load-line analysis and its application to nonlinear circuits.

Chapters 1–4 have focused our attention on resistive circuits. Even so, we have covered most of the circuit theory necessary for the study of this text, as well as for any future courses you may pursue. Chapter 5 will introduce two new devices, the capacitor and the inductor. Both of these devices store energy and thus have memory. To analyze circuits containing these devices requires knowledge of differential equations and their solutions. However, node analysis, mesh analysis, and equivalent circuit analysis techniques are still applicable.

■ **PROBLEMS**

4.1 Measurements at the terminals of the unknown linear resistive circuit shown in Figure P4.1 yield the following data: $V_a = 2\,\mathrm{V}$ when $I_a = 1/2\,\mathrm{A}$ and $V_a = 3\,\mathrm{V}$ when $I_a = 0$. Draw the Thévenin equivalent circuit for this data.

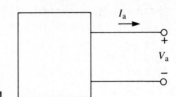

**FIGURE P4.1**

**4.2** Find the Thévenin equivalent between terminals a–a' for the circuit shown in Figure P4.2.

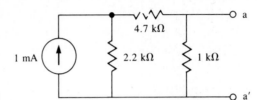

**FIGURE P4.2**

**4.3** Find the Thévenin equivalent between terminals a–a' for the circuit shown in Figure P4.3.

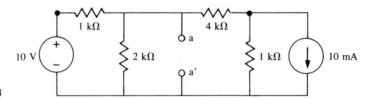

**FIGURE P4.3**

**4.4** For the circuit shown in Figure P4.4,

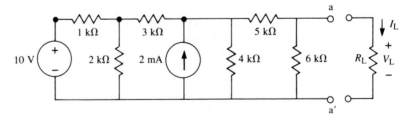

**FIGURE P4.4**

    **a.** Find the Thévenin equivalent seen by the load resistor $R_L$.

    **b.** Find $V_L$, $I_L$, and $P_L$ as functions of $R_L$.

**4.5** Given the following set of V–I data points for a two-terminal black box,

| V (V) | −2.4 | −.3 | .5 | 2.6 | 4.5 | 7.7 | 8.8 | 11.4 | 13.6 |
|-------|------|-----|----|-----|-----|-----|-----|------|------|
| I (mA) | −4 | −3 | −2 | −1 | 0 | 2 | 3 | 4 | 5 |

    **a.** Plot these points on a V–I graph.

    **b.** The data points show a certain amount of scatter because of noise in the measurements. Fit the points with a straight line (use a calculator with a linear regression feature, if available).

    **c.** Derive a circuit model for the data from your straight-line V–I characteristic.

**4.6** Find the Thévenin equivalent seen by $R_L$ in the bridge circuit shown in Figure P4.6.

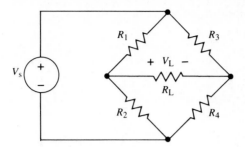

**FIGURE P4.6**

**4.7** For the circuit shown in Figure P4.7,

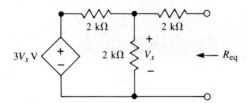

**FIGURE P4.7**

    **a.**   Find the Thévenin equivalent seen by $R_L$.
    **b.**   Find $V_L$, $I_L$, and $P_L$ as functions of $R_L$.

**4.8** Find $R_{eq}$ for the circuit shown in Figure P4.8.

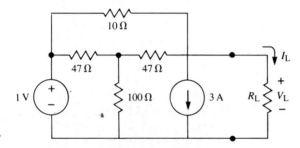

**FIGURE P4.8**

**4.9** Find $R_{eq}$ for the circuit shown in Figure P4.9.

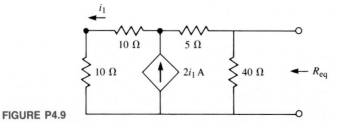

**FIGURE P4.9**

**4.10** Derive the Thévenin equivalent given in this chapter for the op amp circuit shown in Figure 4.16B.

**4.11** For the AC transistor circuit shown in Figure P4.11,

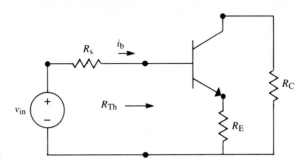

**FIGURE P4.11**

    **a.**   Find $R_{Th}$ if the transistor output resistance ($r_o$) is infinite.

    **b.**   Find $R_{Th}$ if $r_o$ is finite.

    **c.**   Find $i_b$ in terms of $R_{Th}$ for parts a and b.

**4.12** For the AC transistor circuit shown in Figure P4.12 ($r_x = 1\,k\Omega$, $\beta = 100$, and $r_o = 50\,k\Omega$), find the Thévenin equivalent at the output.

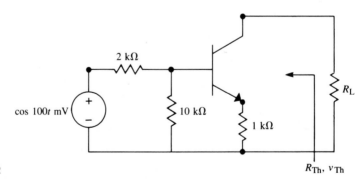

**FIGURE P4.12**

**4.13** Measurements at the terminals of an unknown linear resistive circuit (see Figure P4.1) yield the following data: $I_a = 5\,mA$ when $V_a = 10\,V$ and $I_a = 15\,mA$ when $V_a = 0$. Draw the Norton equivalent circuit for this data.

**4.14** Find the Norton equivalent of the circuit shown in Figure P4.2. Do not transform the Thévenin equivalent.

**4.15** Find the Norton equivalent of the circuit shown in Figure P4.3. Do not transform the Thévenin equivalent.

**4.16** Find the Norton equivalent seen by $R_L$ in the circuit in Figure P4.4. Do not simply transform the Thévenin equivalent.

**4.17** Find the Norton equivalent between terminals a–a′ for the circuit shown in Figure P4.17.

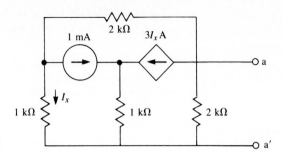

**FIGURE P4.17**

**4.18** Find the Norton equivalent for the transistor circuit of Figure P4.12 for $r_x = 1\,\text{k}\Omega$, $\beta = 100$, and $r_o = 50\,\text{k}\Omega$. Do not transform the Thévenin equivalent.

**4.19** Starting with a Norton equivalent, prove the maximum power transfer relationship (Equation 4.14).

**4.20** Given the FET circuit in Figure P4.20, find the Norton equivalent seen by $R_D$ if $r_d$ is finite.

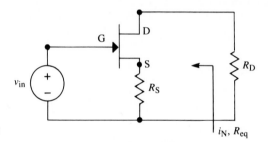

**FIGURE P4.20**

**4.21** For the circuit shown in Figure P4.21,

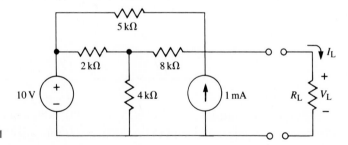

**FIGURE P4.21**

    **a.** Find $R_L$ to maximize power transfer to the load.
    **b.** Find the maximum possible load power.
    **c.** Find the maximum possible load voltage.
    **d.** Find the maximum possible load current.

**4.22** For the independent current source in Figure P3.17,
    **a.** Find the equivalent resistance seen by the source.
    **b.** How much power is delivered to the circuit by the independent current source?

4.23 Find the equivalent resistance seen by the current source in Figure P2.6D. (*Hint:* Use symmetry to find the resistor currents in the cube and then find the voltage between the vertices of interest.)

4.24 Find the equivalent resistance between nodes a and b in the infinite resistor grid shown in Figure P4.24. (*Hint:* Apply a 1 A source between nodes a and b, then use symmetry to find resistor currents, and, finally, find the voltage between nodes a and b.)

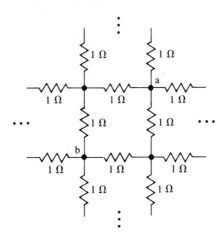

**FIGURE P4.24**

4.25 For the circuit shown in Figure P4.25,

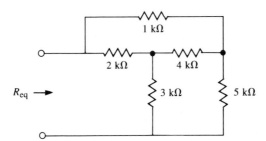

**FIGURE P4.25**

    **a.** Find $R_{eq}$ by replacing the tee ($2\,k\Omega$, $3\,k\Omega$, $4\,k\Omega$) with a pi circuit.

    **b.** Find $R_{eq}$ by replacing the pi ($3\,k\Omega$, $4\,k\Omega$, $5\,k\Omega$) with a tee circuit.

4.26 Use the tee–pi transformation to find $v_{oc}$ and $i_{sc}$ in the circuit shown in Figure P4.21.

4.27 Derive the tee–pi transformation Equations 4.15 and 4.16.

4.28 Find $R_{eq}$ in the circuit shown in Figure P4.8 by manipulating the circuit until you can replace the controlled source with a resistor.

4.29 Find $R_{eq}$ in the circuit shown in Figure P4.9 by manipulating the circuit until you can replace the controlled source with a resistor.

4.30 Find the equivalent resistance seen by the voltage source in the circuit in Figure P4.30. Manipulate the circuit until you can replace the controlled source with a resistor.

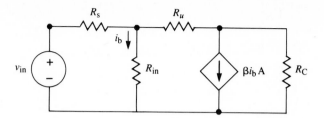

**FIGURE P4.30**

**4.31** A four-bit binary number has the form $b_3b_2b_1b_0$, where $b_i$ is either 0 or 1. The value of the four-bit number is given by $b_32^3 + b_22^2 + b_12 + b_0$. In an electrical circuit, the binary value 1 is represented by an arbitrarily selected voltage (e.g., .5 V), while the binary value 0 is usually represented by 0 V. An op amp circuit can be used to convert the binary representation into an analog voltage. This is known as *digital-to-analog conversion* (DAC). Figure P4.31 shows an op amp DAC (note that for simplicity we use an inverting circuit).

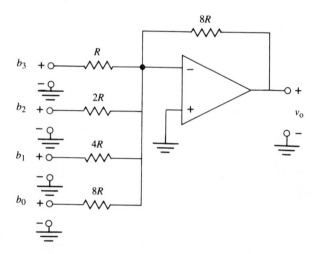

**FIGURE P4.31**

**a.** Use the ideal op amp analysis to find the general expression for $v_o$.

**b.** Assuming that a binary $1 = .5$ V and a binary $0 = 0$ V, find $v_o$ for the following binary numbers: 0001, 0011, 0100, 1010, 1100, and 1111.

**4.32** Using the ideal op amp analysis, find $v_0$ in Figure P4.32.

**a.** Use superposition (analyze for one source at a time).

**b.** Use Thévenin equivalents at the $+$ and $-$ terminals.

**4.33** The answer in Problem 4.32 greatly simplifies if $R_F \| R_1 \| R_2 = R_3 \| R_4$. Show that in this case the answer reduces to

$$v_o = R_F\left(-\frac{v_1}{R_1} - \frac{v_2}{R_2} + \frac{v_3}{R_3} + \frac{v_4}{R_4}\right)$$

**4.34** Find $v_o$ for the op amp circuit shown in Figure P4.34 (*Hint:* Use KCL at the $+$ node to first find $v_+$.)

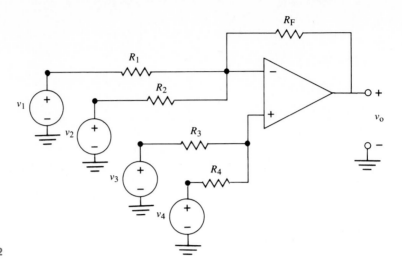

**FIGURE P4.32**

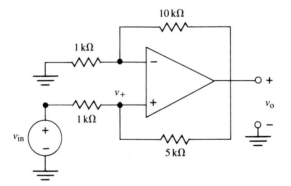

**FIGURE P4.34**

**4.35** Find $v_o$ for the op amp circuit shown in Figure P4.35. (*Hint:* Write node equations only at the − terminals.)

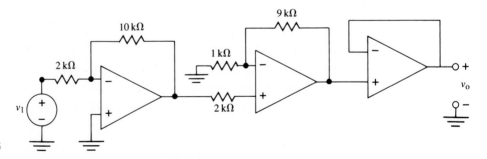

**FIGURE P4.35**

**4.36** In the diode circuit shown in Figure P4.36, the diode current is given by $I_D = 10^{-15}(e^{V_D/.026} - 1)$. Plot the diode *I–V* characteristic and the linear *I–V* load line on the same graph and find $I_D$, $V_D$, and $V_R$.

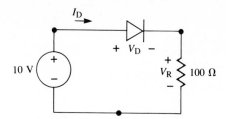

**FIGURE P4.36**

**4.37** The DC behavior of an FET is very nonlinear and usually requires a load-line analysis. A typical FET circuit is shown in Figure P4.37. The $I$–$V$ characteristic for this device is $I_D = I_o(1 - V_{GS}/V_p)^2$, where $I_o$ and $V_p$ are device parameters. Typical values for these parameters are $I_o = 5\,\text{mA}$ and $V_p = -4\,\text{V}$.

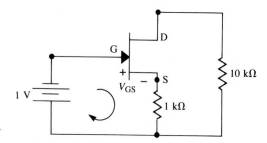

**FIGURE P4.37**

   **a.** Plot $I_D$ vs. $V_{GS}$ for $-4 < V_{GS} < 0$.

   **b.** Write the KVL equation for the loop indicated in Figure P4.37.

   **c.** Plot the straight-line KVL equation on the same graph as part a and find the DC operating point. That is, find $I_D$ and $V_{GS}$.

# Capacitors and Inductors

5

## 5.1 INTRODUCTION

In this chapter, we will examine two new circuit elements, the capacitor and the inductor. These devices differ from resistors in that their *V–I* relationships are described by derivatives and integrals. Since past values of voltage and current determine present behavior in these devices, they are said to have memory.

The circuit laws that you have learned thus far still apply to circuits that contain capacitors and inductors. However, application of the circuit laws to these devices results in differential rather than algebraic equations.

## 5.2 CAPACITANCE AND THE CAPACITOR

Figure 5.1 shows one of the earliest electrical devices, the *Leyden jar*, which was first used in Leyden, The Netherlands, in 1746. It is constructed by lining the interior and exterior of a glass jar with a thin metal conductor. The glass provides electrical insulation between the sheets of conductors. A source of electrical charge is brought in contact with the inner conductor, while the outer conductor is grounded (Figure 5.1A). The inner conductor, in this case, is negatively charged. Because of the insulating glass, none of this charge can reach the outer conductor. However, the negative charge on the inner conductor will repel negative charges from the outer conductor to ground. A net positive charge results on the inner surface of the outer conductor.

When both the source and the ground connections are removed (Figure 5.1B), the charges on the inner and outer conductors remain where they are since both glass and air are insulators. The net charges on the two conductors are equal in magnitude and opposite in sign. The Leyden jar *stores electrical charge*.

Figure 5.1C is a schematic representation of the Leyden jar. The two parallel lines represent the inner and outer conductors, and the space between

**FIGURE 5.1**  *Leyden Jar and Capacitance*

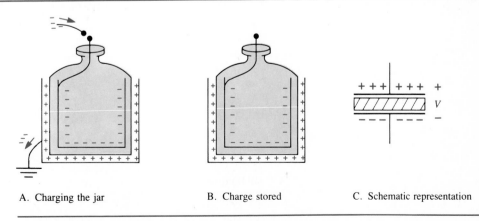

A. Charging the jar          B. Charge stored          C. Schematic representation

FIGURE 5.2    *Capacitor Circuit Symbols*

these lines is filled with an insulator. Chapter 1 showed us that charge separated by a distance creates a voltage between the charges. This voltage is proportional to the charge and has the polarity shown in Figure 5.1C.

It is more common to write charge as a function of voltage:

$$q(t) = Cv(t) \tag{5.1}$$

The constant $C$ is known as **capacitance** and has units of **farads** (F), in honor of the English chemist and physicist Michael Faraday (1791–1867). Capacitance represents the ability of a device to hold charge at a given voltage. The device specifically built to take advantage of this property is the **capacitor.** A 1 F capacitor with 1 V across it will store 1 C of charge, which is equivalent to $6.25 \times 10^{18}$ electrons!

Equation 5.1 assumes that the insulating material between the conductors of the capacitor has infinite resistance. This, of course, cannot be true for a real capacitor. A real capacitor is modeled with an ideal capacitor that obeys Equation 5.1 in parallel with a large resistor. In this textbook, we assume that all capacitors are ideal.

Symbols for the capacitor are shown in Figure 5.2. The symbols are used interchangeably in electrical circuit schematics. Another term for the capacitor is *condenser,* a term still used in automobile ignition systems. The value of the capacitance is dependent on the distance between the plates, the insulating material, and the geometry of the plates. Most capacitors have values in the picofarad (pF) to microfarad (μF) range. A 1 F capacitor is very large.

Example 5.1    *Capacitance*

The capacitance of two equal parallel plates can be determined physically from the following formula:

$$C = \frac{\varepsilon A}{d}$$

where    $A$ = surface area of each plate

$d$ = distance separating the plates

$\varepsilon$ = permittivity of the insulator between the plates

The permittivity of free space is $\varepsilon_0$ and is given by

$$\varepsilon_0 = 8.85 \times 10^{-15} \text{ F/mm}$$

a.  Determine the plate area required to obtain a 1 μF capacitor composed of two parallel plates separated by a distance of 2 mm in free space:

$$A = \frac{Cd}{\varepsilon_0} = \frac{10^{-6} \cdot 2}{8.85 \times 10^{-15}} = .226 \times 10^9 \text{ mm}^2 = 226 \text{ m}^2$$

b.  Determine the permittivity of the insulator required to reduce the surface area of the capacitor in part a to 2.26 cm²:

$$2.26 \text{ cm}^2 = 226 \text{ mm}^2$$

so

$$\varepsilon = \frac{Cd}{A} = \frac{10^{-6} \cdot 2}{226} = 8.85 \times 10^{-9} = 10^6 \varepsilon_0$$

The large surfaces required for some capacitors are created by rolling the conductors and insulating materials concentrically about themselves into tightly wound cylinders.

## Capacitor Current and Voltage

While charge is occasionally used as a circuit variable, we usually are more interested in current. Since current is the derivative of charge with respect to time, we find the following relationship between the capacitor voltage and current, as shown in Figure 5.3:

$$i(t) = \frac{dq}{dt} = \frac{d(Cv)}{dt} = C\frac{dv}{dt} + v\frac{dC}{dt} \tag{5.2}$$

FIGURE 5.3  *Capacitor Current and Voltage*

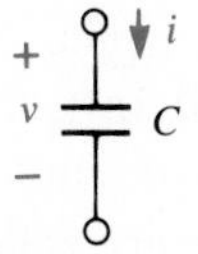

$$i = C\frac{dv}{dt}$$

In some modern switched circuits, the capacitance may be modeled as time varying. However, most capacitors have fixed values, so the relationship expressed in Equation 5.2 reduces to

$$i(t) = C \frac{dv}{dt} \qquad (5.3)$$

Note that the passive sign convention is observed.

The following question may have occurred to you: Since the plates of the capacitor are separated by an insulator, how can charge move from one plate to the other? A physical current does not exist between the plates within the capacitor. Rather, charges move from one plate to another through the circuit to which the capacitor is connected. However, in analyzing circuit behavior, it is convenient to treat capacitor current as existing in the capacitor itself.

Capacitor voltage can be found by integrating both sides of Equation 5.3 to obtain

$$v(t) = \frac{1}{C} \int_{-\infty}^{t} i(t)\, dt \qquad (5.4)$$

or

$$v(t) = \frac{1}{C} \int_{t_0}^{t} v(t)\, dt + v(t_0) \qquad (5.5)$$

Equation 5.4 is compact and is used when writing circuit equations. Equation 5.5 is used to actually find a capacitor voltage since we do not know the behavior of $i(t)$ when $t$ approaches $-\infty$. Note that the initial voltage, $v(t_0)$, is dependent on any charge that is stored on the capacitor before the capacitor is connected to the circuit. The initial voltage $v(t_0)$ is given by

$$v(t_0) = \frac{q(t_0)}{C} \qquad (5.6)$$

The $V$–$I$ relationship for a capacitor is quite different from the $V$–$I$ relationships for the devices that were studied in earlier chapters. Capacitor voltage depends on the past history of capacitor current and any initial charge that may have been stored on the capacitor; that is, the capacitor has memory.

Example 5.2   *Capacitor Currents*

Using Figure 5.4, find $i(t)$ for the following source voltages:

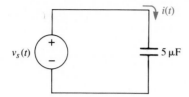

FIGURE 5.4

a.   $v_s(t) = 10$ V

b.   $v_s(t) = 10 \cos 10t$ V

c.   $v_s(t) = 10 \cos 1000t$ V

d.   $v_s(t) = 10t$ V

The capacitor currents are as follows:

a.   $i(t) = 5 \times 10^{-6} \cdot \dfrac{d(10)}{dt} = 0$ A

b.   $i(t) = 5 \times 10^{-6} \cdot \dfrac{d(10 \cos 10t)}{dt} = -.5 \sin 10t$ mA

c.   $i(t) = -50 \sin 10t$ mA

d.   $i(t) = 50 \ \mu$A

Example 5.2 demonstrates some interesting points about capacitor behavior. A most important result of the capacitor $V$–$I$ relationship is that circuit variable waveforms are changed. A voltage cosine is changed to a current sine, and a ramp in voltage is changed to a constant current. Resistors do not have this property; they merely scale the magnitude of the input signal without changing the waveshape.

From part a of Example 5.2, it is clear that if capacitor voltage is constant, capacitor current is zero. Thus, *the capacitor acts as an open circuit for DC voltages*.

Example 5.3   *Capacitor Voltages*

Using Figure 5.5, find $v(t)$ for the following source currents. Note that we assume the current source is turned on at $t = 0$ s and that the capacitor is initially uncharged so that $v(0) = 0$:

(*continues*)

**Example 5.3**  *Continued*

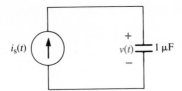

**FIGURE 5.5**

a.  $i_s(t) = 10 \cos 10t \, \mu A$

b.  $i_s(t) = 10 \cos 1000t \, \mu A$

c.  $i_s(t) = 10 \, \mu A$

The capacitor voltages are as follows:

a.  $v(t) = \dfrac{1}{10^{-6}} \displaystyle\int_0^t 10 \times 10^{-6} \cos 10t \, dt = \sin 10t \text{ V}$

b.  $v(t) = .01 \sin 10t \text{ V}$

c.  $v(t) = \dfrac{1}{10^{-6}} \displaystyle\int_0^t 10 \times 10^{-6} \, dt = 10t \text{ V}$

---

Parts a and b of Example 5.3 show that for the same magnitude of sinusoidal current, the magnitude of the voltage decreases with increasing frequency. In fact, as frequency goes to infinity, capacitor voltage goes to zero. Thus, *for sufficiently high-frequency AC signals, the capacitor acts as a short circuit.*

When we first study transistor circuits, we do not fully analyze capacitors. Remember, in Chapter 1, the DC and AC models were described for the transistor. For DC transistor circuits, any capacitor in the circuit is simply replaced with an open circuit. For AC transistor circuits, the assumption is made that the frequency is high enough so that capacitors can be replaced with short circuits. It is usually in more advanced transistor circuit analysis courses that the complete capacitor *V–I* relationship is used.

## Capacitor Power and Energy

As with any electrical device, power into the device is defined as voltage times current:

$$p(t) = v(t) \cdot i(t) = v(t) \cdot C \frac{dv(t)}{dt} = Cv(t) \frac{dv(t)}{dt} \tag{5.7}$$

This relationship does not simplify as it does for the resistor.

Energy is found by integrating power. Thus, for the capacitor, we get

$$w(t) = \int_{-\infty}^{t} p(t)\, dt = \int_{-\infty}^{t} Cv(t) \left( \frac{dv(t)}{dt} \right) dt$$

We can write $(dv/dt) \cdot dt$ as simply $dv$, and we end up integrating with respect to voltage. Note that we must change the limits of integration to get

$$w(t) = C \int_{v(-\infty)}^{v(t)} v(t)\, dv(t) = C \left( \frac{v^2(t)}{2} - \frac{v^2(-\infty)}{2} \right)$$

If we assume that the capacitor is initially uncharged at $t = -\infty$, we get a simple relationship for the net electrical energy absorbed by a capacitor:

$$w(t) = \frac{1}{2} Cv^2(t) \tag{5.8}$$

The electrical energy into a capacitor is always positive or zero, which indicates that the capacitor is a passive device. The energy into a capacitor is stored rather than dissipated. At any given time, the energy stored is a function only of the voltage on the capacitor at that time; that is, past values of voltage and current do not affect the stored energy.

Example 5.4    *Capacitor Power and Energy*

a.    Using Figure 5.6, find $i(t)$, $p(t)$, and $w(t)$ given $v(t) = 5e^{-10t}$ V:

$$i(t) = C \frac{dv(t)}{dt} = -500e^{-10t} \ \mu A$$

$$p(t) = v(t) \cdot i(t) = -2500e^{-20t} \ \mu W$$

$$w(t) = \frac{1}{2} Cv^2(t) = 125e^{-20t} \ \mu J$$

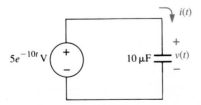

FIGURE 5.6

Note that capacitor voltage is decreasing with time, which indicates that charge is leaving the capacitor and returning to the circuit. Thus, in this example, the capacitor current and power are negative.

(*continues*)

**Example 5.4**   *Continued*

   **b.**  Now, apply two different voltages to the capacitor: $v_1(t) = 20t$ V and $v_2(t) = 20 \cos 2.09t$ V. Find the energy stored in the capacitor at $t = .5$ s for the two voltages:

$$w_1(.5) = \frac{1}{2} \, Cv_1^2(.5) = \frac{10 \cdot 10^{-6}}{2} \, (20 \cdot .5)^2 = 500 \; \mu J$$

$$w_2(.5) = \frac{1}{2} \, Cv_2^2(.5) = \frac{10 \cdot 10^{-6}}{2} \, (20 \cos 1.045)^2 = 500 \; \mu J$$

Although, in general, the two voltages are quite different, at $t = .5$ s, they have the same value. Therefore, the energy stored in the capacitor at $t = .5$ s is the same.

## 5.3   CAPACITORS IN SERIES AND IN PARALLEL

Capacitors connected in series or in parallel can be replaced by equivalent capacitors. However, great care must be taken if the capacitors have an initial charge stored on them. (The derivations of equivalent circuits when there are initial conditions will be pursued in some problems at the end of this chapter.)

### Series Capacitors

Figure 5.7 shows a series connection of capacitors with zero initial conditions. We find the equivalent capacitance $(C_{eq})$ by comparing the V–I relationships for the series connection and for the equivalent circuit:

$$v(t) = v_1(t) + v_2(t) + v_3(t)$$

$$= \frac{1}{C_1} \int_{-\infty}^{t} i(t) \, dt + \frac{1}{C_2} \int_{-\infty}^{t} i(t) \, dt + \frac{1}{C_3} \int_{-\infty}^{t} i(t) \, dt$$

**FIGURE 5.7**   *Capacitors in Series*

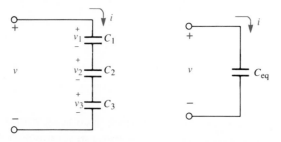

A. Original circuit                    B. Equivalent circuit

Since the current in the series capacitors is the same, we factor out the integral to get

$$v(t) = \left(\frac{1}{C_1} + \frac{1}{C_2} + \frac{1}{C_3}\right) \int_{-\infty}^{t} i(t)\, dt$$

For the equivalent circuit,

$$v(t) = \frac{1}{C_{eq}} \int_{-\infty}^{t} i(t)\, dt$$

Therefore, for series-connected capacitors, we find

$$\frac{1}{C_{eq}} = \frac{1}{C_1} + \frac{1}{C_2} + \frac{1}{C_3}$$

which yields the equivalent series capacitor relationship:

$$C_{eq} = \frac{1}{1/C_1 + 1/C_2 + 1/C_3}$$

or

$$C_{eq} = \frac{1}{\sum_i 1/C_i} \qquad (5.9)$$

Equation 5.9 is analogous to the way in which conductances in series combine.

---

**Example 5.5**   *Series-Connected Capacitors*

In the circuit shown in Figure 5.8A, the capacitors are initially uncharged and

$$v(t) = 10(e^{-5t} - 1) \text{ V}, \qquad t > 0$$

Find $i(t)$ and $v_1(t)$.

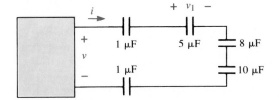
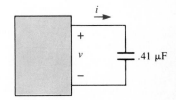

**FIGURE 5.8**   A. Original circuit                          B. Equivalent circuit

*(continues)*

**Example 5.5**   *Continued*

First, we replace capacitors with the equivalent capacitance as shown in Figure 5.8B:

$$C_{eq} = \frac{1}{(1 + .2 + .125 + .1 + 1)10^6} = .41 \,\mu F$$

From the *V–I* characteristic,

$$i = C_{eq}\frac{dv}{dt} = .41 \times 10^{-6}(-50e^{-5t})$$

So,

$$i(t) = -20.5e^{-5t} \,\mu A, \qquad t > 0$$

Returning to the original circuit, we get

$$v_1 = \frac{1}{5 \times 10^{-6}} \int_0^t -20.5 \times 10^{-6}e^{-5t}\, dt = \frac{20.5}{25}(e^{-5t} - 1)$$

So,

$$v_1(t) = .82(e^{-5t} - 1) \text{ V}, \qquad t > 0$$

## Parallel Capacitors

Finding the equivalent capacitance of a number of parallel-connected capacitors is very straightforward if the capacitors are initially uncharged. Parallel capacitors that have different initial conditions require special care in analysis.

Figure 5.9 shows a parallel connection of capacitors with zero initial conditions. We compare the *V–I* relationships for the parallel connection and for the equivalent capacitor:

$$i(t) = C_1\frac{dv}{dt} + C_2\frac{dv}{dt} + C_3\frac{dv}{dt} = (C_1 + C_2 + C_3)\frac{dv}{dt}$$

FIGURE 5.9   *Capacitors in Parallel*

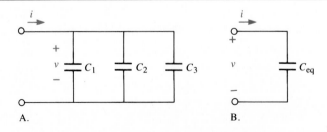

A.                                        B.

and

$$i(t) = C_{eq} \frac{dv}{dt}$$

It is obvious that capacitors in parallel add:

$$C_{eq} = C_1 + C_2 + C_3$$

or

$$C_{eq} = \sum_i C_i \qquad\qquad (5.10)$$

Equation 5.10 is analogous to the way in which conductances in parallel combine.

---

**Example 5.6**   *Parallel–Series-Connected Capacitors*

In the circuit shown in Figure 5.10A, the capacitors are initially uncharged and

$$i(t) = 10e^{-20t} \text{ mA}, \qquad t > 0$$

Find $v(t)$, $v_1(t)$, $v_2(t)$, and $i_2(t)$.

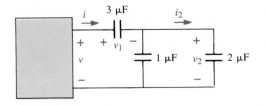

**FIGURE 5.10**    A. Original circuit                                      B. Equivalent circuit

First, we replace capacitors with the equivalent capacitance as shown in Figure 5.10B:

$$C_{eq} = \frac{1}{.33 \times 10^6 + 10^6/(1+2)} = 1.5 \ \mu F$$

From the *V–I* characteristic,

$$v = \frac{1}{1.5 \times 10^{-6}} \int_0^t .01 e^{-20t} \, dt$$

So,

$$v(t) = -333.33(e^{-20t} - 1) \text{ V}, \qquad t > 0$$

*(continues)*

**Example 5.6**  *Continued*

From the original circuit,

$$v_1 = \frac{1}{3 \times 10^{-6}} \int_0^t .01e^{-20t}\, dt$$

So,

$$v_1(t) = -166.67(e^{-20t} - 1)\, V, \qquad t > 0$$

From KVL,

$$v_2(t) = v - v_1 = -166.67(e^{-20t} - 1)\, V, \qquad t > 0$$

From the *V–I* characteristic,

$$i_2(t) = 2 \times 10^{-6}\, \frac{d[-166.67(e^{-20t} - 1)]}{dt} = 6.67e^{-20t}\, mA, \qquad t > 0$$

## 5.4   INDUCTANCE AND THE INDUCTOR

A current in a wire creates a magnetic field that encircles the wire (Figure 5.11A). The flux of the magnetic field is proportional to the current:

$$\phi(t) = Li(t)$$

The scale factor $L$ is known as **inductance** and has units of **henrys** (H), in honor of the American physicist Joseph Henry (1797–1878). If the current is time varying, the magnetic field will also vary with time. This time-varying field will

**FIGURE 5.11**   *Inductance and the Inductor*

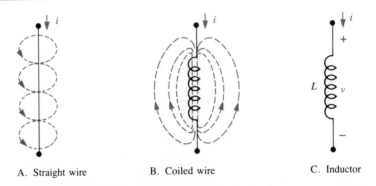

A. Straight wire          B. Coiled wire              C. Inductor

induce a voltage that is proportional to the derivative of magnetic flux:

$$v(t) = \frac{d\phi(t)}{dt} = L\frac{di(t)}{dt} \qquad (5.11)$$

assuming $L$ is constant. For a single small wire at moderate current levels, the induced voltage is small enough to be ignored. If the wire is formed into a coil (Figure 5.11B), the strength of the magnetic field is greatly increased, and the induced voltage becomes significant.

The device that is specifically built to take advantage of this phenomenon is the **inductor.** The circuit symbol for an ideal inductor is a coil (Figure 5.11C). In fact, inductors are often referred to as *coils*. The circuit device in Figure 5.11C is an ideal inductor because we are ignoring the resistance of the wire and the capacitance that exists between adjacent loops of the coil. Typically, inductors have values in the millihenry (mH) to henry (H) range.

## Inductor Current and Voltage

The *V–I* relationships for the inductor are as follows:

$$v(t) = L\frac{di(t)}{dt} \qquad (5.12)$$

and

$$i(t) = \frac{1}{L}\int_{-\infty}^{t} v(t)\,dt = \frac{1}{L}\int_{t_0}^{t} v(t)\,dt + i(t_0) \qquad (5.13)$$

Remember that, as always, the passive sign convention is observed.

---

**Example 5.7** *Inductor Voltages*

For an inductor of 10 mH, find the voltage for each of the following inductor currents:

a. $i(t) = 5\,\text{mA}$

b. $i(t) = 5\sin 10t\,\text{mA}$

c. $i(t) = 5\sin 1000t\,\text{mA}$

The voltages are as follows:

a. $v(t) = .01 \cdot \dfrac{d(.005)}{dt} = 0\,\text{V}$

*(continues)*

**Example 5.7**  *Continued*

b.  $v(t) = .01 \cdot \dfrac{di(t)}{dt} = .5 \cos 10t$ mV

c.  $v(t) = 50 \cos 1000t$ mV

---

Example 5.7 shows us some interesting results. For a constant (DC) current, the inductor voltage is zero. To induce a voltage, the magnetic field must be time varying. Since a constant current produces a steady magnetic field, no voltage is induced. Thus, *an ideal inductor acts as a short circuit for DC current.*

---

**Example 5.8**  *Inductor Currents*

For an inductor of 10 mH and assuming zero initial current at $t = 0$, find the current for each of the following inductor voltages:

a.  $v(t) = 8$ mV

b.  $v(t) = 10 \cos 10t$ mV

c.  $v(t) = 10 \cos 1000t$ mV

The currents are as follows:

a.  $i(t) = \dfrac{1}{.01} \displaystyle\int_0^t .008 \, dt = .8t$ A

b.  $i(t) = .1 \sin 10t$ A

c.  $i(t) = .001 \sin 1000t$ A

---

Parts b and c of Example 5.8 show the frequency dependence of the sinusoidal *V–I* relationships for the inductor. As frequency increases, inductor current decreases. Thus, *for sufficiently high-frequency AC signals, the inductor acts as an open circuit.* In fact, sometimes the term *choke* is used for the inductor because it blocks high-frequency sinusoidal signals.

## Inductor Power and Energy

The power relationship for the inductor is as follows:

$$p(t) = v \cdot i = Li(t) \frac{di(t)}{dt} \tag{5.14}$$

By integrating power, we get

$$w(t) = \int_{-\infty}^{t} p(t)\, dt = \frac{Li^2(t)}{2} - \frac{Li^2(-\infty)}{2}$$

As with capacitor energy, we assume that at $t = -\infty$ the energy stored is zero. Thus, the energy relationship for the inductor becomes

$$w(t) = \frac{1}{2} Li^2(t) \tag{5.15}$$

The inductor stores magnetic energy. Consider a simple example: A current in an inductor creates a magnetic field. When the source of the current is removed, the magnetic field collapses. As the field collapses, a voltage is induced in the inductor, which returns energy to the circuit.

## 5.5 INDUCTORS IN SERIES AND IN PARALLEL

Inductors connected in parallel or in series can be replaced with equivalent inductors. Care must be exercised if initial currents are present, especially if the inductors are connected in series. (These cases will be pursued in some problems at the end of this chapter.)

### Parallel Inductors

Figure 5.12 shows several inductors with zero initial conditions connected in parallel. We find the equivalent inductance ($L_{eq}$) by comparing the $V$–$I$ relation-

FIGURE 5.12 *Inductors in Parallel*

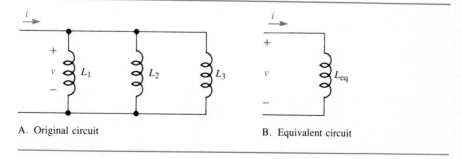

A. Original circuit                    B. Equivalent circuit

ships for the parallel inductors and for the equivalent inductor:

$$i(t) = \frac{1}{L_1} \int_{-\infty}^{t} v(t)\, dt + \frac{1}{L_2} \int_{-\infty}^{t} v(t)\, dt + \frac{1}{L_3} \int_{-\infty}^{t} v(t)\, dt$$

$$= \left( \frac{1}{L_1} + \frac{1}{L_2} + \frac{1}{L_3} \right) \int_{-\infty}^{t} v(t)\, dt$$

and

$$i(t) = \frac{1}{L_{eq}} \int_{-\infty}^{t} v(t)\, dt$$

Therefore, the equivalent parallel inductor relationship is

$$L_{eq} = \frac{1}{1/L_1 + 1/L_2 + 1/L_3}$$

or

$$L_{eq} = \frac{1}{\sum_i 1/L_i} \tag{5.16}$$

## Series Inductors

Figure 5.13 shows several inductors with zero initial conditions connected in series. We find the equivalent inductance as follows:

$$v(t) = L_1 \frac{di}{dt} + L_2 \frac{di}{dt} + L_3 \frac{di}{dt}$$

and

$$v(t) = L_{eq} \frac{di}{dt}$$

We see that inductors in series add:

$$L_{eq} = L_1 + L_2 + L_3$$

or

$$L_{eq} = \sum_i L_i \tag{5.17}$$

**FIGURE 5.13**   *Inductors in Series*

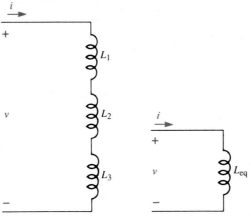

A. Original circuit         B. Equivalent circuit

**Example 5.9**   *Series–Parallel-Connected Inductors*

In the circuit shown in Figure 5.14A, the inductors have zero initial conditions and

$$i(t) = 10(e^{-50t} - 1)\,\text{mA}, \qquad t > 0$$

Find $v(t)$, $v_2(t)$, $v_1(t)$, $i_1(t)$.

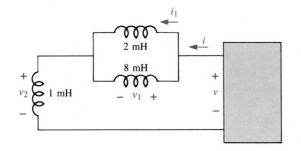
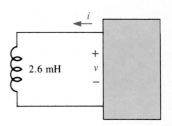

**FIGURE 5.14**   A. Original circuit                         B. Equivalent circuit

First, we replace the inductors with the equivalent inductance as shown in Figure 5.14B:

$$L_{eq} = .001 + \frac{1}{500 + 125} = 2.6\,\text{mH}$$

*(continues)*

**Example 5.9**   *Continued*

From the *V–I* characteristic,

$$v = .0026 \frac{d[.01(e^{-50t} - 1)]}{dt}$$

So,

$$v(t) = -1.3e^{-50t} \text{ mV}, \qquad t > 0$$

From the original circuit, we see that the current in the 1 mH inductor is also $i(t)$. So,

$$v_2(t) = .001 \frac{d[.01(e^{-50t} - 1)]}{dt} = -.5e^{-50t} \text{ mV}, \qquad t > 0$$

From KVL,

$$v_1(t) = v - v_2 = -.8e^{-50t} \text{ mV}, \qquad t > 0$$

From the *V–I* characteristic,

$$i_1(t) = \frac{1}{.002} \int_0^t -.008e^{-50t} \, dt = 8(e^{-50t} - 1) \text{ mA}, \qquad t > 0$$

## 5.6   INTRODUCTION TO COUPLED INDUCTORS AND THE IDEAL TRANSFORMER

Section 5.5 described how a voltage is induced across a coil of wire by the magnetic field around the wire. The magnetic field was created by the current carried in the coil itself. What would happen if the coil of wire were placed in a time-varying magnetic field produced by another source? Of course, a voltage will still be induced. The source of the magnetic field is not important. All that matters is the strength of the magnetic field that actually links with the coil.

In Figure 5.15, a current in coil 2 creates the magnetic field. Even though there is no current in coil 1, a voltage is induced across it. The voltage induced is proportional to the rate of change of flux of the magnetic field that cuts through the coil. The amount of flux linkage is dependent on the total magnetic field strength of the source coil 2, the distance of coil 2 from coil 1, and their geometric relationship. Expressed mathematically,

$$v_1 = M \frac{di_2}{dt} \qquad (\text{when } i_1 = 0)$$

**FIGURE 5.15**  *Coupled Inductors*

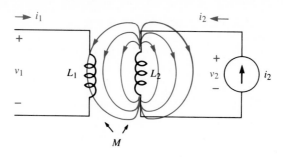

The constant $M$ is known as the **mutual inductance** of the two coils and, like inductance, has units of henrys. If we also provide coil 1 with a current, there will be two sources of magnetic field. Each magnetic source will cause a voltage to be induced in coil 1. The total voltage across coil 1 will be the sum of these two induced voltages. The voltage across coil 2 will also be dependent on the magnetic fields due to coils 1 and 2. These relationships are expressed as follows:

$$v_1 = L_1 \frac{di_1}{dt} + M \frac{di_2}{dt} \qquad \text{(coil 1)} \tag{5.18}$$

$$v_2 = M \frac{di_1}{dt} + L_2 \frac{di_2}{dt} \qquad \text{(coil 2)} \tag{5.19}$$

Technically, $L_1$ and $L_2$ are referred to as **self-inductances** to differentiate them from $M$. When no confusion will result, we drop the term *self*.

The algebraic sign of $M$ depends on whether the two magnetic fields have the same or opposite directions and on the assigned terminal voltages and currents. In circuit diagrams, we indicate the sign of $M$ with the dot notation shown in Figure 5.16. For the current directions and voltage polarities shown, if both $i_1$ and $i_2$ enter or both leave at the dots, then $M$ is positive (Figures 5.16A and 5.16B). If one current enters and the other leaves at the dots, then $M$ is negative (Figures 5.16C and 5.16D).

Mutual inductance is a measure of the coupling between the two coils. Another measure that is often used is the coupling coefficient $k$, where $k$ is given by

$$k = \frac{M}{\sqrt{L_1 L_2}} \tag{5.20}$$

FIGURE 5.16    *Dot Convention*

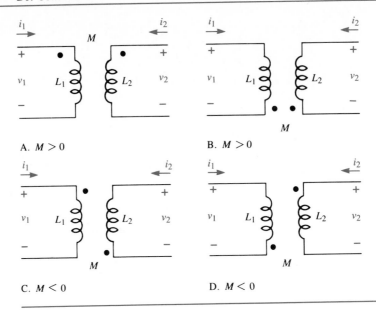

A. $M > 0$                         B. $M > 0$

C. $M < 0$                         D. $M < 0$

    The **transformer** is a device that is built to take advantage of mutual inductance. The simplest transformer is composed of two coils wound around a common iron core. Figure 5.17A is a schematic representation of the transformer; the circuit symbol for the transformer is shown in Figure 5.17B. Note that the dot convention is also used here.

    The transformer is a four-terminal device, which makes it a two-port circuit element. The two-port behavior of the transformer is discussed in detail in Chapter 14. The *V–I* relationships for this device are the same as those given in Equations 5.18 and 5.19 for the coupled coils.

FIGURE 5.17    *Transformer*

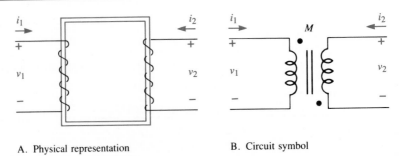

A. Physical representation                    B. Circuit symbol

One of the most important uses for the transformer is to increase or decrease AC voltage. For example, let's assume that $i_2$ in Figure 5.17 is zero—that is, side 2 is open—and find the relationship between $v_1$ and $v_2$. Side 1 is usually referred to as the *primary,* and side 2 as the *secondary.* Note that $M$ is negative in this example circuit. With $i_2 = 0$, the primary and secondary voltages are

$$v_1 = L_1 \frac{di_1}{dt}$$

$$v_2 = M \frac{di_1}{dt}$$

The ratio of $v_2$ to $v_1$ is given by

$$\frac{v_2}{v_1} = \frac{M}{L_1}$$

If we use the coupling coefficient instead of the mutual inductance, we get

$$\frac{v_2}{v_1} = k \frac{\sqrt{L_1 L_2}}{L_1} = k \sqrt{\frac{L_2}{L_1}}$$

If $k = 1$, the transformer is said to be *perfectly coupled.* In most circuits with iron-core transformers, we can usually assume the transformer is perfectly coupled without significant loss of accuracy. The ratio of $v_2$ to $v_1$ for the perfectly coupled transformer is

$$\frac{v_2}{v_1} = \sqrt{\frac{L_2}{L_1}}$$

The ratio of the secondary voltage to the primary voltage is equal to the square root of the ratio of their self-inductances. Since the self-inductance of a coil is proportional to the square of the number of windings that make up the coil, the ratio reduces to

$$\frac{v_2}{v_1} = \frac{n_2}{n_1} \tag{5.21}$$

where $n_2$ and $n_1$ are the number of turns in the secondary and primary, respectively. The ratio of $n_2$ to $n_1$ is known as the *turns ratio.*

If $n_2$ is greater than $n_1$, then the transformer is a *step-up* transformer. That is, the output voltage is greater than the input voltage. If $n_2$ is less than $n_1$, the device is a *step-down* transformer. There are several everyday examples of these devices. The doorbell transformer in a house is used to step down the 120 V line supply to the several volts it takes to ring the bell. The transformers in AC-to-DC converters for hand calculators have a similar function. An induction coil in an automobile is a step-up transformer that increases the low voltage available from the battery to the high voltage required by the spark plugs. Note that since a transformer does not work for DC, a provision is made to turn the battery voltage on and off periodically.

If a perfectly coupled transformer has very large self-inductances and no power losses within the transformer, it is an **ideal transformer.** For an ideal transformer, we are concerned only with the turns ratio and can ignore the self-inductances. If a load is connected to the secondary, as in Figure 5.18, what happens to the current in the ideal transformer? The answer lies in the fact that there is no power loss in an ideal transformer. All the power that is absorbed by the primary is delivered by the secondary so that the total power into the transformer is zero. Expressed mathematically,

$$p_{\text{primary}} + p_{\text{secondary}} = 0$$

so

$$v_1 i_1 + v_2 i_2 = 0$$

giving

$$v_1 i_1 = -v_2 i_2$$

Since $v_1$ and $v_2$ are related by the turns ratio,

$$\frac{i_2}{i_1} = -\frac{v_1}{v_2} = -\frac{n_1}{n_2} \tag{5.22}$$

We see that current is stepped down by the same amount that voltage is stepped up.

A transformer is used in some electronic circuits to change the apparent resistance of a load. For example, let's find the input resistance (primary) of the

**FIGURE 5.18** *Ideal Transformer*

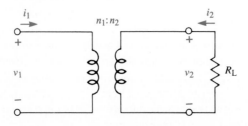

transformer circuit in Figure 5.18. By definition, the input resistance is

$$R_{in} = \frac{v_1}{i_1}$$

If we assume the transformer is ideal, we can substitute for $v_1$ and $i_1$ to get

$$R_{in} = \frac{\frac{1}{n}v_2}{-ni_2}$$

where $n$ is the turns ratio ($n = n_2/n_1$). We get, therefore,

$$R_{in} = -\frac{1}{n^2} \cdot \frac{v_2}{i_2}$$

However, Ohm's law applied at the load in the secondary yields

$$\frac{v_2}{i_2} = -R_L$$

Note the polarity of the voltage and current. We get, finally,

$$R_{in} = \frac{1}{n^2} \cdot R_L \qquad (n = n_2/n_1)$$

(5.23)

---

**Example 5.10**    *Ideal Transformer*

An ideal transformer has $10 \times 10^4$ turns in the primary and 100 turns in the secondary. If the secondary load is 8 $\Omega$, find the equivalent input resistance. The turns ratio $n$ is

$$n = \frac{100}{10 \times 10^4} = \frac{1}{1000}$$

Therefore,

$$R_{in} = 1000^2 \cdot 8 = 8 \text{ M}\Omega$$

---

The simplest way to model the ideal transformer is to use controlled sources. For example, if the turns ratio is $n$, then $v_2 = nv_1$ and $i_1 = -1/ni_2$. One possible controlled source model is shown in Figure 5.19. Caution! This model applies only to time-varying currents and voltages. For DC, no coupling takes place.

FIGURE 5.19    *Ideal Transformer Model (AC Only)*

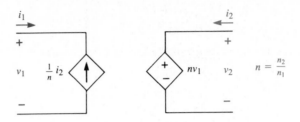

## 5.7  *RLC* CIRCUITS

To analyze circuits that contain capacitors and inductors as well as resistors and sources, we still use the node and mesh analysis techniques of Chapter 3. In later chapters, we will apply equivalent circuit techniques to these networks.

In this section, we will find the equations that describe the behavior of $RLC$ circuits. In Chapter 6, we will solve these equations.

Figure 5.20 shows a single-loop $RC$ circuit in which the output of interest is $v_C$. Usually, we would use KVL to write a mesh equation in terms of $i_C$ for this series circuit. However, since we want to find $v_C$, we will make use of the fact that $i_C = C\, dv_C/dt$ and use $v_C$ as the variable:

$$RC\,\frac{dv_C}{dt} + v_C = v_s \tag{5.24}$$

$$\frac{dv_C}{dt} + \frac{v_C}{RC} = \frac{v_s}{RC} \tag{5.25}$$

A parallel $RL$ circuit is shown in Figure 5.21. In this circuit, the output of interest is $i_L$. Usually, we would apply KCL to write a node equation in terms of $v_L$ for this parallel circuit. However, since we want to find $i_L$, we will make use

FIGURE 5.20    *Series RC Circuit*

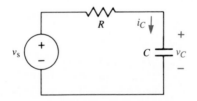

FIGURE 5.21    *Parallel RL Circuit*

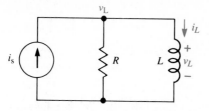

of the fact that $v_L = L \, di_L/dt$ and use $i_L$ as the variable:

$$i_L + \frac{L \, di_L/dt}{R} = i_s \tag{5.26}$$

$$\frac{di_L}{dt} + \frac{R}{L} i_L = \frac{R}{L} i_s \tag{5.27}$$

Example 5.11    *Series RLC Circuit*

Although we usually solve for the current in a series circuit, in the circuit of Figure 5.22 we choose to solve for the capacitor voltage instead. (The reason will become clear in Chapter 6.)

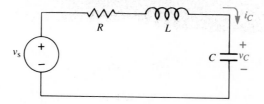

FIGURE 5.22

By KVL,

$$Ri_C + L \frac{di_C}{dt} + v_C = v_s$$

Since the mesh current is $i_C$ and $i_C = C \, dv_C/dt$,

$$RC \frac{dv_C}{dt} + LC \frac{d^2v_C}{dt^2} + v_C = v_s$$

(continues)

**Example 5.11**   *Continued*

Finally, we get

$$\frac{d^2 v_C}{dt^2} + \frac{R}{L}\frac{dv_C}{dt} + \frac{v_C}{LC} = \frac{v_s}{LC}$$

---

**Example 5.12**   *Parallel RLC Circuit*

For reasons that will become clear in Chapter 6, we choose to solve for the inductor current in the circuit of Figure 5.23.

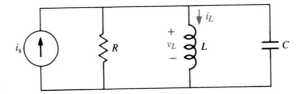

**FIGURE 5.23**

By KCL,

$$\frac{v_L}{R} + i_L + C\frac{dv_L}{dt} = i_s$$

Since the node voltage is $v_L$ and $v_L = L\, di_L/dt$,

$$\frac{L}{R}\frac{di_L}{dt} + i_L + LC\frac{d^2 i_L}{dt^2} = i_s$$

Finally, we get

$$\frac{d^2 i_L}{dt^2} + \frac{1}{RC}\frac{di_L}{dt} + \frac{i_L}{LC} = \frac{1}{LC}i_s$$

---

Analysis of the parallel and series *RLC* circuits in Examples 5.11 and 5.12 results in a single differential equation that can be easily solved. We will solve equations that describe *RLC* circuits in Chapter 6.

In general, we must use node analysis or mesh analysis to find circuit variables in *RLC* networks. The application of these techniques to *RLC* circuits is identical to their application to all resistor circuits. The only differences are the *V–I* characteristics for the inductor and the capacitor.

The analysis of general *RLC* circuits will yield a set of equations that contain integrals and derivatives. These equations are appropriately known as *integro-differential* equations. By differentiating, we can obtain a set of differential equations. A set of differential equations is much more difficult to solve than a single differential equation. The solution of these equations will be discussed in the following chapters.

---

**Example 5.13**   *Node Analysis of RLC Circuit*

Write a set of node equations for the circuit shown in Figure 5.24.

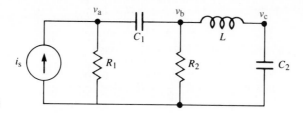

**FIGURE 5.24**

Node a:   $\dfrac{v_a}{R_1} + C_1 \dfrac{d(v_a - v_b)}{dt} = i_s$

Node b:   $C_1 \dfrac{d(v_b - v_a)}{dt} + \dfrac{v_b}{R_2} + \dfrac{1}{L} \displaystyle\int_{-\infty}^{t} (v_b - v_c)\, dt = 0$

Node c:   $\dfrac{1}{L} \displaystyle\int_{-\infty}^{t} (v_c - v_b)\, dt + C_2 \dfrac{dv_c}{dt} = 0$

Differentiating equations at nodes b and c and rearranging yields

Node a:   $\dfrac{v_a}{R_1} + C_1 \dfrac{dv_a}{dt} - C_1 \dfrac{dv_b}{dt} = i_s$

Node b:   $-C_1 \dfrac{d^2 v_a}{dt^2} + C_1 \dfrac{d^2 v_b}{dt^2} + \dfrac{1}{R_2} \dfrac{dv_b}{dt} + \dfrac{1}{L} v_b - \dfrac{1}{L} v_c = 0$

Node c:   $-\dfrac{1}{L} v_b + \dfrac{1}{L} v_c + C_2 \dfrac{d^2 v_c}{dt^2} = 0$

---

**Example 5.14**   *Mesh Analysis of RLC Circuit*

Write a set of mesh equations for the circuit shown in Figure 5.25.

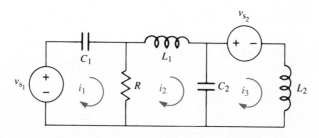

**FIGURE 5.25**

*(continues)*

**Example 5.14** *Continued*

Mesh 1:  $\dfrac{1}{C_1} \displaystyle\int_{-\infty}^{t} i_1 \, dt + R(i_1 - i_2) = v_{s_1}$

Mesh 2:  $R(i_2 - i_1) + L_1 \dfrac{di_2}{dt} + \dfrac{1}{C_2} \displaystyle\int_{-\infty}^{t} (i_2 - i_3) = 0$

Mesh 3:  $\dfrac{1}{C_2} \displaystyle\int_{-\infty}^{t} (i_3 - i_2) \, dt + L_2 \dfrac{di_3}{dt} = -v_{s_2}$

Differentiating and rearranging yields

Mesh 1:  $\dfrac{i_1}{C_1} + R \dfrac{di_1}{dt} - R \dfrac{di_2}{dt} = \dfrac{dv_{s_1}}{dt}$

Mesh 2:  $-R \dfrac{di_1}{dt} + R \dfrac{di_2}{dt} + L_1 \dfrac{d^2 i_2}{dt^2} + \dfrac{1}{C_2} i_2 - \dfrac{1}{C_2} i_3 = 0$

Mesh 3:  $\dfrac{1}{C_2} i_3 - \dfrac{1}{C_2} i_2 + L_2 \dfrac{d^2 i_3}{dt^2} = -\dfrac{dv_{s_2}}{dt}$

---

**Example 5.15** *Coupled-Coil Circuit*

Coupled-coil circuits are most easily analyzed with mesh equations. For the assigned mesh currents in the circuit of Figure 5.26, notice that $i_1$ leaves while $i_2$ enters at the dot. $M$, therefore, is negative in this analysis. Analyze this circuit in two steps. First, write the mesh equations in terms of $v_1$, $i_1$, and $i_2$:

$$v_1 + R_1(i_1 - i_2) = v_s$$
$$R_1(i_2 - i_1) + v_2 + R_2 i_2 = 0$$

Now, replace $v_1$ and $v_2$ with their current relationships:

$$L_1 \dfrac{di_1}{dt} - |M| \dfrac{di_2}{dt} + R_1(i_1 - i_2) = v_s$$

$$R_1(i_2 - i_1) + L_2 \dfrac{di_2}{dt} - |M| \dfrac{di_1}{dt} + R_2 i_2 = 0$$

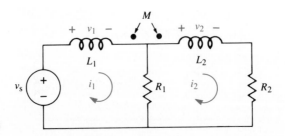

**FIGURE 5.26**

Rearranging, we get

$$L_1 \frac{di_1}{dt} + R_1 i_1 - |M| \frac{di_2}{dt} - R_2 i_2 = v_s$$

$$-|M| \frac{di_1}{dt} - R_1 i_1 + L_2 \frac{di_2}{dt} + (R_1 + R_2) i_2 = 0$$

where $-|M|$ is used to emphasize that $M$ is negative.

## 5.8  SUMMARY

This chapter introduced the capacitor and the inductor. The capacitor stores electrical energy, while the inductor stores magnetic energy. We found the $V$–$I$ characteristics and power and energy relationships for these devices. We saw how the capacitor and the inductor behave at high and low sinusoidal frequency and how they combine in series and in parallel (assuming no initial stored energy). These properties are summarized in Table 5.1.

**TABLE 5.1   *Capacitor and Inductor Relationships***

| Property | Capacitor | Inductor |
|---|---|---|
| $V$–$I$ | $i = C \dfrac{dv}{dt}$ | $v = L \dfrac{di}{dt}$ |
| | $v = \dfrac{1}{C} \displaystyle\int_{-\infty}^{t} i(t)\, dt$ | $i = \dfrac{1}{L} \displaystyle\int_{-\infty}^{t} v(t)\, dt$ |
| | or | or |
| | $v = \dfrac{1}{C} \displaystyle\int_{t_0}^{t} i(t)\, dt + v(t_0)$ | $i = \dfrac{1}{L} \displaystyle\int_{t_0}^{t} v(t)\, dt + i(t_0)$ |
| Series | $\dfrac{1}{C_{eq}} = \dfrac{1}{C_1} + \dfrac{1}{C_2} + \cdots$ | $L_{eq} = L_1 + L_2 + \cdots$ |
| Parallel | $C_{eq} = C_1 + C_2 + \cdots$ | $\dfrac{1}{L_{eq}} = \dfrac{1}{L_1} + \dfrac{1}{L_2} + \cdots$ |
| DC | open | short |
| AC | short (high frequency) | open (high frequency) |
| Power | $Cv \dfrac{dv}{dt}$ | $Li \dfrac{di}{dt}$ |
| Energy | $\dfrac{1}{2} \cdot Cv^2(t)$ | $\dfrac{1}{2} \cdot Li^2(t)$ |

We examined how to write circuit equations for *RC*, *RL*, and *RLC* networks. For an all series or an all parallel network, circuit analysis yields a single differential equation. In Chapters 6 and 7, we will solve these equations. The general *RLC* network results in a simultaneous set of differential equations. The solution to these sets of equations requires some additional mathematical techniques that will be introduced in later chapters.

## ■ PROBLEMS

**5.1** Find the capacitor current if the voltage across a 1 F capacitor is given as follows:

   **a.**  $e^{-5t} \sin 100t$ V

   **b.**  $10t \cos (50t + 30°)$ V

   **c.**  See Figure P5.1A.

   **d.**  See Figure P5.1B.

   **e.**  See Figure P5.1C. (*Hint:* Consider the slopes of the function.)

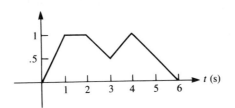
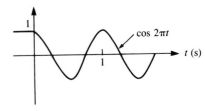

A. Waveform for Problems 5.1 and 5.12c            B. Waveform for Problems 5.1d and 5.12d

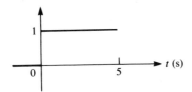

**FIGURE P5.1**   C. Waveform for Problems 5.1e and 5.12e

**5.2** Discuss the consequences of your answer to Problem 5.1e.

**5.3** Find the energy stored in a 10 μF capacitor at $t = 5$ s for each of the voltage waveforms given in Problem 5.1.

**5.4** Find the voltage across a 100 μF capacitor if the capacitor current and initial capacitor voltage are as follows:

   **a.**  $i_C(t) = e^{-5t} \sin 100t$ mA,   $v_C(0) = 0$ V

   **b.**  $i_C(t) = \cos 100t \sin 100t$ mA,   $v_C(0) = 2$ V

   **c.**  $i_C(t) = \sin (10t + 60°)$ mA,   $v_C(2) = 1$ V

   **d.**  See Figure P5.4A.

   **e.**  See Figure P5.4B.

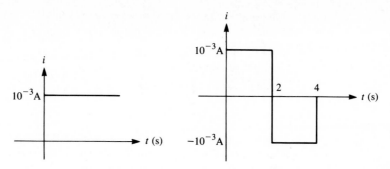

FIGURE P5.4    A. Capacitor current for Problem 5.4d    B. Capacitor current for Problem 5.4e

5.5  For the circuit of uncharged capacitors shown in Figure P5.5,

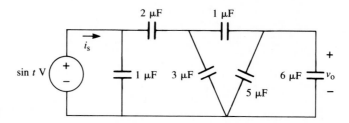

FIGURE P5.5

a.  Find the equivalent capacitance seen by the source.
b.  Find $i_s(t)$ and $v_o(t)$.

5.6  Find $v_o$ as a function of $v_s$ for the uncharged capacitor circuit shown in Figure P5.6. (Your answer should not contain derivatives or integrals.)

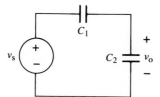

FIGURE P5.6

5.7  Find $i_o$ as a function of $i_s$ for the uncharged capacitor circuit shown in Figure P5.7. (Your answer should not contain derivatives or integrals.)

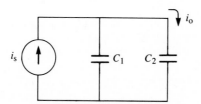

FIGURE P5.7

**5.8** Find the equivalent circuit for the series-connected capacitors with initial conditions shown in Figure P5.8.

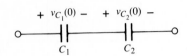

**FIGURE P5.8**

**5.9** The two capacitors shown in Figure P5.9 are connected by a switch at $t = 0$. Just before the switch is closed, there is a voltage stored on each capacitor as shown. Find the equivalent circuit for the parallel-connected capacitors for $t > 0$. (*Hint:* Consider conservation of charge before and after the switch is thrown.)

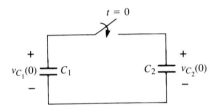

**FIGURE P5.9**

**5.10** For the circuit in Figure P5.9, $C_1 = 2\,\text{F}$ and $C_2 = 4\,\text{F}$. Just before the switch is thrown, $v_{C_1}(0) = 1\,\text{V}$ and $v_{C_2}(0) = .5\,\text{V}$.
   **a.** Find the total energy stored just before the switch is thrown.
   **b.** Find the total energy stored just after the switch is thrown.
   **c.** Compare the answers from parts a and b.

**5.11** For the circuit shown in Figure P5.11, $v(t) = 20e^{-40t}$ mV, $t > 0$, and $v_2(0) = 5$ mV.

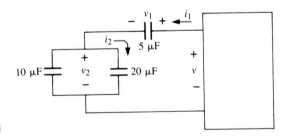

**FIGURE P5.11**

   **a.** Find $v(0)$ and $v_1(0)$.
   **b.** Find $i_1(t)$, $v_1(t)$, $v_2(t)$, and $i_2(t)$.

**5.12** Find the inductor voltage if the current in a 1 H inductor is given as follows:
   **a.** $(10 - 5t)e^{-10t}$ A
   **b.** $5\cos 10t \cos 100t$ A
   **c.** See Figure P5.1A.
   **d.** See Figure P5.1B.
   **e.** See Figure P5.1C. (*Hint:* Consider the slopes of the curve.)

**5.13** Discuss the consequences of your answer to Problem 5.12e.

**5.14** Find the energy stored in a 10 mH inductor at $t = 5$ s, for each of the current waveforms given in Problem 5.12.

**5.15** Find the current in a 100 mH inductor if the voltage across the inductor and the initial current are as follows:

    **a.**  $v_L(t) = 3t^2 + 2t - 5$ V,   $i_L(0) = 0$

    **b.**  $v_L(t) = (20 + 10t)e^{-50t}$ mV,   $i_L(0) = 2$ mA

    **c.**  $v_L(t) = 2t \sin 100t$ V,   $i_L(.02) = 10$ mA

    **d.**  See Figure P5.15A.

    **e.**  See Figure P5.15B.

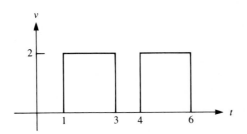

**FIGURE P5.15**    A. Inductor voltage for Problem 5.15d                 B. Inductor voltage for Problem 5.15e

**5.16** For the circuit of uncharged inductors shown in Figure P5.16,

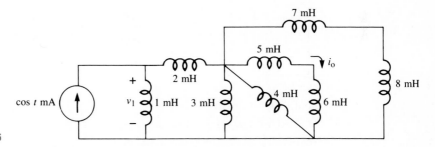

**FIGURE P5.16**

    **a.**  Find the equivalent inductance seen by the source.

    **b.**  Find $v_1(t)$ and $i_o(t)$.

**5.17** Find $i_o$ as a function of $i_s$ for the circuit shown in Figure P5.17. (Your answer should not contain derivatives or integrals.)

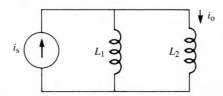

**FIGURE P5.17**

**5.18** Find $v_o$ as a function of $v_s$ for the circuit shown in Figure P5.18. (Your answer should not contain derivatives or integrals.)

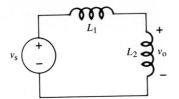

**FIGURE P5.18**

**5.19** Figure P5.19 shows two inductors with initial conditions connected in parallel. Find the equivalent circuit.

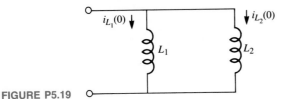

**FIGURE P5.19**

**5.20** For the circuit shown in Figure P5.20, $i_1(t) = 5e^{-10t}$ mA, $t > 0$, and $i_2(0) = 2$ mA.

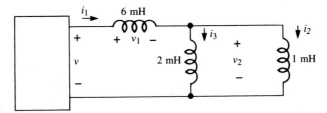

**FIGURE P5.20**

    **a.** Find $i_1(0)$ and $i_3(0)$.
    **b.** Find $v(t)$, $v_1(t)$, $v_2(t)$, $i_2(t)$, and $i_3(t)$.

**5.21** The transformer shown in Figure P5.21 is an ideal transformer. Find $R_{in}$ and $v_o$ if

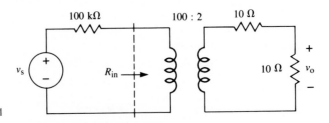

**FIGURE P5.21**

    **a.** $v_s = 10 \cos 1000t$ V
    **b.** $v_s = 10$ V

5.22 Using the ideal op amp analysis, find $v_o$ as a function of $v_{in}$ for
  **a.**  the circuit shown in Figure P5.22A.
  **b.**  the circuit shown in Figure P5.22B.

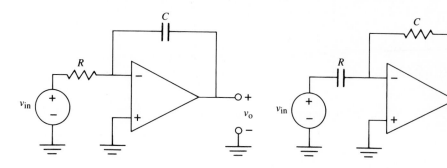

**FIGURE P5.22**   A. Circuit for Problem 5.22a                     B. Circuit for Problem 5.22b

5.23 Refer to the series $RC$ circuit shown in Figure 5.20.
  **a.**  Use KVL to write a differential equation for the mesh current $i_C(t)$.
  **b.**  From your answer to part a, derive the differential equation for $v_C(t)$ (Equation 5.25).
  **c.**  Compare the differential equations in parts a and b.
5.24 Refer to the parallel $RL$ circuit shown in Figure 5.21.
  **a.**  Use KCL to write a differential equation for the node voltage $v_L(t)$.
  **b.**  From your answer to part a, derive the differential equation for $i_L(t)$ (Equation 5.27).
  **c.**  Compare the differential equations in parts a and b.
5.25 Refer to the series $RLC$ circuit of Example 5.11.
  **a.**  Write a differential equation for $i_C(t)$ (use KVL).
  **b.**  Compare the differential equations for $i_C(t)$ and $v_C(t)$.
5.26 Refer to the parallel $RLC$ circuit of Example 5.12.
  **a.**  Write the differential equation for $v_L(t)$ (use KCL).
  **b.**  Compare the differential equations for $v_L(t)$ and $i_L(t)$.
5.27 Write a set of integro-differential node equations for the circuit shown in Figure P5.27.

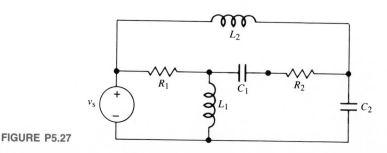

**FIGURE P5.27**

5.28 Write a set of integro-differential mesh equations for the circuit shown in Figure P5.28.

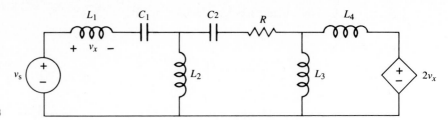

**FIGURE P5.28**

5.29 Write a set of integro-differential mesh equations for the circuit shown in Figure P5.29.

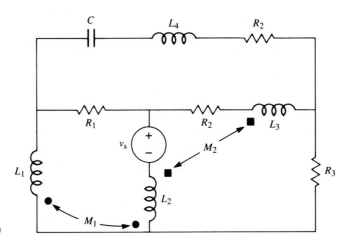

**FIGURE P5.29**

# Analyzing *RLC* Circuits

## 6.1 INTRODUCTION

In Chapter 5, we applied node and mesh analysis to circuits containing resistors, capacitors, inductors, and coupled inductors. For single-capacitor, single-inductor, and series and parallel *RLC* circuits, we were able to derive differential equations in a single variable. Analysis of arbitrary *RLC* circuits will also result in differential equations that have the following general form:

$$a_n \frac{d^n y(t)}{dt^n} + a_{n-1} \frac{d^{n-1} y(t)}{dt^{n-1}} + \cdots + a_1 \frac{dy(t)}{dt} + a_0 y(t) = f(t) \qquad (6.1)$$

where   $n$ = the order of the differential equation

$y(t)$ = a circuit variable such as node voltage or mesh current

$f(t)$ = the forcing function due to independent sources

$a_i$ = the coefficient of the $i$th derivative of $y(t)$

We will standardize all differential equations so that $a_n$ is 1. If all of the coefficients are constant, Equation 6.1 is a linear, time-invariant, ordinary differential equation. While most of the circuits we analyze in this text are described by this type of equation, you should be aware that many circuits and systems are neither linear nor time-invariant. For convenience, we will often drop the $(t)$ for many variables. Unless otherwise noted, all variables will be assumed to be functions of time.

Figure 6.1 shows the *RL* and series *RLC* circuits that we analyzed in Chapter 5. The inductor current in the *RL* circuit shown in Figure 6.1A is described by the first-order differential equation

$$\frac{di_L}{dt} + \frac{R}{L} i_L = \frac{R}{L} i_s$$

which, with substitution, becomes

$$\frac{di_L}{dt} + 5i_L = 5i_s \qquad (6.2)$$

The capacitor voltage in the series *RLC* circuit shown in Figure 6.1B is described by the second-order differential equation

$$\frac{d^2 v_C}{dt^2} + \frac{R}{L} \frac{dv_C}{dt} + \frac{1}{LC} v_C = \frac{1}{LC} v_s$$

which, with substitution, becomes

$$\frac{d^2 v_C}{dt^2} + 10^7 \frac{dv_C}{dt} + 10^7 v_C = 10^7 v_s \qquad (6.3)$$

FIGURE 6.1    *First- and Second-Order Circuits*

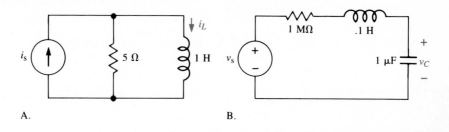

A.                                                                          B.

General *RLC* circuits are described with sets of simultaneous integro-differential equations, which can be difficult to reduce to a differential equation in a single variable. Example 6.1 demonstrates the procedure for a simple *RLC* circuit. Section 6.5 will introduce the differential operator and show how it can be used to simplify the reduction of a set of differential equations.

Example 6.1    *General RLC Circuit*

Write the differential equation for $v_2$ for the circuit shown in Figure 6.2.

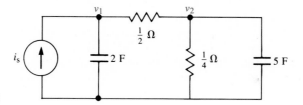

FIGURE 6.2

Write node equations for $v_1$ and $v_2$:

$$2\frac{dv_1}{dt} + 2v_1 - 2v_2 = i_s \tag{1}$$

$$-2v_1 + 6v_2 + 5\frac{dv_2}{dt} = 0 \tag{2}$$

Solve for $v_1$ from equation 2:

$$v_1 = 3v_2 + 2.5\frac{dv_2}{dt} \tag{3}$$

and

$$\frac{dv_1}{dt} = 3\frac{dv_2}{dt} + 2.5\frac{d^2v_2}{dt^2} \tag{4}$$

(*continues*)

**Example 6.1** *Continued*

Substitute equations 3 and 4 into equation 1:

$$6\frac{dv_2}{dt} + 5\frac{d^2v_2}{dt^2} + 6v_2 + 5\frac{dv_2}{dt} - 2v_2 = i_s$$

So,

$$5\frac{d^2v_2}{dt^2} + 11\frac{dv_2}{dt} + 4v_2 = i_s$$

Dividing by 5 yields

$$\frac{d^2v_2}{dt^2} + 2.2\frac{dv_2}{dt} + .8v_2 = .2i_s$$

---

For now, we will assume that an *RLC* circuit has been analyzed and that a differential equation of the form of Equation 6.1 has been found. The solution to this differential equation is composed of two parts: (1) the homogeneous, or *natural*, solution and (2) the particular, or *forced* solution:

$$y(t) = y_N(t) + y_F(t) \tag{6.4}$$

We find the natural solution, $y_N$, by setting the input, $f(t)$ in Equation 6.1, to zero. The forced solution, $y_F$, depends on the form of $f(t)$. In Section 6.2, we will discuss the natural solution, and in the following sections, the forced solution. In each section, the special cases of first- and second-order circuits will be shown. For first-order circuits, solutions can often be derived by inspection of the circuit itself.

## 6.2   THE NATURAL SOLUTION (SOURCE-FREE CIRCUITS)

To find the **natural** solution of a differential equation, we set the input forcing function to zero. If the differential equation is derived from a circuit analysis, then $f(t)$ in Equation 6.1 is a function of the independent energy sources in the circuit. To set $f(t)$ to zero, we turn off all *independent* sources and create a **source-free** circuit. If the current source is turned off in the circuit of Figure 6.1A, for example, then the differential equation describing this source-free circuit is

$$\frac{di_L}{dt} + 5i_L = 0 \tag{6.5}$$

If the voltage source is turned off in the circuit of Figure 6.1B, then the differential equation for this source-free circuit is

$$\frac{d^2v_C}{dt^2} + 10^7 \frac{dv_C}{dt} + 10^7 v_C = 0 \tag{6.6}$$

You might be wondering how any solution can exist when the sources are set to zero. The answer is that initial energy that may be stored in capacitors and inductors is returned to the circuit in the form of branch currents and voltages. The solutions to Equations 6.5 and 6.6 are the natural solutions for $i_L$ and $v_C$, respectively. You should be aware that several terms are used for this type of solution. You may already be familiar from math courses with the term *homogeneous* solution. The term *natural* is used in this text because these solutions depend only on the nature of the source-free circuit. That is, the natural solution depends only on the types of elements in the circuit (excluding independent sources), their interconnections, and any initially stored energy. The term *transient* solution is also used to describe the natural solution. Shortly, you will see why this term is used and why it does not always properly describe the natural solution.

If a circuit has no sources, the voltages and currents that are found will be natural solutions and will also be the total solutions. For circuits with sources, the natural solutions are only part of the total solution.

## First-Order Circuits

The natural solution of a first-order differential equation is easy to find. Consider the *RL* circuit shown in Figure 6.3. The first-order differential equation for this source-free circuit is the same as the one given in Equation 6.5 and is rewritten here:

$$\frac{di_L}{dt} + 5i_L = 0 \tag{6.7}$$

FIGURE 6.3   *Source-Free RL Circuit*

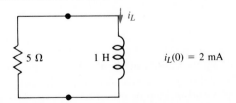

$5\ \Omega$     $1\ H$     $i_L(0) = 2$ mA

One way to solve this equation is to separate the variables $i_L$ and $t$ as follows:

$$\frac{di_L}{dt} = -5i_L$$

$$\frac{di_L}{i_L} = -5\,dt$$

Integrating both sides of this equation yields

$$\ln i_L = -5t + C$$

where $C$ is the constant of integration. If we let $C = \ln K$ (since $C$ and $K$ are both constants, we have not changed anything),

$$\ln i_L = -5t + \ln K$$

$$\ln i_L - \ln K = -5t$$

$$\ln \frac{i_L}{K} = -5t$$

Taking the antilogarithm on both sides of the equation leads to the final equation:

$$i_{L_N} = Ke^{-5t} \tag{6.8}$$

where the subscript N is used to emphasize that we are finding the natural solution.

This solution is easily verified by substituting $i_{L_N}$ into the original differential equation. Note that this solution is valid for any value of $K$. To find $K$, we must know the value of $i_L$ at some time, usually at $t = 0$. Suppose we are told that at $t = 0$, $i_L = 2\,\text{mA}$. Since this is a source-free circuit, $i_{L_N}$ is the total solution and must equal $2\,\text{mA}$ at $t = 0$. Therefore,

$$i_{L_N}(0) = i_L(0) = 2\,\text{mA} = Ke^{-5\cdot0} = K$$

so

$$K = 2\,\text{mA}$$

and

$$i_L(t) = 2e^{-5t}\,\text{mA}, \qquad t \geqslant 0 \tag{6.9}$$

Note that we do not know the behavior of $i_L(t)$ for $t < 0$. All we know is the value of $i_L(t)$ at $t = 0$. Therefore, we can determine $i_L(t)$ only from this time on. This value of $i_L(t)$ is expressed in Equation 6.9 with $t \geqslant 0$. The value of $i_L(t)$ at $t = 0$ is known as the **initial value** of $i_L(t)$.

You can see that the natural solution of a first-order differential equation is an *exponential*. Let's look at this very important function of time in more detail (Figure 6.4). The general form for the exponential is

$$y(t) = Ke^{\sigma t} \tag{6.10a}$$

or

$$y(t) = Ke^{-t/\tau} \tag{6.10b}$$

The constant $K$ gives the value of the exponential at $t = 0$. Equations 6.10a and 6.10b show the two ways of expressing the exponential. In both cases, the power of $e$ must be dimensionless. That is, $a$ must have units of inverse seconds $(s^{-1})$, and $\tau$ must have units of seconds (s). A variable with units of $s^{-1}$ is a frequency. The variable $a$ is known as the **neper frequency** of the exponential. The variable $\tau$ is measured in seconds and is known as the **time constant** of the exponential. It is obvious that

$$\sigma = \frac{-1}{\tau} \tag{6.11}$$

Figure 6.4 shows the behavior of the exponential for $\sigma < 0$ (Figure 6.4A), $\sigma = 0$ (Figure 6.4B), and $\sigma > 0$ (Figure 6.4C). For a negative neper frequency, the exponential decays with time. If the neper frequency is 0, then the exponential is constant. For a positive neper frequency, the exponential grows toward infinity. You can see that the solution we found for the parallel $RL$ circuit (Equation 6.9) was an exponential with a neper frequency of $-5\,s^{-1}$ and a time constant of .2 s. Since the neper frequency is negative, $i_L(t)$ decays with time. It is transient in its behavior. Now you can see why the natural solution is also known as the transient solution.

How fast does an exponential decay? To find out, let's return to the general

FIGURE 6.4　*The Exponential*

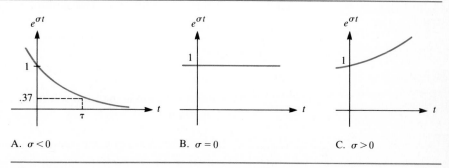

A. $\sigma < 0$　　　　B. $\sigma = 0$　　　　C. $\sigma > 0$

exponential of Equation 6.10b and find $y(t)$ for various values of $t$. The easiest values of $t$ to use are integer multiples of the time constant $\tau$. For values of $t$ from 0 to $5\tau$,

$$y(0) = Ke^{0} = K$$

$$y(\tau) = Ke^{-1} = .37K$$

$$y(2\tau) = Ke^{-2} = .14K$$

$$y(3\tau) = Ke^{-3} = .05K \tag{6.12}$$

$$y(4\tau) = Ke^{-4} = .02K$$

$$y(5\tau) = Ke^{-5} = .007K$$

The initial value of $y$ is $K$. The exponential decays to 37% of its initial value in one time constant. Therefore, $\tau$ may be measured from experimental data as shown in Figure 6.4A. The exponential decays to 2% of its initial value in four time constants and to 0.7% of its initial value in five time constants. A rule of thumb that most engineers use is that an exponential dies out to approximately zero in four-to-five time constants.

An important property of the exponential is that its derivative is also an exponential with the same time constant. That is,

$$\frac{dy(t)}{dt} = \frac{d(Ke^{-t/\tau})}{dt} = -\frac{K}{\tau}e^{-t/\tau} \tag{6.13}$$

This property is the basis for the solution technique for high-order source-free circuits.

If you return to the natural solution of the differential equation (Equation 6.7) and examine the final result (Equation 6.8), you can see that the answer can be derived by inspection. This simple procedure is done as follows. First, write the differential equation in the form of

$$\frac{dy(t)}{dt} + ay(t) = 0 \tag{6.14}$$

Note that the coefficient of $dy/dt$ is 1. The natural solution is given simply by

$$y_{N}(t) = Ke^{-at} \tag{6.15}$$

If, *and only if*, the natural solution is the total solution, and the initial value of $y$ is given (e.g., $y(0) = Y_0$), then

$$y(t) = y_{N}(t) = Y_{0}e^{-at}, \qquad t \geqslant 0 \tag{6.16}$$

**Example 6.2**  *First-Order Examples*

a.  Given $dy/dt + 6y = 0$ and $y(0) = .5$, find $y(t)$.

$$y(t) = .5e^{-6t}, \qquad t \geq 0$$

b.  Given $3\,dy/dt + 5y = 0$ and $y(0) = -1$, find $y(t)$.

$$\frac{dy}{dt} + \frac{5}{3}\,y = 0$$

$$y(t) = -1e^{-5/3t}, \qquad t \geq 0$$

c.  Given the *RC* circuit in Figure 6.5A, find $v_C(t)$.

$$C\,\frac{dv_C}{dt} + \frac{v_C}{R} = 0$$

$$\frac{dv_C}{dt} + \frac{v_C}{RC} = 0$$

$$v_{C_N} = Ke^{-t/RC}$$

$$v_C(t) = V_0 e^{-t/RC} \text{ V}, \qquad t \geq 0$$

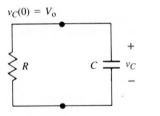
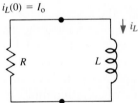

**FIGURE 6.5**  A. *RC* circuit          B. *RL* circuit

d.  Given the *RL* circuit in Figure 6.5B, find $i_L(t)$.

$$L\,\frac{di_L}{dt} + Ri_L = 0$$

$$\frac{di_L}{dt} + \frac{R}{L}\,i_L = 0$$

$$i_{L_N} = Ke^{-Rt/L}$$

$$i_L(t) = I_0 e^{-Rt/L} \text{ A}, \qquad t \geq 0$$

The example circuits of Figure 6.5 show how easy it is to find the natural solution of an *RC* or an *RL* circuit. For an *RC* circuit, the time constant is simply *RC*. For an *RL* circuit, the time constant is *L/R*. That is,

**RC Circuit**

$$\tau = RC \tag{6.17}$$

**RL Circuit**

$$\tau = \frac{L}{R} \tag{6.18}$$

In fact, any circuit that contains just one energy storage device, either a capacitor or an inductor, can be analyzed without writing differential equations. Consider, for example, the circuit in Figure 6.6A.

If we want to find the solution for $v_C$, we can write node or mesh equations, or, better still, we can use the Thévenin equivalent approach. Figure 6.6B shows the result of taking a Thévenin equivalent across the terminals of the capacitor. We then have a simple *RC* circuit and can immediately write the response of $v_C$:

$$v_C(t) = v_{C_N} = 2e^{-t/10} \text{ V}, \qquad t \geq 0$$

since $\tau = 2 \cdot 5 = 10$ s and $v_C(0) = 2$ V.

Finding the voltage across a capacitor in a source-free first-order circuit, therefore, requires finding the Thévenin resistance that the capacitor "sees". The time constant for such a circuit is just $\tau = R_{eq}C$.

What if we want to find the natural solution for a circuit variable other than the capacitor voltage? For example, to find $v_a$ in the circuit of Figure 6.6A, we could write a differential equation for $v_a$. We would then need a technique to find

**FIGURE 6.6    *First-Order RC Circuit Analysis***

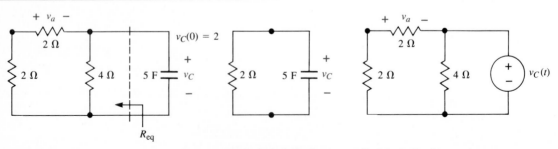

A.  Original circuit                          B.  Equivalent circuit          C.  Substitution for capacitor

$v_a(0)$ since it is not given. This procedure is examined in problems at the end of the chapter.

We will, however, always first find the capacitor voltage from the equivalent circuit (Figure 6.6B). Since we now know the value of $v_C(t)$ for all $t > 0$, we can replace the capacitor in the original circuit with a voltage source with a value of $v_C(t)$. This technique is known formally as the **substitution theorem.** We then proceed to analyze the all-resistor circuit for the variable of interest. This type of analysis is the basis for a modern technique known as **state–space analysis,** in which capacitor voltages and inductor currents are the variables used to write circuit equations.

In the circuit of Figure 6.6C, we have substituted the $2e^{-t/10}$ V voltage source for the 5 F capacitor (be careful of polarity). We use the voltage divider relationship to find

$$v_a(t) = \frac{2}{4} \cdot 2e^{-t/10} = e^{-t/10} \text{ V}, \qquad t \geq 0$$

You can see that both $v_C$ and $v_a$ have the same time constant. All passive-element circuit variables in a given first-order circuit have the same time constant. This statement may not hold for circuits that contain controlled sources.

In this circuit, the 4 Ω resistor does not affect the magnitude of $v_a$ since it is in parallel with the capacitor voltage source. When you use the substitution theorem in this way, look for such occurrences. They can help to simplify analysis.

For circuits that contain only a single inductor, we find the Norton equivalent that the inductor "sees". The use of the Norton equivalent will become apparent when we consider circuits with sources. Of course, the Norton equivalent resistance is exactly the same as the Thévenin equivalent resistance. To find other branch variables, we replace the inductor with a *current source* equal to $i_L(t)$ and use resistive circuit techniques.

---

**Example 6.3** *Circuit with Single Inductor*

Find $i_L(t)$ and $i_a(t)$ for the circuit shown in Figure 6.7A.

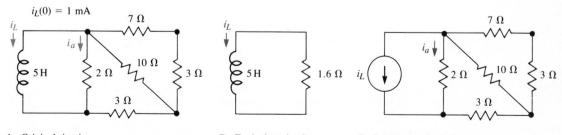

A. Original circuit  B. Equivalent circuit  C. Substitution for inductor

**FIGURE 6.7**

*(continues)*

**Example 6.3** *Continued*

The equivalent resistance seen by the inductor is

$$R_{eq} = 1.6 \, \Omega$$

From the equivalent circuit shown in Figure 6.7B,

$$i_{L_N} = Ke^{-.32t}$$

where $\tau = 5/1.6 = 3.13 \, \text{s}$.

Since $i_L(0) = 1 \, \text{mA}$,

$$i_L(t) = e^{-.32t} \, \text{mA}, \qquad t \geqslant 0$$

To find $i_a$, replace the inductor with a current source as shown in Figure 6.7C:

$$i_a(t) = -.8 i_L(t) = -.8 e^{-.32t} \, \text{mA}, \qquad t \geqslant 0$$

---

## Second-Order Circuits

The analysis of circuits that contain more than one energy storage device usually results in higher-order differential equations. Figure 6.1B shows such a circuit. Its second-order differential equation is given in Equation 6.3; the unforced equation for this circuit is given in Equation 6.6.

How can we find the natural solution to a second-order unforced differential equation? While separation of variables works easily for first-order differential equations, it does not work for higher-order equations. There is, however, a general technique for finding the natural solution of differential equations. It is based on the fact that to satisfy a linear homogeneous differential equation, a function and its derivatives must be of the same form. The exponential $e^{st}$ meets this requirement.

Consider the following second-order differential equation:

$$\frac{d^2y}{dt^2} + 5\frac{dy}{dt} + 6y = 0 \qquad \qquad (6.19)$$

Let's assume that the natural solution has the form

$$y_N = Ke^{st}$$

where $s$ is to be determined. We have to take the first and second derivatives of $y_N$ and substitute them into the differential equation:

$$\frac{dy_N}{dt} = sKe^{st}$$

and

$$\frac{d^2y_N}{dt^2} = s^2Ke^{st}$$

Again, note that the derivative of an exponential is also an exponential. It is this fact that permits us to find a solution. Substituting these relationships into the differential equation gives

$$s^2Ke^{st} + 5sKe^{st} + 6Ke^{st} = 0$$

Factoring out $Ke^{st}$, which appears in each term, yields

$$(s^2 + 5s + 6)Ke^{st} = 0$$

Since $Ke^{st} \neq 0$ for nonzero $K$, we can divide both sides by this term to get

$$s^2 + 5s + 6 = 0 \tag{6.20}$$

This equation is known as the **characteristic equation** of the circuit. The roots of Equation 6.20 are

$$s_1 = -2 \quad \text{and} \quad s_2 = -3$$

There are two roots since this equation is a second-order polynomial equation. These roots are known as the **natural frequencies** of the system. So, $y_N = K_1e^{-2t}$ and $y_N = K_2e^{-3t}$ must satisfy the unforced equation (Equation 6.19). The total natural solution is, therefore,

$$y_N = K_1e^{-2t} + K_2e^{-3t} \tag{6.21}$$

You should confirm the correctness of this answer by substituting $y_N$ into Equation 6.19.

It is not necessary to go through all of the steps just described to find the characteristic equation of the system. If you compare the differential equation (Equation 6.19) to the characteristic equation (Equation 6.20), you will see a very simple relationship. To get the characteristic equation from a differential equation,

$$\begin{array}{ll}
\text{replace:} & \text{with:} \\
y & s^0 = 1 \\
dy/dt & s^1 = s \\
d^2y/dt^2 & s^2 \\
\vdots & \vdots \\
d^ny/dt^n & s^n
\end{array}$$

Several examples are shown in Table 6.1. Note that the first-order differential equation has a single, easily found, natural frequency (see example 1, Table 6.1). We already know that the natural solution to this first-order equation is $y_N = Ke^{-4t}$. We have merely confirmed that the characteristic equation technique applies to first-order differential equations.

The second-order differential equation has two natural frequencies that can be distinct, identical, or complex conjugates (see examples 2, 3, and 4, Table 6.1).

**TABLE 6.1**  *Examples of Differential Equations and Their Characteristic Equations*

| Differential Equation | Characteristic Equation | Natural Frequencies |
|---|---|---|
| (1) $\dfrac{dy}{dt} + 4y = 0$ | $s + 4 = 0$ | $s = -4$ |
| (2) $\dfrac{d^2y}{dt^2} + 8\dfrac{dy}{dt} + 15y = 0$ | $s^2 + 8 + 15 = 0$ <br> or <br> $(s+3)(s+5) = 0$ | $s_{1,2} = -3, -5$ |
| (3) $\dfrac{d^2y}{dt^2} + 6\dfrac{dy}{dt} + 9y = 0$ | $s^2 + 6s + 9 = 0$ <br> or <br> $(s+3)(s+3) = 0$ | $s_{1,2} = -3, -3$ |
| (4) $\dfrac{d^2y}{dy^2} + 6\dfrac{dy}{dt} + 10y = 0$ | $s^2 + 6s + 10 = 0$ <br> or <br> $(s+3-j)(s+3+j) = 0$ | $s_1 = -3 + j$ <br> $s_2 = -3 - j$ |
| (5) $\dfrac{d^3y}{dt^3} + 5\dfrac{d^2y}{dt^2} + 11\dfrac{dy}{dt} + 15y = 0$ | $s^3 + 5s^2 + 11s + 15 = 0$ <br> or <br> $(s+3)(s^2 + 2s + 5) = 0$ | $s_1 = -3$ <br> $s_2 = -1 + j2$ <br> $s_3 = -1 - j2$ |

We use the letter $j$ instead of $i$ to represent $\sqrt{-1}$, since $i$ is reserved for currents. Appendix C has a review of complex arithmetic, which will be used extensively from now on. The second-order differential equation is important for two reasons. First, it provides the building block, along with the first-order equation, for higher-order equations (see example 5, Table 6.1). Second, even though many systems are of higher order, they often can be reasonably approximated with a second-order system.

The general second-order differential equation can be written using several standard forms. Two of them are as follows:

$$\frac{d^2y}{dt^2} + 2\alpha \frac{dy}{dt} + \omega_n^2 y = 0 \tag{6.22a}$$

and

$$\frac{d^2y}{dt^2} + 2\zeta\omega_n \frac{dy}{dt} + \omega_n^2 y = 0 \tag{6.22b}$$

where $\omega_n$ = undamped natural frequency

$\alpha$ = damping coefficient

$\zeta$ = damping ratio

Obviously, $\alpha = \zeta\omega_n$. The characteristic equation for the second formulation (Equation 6.22b) is

$$s^2 + 2\zeta\omega_n s + \omega_n^2 = 0 \tag{6.23}$$

The characteristic equation is factored to get the natural frequencies:

$$s_1 = -\zeta\omega_n + \omega_n\sqrt{\zeta^2 - 1} \tag{6.24a}$$

and

$$s_2 = -\zeta\omega_n - \omega_n\sqrt{\zeta^2 - 1} \tag{6.24b}$$

Table 6.1 showed examples of second-order characteristic equations that had two distinct real roots, two multiple real roots, and two complex roots. If you examine the solutions in Equations 6.24a and 6.24b, you can see that the three cases depend entirely on the damping ratio, $\zeta$.

**Case 1:** If $\zeta > 1$, the term inside the radical is positive, and we get the two real roots given in Equations 6.24a and 6.24b. For reasons that will be explained later, we say that this is an **overdamped** response.

**Case 2:**  If $\zeta = 1$, the radical term is zero, and $s$ has two real roots at

$$s_{1,2} = -\zeta\omega_n \tag{6.25}$$

This is known as a **critically damped** response.

**Case 3:**  If $\zeta < 1$, the term inside the radical is negative, and we get complex roots. If we write $\sqrt{\zeta^2 - 1}$ as $\sqrt{-(1 - \zeta^2)} = j\sqrt{1 - \zeta^2}$, then

$$s_1 = -\zeta\omega_n + j\omega_n\sqrt{1 - \zeta^2}$$

and

$$s_2 = -\zeta\omega_n - j\omega_n\sqrt{1 - \zeta^2} \tag{6.26}$$

This is known as an **underdamped** response.

For the circuits that we will be analyzing—in fact, for any system with real coefficients in the differential equation—complex roots always occur as complex conjugate pairs. Since the form of the natural solution depends on the roots of the characteristic equation, we will examine the three cases one at a time. We will not evaluate the constants that appear in the natural solution until after we have presented all three cases. We have, of course, already found the natural solution for a characteristic equation that has two real roots (Equation 6.21). This case is included again for completeness.

**Case 1:**  *Distinct Real Roots (Overdamped).*

$$\frac{d^2y}{dt^2} + 10\frac{dy}{dt} + 16y = 0 \tag{6.27}$$

The characteristic equation is

$$s^2 + 10s + 16 = 0$$

For this system, $\omega_n = 4$ and $\zeta = 1.25$, which is greater than 1. The natural frequencies of the characteristic equation must, therefore, be real and distinct. They are as follows:

$$s_1 = -8 \quad \text{and} \quad s_2 = -2$$

The natural solution is

$$y_N = K_1 e^{-8t} + K_2 e^{-2t}$$

We confirm that this answer is correct by substituting $y_N$ into Equation 6.27:

$$\frac{dy_N}{dt} = -8K_1e^{-8t} - 2K_2e^{-2t}$$

and

$$\frac{d^2y_N}{dt^2} = 64K_1e^{-8t} + 4K_2e^{-2t}$$

Substituting in Equation 6.27, we get

$$64K_1e^{-8t} + 4K_2e^{-2t} - 80K_1e^{-8t} - 20K_2e^{-2t} + 16K_1e^{-8t} + 16K_2e^{-2t}$$

$$= (64 - 80 + 16)K_1e^{-8t} + (4 - 20 + 16)K_2e^{-2t}$$

$$= 0 + 0 = 0$$

Figure 6.8 shows what the natural solution of the overdamped case looks like for several values of $K_1$ and $K_2$. You can see that $y_N$ starts at $K_1 + K_2$ and goes to zero as time increases. As in the first-order solution, the natural solution here is transient. Here, there are two exponentials, so there are two time constants: $\tau_1 = 1/8$ s and $\tau_2 = 1/2$ s. The natural response will completely die out only when both exponentials have died out. Thus, the natural solution dies out in $5\tau_2 = 2.5$ s.

In general, the natural solution for a second-order differential equation with two distinct real roots, $s_1$ and $s_2$, is

*Case 1*

$$y_N = K_1e^{s_1t} + K_2e^{s_2t}$$

(6.28)

FIGURE 6.8    ***Overdamped Second-Order Response***

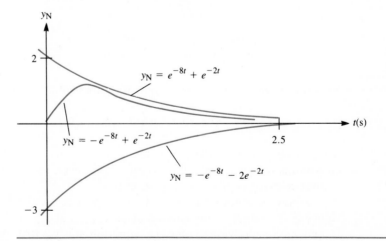

**Case 2:** *Multiple Real Roots (Critically Damped).*

$$\frac{d^2y}{dt^2} + 10\frac{dy}{dt} + 25y = 0 \tag{6.29}$$

The characteristic equation is

$$s^2 + 10s + 25 = 0$$

For this system, $\omega_n = 5$ and $\zeta = 1$. This characteristic equation has two roots at $s = -5$. If we try to apply the previous rule to this case, we get

$$y_N = K_1 e^{-5t} + K_2 e^{-5t}$$

which equals

$$y_N = (K_1 + K_2)e^{-5t}$$

Since the sum of $K_1 + K_2$ is just another constant, we have found only one solution to the homogeneous equation. A second-order differential equation must have two independent solutions. It can be shown that the second solution is found by multiplying one of the exponentials by $t$. That is,

$$y_N = K_1 e^{-5t} + K_2 t e^{-5t} = (K_1 + K_2 t)e^{-5t} \tag{6.30}$$

We already know that $K_1 e^{-5t}$ satisfies Equation 6.29. Let's see whether $K_2 t e^{-5t}$ works. Differentiating the product of two functions of time yields

$$\frac{dK_2 t e^{-5t}}{dt} = (K_2 - 5K_2 t)e^{-5t}$$

and

$$\frac{d^2 K_2 t e^{-5t}}{dt^2} = (-5K_2 - 5K_2 + 25K_2 t)e^{-5t} = (-10 + 25t)K_2 e^{-5t}$$

Substituting these relationships into the differential equation, we get

$$K_2(-10 + 25t + 10 - 50t + 25t)e^{-5t} = 0$$

Since the terms inside the parentheses sum to zero, the homogeneous differential equation is satisfied, and we have found the second solution. Therefore, the total natural solution is given by Equation 6.30.

Figure 6.9 shows the natural solution of the critically damped case for two values of $K_1$ and $K_2$. The solution always starts at $K_1$ and goes to zero with time. It may not be obvious from the equation for $y_N$ that this natural solution is

FIGURE 6.9   *Critically Damped Second-Order Response*

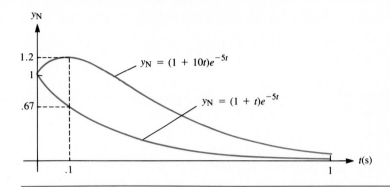

transient since it contains the term $t$. However, the exponential dies out faster than $t$ grows, so the total solution does, indeed, die out. Case 2 natural solutions have only one exponential, so there is only one time constant: $\tau = 1/5$ s. This solution dies out in approximately 1 s.

In general, the natural solution for a second-order differential equation with two identical real roots, $s_1$, is

*Case 2*

$$y_N = (K_1 + K_2 t)e^{s_1 t} \tag{6.31}$$

**Case 3:** *Complex Roots (Underdamped).*

$$\frac{d^2 y}{dt^2} + 4\frac{dy}{dt} + 148y = 0 \tag{6.32}$$

The characteristic equation is

$$s^2 + 4s + 148 = 0$$

For this system, $\omega_n = \sqrt{148}$ and $\zeta = 2/\sqrt{148}$, which is less than 1. The roots of the characteristic equation are, therefore, complex and given by

$$s_1 = -2 + j12 \quad \text{and} \quad s_2 = -2 - j12$$

Again, note that $j$ is the square root of $-1$ and that the two roots are complex conjugates. Even though the roots are complex, they are distinct and can be treated as in case 1:

$$y_N = K_1 e^{(-2+j12)t} + K_2 e^{(-2-j12)t}$$

Let's see whether this solution works:

$$148y_N = 148K_1e^{(-2+j12)t} + 148K_2e^{(-2-j12)t}$$

$$+4\frac{dy_N}{dt} = 4(-2+j12)K_1e^{(-2+j12)t} + 4(-2-j12)K_2e^{(-2-j12)t}$$

$$+\frac{d^2y_N}{dt^2} = (-140-j48)K_1e^{(-2+j12)t} + (-140+j48)K_2e^{(-2-j12)t}$$

$$0 = 0 + 0$$

Complex conjugate roots do obey case 1 rules. Why should another case be introduced then? The solution to a real differential equation must be real. As presented, the case 3 solution does not appear to be real and is not easily visualized. However, we can manipulate the complex exponential to derive a recognizable real function. The basis for finding a case 3 solution is a formula derived by the German mathematician Leonhard Euler (1707–1783):

$$e^{j\omega t} = \cos \omega t + j \sin \omega t$$

or                                                                                        (6.33)

$$e^{-j\omega t} = \cos \omega t - j \sin \omega t$$

Let's return to the case 3 exponential solution:

$$y_N = K_1e^{(-2+j12)t} + K_2e^{(-2-j12)t}$$

Factoring out the $e^{-2t}$ term gives

$$y_N = K_1e^{-2t}e^{j12t} + K_2e^{-2t}e^{-j12t} = e^{-2t}(K_1e^{j12t} + K_2e^{-j12t})$$

Now, we apply Euler's formula to both complex exponentials:

$$y_N = e^{-2t}[K_1(\cos 12t + j \sin 12t) + K_2(\cos 12t - j \sin 12t)]$$

$$= e^{-2t}[(K_1 + K_2)\cos 12t + j(K_1 - K_2)\sin 12t]$$

We can introduce two new constants at this point:

$$A = K_1 + K_2 \quad \text{and} \quad B = j(K_1 - K_2)$$

It may appear that $B$ is a complex number. However, it can be shown that $K_1$ and $K_2$ are complex conjugates so that both $A$ and $B$ are real numbers. In any case, they are just another set of constants that need to be evaluated at a later time. Using these new constants, we get the natural solution:

$$y_N = e^{-2t}(A \cos 12t + B \sin 12t) \tag{6.34}$$

Note that from the characteristic equation, $\zeta\omega_n = 2$ and $\omega_n\sqrt{1 - \zeta^2} = 12$.

Figure 6.10 shows the natural solution of the underdamped case. Once again, the natural solution is a transient solution. This function is a damped sinusoid. If the peaks of the sinusoid are connected with a dashed line, what is known as the *envelope* of the damped sinusoid is formed. The envelope here is an exponential with the neper frequency of $-\zeta\omega_n$ s$^{-1}$, while the sinusoid has a radian frequency of $\omega_n\sqrt{1 - \zeta^2}$ rad/s. Case 3 natural solutions have only a single real exponential, so there is one time constant, which is given by $\tau = 1/\zeta\omega_n$. The natural solution here dies out in $5/(\zeta\omega_n)$ s.

In general, the natural solution for a second-order differential equation with complex natural frequencies $-\zeta\omega_n \pm j\omega_n\sqrt{1 - \zeta^2}$ is

*Case 3*

$$y_N = e^{-\zeta\omega_n t}(A \cos \omega_n\sqrt{1 - \zeta^2}t + B \sin \omega_n\sqrt{1 - \zeta^2}t) \tag{6.35}$$

A good example of a second-order mechanical system is the suspension system of an automobile. This system is composed of the mass of the car, the springs, and the shock absorbers. The mass and springs store mechanical energy and, hence, are analogous to the capacitors and inductors in electrical circuits. The shock absorbers dissipate energy and are analogous to the resistors. Mechanical devices that dissipate energy are often known as *dampers*.

**FIGURE 6.10**   *Underdamped Second-Order Response*

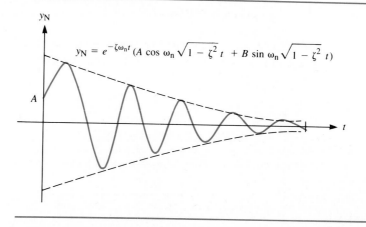

An automobile with bad shock absorbers will oscillate if the springs are compressed and then released. The shock absorbers do not provide enough damping to the suspension system; it is underdamped. If the springs on a car with good shock absorbers are compressed, the car slowly returns to its resting position without overshooting or oscillating. This system has a lot of damping; it is overdamped. Critical damping is simply the boundary between under- and overdamping. An undamped system ($\zeta = 0$) has no means for dissipating energy and will oscillate forever when disturbed.

---

**Example 6.4**     *Undamped Second-Order System*

Find $\zeta$, $\omega_n$, and the natural solution for

$$\frac{d^2 y}{dt^2} + 16y = 0$$

The characteristic equation is

$$s^2 + 16 = 0$$

So,

$$\omega_n = 4 \qquad \text{and} \qquad \zeta = 0$$

Since the damping coefficient is zero, this is an undamped system. The roots of this system are

$$s_1 = j4 \qquad \text{and} \qquad s_2 = -j4$$

and the natural solution is

$$y_N = A \cos 4t + B \sin 4t$$

The natural solution is shown in Figure 6.11. Notice from the plot of $y_N$ that this solution is not transient. The natural response is a pure sinusoid

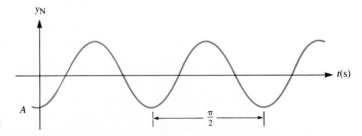

**FIGURE 6.11**

with a radian frequency of $\omega_n = 4\,\text{rad/s}$. We can now see that an undamped second-order system will oscillate at its undamped natural frequency, $\omega_n$. Zero damping occurs in electrical circuits that have no net resistance.

**Example 6.5**    *Parallel RLC circuit*

Find the values of $R$ for which $i_L(t)$ in Figure 6.12 will be over, critically and underdamped.

$$\frac{d^2 i_L}{dt^2} + \frac{1}{RC}\frac{di_L}{dt} + \frac{i_L}{LC} = 0$$

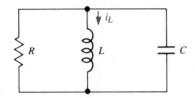

**FIGURE 6.12**

The characteristic equation is

$$s^2 + \frac{1}{RC}s + \frac{1}{LC} = 0$$

So,

$$\omega_n = \sqrt{\frac{1}{LC}} \qquad \text{and} \qquad \zeta = \frac{1}{2R}\sqrt{\frac{L}{C}}$$

The undamped natural frequency, $\omega_n$, is a function of $L$ and $C$. If $L$ and $C$ are given, then the damping ratio, $\zeta$, depends on $R$.

Case 1:   If $\zeta > 1$, then $R < (1/2)\sqrt{L/C}$.
Case 2:   If $\zeta = 1$, then $R = (1/2)\sqrt{L/C}$.
Case 3:   If $\zeta < 1$, then $R > (1/2)\sqrt{L/C}$.

The parallel *RLC* circuit is undamped if the resistor is removed $(R \to \infty)$.

**Example 6.6**    *Series RLC Circuit*

Find the values of $R$ for which $v_C(t)$ in Figure 6.13 will be over, critically and underdamped.

$$\frac{d^2 v_C}{dt^2} + \frac{R}{L}\frac{dv_C}{dt} + \frac{1}{LC}v_C = 0$$

*(continues)*

Example 6.6    *Continued*

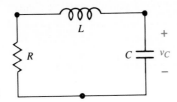

FIGURE 6.13

The characteristic equation is

$$s^2 + \frac{R}{L}s + \frac{1}{LC} = 0$$

So,

$$\omega_n = \sqrt{\frac{1}{LC}} \quad \text{and} \quad \zeta = \frac{R}{2}\sqrt{\frac{C}{L}}$$

The undamped natural frequency, $\omega_n$, is a function of $L$ and $C$. If $L$ and $C$ are given, then the damping ratio, $\zeta$, depends on $R$.

Case 1:   If $\zeta > 1$, then $R > 2\sqrt{L/C}$.
Case 2:   If $\zeta = 1$, then $R = 2\sqrt{L/C}$.
Case 3:   If $\zeta < 1$, then $R < 2\sqrt{L/C}$.

The series *RLC* circuit is undamped if the resistor is shorted ($R = 0$). As in the parallel *RLC* circuit, the frequency of the undamped sinusoid is $\sqrt{1/LC}$. This should not be surprising, since the series and parallel *LC* circuits are exactly the same when their resistances are removed.

Figure 6.14 shows how the natural response of a second-order system varies

FIGURE 6.14    ***Second-Order Response as Function of Damping Ratio***

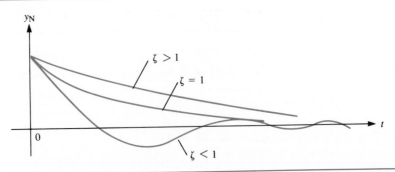

when $\omega_n$ is fixed and $\zeta$ is varied. You can see that as $\zeta$ is decreased, the response decays faster. When $\zeta$ is less than 1, the response overshoots its final value ($y_N = 0$) and oscillates about that value with decreasing amplitude. These responses are important considerations in designing a second-order system. Often, we want the response to reach its final value quickly but do not want much overshoot or ringing (oscillation). Therefore, a trade-off must be made in choosing the value for $\zeta$.

---

**Example 6.7** *Design Problem*

Design a parallel *RLC* circuit that has an undamped natural frequency of 10 kHz and a damping coefficient of 0.707.

First, convert Hz to radian frequency:

$$\omega_n = 2\pi \cdot 10 \times 10^3 = 62{,}832 \text{ rad/s}$$

Next, since $\omega_n = \sqrt{1/LC} = 62{,}832$, choose $L = 1$ mH and $C = .253\ \mu$F. Note that the choices for $L$ and $C$ are arbitrary, provided their product equals $1/62{,}832^2$.

Finally, since, for a parallel *RLC* circuit, $\zeta = (1/2R)\sqrt{L/C}$,

$$R = \frac{1}{2\zeta}\sqrt{\frac{L}{C}} = 44.5\ \Omega$$

Keep in mind that in most circuit designs, we must use element values for inductors, capacitors, and resistors that are commercially available. This limitation will lead to designs that only approximate the desired $\omega_n$ and $\zeta$.

---

You may have noticed that we have not yet evaluated any of the constants in the second-order natural solutions. Finding these constants is not as easy as in the first-order case. If we are solving a purely mathematical second-order differential equation, then finding the constants is fairly straightforward. For example, let's solve the following differential equation with the given initial conditions (note that both $y(0)$ and $dy(0)/dt$ are given):

$$\frac{d^2y}{dt^2} + 6\frac{dy}{dt} + 8y = 0, \qquad y(0) = 1$$

$$\frac{dy(0)}{dt} = 1$$

The solution to this equation is

$$y(t) = K_1 e^{-2t} + K_2 e^{-4t} \tag{6.36}$$

We can use the initial condition on $y(t)$ to find

$$y(0) = 1 = K_1 + K_2 \tag{6.37}$$

We have one equation but two unknowns. We can use the initial condition for the derivative, $dy/dt$, to obtain the necessary second equation. We first take the derivative of $y$ as given in Equation 6.36 and then evaluate at $t = 0$:

$$\frac{dy(t)}{dt} = -2K_1 e^{-2t} - 4K_2 e^{-4t}$$

$$\frac{dy(0)}{dt} = 1 = -2K_1 - 4K_2 \tag{6.38}$$

Next, we use Equations 6.37 and 6.38 to find $K_1$ and $K_2$:

$$K_1 + K_2 = 1$$
$$-2K_1 - 4K_2 = 1$$

So,

$$K_1 = 2.5 \quad \text{and} \quad K_2 = 1.5$$

Thus,

$$y(t) = 2.5e^{-2t} + 1.5e^{-4t}, \quad t \geq 0$$

---

**Example 6.8**   *Second-Order Equation with Initial Conditions (Case 2, Critically Damped)*

Find $y(t)$ if

$$\frac{d^2y}{dt^2} + 8\frac{dy}{dt} + 16y = 0, \qquad y(0) = 1$$

$$\frac{dy(0)}{dt} = 0$$

The characteristic equation is

$$s^2 + 8s + 16 = 0$$

So,

$$s = -4, -4 \quad \text{(case 2, critically damped)}$$
$$y(t) = (K_1 + K_2 t)e^{-4t}$$

$$\frac{dy}{dt} = (K_2 - 4K_1 - 4K_2 t)e^{-4t}$$

$$y(0) = 1 = K_1$$

$$\frac{dy(0)}{dt} = 0 = K_2 - 4K_1 = K_2 - 4$$

$$K_2 = 4$$

The solution is

$$y(t) = (1 + 4t)e^{-4t}, \qquad t \geqslant 0$$

**Example 6.9**    *Second-Order Equation with Initial Conditions (Case 3, Underdamped)*

Find $y(t)$ if

$$\frac{d^2y}{dt^2} + 4\frac{dy}{dt} + 8y = 0, \qquad y(0) = 2$$

$$\frac{dy(0)}{dt} = 1$$

The characteristic equation is

$$s^2 + 4s + 8 = 0$$

So,

$$s = -2 + j2, \ -2 - j2 \qquad \text{(case 3, underdamped)}$$

$$y(t) = e^{-2t}(A \cos 2t + B \sin 2t)$$

$$\frac{dy}{dt} = -2e^{-2t}(A \cos 2t + B \sin 2t) + e^{-2t}(-2A \sin 2t + 2B \cos 2t)$$

$$y(0) = 2 = A$$

$$\frac{dy(0)}{dt} = 1 = -2A + 2B = -4 + 2B$$

$$B = 2.5$$

The solution is

$$y(t) = e^{-2t}(2 \cos 2t + 2.5 \sin 2t), \qquad t \geqslant 0$$

**Source-Free Series *RLC* Circuit.**    Since the initial conditions were provided in Examples 6.8 and 6.9, solving the differential equations was straightforward. If the differential equation that we are solving is obtained by analysis of an electrical circuit, then we must find the appropriate initial conditions. As an example, let's analyze the source-free series *RLC* circuit shown in Figure 6.15.

The differential equation for $v_C$ has already been derived in Chapter 5. For practice, we derive it again here. We write the mesh equation in terms of $v_C$ ($i_L = i_C = C \, dv_C/dt$):

$$RC \frac{dv_C}{dt} + LC \frac{d^2v_C}{dt^2} + v_C = 0$$

so

$$\frac{d^2v_C}{dt^2} + \frac{R}{L} \frac{dv_C}{dt} + \frac{1}{LC} v_C = 0 \qquad (6.39)$$

For the values of $R$, $L$, and $C$ given, Equation 6.39 becomes

$$\frac{d^2v_C}{dt^2} + 15 \frac{dv_C}{dt} + 50v_C = 0$$

The solution proceeds as follows:

$$s^2 + 15s + 50 = 0$$

$$s = -5, -10 \qquad \text{(overdamped)}$$

$$v_C = K_1 e^{-5t} + K_2 e^{-10t} \qquad (6.40)$$

$$\frac{dv_C}{dt} = -5K_1 e^{-5t} - 10K_2 e^{-10t} \qquad (6.41)$$

The solution to this point has been straightforward. We run into a complication, however, when we try to evaluate $K_1$ and $K_2$. You can see that although we know $v_C(0)$, we do not yet know $dv_C(0)/dt$. The only initial conditions generally given in circuits are capacitor voltages and inductor currents. Thus, we have some work to do. The key to the solution is to remember the relationship between

FIGURE 6.15   *Source-Free Series RLC Circuit*

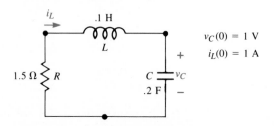

capacitor voltage and current:

$$i_C = C \frac{dv_C}{dt}$$

Therefore,

$$\frac{dv_C}{dt} = \frac{i_C}{C}$$

so

$$\frac{dv_C(0)}{dt} = \frac{i_C(0)}{C} \tag{6.42}$$

Equation 6.42 indicates that we can find $dv_C(0)/dt$ if we first find $i_C(0)$. In a general circuit, finding $i_C(0)$ can be a difficult undertaking. However, for the series $RLC$ circuit that we are analyzing, it is fairly easy to find $i_C(0)$. In this circuit, we see immediately that $i_C = i_L$. Since $i_L(0)$ is given, we know $i_C(0)$:

*Series RLC*

$$\frac{dv_C(0)}{dt} = \frac{i_C(0)}{C} = \frac{i_L(0)}{C}$$

For this example circuit, therefore,

$$\frac{dv_C(0)}{dt} = \frac{1}{.2} = 5$$

We can now return to Equations 6.40 and 6.41 and complete the solution:

$$v_C(0) = 1 = K_1 + K_2$$

$$\frac{dv_C(0)}{dt} = 5 = -5K_1 - 10K_2$$

so

$$K_1 = 3 \qquad \text{and} \qquad K_2 = -2$$

Thus,

$$v_C(t) = 3e^{-5t} - 2e^{-10t}, \qquad t \geqslant 0$$

**Source-Free Parallel *RLC* Circuit.**   The parallel $RLC$ circuit shown in Figure 6.16 was also analyzed in Chapter 5. Here, we write a node equation in terms of $i_L$ ($v_C = v_L = L\,di_L/dt$):

$$\frac{L}{R} \frac{di_L}{dt} + i_L + CL \frac{d^2 i_L}{dt^2} = 0$$

FIGURE 6.16     *Source-Free Parallel RLC Circuit*

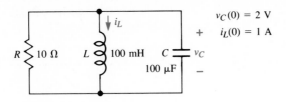

so

$$\frac{d^2 i_L}{dt^2} + \frac{1}{RC}\frac{di_L}{dt} + \frac{1}{LC} i_L = 0 \tag{6.43}$$

For the values of $R$, $L$, and $C$ given, Equation 6.43 becomes

$$\frac{d^2 i_L}{dt^2} + 10^3 \frac{di_L}{dt} + 10^5 i_L = 0$$

The solution proceeds as follows:

$$s^2 + 10^3 s + 10^5 = 0$$

$$s = -112.7,\ -887.3 \qquad \text{(overdamped)}$$

$$i_L(t) = K_1 e^{-112.7t} + K_2 e^{-887.3t} \tag{6.44}$$

$$\frac{di_L(t)}{dt} = -112.7 K_1 e^{-112.7t} - 887.3 K_2 e^{-887.3t} \tag{6.45}$$

We know $i_L(0)$, but we do not know $di_L(0)/dt$. We can find $di_L/dt$ by remembering that

$$v_L = L \frac{di_L}{dt}$$

Therefore,

$$\frac{di_L}{dt} = \frac{v_L}{L}$$

so

$$\frac{di_L(0)}{dt} = \frac{v_L(0)}{L} \tag{6.46}$$

To find $di_L(0)/dt$, we must first find $v_L(0)$, which is not necessarily a simple task. However, for the parallel circuit that we are analyzing, $v_L(t)$ equals $v_C(t)$.

So,

*Parallel RLC*

$$\frac{di_L(0)}{dt} = \frac{v_L(0)}{L} = \frac{v_C(0)}{L}$$

For this example circuit, therefore,

$$\frac{di_L(0)}{dt} = \frac{2}{.1} = 20$$

We are given $i_L(0) = 1$, and we just found that $di_L(0)/dt = 20$. We now use these values in Equations 6.44 and 6.45 and complete the solution:

$$i_L(0) = 1 = K_1 + K_2$$

$$\frac{di_L(0)}{dt} = 20 = -112.7K_1 - 887.3K_2$$

so

$$K_1 = 1.17 \qquad \text{and} \qquad K_2 = -.17$$

Thus,

$$i_L(t) = 1.17e^{-112.7t} - .17e^{-887.3t} \text{ A}, \qquad t \geqslant 0$$

**General *RLC* Circuits.**  Thus far we have analyzed only series and parallel *RLC* circuits. General *RLC* circuits are more difficult to analyze. They usually require the solution of simultaneous sets of differential equations. Initial conditions are also harder to find. These circuits are more easily analyzed with the differential operator technique that will be described in Section 6.5.

---

**Example 6.10**  *General Second-Order RLC Circuit*

Given the circuit in Figure 6.17, with $v_1(0) = 1$ V and $v_2(0) = .2$ V, solve for $v_2(t)$.

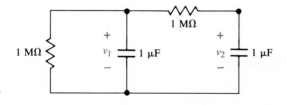

**FIGURE 6.17**

*(continues)*

**Example 6.10**   *Continued*

Write the node equations:

$$v_1\left(\frac{1}{10^6} + \frac{1}{10^6}\right) + 10^{-6}\frac{dv_1}{dt} - \frac{v_2}{10^6} = 0 \qquad (1)$$

$$-\frac{v_1}{10^6} + \frac{v_2}{10^6} + 10^{-6}\frac{dv_2}{dt} = 0 \qquad (2)$$

Multiply equations 1 and 2 by $10^6$:

$$2v_1 + \frac{dv_1}{dt} - v_2 = 0 \qquad (3)$$

$$-v_1 + v_2 + \frac{dv_2}{dt} = 0 \qquad (4)$$

From equation 4,

$$v_1 = v_2 + \frac{dv_2}{dt} \qquad (5)$$

and

$$\frac{dv_1}{dt} = \frac{dv_2}{dt} + \frac{d^2v_2}{dt^2} \qquad (6)$$

Substitute equations 5 and 6 into equation 3:

$$2v_2 + 2\frac{dv_2}{dt} + \frac{dv_2}{dt} + \frac{d^2v_2}{dt^2} - v_2 = 0$$

which gives

$$\frac{d^2v_2}{dt^2} + 3\frac{dv_2}{dt} + v_2 = 0$$

The characteristic equation is

$$s^2 + 3s + 1 = 0$$

So,

$$s = -2.6, -.38$$

The natural solution is

$$v_2 = K_1e^{-.38t} + K_2e^{-2.6t}$$

To find $K_1$ and $K_2$, we need $v_2(0)$ and $dv_2(0)/dt$. We are given $v_2(0)$, but we must find $dv_2(0)/dt$.

Note that $dv_2(0)/dt = i_2(0)/C_2 = i_2(0)/10^{-6}$. In this circuit,

$$i_2(0) = \frac{v_1 - v_2}{10^6}$$

so

$$i_2(0) = [v_1(0) - v_2(0)] \cdot 10^{-6}$$

and

$$\frac{dv_2(0)}{dt} = \frac{i_2(0)}{10^{-6}} = (1 - .2) \cdot \frac{10^{-6}}{10^{-6}} = .8$$

Use the initial values to find $K_1$ and $K_2$:

$$v_2(0) = .2 = K_1 + K_2$$

$$\frac{dv_2(0)}{dt} = .8 = -.38 K_1 - 2.6 K_2$$

$$K_1 = -.59 \qquad \text{and} \qquad K_2 = -.39$$

Thus,

$$v_2(t) = -.59 e^{-.38t} - .39 e^{-2.6t} \text{ V}, \qquad t \geqslant 0$$

## High-Order Source-Free Circuits

The natural solutions to high-order systems are simply combinations of the cases just described. A characteristic equation of any order can be factored into first- and second-order terms. The difficulties in solving high-order equations lie in factoring them and in finding the necessary initial conditions. A computer or calculator program is usually required to perform the factoring. Consider the following examples.

Example 1: $\dfrac{d^3 y}{dt^3} + 12 \dfrac{d^2 y}{dt^2} + 41 \dfrac{dy}{dt} + 30 y = 0$

Characteristic equation: $s^3 + 12 s^2 + 41 s + 30 = (s + 1)(s + 5)(s + 6) = 0$

Natural frequencies: $s_1 = -1; s_2 = -5; s_3 = -6$

Natural response: $y_N = K_1 e^{-t} + K_2 e^{-5t} + K_3 e^{-6t}$

Example 2: $\dfrac{d^3y}{dt^3} + 3\dfrac{d^2y}{dt^2} + 3\dfrac{dy}{dt} + 2y = 0$

Characteristic equation: $s^3 + 3s^2 + 3s + 2 = (s + 2)(s^2 + s + 1) = 0$

Natural frequencies: $s_1 = -2;\ s_2 = -.5 + j.5\sqrt{3};\ s_3 = -.5 - j.5\sqrt{3}$

Natural response: $y_N = K_1e^{-2t} + e^{-.5t}(A\cos.5\sqrt{3}t + B\sin.5\sqrt{3}t)$

Example 3: $\dfrac{d^3y}{dt^3} + 3\dfrac{d^2y}{dt^2} + 3\dfrac{dy}{dt} + y = 0$

Characteristic equation: $s^3 + 3s^2 + 3s + 1 = (s + 1)^3 = 0$

Natural frequencies: $s_{1,2,3} = -1$

Natural response: $y_N = (K_1 + K_2t + K_3t^2)e^{-t}$

Notice that in example 3, there are three roots at the same value. Previously, when there were two roots at the same value, we added the term $K_2t$ to the constant $K_1$ before multiplying by the exponential. The natural response in example 3 is given in a similar manner. You should verify that this solution is correct. Remember: *If a real root has multiplicity m, multiply the exponential by a polynomial in t, with the highest order of t equaling m − 1.*

Example 4: $\dfrac{d^4y}{dt^4} + 7\dfrac{d^3y}{dt^3} + 18\dfrac{d^2y}{dt^2} + 22\dfrac{dy}{dt} + 12 = 0$

Characteristic equation: $s^4 + 7s^3 + 18s^2 + 22s + 12 = 0$

$(s + 2)(s + 3)(s^2 + 2s + 2) = 0$

Natural frequencies: $s_1 = -2;\ s_2 = -3;\ s_3 = -1 + j;\ s_4 = -1 - j$

Natural response: $y_N = K_1e^{-2t} + K_2e^{-3t} + e^{-t}(A\cos t + B\sin t)$

Example 5: $\dfrac{d^4y}{dt^4} + 4\dfrac{d^3y}{dt^3} + 8\dfrac{d^2y}{dt^2} + 8\dfrac{dy}{dt} + 4y = 0$

Characteristic equation: $s^4 + 4s^3 + 8s^2 + 8s + 4 = (s^2 + 2s + 2)^2 = 0$

Natural frequencies: $s_{1,3} = -1 + j;\ s_{2,4} = -1 - j$

Natural response: $y_N = (K_1 + K_2t)[e^{-t}(A\cos t + B\sin t)]$

In this last example, we have a case of a multiple complex root. As with multiple real roots, we merely multiply the single complex root response by a polynomial in *t*.

## 6.3 THE FORCED AND TOTAL SOLUTIONS

### The Forced Solution

The natural solution is only one part of the total solution to a differential equation. The rest of the solution depends on the forcing function (independent sources) and is the particular, or **forced,** solution. Forced solutions that do not decay to zero with time (for example, constants and periodic functions) are also known as **steady-state** solutions.

Several techniques can be used to find the forced solution of linear, time-invariant, differential equations. The method that will be described here is based on the **method of undetermined coefficients.** We will first examine the mathematical groundwork that demonstrates the applicability of this method and then describe its practical use.

The forced solution, $y_F$, must satisfy the following differential equation:

$$\frac{d^n y_F}{dt^n} + a_{n-1} \frac{d^{n-1} y_F}{dt^{n-1}} + \cdots + a_1 \frac{dy_F}{dt} + a_0 y_F = f(t) \qquad (6.47)$$

where we have assumed that $a_n = 1$. For most of the forcing functions, $f(t)$, that electrical engineers encounter, the forced solution is given by

$$y_F = b_0 f(t) + b_1 \frac{df(t)}{dt} + \cdots + b_m \frac{d^m f(t)}{dt^m} \qquad (6.48)$$

That is, the forced solution is a linear combination of the first $m$ derivatives of the forcing function [remember, $d^0 f(t)/dt^0 = f(t)$]. For this method to work, $m$ must be finite. Functions of time that are not continuous do not satisfy this criterion. We will discuss such functions in Chapter 7.

How do we find $m$? As you will see, for the simple functions of time that we normally use, the answer is obvious. However, in general, we find $m$ when the next highest derivative is a linear combination of the previous derivatives. That is,

$$\frac{d^{m+1} f(t)}{dt^{m+1}} = K_0 f(t) + K_1 \frac{df(t)}{dt} + \cdots + K_m \frac{d^m f(t)}{dt^m} \qquad (6.49)$$

For our purposes, the values of the $K_i$'s are not important. We are interested only in $m$.

---

**Example 6.11**     *Finding m*

For each of the following functions, find the value of $m$ that satisfies Equation 6.49.

a.   $f(t) = 100$

$$\frac{df(t)}{dt} = 0 = 0 \cdot f(t)$$

So,     $m = 0$

*(continues)*

**Example 6.11**   *Continued*

   **b.**  $f(t) = e^{-10t}$

$$\frac{df(t)}{dt} = -10e^{-10t} = -10f(t)$$

So,     $m = 0$

   **c.**  $f(t) = \sin 50t$

$$\frac{df(t)}{dt} = 50 \cos 50t$$

$$\frac{d^2f(t)}{dt^2} = -250 \sin 50t = -250f(t)$$

So,     $m = 1$

   **d.**  $f(t) = e^{-t} \cos 5t$

$$\frac{df(t)}{dt} = -e^{-t} \cos 5t - 5e^{-t} \sin 5t$$

$$\frac{d^2f(t)}{dt^2} = -24e^{-t} \cos 5t + 10e^{-t} \sin 5t = -26f(t) - 2\,\frac{df(t)}{dt}$$

So,     $m = 1$

   **e.**  $f(t) = t^4 + 3t^3 + 4t^2 + t + 5$

$$\frac{df(t)}{dt} = 4t^3 + 9t^2 + 8t + 1$$

$$\frac{d^2f(t)}{dt^2} = 12t^2 + 18t + 8$$

$$\frac{d^3f(t)}{dt^3} = 24t + 18$$

$$\frac{d^4f(t)}{dt^4} = 24$$

$$\frac{d^5f(t)}{dt^5} = 0 = 0f + 0\,\frac{df}{dt} + 0\,\frac{d^2f}{dt^2} + 0\,\frac{d^3f}{dt^3} + 0\,\frac{d^4f}{dt^4}$$

So,     $m = 4$

Now that we know how to find $m$, we can demonstrate that the linear combination of the first $m$ derivatives of $f(t)$ is the forced solution to the differential equation (Equation 6.47). The validity of Equation 6.48 is demonstrated for a second-order differential equation, where $m = 1$. The general proof is left to you.

Given

$$\frac{d^2 y_F}{dt^2} + a_1 \frac{dy_F}{dt} + a_0 y_F = f(t) \tag{6.50}$$

where $m = 1$,

$$y_F = b_0 f + b_1 \frac{df}{dt} \tag{6.51}$$

So,

$$\frac{dy_F}{dt} = b_0 \frac{df}{dt} + b_1 \frac{d^2 f}{dt^2}$$

However, since $m = 1$, we get from Equation 6.49

$$\frac{d^2 f}{dt^2} = K_0 f + K_1 \frac{df}{dt}$$

So,

$$\frac{dy_F}{dt} = b_1 K_0 f + (b_0 + b_1 K_1) \frac{df}{dt} \tag{6.52}$$

and

$$\frac{d^2 y_F}{dt^2} = b_1 K_0 \frac{df}{dt} + (b_0 + b_1 K_1) \frac{d^2 f}{dt^2}$$

Substituting for $d^2 f/dt^2$ gives

$$\frac{d^2 y_F}{dt^2} = (b_0 + b_1 K_1) K_0 f + (b_1 K_0 + b_0 K_1 + b_1 K_1^2) \frac{df}{dt} \tag{6.53}$$

Substitution of Equations 6.51, 6.52, and 6.53 into Equation 6.50 yields

$$(b_0 + b_1 K_1) K_0 f + (b_1 K_0 + b_0 K_1 + b_1 K_1^2) \frac{df}{dt} + a_1 b_1 K_0 f$$

$$+ a_1 (b_0 + b_1 K_1) \frac{df}{dt} + a_0 b_0 f + a_0 b_1 \frac{df}{dt} = f \tag{6.54}$$

Remember that $a_0$, $a_1$, $K_0$, and $K_1$ in Equation 6.54 are known. The $a_i$'s are given, and we have solved for the $K_i$'s. Thus, the coefficients $b_0$ and $b_i$ are *undetermined coefficients*. We find the $b_i$'s by comparing the coefficients of $f$ and $df/dt$ on both sides of Equation 6.54. Factoring results in

$$(a_0 + K_0)b_0 + (K_0 K_1 + a_1 K_0)b_1 = 1$$

$$(K_1 + a_1)b_0 + (K_0 + K_1^2 + a_1 K_1 + a_0)b_1 = 0$$

since the coefficient of $f(t)$ on the right side of Equation 6.54 is 1 and the coefficient of $df(t)/dt$ is 0.

We now have two independent equations in two unknowns, and we can solve for $b_0$ and $b_1$. We have, therefore, found the forced solution that satisfies the differential equation. As the next few examples demonstrate, this technique is much easier to use in practice than is implied by this derivation.

---

**Example 6.12**    *Forced Solution (Constant Input)*

Find the forced solution for

$$\frac{dy}{dt} + 100y = 20$$

We know from Example 6.11 that $m = 0$ for a constant input. Therefore,

$$y_F = b_0 f(t) = 20b_0$$

For simplicity, we introduce a new constant $C = 20b_0$, so

$$y_F = C$$

and

$$\frac{dy_F}{dt} = 0$$

Substituting in the differential equation gives

$$0 + 100C = 20$$

So,

$$C = .2$$

Finally,

$$y_F = .2$$

---

**Example 6.13**    *Forced Solution (Exponential Input)*

Find the forced solutions to the differential equations in part a and part b.

a.    $\dfrac{dy}{dt} + 5y = 4e^{-10t}$

We know from Example 6.11 that $m = 0$ for the exponential. Therefore,

$$y_F = b_0 e^{-10t}$$

and

$$\frac{dy_F}{dt} = -10b_0 e^{-10t}$$

To be consistent with the notation used in Example 6.12, we will let

$$y_F = Ce^{-10t}$$

and

$$\frac{dy_F}{dt} = -10Ce^{-10t}$$

Substituting into the differential equation gives

$$-10Ce^{-10t} + 5e^{-10t} = 4e^{-10t}$$
$$-5Ce^{-10t} = 4e^{-10t}$$
$$-5C = 4$$

So,

$$C = -.8$$

and

$$y_F = -.8e^{-10t}$$

b.    $\dfrac{d^2y}{dt^2} + 4\dfrac{dy}{dt} + 3y = 4e^{-10t}$

Again, for the exponential, $m = 0$, and, therefore,

$$y_F = Ce^{-10t}$$

*(continues)*

**Example 6.13**     *Continued*

and

$$\frac{dy_F}{dt} = -10Ce^{-10t}$$

and

$$\frac{d^2y_F}{dt^2} = 100Ce^{-10t}$$

Substituting into the differential equation gives

$$100Ce^{-10t} - 40Ce^{-10t} + 3Ce^{-10t} = 4e^{-10t}$$

$$63Ce^{-10t} = 4e^{-10t}$$

So,

$$C = \frac{4}{63} = .063$$

and

$$y_F = .063e^{-10t}$$

These two example equations demonstrate an important point. The *form* of the particular solution does not depend on the order of the differential equation. A higher-order differential equation just requires taking more derivatives of $y_F$.

---

**Example 6.14**     *Forced Response (Sinusoidal Input)*

Find the forced response for

$$\frac{dy}{dt} + 10y = 5\cos 20t$$

We know from Example 6.11 that $m = 1$ for the cosine. Therefore,

$$y_F = b_0 f(t) + b_1 \frac{df(t)}{dt} = b_0 \cos 20t - 20b_1 \sin 20t$$

We introduce new constants to simplify $y_F$. Let $C_1 = b_0$ and $C_2 = -20b_1$. We use

$$y_F = C_1 \cos 20t + C_2 \sin 20t$$

and

$$\frac{dy_F}{dt} = -20C_1 \sin 20t + 20C_2 \cos 20t$$

Substituting into the differential equation gives

$$-20C_1 \sin 20t + 20C_2 \cos 20t + 10C_1 \cos 20t + 10C_2 \sin 20t = 5 \cos 20t$$

$$(10C_1 + 20C_2) \cos 20t + (-20C_1 + 10C_2) \sin 20t = 5 \cos 20t$$

$$10C_1 + 20C_2 = 5$$

$$-20C_1 + 10C_2 = 0 \qquad \text{(no sine term on right side)}$$

So,

$$C_1 = .1 \qquad \text{and} \qquad C_2 = .2$$

and

$$y_F = .1 \cos 20t + .2 \sin 20t$$

---

**Example 6.15**  *Forced Solution (Polynomial in t)*

Find the forced solution for

$$\frac{d^2y}{dt^2} + 10 \frac{dy}{dt} + 5y = t^2 + 3t + 2$$

It is fairly easy to demonstrate that in this case $m = 2$. Therefore,

$$y_F = b_0(t^2 + 3t + 2) + b_1(2t + 3) + b_2(2)$$

$$= b_0 t^2 + (3b_0 + 2b_1)t + 2b_0 + 2b_2$$

To simplify, we let

$$y_F = C_1 + C_2 t + C_3 t^2$$

and

$$\frac{dy_F}{dt} = C_2 + 2C_3 t$$

$$\frac{d^2y_F}{dt^2} = 2C_3$$

*(continues)*

**Example 6.15**   *Continued*

Substituting into the differential equation gives

$$2C_3 + 10C_2 + 20C_3t + 5C_1 + 5C_2t + 5C_3t^2 = t^2 + 3t + 2$$
$$2C_3 + 10C_2 + 5C_1 = 2$$
$$20C_3 + 5C_2 = 3$$
$$5C_3 = 1$$

So,

$$C_1 = .72, \qquad C_2 = -.2, \qquad C_3 = .2$$

and

$$y_F = .72 - .2t + .2t^2$$

The most common input forcing functions and their respective forced solutions are shown in Table 6.2. Example 6.16 provides a circuit example that shows the use of this table and the superposition principle to find the forced solution.

**TABLE 6.2**   *Common Input Forcing Functions and Their Forced Solutions*

| *Input* | *Forced Solution* |
|---|---|
| Constant | $C$ |
| Ramp $t$ | $C_1 + C_2t$ |
| Parabola $t^2$ | $C_1 + C_2t + C_3t^2$ |
| . | . |
| . | . |
| . | . |
| $t^n$ | $C_1 + C_2t + \cdots + C_nt^n$ |
| $e^{at}$* | $Ce^{at}$ |
| $\cos \omega t$* | $C_1 \cos \omega t + C_2 \sin \omega t$ |
| $\sin \omega t$* | $C_1 \cos \omega t + C_2 \sin \omega t$ |
| $e^{at} \cos \omega t$* | $e^{at}(C_1 \cos \omega t + C_2 \sin \omega t)$ |
| $e^{at} \sin \omega t$* | $e^{at}(C_1 \cos \omega t + C_2 \sin \omega t)$ |

*If the natural solution has the same frequencies as the input, then these results must be modified. See text.

**Example 6.16**  *Forced Solution (Circuit Example)*

Given the circuit in Figure 6.18A, find the forced solution for $v_o$.

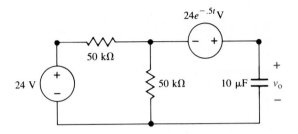

**FIGURE 6.18**  A. Original circuit           B. Equivalent circuit

First find the Thévenin equivalent seen by the capacitor:

$$v_{Th} = 12 + 24e^{-.5t} \text{ V}$$
$$R_{Th} = 25 \text{ k}\Omega$$

The equivalent circuit is shown in Figure 6.18B. The differential equation for $v_o$ is

$$\frac{dv_o}{dt} + 4v_o = 48 + 96e^{-.5t}$$

Now use superposition to find the total forced solution. Find $v_{o_F}$ for the constant term and separately for the exponential:

$$\text{Constant:} \quad v_{o_{F1}} = C_1$$
$$4C_1 = 48$$
$$C_1 = 12$$
$$\text{Exponential:} \quad v_{o_{F2}} = C_2 e^{-.5t}$$

$$\frac{dv_{o_{F2}}}{dt} = -.5C_2 e^{-.5t}$$
$$-.5C_2 + 4C_2 = 96$$
$$C_2 = 27.43$$

The total forced solution is

$$v_{o_F} = 12 + 27.43e^{-2t} \text{ V}$$

The footnote to Table 6.2 states that some of the forced solution forms will not always work. Let's try the following example:

$$\frac{dy}{dt} + 5y = 6e^{-5t}$$

The forced solution, according to the table, is $Ce^{-5t}$. If we substitute this solution and its derivative into the differential equation, we get

$$-5Ce^{-5t} + 5Ce^{-5t} = 6e^{-5t}$$

$$(-5C + 5C)e^{-5t} = 6e^{-5t}$$

$$0 = 6$$

Obviously, our simple rule did not work.

To find out what happened, let's obtain the natural solution for this example:

$$s + 5 = 0$$

so

$$s = -5$$

and

$$y_N = Ke^{-5t}$$

The natural solution has the same frequency as the input. All of the exceptions to the forced solutions given in Table 6.2 occur when the input has the same frequency (neper or radian) as the natural solution(s). What do we do in this case? The critically damped natural solution to a second-order differential equation provides the answer. To find the second natural solution to a differential equation with multiple real roots, we had to multiply the first solution by powers of $t$. We do the same here. Let's try

$$y_F = Cte^{-5t}$$

$$\frac{dy_F}{dt} = -5Cte^{-5t} + Ce^{-5t}$$

so

$$(-5Ct + C)e^{-5t} + 5Cte^{-5t} = 6e^{-5t}$$

$$(-5C + 5C)t + C = 6$$

$$C = 6$$

The forced solution is, therefore,

$$y_F = 6te^{-5t}$$

In general, the forced solution for a circuit driven by a source that has some frequencies in common with the circuit's natural frequencies is given by the following statement: *If the input has the same frequency as m identical natural solutions, multiply the forced solution given in Table 6.2 by the polynomial in t,* $C_m t^m + \cdots + C_1 t.$

## The Total Solution

The **total** solution to a differential equation is the sum of the natural solution and the forced solution. The most common error that is made in finding the total solution is the misuse of the initial conditions. Always remember that the initial conditions apply to the total solution and not to the natural solution.

---

**Example 6.17**    *Total Solution (Constant Input)*

Find $y(t)$ for

$$\frac{dy}{dt} + 100y = 20, \qquad y(0) = 1$$

We found the forced solution to this equation in Example 6.12:

$$y_F = .2$$

The natural solution is found by inspection to be

$$y_N = Ke^{-100t}$$

Note that $K$ is *not* 1 in this example. We must first find the total solution:

$$y(t) = y_N + y_F = Ke^{-100t} + .2$$

Only now can we use the initial condition:

$$y(0) = 1 = K + .2$$

so

$$K = .8$$

and

$$y(t) = .8e^{-100t} + .2, \qquad t \geq 0$$

---

**Example 6.18**    *Total Solution (Exponential Input)*

Find $y(t)$ for

$$\frac{d^2y}{dt^2} + 4\frac{dy}{dt} + 3y = 4e^{-10t}, \qquad y(0) = 1$$

$$\frac{dy(0)}{dt} = 0$$

We found the forced solution to this equation in part b of Example 6.13:

$$y_F = .063e^{-10t}$$

The natural solution is found from

$$\frac{d^2y}{dt^2} + 4\frac{dy}{dt} + 3y = 0$$

$$s^2 + 4s + 3 = 0$$

$$s = -1, -3$$

$$y_N = K_1 e^{-t} + K_2 e^{-3t}$$

$$y(t) = y_N + y_F = K_1 e^{-t} + K_2 e^{-3t} + .063e^{-10t}$$

$$\frac{dy}{dt} = -K_1 e^{-t} - 3K_2 e^{-3t} - .63e^{-10t}$$

Now we use the initial conditions:

$$y(0) = 1 = K_1 + K_2 + .063$$

$$\frac{dy(0)}{dt} = 0 = -K_1 - 3K_2 - .63$$

so

$$K_1 = 1.72 \qquad \text{and} \qquad K_2 = -.783$$

and

$$y(t) = 1.72e^{-t} - .783e^{-3t} + .063e^{-10t}, \qquad t \geq 0$$

**Example 6.19**    *Total Solution (Circuit Example)*

Given the circuit in Figure 6.19A and $v_C(0) = 0$ V, find $v_o(t)$.

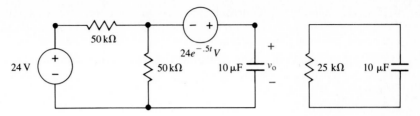

**FIGURE 6.19**    A. Original circuit                                                   B. Source-free equivalent circuit

The circuit in Figure 6.19A is the same as the circuit in Figure 6.18A. Therefore, the forced solution from Example 6.16 is

$$v_{o_F} = 12 + 27.43e^{-2t}$$

The natural solution is found by turning off the independent sources ($v_{Th} = 0$) to obtain a source-free first-order $RC$ circuit, where $R_{eq} = 25\,k\Omega$ (see Figure 6.19B). The natural solution is, therefore,

$$v_{o_N} = v_{C_N} = Ke^{-t/R_{eq}C} = Ke^{-t/.25}$$

Note that, in general, $K \neq v_C(0)$. We must first find the total voltage:

$$v_o(t) = v_C(t) = Ke^{-t/.25} + 12 + 27.43e^{-2t}$$

Now we use the initial condition:

$$v_o(0) = 0 = K + 12 + 27.43$$

so

$$K = -39.43$$

and

$$v_o(t) = -39.43e^{-t/.25} + 12 + 27.43e^{-2t}\ \text{V}, \qquad t \geq 0$$

## 6.4   DC STEADY-STATE SOLUTIONS

If an $RLC$ circuit is driven only by constant sources, there is a very simple method for finding the forced solution of any variable. You do not need to write differential equations for the circuit. We have just seen that, when the input to a linear system is a constant, the forced response is also a constant. Thus, all voltages and currents in a circuit will be constants after the transient natural solutions have died out. The circuit is then in the DC steady state.

The key questions are (1) what is the current in a capacitor when the voltage across it is a constant, and (2) what is the voltage across an inductor when the current in it is constant? The answers, as we have already seen, are

$$i_C = C\,\frac{dv_C}{dt} = C\,\frac{d(\text{constant})}{dt} = 0$$

and

$$v_L = L\,\frac{di_L}{dt} = L\,\frac{d(\text{constant})}{dt} = 0$$

For constant inputs, currents in all capacitors and voltages across all inductors go to zero as the circuit approaches steady-state operating conditions—that is, as $t$ approaches $\infty$. In circuit model terms, capacitors act as open circuits ($i_C = 0$), while inductors act as short circuits ($v_L = 0$). To find the forced solution for any variable in a circuit driven only by constant sources, we replace all capacitors with open circuits and all inductors with short circuits. We can then analyze the resultant resistive circuit for circuit variable(s) of interest.

To demonstrate this technique, let's first analyze the simple *RC* circuit of Figure 6.20A by using differential equations. We get

$$\frac{dv_C}{dt} + 1000v_C = 5000$$

so

$$v_{C_F} = 5\text{ V}$$

Now, without writing a differential equation, let's find the forced solution by replacing the capacitor with an open circuit, as in Figure 6.20B. Since there is no closed path, the current in the circuit must be zero. The voltage across the resistor

**FIGURE 6.20**   *Constant-Input Forced Response of RC Circuit*

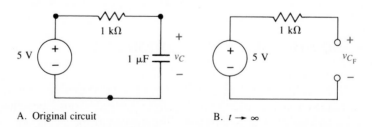

A.  Original circuit                         B. $t \rightarrow \infty$

must then also be zero. Therefore,

$$v_{C_F} = 5 - 0 = 5 \text{ V}$$

Another simple example is shown in Figure 6.21A. In this circuit, we replace the inductor with a short circuit to get the circuit shown in Figure 6.21B. We find $i_{L_F}$ to be

$$i_{L_F} = 10 - 0 = 10 \text{ A}$$

Before we examine a more complicated example, consider the significance of the two circuits that we just analyzed. The first circuit (Figure 6.20A) represents all first-order capacitor circuits if we consider the resistor and source as the Thévenin equivalent seen by the capacitor. Likewise, the second circuit (Figure 6.21A) represents all first-order inductor circuits if we consider the resistor and source as the Norton equivalent seen by the inductor.

Chapter 7, Section 7.3 will provide you with a formula for writing the total solution of a first-order circuit with constant sources. The response of any first-order DC circuit can be written once the time constant of the circuit, steady-state value, and initial value are known.

As a final example, let's analyze the higher-order circuit shown in Figure 6.22A for the forced solutions of the indicated circuit variables.

First, we replace the capacitors with open circuits and the inductors with short circuits, as shown in Figure 6.22B. We are left with a purely resistive circuit. We can solve for $i_F$ and $v_F$ by using any of the circuit analysis techniques described in earlier chapters. One of the simplest ways to analyze this circuit is to observe that the $1 \, \Omega$ resistor is in series with the current source, which immediately implies that

$$i_F = -2 \text{ A}$$

We can use superposition and voltage divider relationships to find $v_F$:

$$v_F = \left( \frac{4}{5} \cdot 2.5 \right) + \left( \frac{1}{5} \cdot 2i_F \right) = 2 - \frac{4}{5} = 1.2 \text{ V}$$

FIGURE 6.21  *Constant-Input Forced Response of RL Circuit*

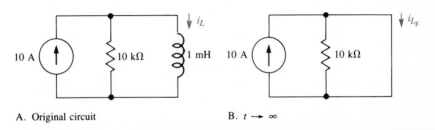

A. Original circuit          B. $t \rightarrow \infty$

FIGURE 6.22    *Constant-Input Forced Response of General Circuit*

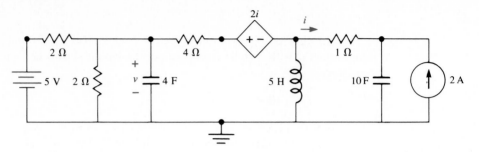

A.  Original circuit

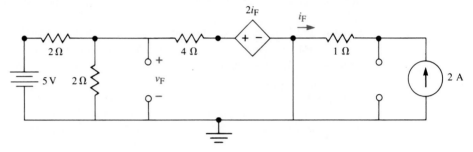

B. $t \rightarrow \infty$

## *6.5  DIFFERENTIAL OPERATORS

In the late 1800s, Oliver Heaviside (1850–1921) devised a technique for replacing the derivatives in linear, time-invariant, mathematical equations with algebraic operators. This technique is very simple to use and is extremely powerful. It allows us to reduce a set of simultaneous integro-differential equations to a single differential equation in one variable. It also leads directly to the characteristic equation of a system and, for certain inputs, allows very easy calculation of the forced response.

We start with the definitions of the differential operator $D$ and the powers of $D$:

$$Dy = \frac{dy}{dt}$$

$$D^2y = D \cdot Dy = \frac{d(dy/dt)}{dt} = \frac{d^2y}{dt^2}$$

$$\vdots$$

$$D^ny = \frac{d^ny}{dt^n}$$

From these definitions, it is clear that multiplying by $D$ represents taking the first derivative. Multiplying by powers of $D$ represents taking successive derivatives.

An example of the use of the differential operator is

$$\frac{d^3y}{dt^3} + 4\frac{d^2y}{dt^2} + 2\frac{dy}{dt} + 5y = 4\frac{dx}{dt} + x$$

which can be represented as

$$D^3y + 4D^2y + 2Dy + 5y = 4Dx + x$$

Given a differential equation, we can write the algebraic equation in $D$. Given the operator equation, we can write the differential equation it represents.

The usefulness of the differential operator is that it can be treated as any other algebraic quantity. The variables $y$ and $x$ in the preceding example equation can be factored out to produce

$$(D^3 + 4D^2 + 2D + 5)y = (4D + 1)x$$

To see how valuable a tool the differential operator is, let's analyze the circuit shown in Figure 6.23. Analysis of this circuit results in two simultaneous differential node equations that are difficult to solve. The two node equations are as follows:

$$\frac{dv_a}{dt} + 5v_a - 2v_b = \cos 5t$$

$$-2v_a + 3\frac{dv_b}{dt} + 6v_b = e^{-4t}$$

To solve these equations using the techniques of this chapter, we would reduce the two equations in two unknowns to two equations with one unknown each. This technique is known as *separation of variables*. From the second

---

**FIGURE 6.23**    *Node Analysis Using Differential Operators*

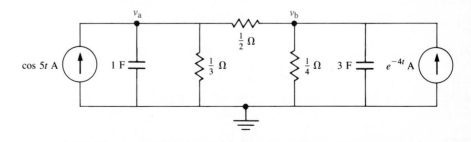

equation,

$$v_a = 1.5 \frac{dv_b}{dt} + 3v_b - .5e^{-4t}$$

and

$$\frac{dv_a}{dt} = 1.5 \frac{d^2v_b}{dt^2} + 3 \frac{dv_b}{dt} + 2e^{-4t}$$

Substitution in the first equation gives

$$1.5 \frac{d^2v_b}{dt^2} + 3 \frac{dv_b}{dt} + 2e^{-4t} + 7.5 \frac{dv_b}{dt} + 15v_b - 2.5e^{-4t} - 2v_b = \cos 5t$$

which reduces to

$$1.5 \frac{d^2v_b}{dt^2} + 10.5 \frac{dv_b}{dt} + 13v_b = \cos 5t + .5e^{-4t}$$

Thus,

$$\frac{d^2v_b}{dt^2} + 7 \frac{dv_b}{dt} + 8.67v_b = .67 \cos 5t + .33e^{-4t} \qquad (6.55)$$

We can now solve for $v_b$ and then find $v_a$, which is a lot of work, even for a simple set of differential equations.

Let's use differential operators to get the same result. The two differential equations become

$$Dv_a + 5v_a - 2v_b = \cos 5t$$

and

$$-2v_a + 3Dv_b + 6v_b = e^{-4t}$$

Factoring, we get

$$(D + 5)v_a - 2v_b = \cos 5t$$
$$-2v_a + (3D + 6)v_b = e^{-4t}$$

Here, we have a simultaneous set of algebraic equations that can be readily solved. For example, we can use Cramer's rule to solve for $v_b$:

$$v_b = \frac{\begin{vmatrix} D+5 & \cos 5t \\ -2 & e^{-4t} \end{vmatrix}}{\begin{vmatrix} D+5 & -2 \\ -2 & 3D+6 \end{vmatrix}} = \frac{De^{-4t} + 5e^{-4t} + 2\cos 5t}{3D^2 + 21D + 26}$$

Multiplying both sides of the equation by the denominator gives

$$3D^2v_b + 21Dv_b + 26v_b = De^{-4t} + 5e^{-4t} + 2\cos 5t$$

We transform $D^n(v_b)$ to $d^n(v_b)/dt^n$ and get

$$3\frac{d^2v_b}{dt^2} + 21\frac{dv_b}{dt} + 26v_b = -4e^{-4t} + 5e^{-4t} + 2\cos 5t$$

Thus,

$$\frac{d^2v_b}{dt^2} + 7\frac{dv_b}{dt} + 8.67v_b = .67\cos 5t + .33e^{-4t} \tag{6.56}$$

This result is the same as in Equation 6.55. The advantage in using the differential operator is that a very logical algebraic approach can be used.

Let's use operators to find a differential equation for $v_a$ in the circuit of Figure 6.24. We note that the current-controlled current source can be written as

$$2i = 2 \cdot 3 \frac{dv_a}{dt}$$

The node equations are as follows:

$$3\frac{dv_a}{dt} + 5v_a + 2\int_{-\infty}^{t}(v_a - v_b)\,dt = 5t$$

$$2\int_{-\infty}^{t}(v_b - v_a)\,dt + 10v_b + 4\int_{-\infty}^{t}v_b\,dt + 6\frac{dv_a}{dt} = 0$$

To solve this set of equations in terms of derivatives, we first differentiate to get rid of the integrals:

$$3\frac{d^2v_a}{dt} + 5\frac{dv_a}{dt} + 2(v_a - v_b) = 5$$

$$2(v_b - v_a) + 10\frac{dv_b}{dt} + 4v_b + 6\frac{d^2v_a}{dt^2} = 0$$

FIGURE 6.24    *Another Node Analysis Using Differential Operators*

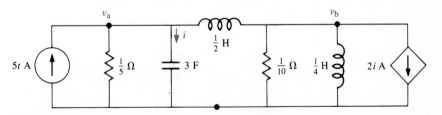

We then use the differential operator to transform the simultaneous equations:

$$(3D^2 + 5D + 2)v_a - 2v_b = 5$$

$$(6D^2 - 2)v_a + (10D + 6)v_b = 0$$

Solving for $v_a$ yields

$$v_a = \frac{\begin{vmatrix} 5 & -2 \\ 0 & 10D + 6 \end{vmatrix}}{\begin{vmatrix} 3D^2 + 5D + 2 & -2 \\ 6D^2 - 2 & 10D + 6 \end{vmatrix}} = \frac{(10D + 6) \cdot 5}{30D^3 + 80D^2 + 50D + 8}$$

Note that in the numerator, the constant 5 is multiplied by a polynomial in $D$. Cross multiplying, we get

$$(30D^3 + 80D^2 + 50D + 8)v_a = (10D + 6) \cdot 5$$

and

$$\left( D^3 + \frac{8D^2}{3} + \frac{5D}{3} + \frac{8}{30} \right)v_a = \left( \frac{D}{3} + \frac{1}{5} \right) \cdot 5 \tag{6.57}$$

On the right side of Equation 6.57, the constant 5 is multiplied by $D/3$. Remembering that the derivative of a constant is zero, we get the following differential equation:

$$\frac{d^3v_a}{dt^3} + \frac{8}{3}\frac{d^2v_a}{dt^2} + \frac{5}{3}\frac{dv_a}{dt} + \frac{8}{30}v_a = 1 \tag{6.58}$$

To solve the differential equation, we first find the natural solution as follows:

$$s^3 + \frac{8}{3}s^2 + \frac{5}{3}s + \frac{8}{30} = 0 \tag{6.59}$$

We then solve for the roots of this characteristic equation. You may have noticed that the differential operator equation for this circuit (Equation 6.57) is identical to the characteristic equation (Equation 6.59) if the input is turned off. That is,

$$\left( D^3 + \frac{8}{3}D^2 + \frac{5}{3}D + \frac{8}{30} \right)v_a = 0$$

So,

$$D^3 + \frac{8}{3}D^2 + \frac{5}{3}D + \frac{8}{30} = 0$$

The only difference in the characteristic equation and the differential operator equation here is the algebraic symbol used.

**Example 6.20**   *Differential Operators*

For the circuit in Figure 6.25, find the forced solution for $i_1(t)$.

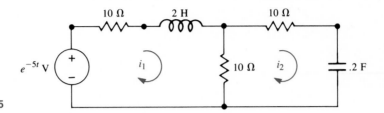

**FIGURE 6.25**

The two mesh equations are as follows:

Mesh 1:   $20i_1 + 2\dfrac{di_1}{dt} - 10i_2 = e^{-5t}$   (1)

Mesh 2:   $-10i_1 + 20i_2 + 5\displaystyle\int_{-\infty}^{t} i_2\, dt = 0$   (2)

Take the derivative of the mesh 2 equation:

$$-10\dfrac{di_1}{dt} + 20\dfrac{di_2}{dt} + 5i_2 = 0 \qquad (2)$$

Use the differential operators for equations 1 and 2:

$$(2D + 20)i_1 - 10i_2 = e^{-5t}$$
$$-10Di_1 + (20D + 5)i_2 = 0$$

Use Cramer's rule:

$$i_1 = \dfrac{\begin{vmatrix} e^{-5t} & 5 \\ 0 & 20D + 5 \end{vmatrix}}{\begin{vmatrix} 2D + 20 & -10 \\ -10D & 20D + 5 \end{vmatrix}} = \dfrac{(20D + 5)e^{-5t}}{40D^2 + 310D + 100} \qquad (3)$$

The differential operator equation for $i_1$ is

$$(40D^2 + 310D + 100)i_1 = (20D + 5)e^{-5t} = 20De^{-5t} + 5e^{-5t} \qquad (4)$$

The differential equation for $i_1$ is

$$40\dfrac{d^2 i_1}{dt^2} + 310\dfrac{di_1}{dt} + 100i_1 = (-100 + 5)e^{-5t} = -95e^{-5t} \qquad (5)$$

*(continues)*

**Example 6.20**    *Continued*

The forced solution for $i_1$ is

$$i_{1_F} = Ce^{-5t}$$

so

$$(1000C - 1550C + 100C)e^{-5t} = -95e^{-5t} \qquad (6)$$

and

$$C = \frac{-95}{-450} = .21$$

Thus,

$$i_{1_F} = .21e^{-5t} \text{ A} \qquad (7)$$

---

The differential operator can also be used to develop some shortcuts to finding the forced solution, especially when the forcing function is an exponential. For example, if we take equation 3 from Example 6.20,

$$i_1 = \frac{(20D + 5)e^{-5t}}{40D^2 + 310D + 100}$$

and set $D$ equal to the neper frequency of the forcing function $(-5)$, we find

$$\frac{[(-20 \cdot 5) + 5]e^{-5t}}{(40 \cdot 25) - (310 \cdot 5) + 100} = .21e^{-5t} = i_{1_F}$$

This result is the same as the forced response that we found in Example 6.20 by analyzing the differential equation.

For exponential inputs, we can find the forced solution of a circuit variable from the differential operator equation for that variable. We simply set $D$ equal to the input neper frequency and solve. That is, for exponential input $Me^{at}$, if $y = H(D) \cdot Me^{at}$, where $H(D) =$ a ratio of polynomials in $D$, the forced solution is

$$\boxed{y_F = H(a) \cdot Me^{at}}$$

Caution! This technique will not work if the input frequency is the same as a natural solution frequency. In this case, use the method of undetermined coefficients to find the forced solution.

## 6.6 SUMMARY

In this chapter, we found out how to solve differential equations that result from the analysis of $RLC$ circuits. The solution to a differential equation is the addition of the natural and the forced solutions.

The natural solution is found by setting the forcing function (input) to zero. The $n$th derivative of $y$ is then replaced by $s^n$, and the roots of the resulting characteristic equation are found. These roots are the natural frequencies of the circuit variable. For a first-order circuit, the magnitude of the neper frequency of the natural solution is the inverse of the circuit time constant ($\tau$), which is $R_{eq}C$ for a capacitor circuit and $L/R_{eq}$ for an inductor circuit. For a second-order circuit, the damping ratio ($\zeta$) and the undamped natural frequency ($\omega_n$) determine the natural response. The following list summarizes the natural solutions to first- and second-order characteristic equations.

First order:   For $s + a = 0$,      $y_N = Ke^{-at} = Ke^{-t/\tau}$

Second order:   For $s^2 + 2\zeta\omega_n + \omega_n^2 = 0$

Case 1:   if $\zeta > 1$, $y_N = K_1 e^{(-\zeta\omega_n + \omega_n\sqrt{\zeta^2-1})t} + K_2 e^{(-\zeta\omega_n - \omega_n\sqrt{\zeta^2-1})t}$

Case 2:   if $\zeta = 1$, $y_N = (K_1 + K_2 t)e^{-\zeta\omega_n t}$

Case 3:   if $\zeta < 1$, $y_N = e^{-\zeta\omega_n t}(A \cos \omega_n\sqrt{1 - \zeta^2}t + B \sin \omega_n\sqrt{1 - \zeta^2}t)$

The forced solution usually has the same form as the input. For constant inputs, we found the forced solution directly from the circuit. In this case, we simply open all capacitors and short all inductors and analyze the resultant resistive circuit for the variable of interest.

We then examined the differential operator, which helps solve simultaneous differential equations. The differential operator simplifies the separation of variables of simultaneous sets of differential equations. The differential operator can also be used to find a circuit variable's characteristic equation and forced response to an exponential input.

## ■ PROBLEMS

---

**6.1**  Given $dy/dt + 10y = 0$, and $y(0) = 5$.
  **a.**  Find and sketch $y(t)$.
  **b.**  Find $y(.1)$.
  **c.**  Find the time $t_0$ at which $y(t) = 2.5$.

**6.2**  Given $10 \, dy/dt + 5y = 0$, and $y(0) = 1$. Find and sketch $y(t)$.

**6.3**  Given $d^2y/dt^2 + 2\alpha \, dy/dt + 100y = 0$, $y(0) = 1$, and $dy(0)/dt = 0$. Find and sketch $y(t)$ for the following values of $\alpha$:
  **a.**  $\alpha = 20$
  **b.**  $\alpha = 10$
  **c.**  $\alpha = 2$

**6.4** For each part of Problem 6.3, find the first time at which $y(t) = 0.5$.

**6.5** Given $d^2y/dt^2 + 2\zeta\omega_n \, dy/dt + \omega_n^2 y = 0$, $y(0) = 1$, and $dy(0)/dt = 0$.

   **a.** Assume $\zeta < 1$, and find and sketch $y(t)$.

   **b.** Find the time at which the first negative peak occurs $(t_P)$.

   **c.** Find $y(t_P)$.

**6.6** Find and sketch $v_0(t)$ and $i_1(t)$ for the circuit shown in Figure P6.6 if

   **a.** $\beta = 0$

   **b.** $\beta = 3$

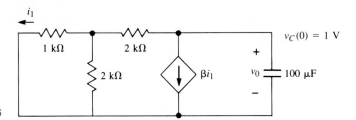

**FIGURE P6.6**

**6.7** Find and sketch $i(t)$ and $v_1(t)$ for the circuit shown in Figure P6.7 if

   **a.** $\alpha = 0$

   **b.** $\alpha = 3$

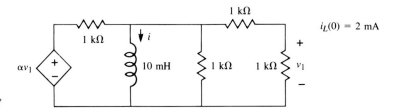

**FIGURE P6.7**

**6.8** The capacitor in the simple *RC* circuit of Figure P6.8 has $2 \, \mu\text{J}$ of energy stored at $t = 0\,\text{s}$.

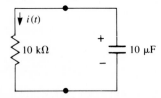

**FIGURE P6.8**

   **a.** Find $i(t)$ for $t > 0$.

   **b.** Find $p_R(t)$ for $t > 0$.

   **c.** Find $w_C(t)$ for $t > 0$.

   **d.** Find the total energy dissipated by the resistor as $t$ approaches $\infty$. Discuss your result.

**6.9** The inductor in the simple *RL* circuit of Figure P6.9 has 10 mJ of energy stored at $t = 0$ s.

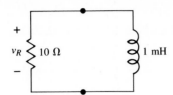

**FIGURE P6.9**

    **a.** Find $v_R(t)$ for $t > 0$.
    **b.** Find $p_R(t)$ for $t > 0$.
    **c.** Find $w_L(t)$ for $t > 0$.
    **d.** Find the total energy dissipated by the resistor as $t$ approaches $\infty$. Discuss your result.

**6.10** Figure P6.10 shows the capacitor voltage response of a simple *RC* circuit. Find *R* and *C* if $w_C(0) = 100\ \mu$J.

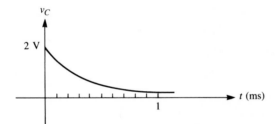

**FIGURE P6.10**

**6.11** Figure P6.11 shows the inductor current response of a simple *RL* circuit. Find *R* and *L* if $w_L(0) = 1$ mJ.

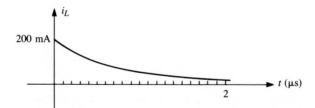

**FIGURE P6.11**

**6.12** The inductor current in a simple unforced parallel *RLC* circuit is measured and found to be $i_L(t) = 5e^{-5t} - 2e^{-10t}$ A. If the resistance in the circuit is known to be 2 Ω, find *L*, *C*, and their initial conditions. Draw the circuit.

**6.13** The capacitor voltage in a simple series *RLC* circuit is measured and found to be $v_C(t) = (8 - 10t)e^{-1000t}$ V. If the resistance in the circuit is known to be 1 kΩ, find *L*, *C*, and their initial conditions. Draw the circuit.

**6.14** Find the values of $\beta$ for which the circuit shown in Figure P6.14 is

    **a.** overdamped.
    **b.** critically damped.

**c.** underdamped.

**d.** undamped.

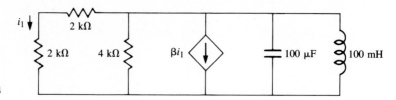

FIGURE P6.14

**6.15** Find the values of $\alpha$ for which the circuit shown in Figure P6.15 is
**a.** overdamped.
**b.** critically damped.
**c.** underdamped.
**d.** undamped.

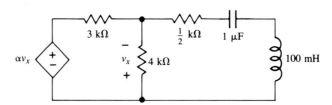

FIGURE P6.15

**6.16** The circuit shown in Figure P6.14 has initial conditions of $v_C(0) = 2\,\text{V}$ and $i_L(0) = .1\,\text{A}$. Find $i_L(t)$ and $v_C(t)$ if
**a.** $\beta = 400$
**b.** $\beta = 40$

**6.17** The circuit shown in Figure P6.15 has initial conditions of $v_C(0) = 1\,\text{V}$ and $i_L(0) = 1\,\text{mA}$. Find $i_L(t)$ and $v_C(t)$ if
**a.** $\alpha = -1.8$
**b.** $\alpha = 2$

**6.18** Given $d^2y/dt^2 + 2\zeta\omega_n\,dy/dt + \omega_n^2 y = \omega_n^2$, $y(0) = 0$, and $dy(0)/dt = 0$, and assume $\zeta < 1$ (underdamped).
**a.** Find $y(t)$ as a function of $\zeta$ and $\omega_n$.
**b.** Find the time at which the first (and largest) peak in $y(t)$ occurs.
**c.** Find $y_P$, which is the peak value of $y(t)$.
**d.** Find the overshoot $\gamma$, which is given by $\gamma = y_P - 1$.
**e.** Plot $\gamma$ as a function of $\zeta$, for $0 < \zeta < 1$.

**6.19** Given the differential equation of Problem 6.18, with general initial conditions of $y(0) = Y_0$ and $dy(0)/dt = Y_0'$.
**a.** Find $y(t)$.
**b.** Find $Y_0$ and $Y_0'$ so that $y(t) = 1$, $t > 0$.

6.20 Find the natural solutions for each of the following differential equations. Any initial condition not given is zero.

a. $\dfrac{d^3y}{dt^3} + 31\dfrac{d^2y}{dt^2} + 230\dfrac{dy}{dt} + 200y = 0,$   $y(0) = 2$

b. $\dfrac{d^3y}{dt^3} + 13\dfrac{d^2y}{dt^2} + 130\dfrac{dy}{dt} + 300y = 0,$   $\dfrac{dy(0)}{dt} = 1$

c. $\dfrac{d^3y}{dt^3} + 30\dfrac{d^2y}{dt^2} + 300\dfrac{dy}{dt} + 1000y = 0,$   $\dfrac{d^2y(0)}{dt^2} = 5$

6.21 Find $m$ as defined in Equation 6.49 for

a. $f(t) = te^{-5t}$
b. $f(t) = t\cos 10t$
c. $f(t) = 10\cos 100t + 5\sin 100t$

6.22 Given $d^2y/dt^2 + 10\,dy/dt + 16y = f(t)$, $y(0) = 1$, and $dy(0)/dt = 0$. Find $y_F$ and $y(t)$ if

a. $f(t) = 50$
b. $f(t) = 5e^{-10t}$
c. $f(t) = \sin 5t$
d. $f(t) = e^{-t}\cos 10t$
e. $f(t) = e^{-8t}$

6.23 The circuit shown in Figure P6.23 is in steady state; all transients have died out. Find the steady-state values of $v_C$, $v_1$, and $i_1$.

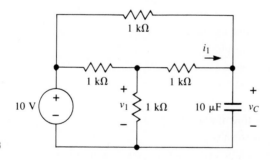

**FIGURE P6.23**

6.24 The circuit shown in Figure P6.24 is in steady state. Find the steady-state values of $i_L$, $i_1$, and $v_1$.

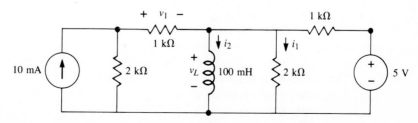

**FIGURE P6.24**

**6.25** The circuit shown in Figure P6.25 is in steady state. Find the steady-state values of $i_1$, $i_2$, $v_1$, and $v_2$.

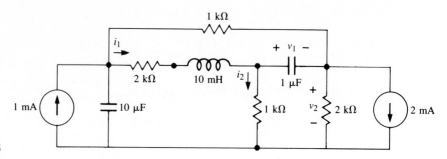

**FIGURE P6.25**

**6.26** Given the sinusoidally driven *RC* circuit shown in Figure P6.26.

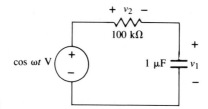

**FIGURE P6.26**

   **a.**  Find $v_{1_F}$ as a function of $\omega$ ($v_{1_F} = A \cos \omega t + B \sin \omega t$).
   **b.**  Plot $A$ as a function of $\omega$.
   **c.**  Plot $B$ as a function of $\omega$.
   **d.**  Discuss your results.

**6.27** Given the circuit shown in Figure P6.26.
   **a.**  Find $v_{2_F}$ as a function of $\omega$ ($v_{2_F} = C \cos \omega t + D \sin \omega t$).
   **b.**  Plot $C$ as a function of $\omega$.
   **c.**  Plot $D$ as a function of $\omega$.
   **d.**  Discuss your results.

**6.28** Given the sinusoidally driven *RL* circuit shown in Figure P6.28.

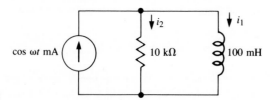

**FIGURE P6.28**

   **a.**  Find $i_{1_F}$ as a function of $\omega$ ($i_{1_F} = A \cos \omega t + B \sin \omega t$).
   **b.**  Plot $A$ as a function of $\omega$.
   **c.**  Plot $B$ as a function of $\omega$.
   **d.**  Discuss your results.

**6.29** Given the circuit shown in Figure P6.28.
  **a.**  Find $i_{2_F}$ as a function of $\omega$ ($i_{2_F} = C \cos \omega t + D \sin \omega t$).
  **b.**  Plot $C$ as a function of $\omega$.
  **c.**  Plot $D$ as a function of $\omega$.
  **d.**  Discuss your results.

**6.30** A transistor AC circuit is shown in Figure P6.30.

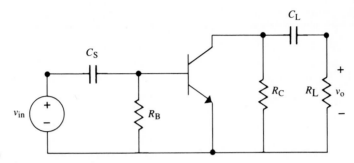

**FIGURE P6.30**

  **a.**  Redraw the circuit using the AC model for the transistor ($r_o = \infty$).
  **b.**  Find the differential equation for $v_o$.
  **c.**  What can you say about the damping of this circuit?

**6.31** Given the following set of simultaneous differential equations, use the differential operator to find a differential equation in $y_2$:

$$\frac{dy_1}{dt} + 5y_1 - \frac{dy_2}{dt} = 10t$$

$$-\frac{dy_1}{dt} + 10\frac{dy_2}{dt} + 10y_2 = 0$$

**6.32** Given the following set of simultaneous integro-differential equations, use the differential operator to find a differential equation in $y_1$:

$$\frac{d^2y_1}{dt^2} + 10y_1 - \int_{-\infty}^{t} y_2 \, dt - \frac{dy_3}{dt} = e^{-5t}$$

$$\frac{dy_2}{dt} + 5y_2 - 10y_3 = 5$$

$$-y_1 - \frac{dy_1}{dt} - \frac{dy_2}{dt} + 5\int_{-\infty}^{t} y_3 \, dt = 10$$

**6.33** The circuit shown in Figure P6.33 is a high-frequency transistor circuit model.

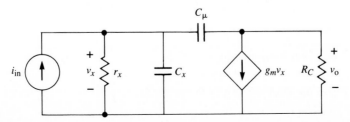

**FIGURE P6.33**

**a.** Find a differential equation for $v_o(t)$.

**b.** What are $\omega_n$ and $\zeta$ for this circuit?

**6.34** Use node analysis and the differential operator to find a differential equation for $v_C(t)$ in the circuit shown in Figure P6.23. The capacitor is initially uncharged.

**a.** Find $v_C(t)$ from your differential equation.

**b.** If the voltage source is changed from a constant 10 V to an exponential $e^{-100t}$ V, find $v_C(t)$.

**6.35** Use mesh analysis and the differential operator to find a differential equation for $i_L(t)$ in the circuit shown in Figure P6.24. The initial inductor current is zero.

**a.** Find $i_L(t)$ from your differential equation.

**b.** If the voltage source is changed from a constant 5 V to an exponential $e^{-1000t}$ V and the current source is changed from a constant 10 mA to $e^{-5000t}$ mA, find $i_L(t)$.

**6.36** We usually solve for $v_C(t)$ or $i_L(t)$ when we are analyzing simple *RC*, simple *RL*, series *RLC*, and parallel *RLC* circuits. We then use the substitution theorem to find any other desired circuit variable. One reason for using this procedure is that we are usually given the initial conditions on capacitors and inductors. It is possible, however, to find directly any circuit variable's initial condition. We do this by replacing all capacitors with batteries whose values equal the capacitors' initial voltages and by replacing all inductors with current sources whose values equal the inductors' initial currents. The initial value of any circuit variable may now be found. Then, with the aid of the differential operator, any first-order circuit variable's differential equation can be found and solved.

**a.** For the circuit in Figure P6.6, set $\beta = 0$ and find $i_1(0)$.

**b.** Use differential operators and the answer to part a to find directly $i_1(t)$.

**6.37 a.** For the circuit in Figure P6.7, set $\alpha = 0$ and find $v_1(0)$.

**b.** Use differential operators to find $v_1(t)$.

**6.38** Second-order circuits require additional work to find appropriate initial conditions. Consider, for example, the two-capacitor circuit shown in Figure P6.38. We wish to find directly the current $i_R$. We will thus need to find $i_R(0)$ and also $di_R(0)/dt$. You know how to find $i_R(0)$. To find $di_R(0)/dt$, you must first write $i_R$ as a function of the capacitor voltages and then differentiate.

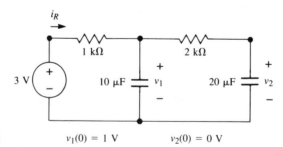

**FIGURE P6.38**            $v_1(0) = 1$ V            $v_2(0) = 0$ V

**a.** Find $i_R(0)$ and $di_R(0)/dt$.

**b.** Use differential operators to find $i_R(t)$.

**6.39** Use the circuit shown in Figure P6.39.

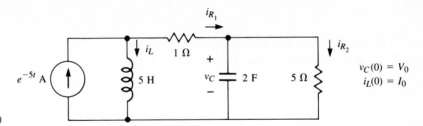

**FIGURE P6.39**

**a.** Find $i_{R_1}(0)$, $i_{R_2}(0)$, $di_{R_1}(0)/dt$, and $di_{R_2}(0)/dt$ in terms of $V_0$ and $I_0$. (*Hint:* Write $i_{R_1}$ and $i_{R_2}$ in terms of $v_C$ and $i_L$ and then differentiate.)

**b.** Find $i_{R_1}(t)$ and $i_{R_2}(t)$ if $V_0 = 1$ V and $I_0 = 2$ A.

# Switched Circuits

## 7.1 INTRODUCTION

In Chapter 6, you learned how to solve the differential equations for *RLC* circuits, given the initial capacitor voltages and inductor currents. In many circuits, you must find the initial conditions yourself. This chapter will show you how to find initial conditions in circuits that contain switches. It will then introduce some new functions of time that can be used to represent switching. The chapter will conclude with a brief discussion of an important modern circuit that uses switched capacitors.

## 7.2 CONTINUITY OF CAPACITOR VOLTAGE AND INDUCTOR CURRENT

Figure 7.1 shows an *RC* circuit that is connected by a switch to a current source at $t = 0$. We assume in this textbook that all switches are ideal and open or close instantaneously. It should seem reasonable to you that the branch variables $v_C$, $i_C$, and $i_R$ might have different values just before and just after switch $S_1$ is closed—and indeed they may. Therefore, we need some notation to indicate a variable's value just before and just after a switch opens or closes. If a switch changes positions at $t = t_0$, then at the instant before the change $t = t_0^-$ and at the instant after the change $t = t_0^+$. It is important to note that there is a vanishingly small time between $t_0^-$ and $t_0^+$. Since the switch in Figure 7.1 is thrown at $t = 0$, then the circuit variables at the switching time are $i_R(0^-)$, $i_C(0^-)$, $v_C(0^-)$, and $i_R(0^+)$, $i_C(0^+)$, $v_C(0^+)$.

If the value of a circuit variable at $t = t_0^-$ equals its value at $t = t_0^+$, then the variable is *continuous* at $t = t_0$; otherwise, the variable is *discontinuous* at $t = t_0$. With the exception of capacitor voltage and inductor current, circuit variables are usually discontinuous at switching instants. To see what happens to capacitor voltages, let's examine Figure 7.2.

In Figure 7.2A, we assume that a capacitor voltage is discontinuous and jumps from a value of 0 to a value of 1 at $t = 0$ as a result of a switch being

FIGURE 7.1　*Switched RC Circuit*

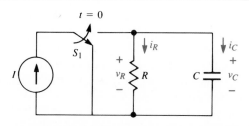

FIGURE 7.2   *Discontinuous Capacitor Voltage and Current*

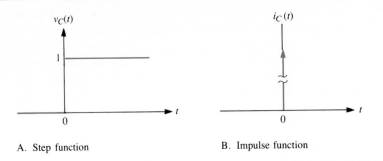

A. Step function                           B. Impulse function

thrown:

$$v_C(t) = \begin{cases} 0, & t < 0 \\ 1, & t > 0 \end{cases}$$

Note that $v_C$ equals 0 at $t = 0^-$, equals 1 at $t = 0^+$, and is not defined at $t = 0$. This function of time is known as a **step function.** What capacitor current would correspond to such a voltage discontinuity?

For $t < 0$ and $t > 0$, capacitor voltage is a constant, so its derivative is zero. However, at $t = 0$, capacitor voltage is changing instantaneously. The derivative of a function at any time is equal to the slope of the function at that time. At $t = 0$, the slope of $v_C(t)$ is infinite (unit change in zero time), so

$$i_C(t) = \begin{cases} 0, & t < 0 \\ \infty, & t = 0 \\ 0, & t > 0 \end{cases}$$

The capacitor current is shown in Figure 7.2B. Since $i_C(t)$ has a value of infinity at $t = 0$, we cannot actually plot it. Figure 7.2B is merely a representation of $i_C(t)$. We will discuss this very important function, known as an **impulse function,** in Chapter 8.

What has been demonstrated with this analysis is that a step change in capacitor voltage requires an infinite current at the time of the change. Since infinite currents are not available in the real world, we can say:

Capacitor voltage cannot change instantaneously.

The same relationship that was just demonstrated between capacitor voltage and current also holds for inductor current and voltage. A step change in inductor current requires an infinite voltage at the time of the change. Since this situation too is, in general, not possible, we state:

Inductor current cannot change instantaneously.

All other circuit variables, such as capacitor current, inductor voltage, and resistor current and voltage, can and do change instantaneously. The capacitor voltage and inductor current relationships just given can be succinctly stated as follows:

$$v_C(t_0^-) = v_C(t_0^+) \tag{7.1}$$

and

$$i_L(t_0^-) = i_L(t_0^+) \tag{7.2}$$

We again use the notation that $t_0^-$ is the time just before the discontinuity and $t_0^+$ is the time just after the discontinuity. Remember that $t_0^-$ and $t_0^+$ are vanishingly close to each other.

---

**Example 7.1** *Discontinuities*

Figure 7.3 shows an example of a function with discontinuities. We see that, at the discontinuities,

$$f(10^-) = 1, \qquad f(10^+) = 3$$

and

$$f(20^-) = 2, \qquad f(20^+) = .5$$

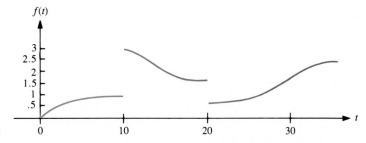

**FIGURE 7.3**

---

## 7.3 FIRST-ORDER SWITCHED CIRCUITS

The circuit in Figure 7.4A has a single switch that has been closed for a long time and then is opened at $t = 0$. Figure 7.4B shows the circuit with the switch open—that is, for $t > 0$. The analysis of switched circuits is simplified if you draw a separate circuit for each switch state.

FIGURE 7.4   *First-Order Switched RC Circuit*

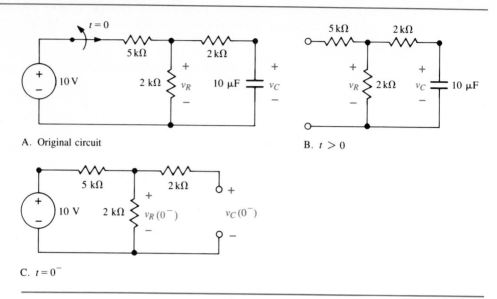

A. Original circuit

B. $t > 0$

C. $t = 0^-$

Let's find $v_C(t)$ and $v_R(t)$ for $t > 0$. Figure 7.4B shows that we are analyzing a source-free $RC$ circuit for $t > 0$. First, we find the equivalent resistance seen by the capacitor:

$$R_{eq} = 2 \text{ k}\Omega + 2 \text{ k}\Omega = 4 \text{ k}\Omega$$

The 5 kΩ resistor is in an open pathway and so is not included. The time constant of the natural solution is $R_{eq}C = 40$ ms. Since the total solution in a source-free circuit equals the natural solution, we get

$$v_C(t) = Ke^{-t/\tau} = Ke^{-t/R_{eq}C} = Ke^{-t/.04}, \qquad t > 0$$

Note that this expression is valid only for $t > 0$, as is explicitly stated.

To find $v_C(t)$, we need to know the initial condition on the capacitor. Since the equation for $v_C(t)$ is valid only for $t > 0$, we need the value of $v_C(t)$ for some $t > 0$. As in Chapter 6, we will use the initial value, which in this circuit occurs at $t = 0^+$—that is, $v_C(0^+)$. However, unlike the circuit problems in Chapter 6, the initial condition is not given. We find it by examining the circuit just before the switch opens—that is, at $t = 0^-$.

We do not need to write differential equations for Figure 7.4A. All we have to do is to recognize that the switch has been closed for a long time. We thus assume that all of the transients in the circuit have died out and that the circuit is in the DC steady state. We need to know only the forced solution just prior to the time when the switch opens.

Since the input is a constant for $t < 0$, the forced solution must also be a constant. As you may recall from Chapter 6, to find the forced solution to a constant input, we open all capacitors and short all inductors. Having done this in Figure 7.4C, which shows the circuit at $t = 0^-$, we find that

$$v_C(0^-) = v_R(0^-) = \frac{2}{7} \cdot 10 = \frac{20}{7} \text{ V}$$

So, we have now found $v_C(0^-)$ and $v_R(0^-)$. However, to find $v_C(t)$ and $v_R(t)$, we must know their values at $t = 0^+$. We use Equation 7.1 to immediately determine that

$$v_C(0^+) = v_C(0^-) = \frac{20}{7} \text{ V}$$

To find $v_R(0^+)$, we analyze the circuit shown in Figure 7.4B at $t = 0^+$. We will describe this procedure later. For now, we use $v_C(0^+)$ to find that

$$v_C(0^+) = \frac{20}{7} = Ke^0 = K$$

so

$$v_C(t) = \frac{20}{7} e^{-t/.04} \text{ V}, \qquad t > 0$$

How do we find $v_R(t)$? We use the substitution theorem. That is, we replace the capacitor with a voltage source equal to $(20/7)e^{-t/.04}$ and then find $v_R(t)$. In this simple circuit, we see directly from Figure 7.4B that

$$v_R(t) = \frac{2}{4} \cdot v_C(t)$$

so

$$v_R(t) = \frac{10}{7} e^{-t/.04} \text{ V}, \qquad t > 0$$

We use this general procedure for all first-order circuits. It does not matter which variable you want to find. First find the capacitor voltage or inductor current, and then use the substitution theorem to find the variable of interest. In the problems at the end of the chapter, we will discuss how to solve directly for the general circuit variable.

Remember, only capacitor voltage and inductor current cannot change instantaneously. We had found that $v_R(0^-) = 20/7$ V. We can now see that $v_R(0^+) = 10/7$ V. Resistor voltages can and do show discontinuities. Figure 7.5 shows the plots of the two voltages we have found.

The procedure for analyzing any switched (at $t = 0$) first-order capacitor or inductor circuit is now evident. First, redraw the circuit for $t = 0^-$. If we assume

FIGURE 7.5    *First-Order Switched RC Circuit Responses*

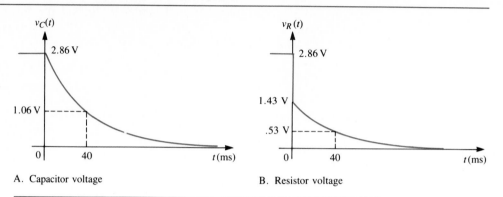

A. Capacitor voltage                    B. Resistor voltage

DC conditions prior to switching, all capacitors are opened and all inductors are shorted in the circuit at $t = 0^-$. Then, analyze this circuit for the appropriate capacitor voltage or inductor current initial condition. Next, redraw the circuit for $t > 0$ and find the capacitor voltage or inductor current using the appropriate initial condition. Finally, use the substitution theorem to find any circuit variable of interest.

**Example 7.2**    *First-Order Capacitor Example*

In the circuit in Figure 7.6A, the switch has been at position 1 for a long time. At $t = 0$, the switch is thrown instantaneously to position 2. Find $v_C(t)$ for $t > 0$.

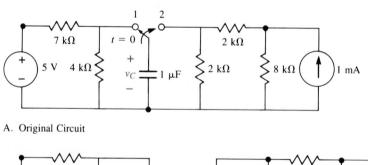

A. Original Circuit

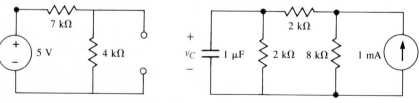

FIGURE 7.6    B. $t = 0^-$                    C. $t > 0$

First, draw the circuit for $t = 0^-$. Since the source in this circuit is constant, the capacitor can be replaced with an open circuit, as shown in Figure 7.6B.

$$v_C(0^-) = 5 \cdot 4000/11{,}000 = 1.82 \text{ V}$$

so

$$v_C(0^+) = 1.82 \text{ V}$$

The circuit is now redrawn for $t > 0$, as shown in Figure 7.6C. Find the natural solution:

$$R_{eq} = 2000 \parallel (2000 + 8000) = 1.67 \text{ k}\Omega$$

$$\tau = R_{eq}C = 1.67 \text{ ms}$$

$$v_{C_N} = Ke^{-t/.00167}$$

Find the forced solution by assuming the capacitor is open. Use source transformation and the voltage divider relationship to find

$$v_{C_F} = \frac{2000}{12{,}000} \cdot 8 = 1.33 \text{ V}$$

and

$$v_C(t) = Ke^{-t/.00167} + 1.33$$

Find $K$:

$$v_C(0^+) = 1.82 = K + 1.33$$

$$K = .49$$

so

$$v_C(t) = .49e^{-t/.00167} + 1.33 \text{ V}, \qquad t > 0$$

---

A first-order differential equation with a constant forcing function always has the same general form of response. The response starts at the initial value of the solution and exponentially approaches the DC steady-state value. Examples of such responses are shown in Figure 7.7.

A general expression can be derived for any first-order DC circuit response:

$$x(t) = x(\infty) + [x(0^+) - x(\infty)]e^{-t/\tau}, \qquad t > 0 \tag{7.3}$$

FIGURE 7.7    *First-Order Step Responses*

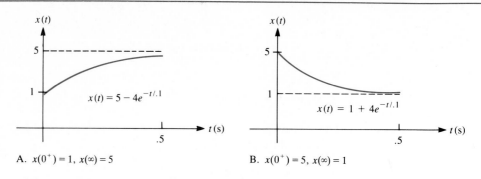

A. $x(0^+) = 1$, $x(\infty) = 5$        B. $x(0^+) = 5$, $x(\infty) = 1$

where    $x(\infty) =$ final value (forced solution)

$x(0^+) =$ initial value

$\tau =$ time constant of the circuit variable

You can see in Equation 7.3 that as $t$ approaches $\infty$, $x(t)$ approaches $x(\infty)$ and that when $t = 0^+$, $x(t) = x(\infty) + [x(0^+) - x(\infty)] = x(0^+)$. We will use Equation 7.3 in the two examples that follow.

**Example 7.3**    *Another First-Order Capacitor Example*

In the circuit in Figure 7.8A, the switch has been at position 1 for a long time. At $t = 0$, the switch is moved to position 2. Find $v_R(t)$ for $t > 0$.

First, find $v_C(t)$. Remember, just before the switch opens ($t = 0^-$), the inductor can be replaced with a short circuit and the capacitors with an open circuit, as shown in Figure 7.8B.

$$v_C(0^-) = 30 \text{ V}$$

so

$$v_C(0^+) = 30 \text{ V}$$

Now analyze the circuit for $t > 0$. Since this is a DC circuit, use Equation 7.3. Both $R_{eq}$ and $v_C(\infty)$ are found by determining the Thévenin equivalent circuit seen by the capacitor (Figure 7.8C). Remember, $v_{C_F} = v_{Th}$:

$$R_{eq} = 8000 \parallel [3000 + (10{,}000 \parallel 10{,}000)] = 4 \text{ k}\Omega$$

$$\tau = R_{eq}C = .04 \text{ s}$$

$$v_C(\infty) = 2.5 \text{ V}$$

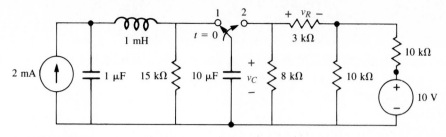

A. Original circuit

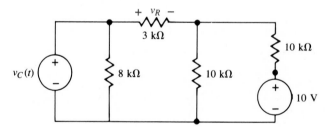

B. $t = 0^-$

C. $t > 0$

Thévenin equivalent

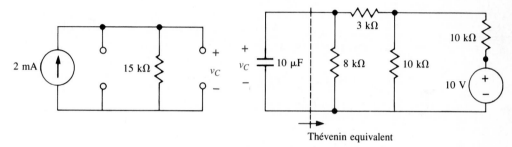

**FIGURE 7.8** D. Substitution theorem

Since $v_C(0^+) = 30$,

$$v_C(t) = 2.5 + (30 - 2.5)e^{-t/.04}$$

so

$$v_C(t) = 2.5 + 27.5e^{-t/.04} \text{ V}, \qquad t > 0$$

To find $v_R(t)$, use the substitution theorem and replace the capacitor with a voltage source (see Figure 7.8D). Note that the 8 kΩ resistor is in parallel with the capacitor voltage source and can be ignored. Finally, use

(*continues*)

Example 7.3    *Continued*

superposition and voltage dividers to find

$$v_R(t) = \frac{3}{8} v_C(t) - \frac{15}{8}$$

so

$$v_R(t) = 10.31e^{-t/.04} - .94 \text{ V}, \qquad t > 0$$

Example 7.4    *First-Order Inductor Circuit*

The switch in the circuit of Figure 7.9A is known as a *make-before-break* switch. This switch is always in contact with position 1 or position 2 and is never open. A make-before-break switch is required to keep the inductor current from instantaneously falling to 0 A. The switch has been at position

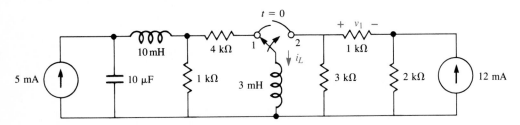

A. Original circuit

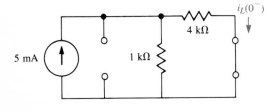

B. *t* = 0

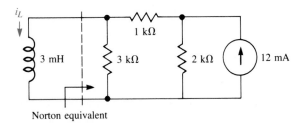

Norton equivalent

C. *t* > 0

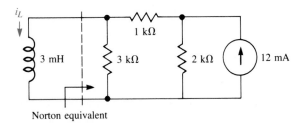

D. Substitution theorem

FIGURE 7.9

1 for a long time. At $t = 0$, the switch is thrown to position 2. Find $i_L(t)$ and $v_1(t)$ for $t > 0$.

From the circuit in Figure 7.9B,

$$i_L(0^-) = \frac{1}{5} \cdot 5 \text{ mA} = 1 \text{ mA}$$

$$i_L(0^+) = i_L(0^-) = 1 \text{ mA}$$

In a DC circuit, to find $i_L(\infty)$ and $R_{eq}$, find the Norton equivalent seen by the inductor (Figure 7.9C):

$$R_{eq} = 3000 \parallel 3000 = 1.5 \text{ k}\Omega$$

$$\tau = \frac{L}{R_{eq}} = 2 \text{ μs}$$

$$i(\infty) = \frac{2000 \cdot .012}{3000} = 8 \text{ mA}$$

Thus,

$$i_L(t) = 8 + (1 - 8)e^{-t/2 \times 10^{-6}} \text{ mA}$$

so

$$i_L(t) = 8 - 7e^{-t/2 \times 10^{-6}} \text{ mA}, \qquad t > 0$$

Now use the substitution theorem and replace the inductor with a current source equal to $i_L$ to find $v_1$ (see Figure 7.9D):

$$v_1(t) = 3.5e^{-t/2 \times 10^{-6}} - 8 \text{ V}, \qquad t > 0$$

Note in this example that $v_1(0^-) = 4 \text{ V}$, while $v_1(0^+) = -4.5 \text{ V}$.

---

Caution! Do not use Equation 7.3 if the sources in the $t > 0$ circuit are not constants. The following example shows such a case.

---

**Example 7.5**    *First-Order Inductor Circuit (Exponential Input)*

In the circuit in Figure 7.10A, the switch has been at position 1 for a long time. At $t = 0$, the switch is thrown to position 2. Find $i_L(t)$ for $t > 0$.

For $t < 0$, replace the inductor with a short circuit, as shown in Figure 7.10B.

*(continues)*

**Example 7.5**   *Continued*

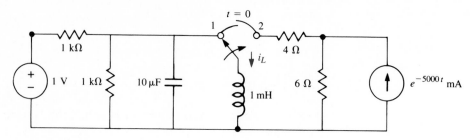

A. Original circuit

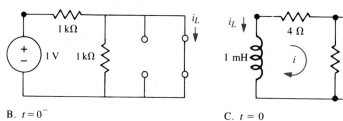

FIGURE 7.10   B. $t = 0^-$                                    C. $t = 0$

$$i_L(0^-) = \frac{1}{1000} = 1 \text{ mA}$$

$$i_L(0^+) = i_L(0^-) = 1 \text{ mA}$$

For $t > 0$, the circuit shown in Figure 7.10C is driven by an exponential source and the initial energy stored in the inductor. Use mesh analysis to write a differential equation for $i_L$:

$$10^{-3} \frac{di}{dt} + 4i + 6(i + .001e^{-5000t}) = 0$$

so

$$\frac{di}{dt} + 10,000i = -6e^{-5000t}$$

Now, since $i_L = -i$,

$$\frac{di_L}{dt} + 10,000i_L = 6e^{-5000t}$$

Thus,

$$i_L(t) = Ke^{-10,000t} + .0012e^{-5000t}$$

and

$$i_L(0^+) = .001 = K + .0012$$

$$K = -.0002$$

so

$$i_L(t) = -.2e^{-10,000t} + 1.2e^{-5000t} \text{ mA}, \qquad t > 0$$

---

## 7.4 SECOND-ORDER SWITCHED CIRCUITS

### Parallel *RLC* Circuit

Figure 7.11A shows a switched circuit that becomes a parallel *RLC* circuit for $t > 0$ (Figure 7.11C). Analysis of the circuit begins when we analyze the circuit at $t = 0^-$ (Figure 7.11B) to find initial capacitor voltage and inductor current.

We analyze the $t = 0^-$ circuit just as we did in Section 7.3. Since all capacitors are opened and all inductors are shorted at $t = 0^-$, the order of the circuit does not matter. From this circuit, we find that

$$v_C(0^-) = 0 \text{ V}$$

and

$$i_L(0^-) = 1.5 \text{ mA}$$

The parallel *RLC* circuit (Figure 7.11C) has already been analyzed in Chapter 6. The differential equation for $i_L(t)$ is

$$\frac{d^2 i_L}{dt^2} + \frac{1}{RC} \frac{di_L}{dt} + \frac{1}{LC} i_L(t) = \frac{1}{LC} i_s$$

so

$$\frac{d^2 i_L}{dt^2} + 500 \frac{di_L}{dt} + 60,000 i_L = 600$$

The natural solution is found from

$$s^2 + 500s + 60,000 = 0$$

$$s_1 = -200 \qquad \text{and} \qquad s_2 = -300$$

$$i_{L_N} = K_1 e^{-200t} + K_2 e^{-300t}$$

**FIGURE 7.11**    *Parallel Switched RLC Circuit*

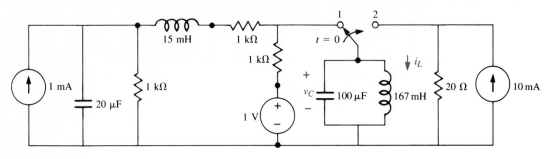

A. Original circuit

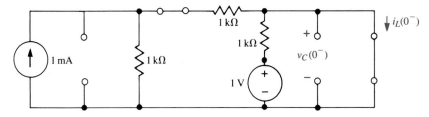

B. $t = 0^-$

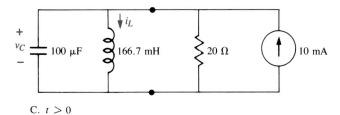

C. $t > 0$

The forced solution is

$$i_{L_F} = 10 \text{ mA}$$

The total solution is

$$i_L(t) = K_1 e^{-200t} + K_2 e^{-300t} + .01$$

$$\frac{di_L(t)}{dt} = -200 K_1 e^{-200t} - 300 K_2 e^{-300t}$$

To evaluate $K_1$ and $K_2$, we need to know the values of $i_L(0^+)$ and $di_L(0^+)/dt$. We already know $i_L(0^+)$, so $di_L(0^+)/dt$ remains to be found. Also, remember that $di_L/dt = v_L/L$; thus, we must find $v_L(0^+)$ for this circuit. In this parallel circuit,

$v_L = v_C$, so we have

$$\frac{di_L(0^+)}{dt} = \frac{v_L(0^+)}{L} = \frac{v_C(0^+)}{L} = \frac{0}{.167} = 0$$

We can now evaluate the constants:

$$i_L(0^+) = .0015 = K_1 + K_2 + .01$$

$$\frac{di_L(0^+)}{dt} = 0 = -200K_1 - 300K_2$$

so

$$K_1 = -.0255 \quad \text{and} \quad K_2 = .017$$

The total solution is

$$i_L(t) = -.0255e^{-200t} + .017e^{-300t} + .01 \text{ A}, \qquad t > 0$$

---

**Example 7.6**    *Parallel RLC Circuit*

In the circuit in Figure 7.12A, $S_1$ opens and $S_2$ closes at $t = 0$. Find $v_C(t)$ for $t > 0$.

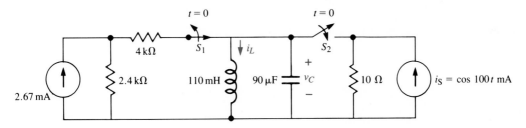

A. Original circuit

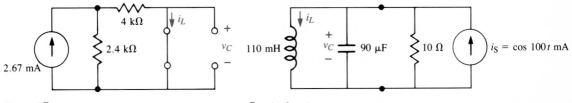

B. $t = 0^-$                              C. $t > 0$

**FIGURE 7.12**

*(continues)*

**Example 7.6**   *Continued*

We could find $v_C(t)$ by first recognizing that $v_C(t) = v_L(t) = L\, di_L/dt$ and then finding $i_L(t)$ as was done in the preceding discussion. It is more instructive, however, to find $v_C(t)$ directly. In either case, we must first find the initial inductor current and capacitor voltage.

From the circuit in Figure 7.12B,

$$v_C(0^-) = v_C(0^+) = 0 \text{ V}$$

and

$$i_L(0^-) = i_L(0^+) = \frac{2.4}{6.4} \cdot 2.67 = 1 \text{ mA}$$

Writing a node equation for the $t > 0$ circuit shown in Figure 7.12C in terms of $v_C$ gives

$$\frac{1}{L}\int_{-\infty}^{t} v_C(t)\, dt + C\,\frac{dv_C}{dt} + \frac{v_C}{R} = i_s$$

so

$$\frac{d^2 v_C}{dt^2} + \frac{1}{RC}\frac{dv_C}{dt} + \frac{1}{LC}v_C = \frac{1}{C}\frac{di_s}{dt}$$

$$\frac{d^2 v_C}{dt^2} + 1111\frac{dv_C}{dt} + 101{,}010 = -1111 \sin 100t$$

The solution to the characteristic equation is

$$s^2 + 1111s + 101{,}010 = 0$$

$$s = -100,\ -1011$$

The forced solution is

$$v_{C_F} = A \cos 100t + B \sin 100t$$

Substituting $v_{C_F}$ into the differential equation yields

$$A = .006 \qquad \text{and} \qquad B = -.005$$

so

$$v_C(t) = K_1 e^{-100t} + K_2 e^{-1011t} + .006 \cos 100t - .005 \sin 100t$$

$$\frac{dv_C(t)}{dt} = -100 K_1 e^{-100t} - 1011 K_2 e^{-1011t} - .6 \sin 100t - .5 \cos 100t$$

Since $v_C(0^+)$ is already known, $dv_C(0^+)/dt$ must now be found. Recognizing that

$$\frac{dv_C}{dt} = \frac{i_C}{C}$$

we need to find $i_C(0^+)$. In this parallel circuit, we see immediately that

$$i_C = i_s - i_L - i_R = i_s - i_L - \frac{v_C}{R}$$

so

$$i_C(0^+) = i_s(0^+) - i_L(0^+) - \frac{v_C(0^+)}{R} = .001 - .001 - \frac{0}{10^5} = 0$$

Thus,

$$\frac{dv_C(0^+)}{dt} = \frac{i_C(0^+)}{C} = 0$$

We can now evaluate the constants:

$$v_C(0^+) = 0 = K_1 + K_2 + .006$$

$$\frac{dv_C(0^+)}{dt} = 0 = -100K_1 - 1011K_2 - .5$$

so

$$K_1 = -.0061 \qquad \text{and} \qquad K_2 = .0001$$

The total solution is

$$v_C(t) = -6.1e^{-100t} + .1e^{-1011t} + 6\cos 100t - 5\sin 100t \text{ mV}, \qquad t > 0$$

---

### Series *RLC* Circuit

Figure 7.13A shows a switched circuit that becomes a series $RLC$ circuit for $t > 0$ (Figure 7.13C). We begin the analysis as we did for the circuit in Figure 7.11: We analyze the circuit at $t = 0^-$ (Figure 7.13B) to find the initial conditions.

From the $t = 0^-$ circuit, we find that

$$v_C(0^-) = 20 \text{ V}$$

and

$$i_L(0^-) = 0 \text{ A}$$

**FIGURE 7.13** *Series Switched RLC Circuit*

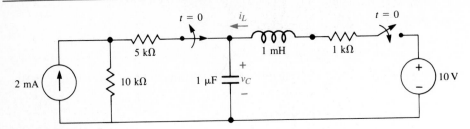

A. Original circuit

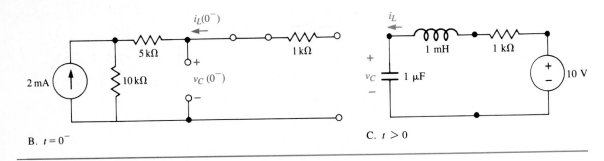

B. $t = 0^-$

C. $t > 0$

The series $RLC$ circuit (Figure 7.13C) has already been analyzed in Chapter 6. The differential equation for $v_C(t)$ is

$$\frac{d^2 v_C}{dt^2} + \frac{R}{L}\frac{dv_C}{dt} + \frac{1}{LC}v_C = \frac{1}{LC}v_s$$

so

$$\frac{d^2 v_C}{dt^2} + 10^6 \frac{dv_C}{dt} + 10^9 v_C = 10^{10}$$

The natural solution is

$$s^2 + 10^6 s + 10^9 = 0$$

$$s_1 \approx -10^3 \quad \text{and} \quad s_2 \approx -10^6$$

$$v_{C_N} = K_1 e^{-t/10^{-3}} + K_2 e^{-t/10^{-6}}$$

The forced solution is

$$v_{C_F} = 10 \text{ V}$$

The total solution is

$$v_C(t) = K_1 e^{-t/10^{-3}} + K_2 e^{-t/10^{-6}} + 10$$

$$\frac{dv_C(t)}{dt} = -10^3 K_1 e^{-t/10^{-3}} - 10^6 K_2 e^{-t/10^{-6}}$$

We already know $v_C(0^+)$ and now need to find $dv_C(0^+)/dt$. We do this by remembering that $dv_C/dt = i_C/C$. In this series circuit, $i_C = i_L$, so we have

$$\frac{dv_C(0^+)}{dt} = \frac{i_C(0^+)}{C} = \frac{i_L(0^+)}{C} = \frac{0}{1\,\mu F} = 0$$

We can now evaluate the constants:

$$v_C(0^+) = 20 = K_1 + K_2 + 10$$

$$\frac{dv_C(0^+)}{dt} = 0 = -10^3 K_1 - 10^6 K_2$$

so

$$K_1 \approx 10 \quad \text{and} \quad K_2 \approx -10^{-2}$$

The total solution is

$$v_C(t) = 10 e^{-t/10^{-3}} - .01 e^{-t/10^{-6}} + 10\ \text{V}, \qquad t > 0$$

---

**Example 7.7**   *Series RLC Circuit*

In the circuit in Figure 7.14A, $S_1$ opens and $S_2$ closes at $t = 0$. Find $i_L(t)$ for $t > 0$.

We could find $i_L(t)$ by first finding $v_C(t)$ as just described and then recognizing that $i_L = i_C = C(dv_C/dt)$ for $t > 0$. However, we can gain some additional experience by solving for $i_L(t)$ directly. In either case, we must first find the initial conditions.

From the circuit in Figure 7.14B,

$$v_C(0^-) = 0\ \text{V}$$

and

$$i_L(0^-) = .57\ \text{mA}$$

(*continues*)

**Example 7.7**   *Continued*

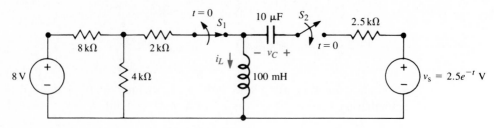

A. Original circuit

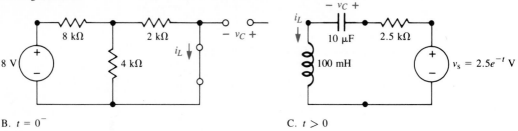

B. $t = 0^-$                                           C. $t > 0$

**FIGURE 7.14**

For the $t > 0$ circuit shown in Figure 7.14C, the differential equation for $i_L(t)$ is

$$\frac{d^2 i_L}{dt^2} + \frac{R}{L}\frac{di_L}{dt} + \frac{1}{LC}i_L = \frac{1}{L}\frac{dv_s}{dt}$$

so

$$\frac{d^2 i_L}{dt^2} + (25 \times 10^3)\frac{di_L}{dt} + 10^6 i_L = -25e^{-t}, \qquad t > 0$$

$$i_L(t) = K_1 e^{-24,960t} + K_2 e^{-40t} - (25.6 \times 10^{-6})e^{-t}, \qquad t > 0$$

$$i_L(0^+) = K_1 + K_2 - 25.6 \times 10^{-6}$$

and

$$\frac{di_L(0^+)}{dt} = -24,960K_1 - 40K_2 + 25.6 \times 10^{-6}$$

We already know that $i_L(0^+) = i_L(0^-) = .57\,\text{mA}$. We need to find $di_L/dt$. Remembering that

$$\frac{di_L}{dt} = \frac{v_L}{L}$$

we see that

$$v_L = v_s - v_C - v_R = v_s - v_C - i_L R$$

so

$$v_L(0^+) = v_s(0^+) - v_C(0^+) - i_L(0^+)R = 2.5 - 0 - 1.43 = 1.07$$

Thus,

$$\frac{di_L(0^+)}{dt} = \frac{v_L(0^+)}{L} = \frac{1.07}{.1} = 10.7$$

We can now evaluate the constants:

$$.57 \times 10^{-3} = K_1 + K_2 - 25.6 \times 10^{-6}$$
$$10.7 = -24,960 K_1 - 40 K_2 + 25.6 \times 10^{-6}$$

so

$$K_1 = -.43 \times 10^{-3} \qquad \text{and} \qquad K_2 = 10^{-3}$$

The final solution is

$$i_L(t) = -.43 e^{-24,960t} + e^{-40t} - .0256 e^{-t} \text{ mA}, \qquad t > 0$$

---

## General Second-Order Circuit

In the circuit shown in Figure 7.15A, $S_1$ opens and $S_2$ closes at $t = 0$. Let's find $v_{C_1}$ and $v_{C_2}$. From the $t = 0^-$ circuit shown in Figure 7.15B, we find that

$$v_{C_1}(0^-) = 2 \text{ V} \qquad \text{and} \qquad v_{C_2}(0^-) = .5 \text{ V}$$

so

$$v_{C_1}(0^+) = 2 \text{ V} \qquad \text{and} \qquad v_{C_2}(0^+) = .5 \text{ V}$$

We use node analysis in the $t > 0$ circuit shown in Figure 7.15C to derive differential equations for $v_{C_1}$ and $v_{C_2}$ (the differential operator technique may also be used). After manipulation of the node equations, we get

$$3 \frac{dv_{C_1}}{dt} + 10^3 v_{C_1} - 10^3 v_{C_2} = 0$$

$$-10^3 v_{C_1} + 6 \frac{dv_{C_2}}{dt} + (4 \times 10^3) v_{C_2} = 15 \times 10^3$$

FIGURE 7.15    *General Second-Order Circuit*

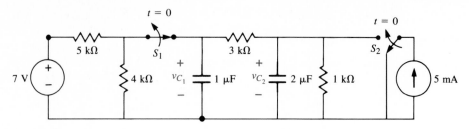

A. Original circuit

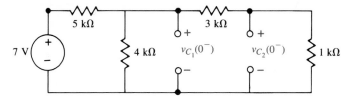

B. $t = 0^-$

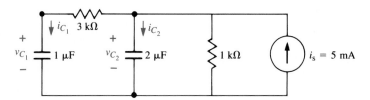

C. $t > 0$

From the first node equation, we find

$$v_{C_2} = (3 \times 10^{-3}) \frac{dv_{C_1}}{dt} + v_{C_1}$$

and

$$\frac{dv_{C_2}}{dt} = (3 \times 10^{-3}) \frac{d^2 v_{C_1}}{dt^2} + \frac{dv_{C_1}}{dt}$$

We substitute these relationships into the second node equation:

$$.018 \frac{d^2 v_{C_1}}{dt^2} + 18 \frac{dv_{C_1}}{dt} + (3 \times 10^3) v_{C_1} = 15 \times 10^3$$

so

$$\frac{d^2 v_{C_1}}{dt^2} + 1000 \frac{d v_{C_1}}{dt} + (.167 \times 10^6) v_{C_1} = .833 \times 10^6$$

In a similar manner, we derive the following equation for $v_{C_2}$:

$$\frac{d^2 v_{C_2}}{dt^2} + 1000 \frac{d v_{C_2}}{dt} + (.167 \times 10^6) v_{C_2} = .833 \times 10^6$$

In this particular problem, the mathematical forcing function for the two equations turns out to be the same, which is just coincidental. We get the following answers for the capacitor voltages and their derivatives:

$$v_{C_1}(t) = K_1 e^{-800t} + K_2 e^{-200t} + 5$$

$$\frac{d v_{C_1}}{dt} = -800 K_1 e^{-800t} - 200 K_2 e^{-200t}$$

and

$$v_{C_2}(t) = K_3 e^{-800t} + K_4 e^{-200t} + 5$$

$$\frac{d v_{C_2}}{dt} = -800 K_3 e^{-800t} - 200 K_4 e^{-200t}$$

We now need the initial conditions. We already found that $v_{C_1}(0^+) = 2$ and that $v_{C_2}(0^+) = .5$. We find $d v_{C_1}(0^+)/dt$ and $d v_{C_2}(0^+)/dt$ by solving for $i_{C_1}$ and $i_{C_2}$:

$$i_{C_1}(t) = C_1 \frac{d v_{C_1}}{dt} = \frac{v_{C_2} - v_{C_1}}{3000}$$

and

$$i_{C_2}(t) = C_2 \frac{d v_{C_2}}{dt} = i_s - \frac{v_{C_2}}{1000} - \frac{v_{C_2} - v_{C_1}}{3000}$$

so

$$\frac{d v_{C_1}(0^+)}{dt} = \frac{i_{C_1}(0^+)}{C_1} = -500$$

and

$$\frac{d v_{C_2}(0^+)}{dt} = \frac{i_{C_2}(0^+)}{C_2} = 2500$$

We use the initial conditions to find the unknown coefficients, $K_1$, $K_2$, $K_3$, and $K_4$. The final answers are

$$v_{C_1}(t) = \frac{11}{6} e^{-800t} - \frac{29}{6} e^{-200t} + 5 \text{ V}, \qquad t > 0$$

and

$$v_{C_2}(t) = -\frac{11}{6} e^{-800t} - \frac{16}{6} e^{-200t} + 5 \text{ V}, \qquad t > 0$$

## 7.5  THE UNIT STEP FUNCTION

The unit step function is shown in Figure 7.16. It is the same function of time that was shown in Figure 7.2. Because of its importance, this function is given its own symbol, $u(t)$, and is defined as follows:

$$u(t) = \begin{cases} 1, & t > 0 \\ 0, & t < 0 \end{cases} \qquad\qquad (7.4)$$

At this point, the unit step is not defined at $t = 0$. Some advanced mathematical procedures do require that the unit step be defined at $t = 0$. However, circuit analysis does not generally require it.

The unit step is useful for many reasons. For example, a function of time that starts at $t = 0$ can be represented by multiplying the original function by the unit step. Figure 7.17 shows a sinusoidal function that has been on for all time (Figure 7.17A) and the same function that starts at $t = 0$ (Figure 7.17B). For Figure 7.17A, the function of time is given by

$$f(t) = 5 \sin 100t$$

For Figure 7.17B, the function of time is given by

$$g(t) = \begin{cases} 5 \sin 100t, & t > 0 \\ 0, & t < 0 \end{cases}$$

**FIGURE 7.16**  *Unit Step Function*

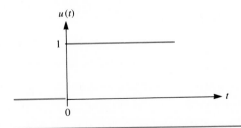

**FIGURE 7.17**  *Switched Sinusoid*

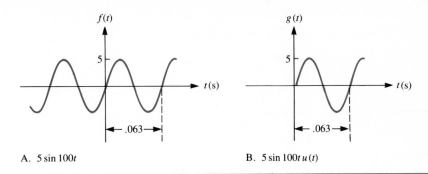

A.  $5 \sin 100t$                                          B.  $5 \sin 100t \, u(t)$

We can write a more compact and mathematically tractable description for $g(t)$ by using the unit step—that is, by multiplying $f(t)$ by the unit step:

$$g(t) = f(t)u(t) = 5 \sin 100t \, u(t)$$

Since the unit step equals 0 for $t < 0$, then any function multiplied by the unit step is also 0 for $t < 0$. For $t > 0$, the unit step equals 1, so the product yields just $f(t)$.

Figure 7.18 shows how the unit step function can be used to represent switched sources. Figure 7.18A shows a current source that is switched in at $t = 0$ and its equivalent source, $i_s(t)u(t)$. Figure 7.18B shows a voltage source that is switched in at $t = 0$ and its equivalent source, $v_s(t)u(t)$.

A common error in using the unit step to represent switched sources is shown in Figure 7.18C. For $t < 0$, that the series switch is open does *not* imply that the terminal voltage shown is zero. The value of the terminal voltage will depend on the circuit to which the terminals are connected. In this case, $v_s'(t) \neq v_s(t)u(t)$.

Figure 7.19A shows a simple example of the use of step functions to represent switching operations in circuits. For $t < 0$, the current source is open, as shown in Figure 7.19B. Since there is no source, the capacitor voltage is zero: $v_C(0^-) = v_C(0^+) = 0$. For $t > 0$, the source is a constant 5 mA, as shown in Figure 7.19C. Therefore, $v_{C_F} = 25$ V and

$$v_C(t) = \begin{cases} 0 \text{ V}, & t < 0 \\ 25 - 25e^{-t/.05} \text{ V}, & t > 0 \end{cases}$$

or

$$v_C(t) = (25 - 25e^{-t/.05})u(t) \text{ V}$$

Caution! You can only use the unit step notation for a circuit variable response if that variable is known to be *zero for all $t < 0$*. Otherwise, you indicate that the solution is valid only for $t > 0$.

FIGURE 7.18    *Switched Sources*

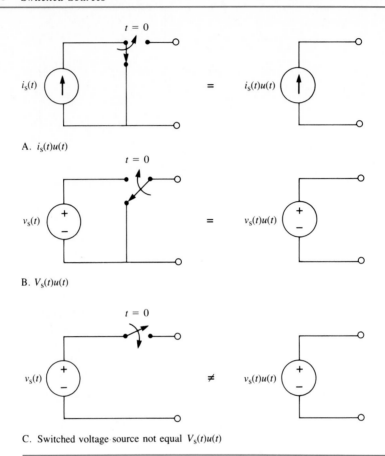

A. $i_s(t)u(t)$

B. $V_s(t)u(t)$

C. Switched voltage source not equal $V_s(t)u(t)$

The unit step is a versatile function. A more general description of it is as follows:

$$u(x) = \begin{cases} 0, & x < 0 \\ 1, & x > 0 \end{cases} \tag{7.5}$$

where $x$ is an arbitrary function known as the *argument* of the unit step. The unit step is 0 when its argument is negative, 1 when its argument is positive, and, for now, undefined when its argument is 0.

Figure 7.20 shows several examples of step functions. In Figure 7.20A, the unit step is delayed by $T$ seconds. Delay implies that the function of time is shifted to the right on the time axis. In this case, the argument of $u(x)$ should be negative for $t < T$ and positive for $t > T$. This function is given by

$$g_1(t) = u(t - T)$$

FIGURE 7.19   *First-Order Step Response Analysis*

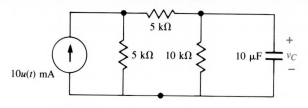

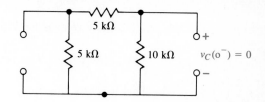

A. Original circuit

B. $t = 0^-$

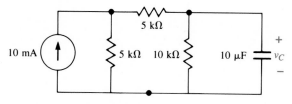

C. $t > 0$

FIGURE 7.20   *Step Functions*

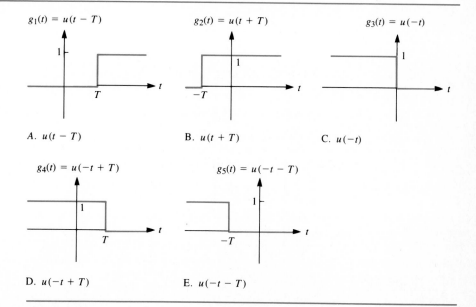

$g_1(t) = u(t - T)$

A. $u(t - T)$

$g_2(t) = u(t + T)$

B. $u(t + T)$

$g_3(t) = u(-t)$

C. $u(-t)$

$g_4(t) = u(-t + T)$

D. $u(-t + T)$

$g_5(t) = u(-t - T)$

E. $u(-t - T)$

Figure 7.20B shows a step function shifted to the left on the time axis. This function is an advance (or negative delay) of the unit step and is given by

$$g_2(t) = u(t + T)$$

The argument $t + T$ is negative for $t < -T$ and positive for $t > -T$.

Figure 7.20C shows a step function that is rotated about the vertical axis. This function turns off at $t = 0$ and is given by

$$g_3(t) = u(-t)$$

Here, the argument is negative for $t > 0$ and positive for $t < 0$.

Figures 7.20D and 7.20E show the delayed and advanced versions, respectively, of $u(-t)$. These functions are given by

$$g_4(t) = u(-t + T)$$

and

$$g_5(t) = u(-t - T)$$

Check the arguments of these functions to verify their correctness.

---

**Example 7.8**    *Switched Sources*

For the circuit in Figure 7.21A, find $v_C(t)$ for $t > 0$.

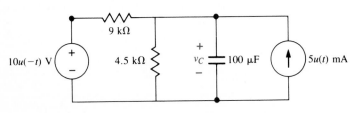

A. Original circuit

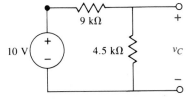

B. $t = 0^-$

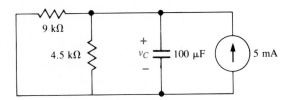

C. $t > 0$

**FIGURE 7.21**

For $t<0$, the voltage source is on and the current source is off. At $t = 0^-$ (Figure 7.21B),

$$v_C(0^-) = 3.33 \text{ V}$$

For $t>0$, the voltage source is off and the current source is on (Figure 7.21C):

$$v_C(t) = 15 + Ke^{-t/.3}$$

Using the initial condition gives

$$v_C(t) = 15 - 11.67e^{-t/.3} \text{ V}, \qquad t>0$$

Although it is common to write this answer as $(15 - 11.67e^{-t/.3})u(t)$, that is not strictly correct. Using the step function in the answer says that $v_C(t)$ is zero for all $t<0$, which is not true here.

## 7.6 SEQUENTIAL SWITCHING

In many switched circuits, switches are opened and closed more than once. If these circuits are driven by time-varying sources, their analysis is very difficult and often accomplished with the aid of computer programs. However, if the inputs are constant, then analysis is reasonably straightforward. It is merely a matter of repeated application of techniques you already know.

The first step in analyzing sequentially switched circuits is the understanding of circuit response to a delayed input. The simple circuit shown in Figure 7.22 is driven by a delayed unit step.

The differential equation for this circuit is

$$\frac{dv(t)}{dt} + .1v(t) = u(t - 2)$$

**FIGURE 7.22** *Delayed Step Input*

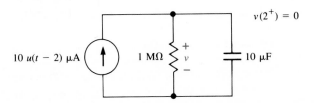

A formal solution to this equation is straightforward if we first make the following variable substitution:

$$m = t - 2$$

so

$$t = m + 2$$

and

$$dt = dm$$

The differential equation then becomes

$$\frac{dv(m + 2)}{dm} + .1v(m + 2) = u(m)$$

The voltage in this differential equation is the same voltage as in the original equation. We have changed only the time label.

The solution for $v(m + 2)$ is a simple step response whose forced response is 10. We get, therefore,

$$v(m + 2) = (10 + Ke^{-.1m})u(m)$$

Since $m = t - 2$, we can write

$$v(t) = (10 + Ke^{-.1(t-2)})u(t - 2)$$

Given the value of $v(t)$ for $t = 2$, as shown in Figure 7.22, use this value to find $K$:

$$v(2) = 0 = 10 + K$$
$$K = -10$$

We get the final solution as follows:

$$v(t) = (10 - 10e^{-.1(t-2)})u(t - 2) \text{ V} \tag{7.6}$$

It is instructive at this point to compare this answer to the response that we get if the input turns on at $t = 0$. That is, if the current source in the circuit of Figure 7.22 equals $10u(t)$ and we are given $v(0^+) = 0$, then the response is

$$v(t) = (10 - 10e^{-.1t})u(t) \text{ V} \tag{7.7}$$

The response to the unit step (Equation 7.7) is plotted in Figure 7.23A, and the response to the delayed unit step (Equation 7.6) is plotted in Figure 7.23B. You can see that the only difference in the two responses is a time delay of 2 s. The output shown in Figure 7.23B is delayed by the same amount as the input to

FIGURE 7.23   *Comparison of Step and Delayed Step Responses*

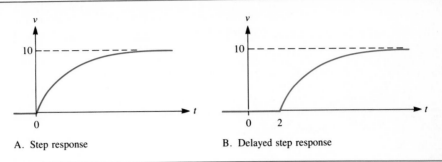

A. Step response

B. Delayed step response

the circuit. This response is a demonstration of time invariance. In a time-invariant circuit, a delay in the input causes the same delay in the output but does not change the waveshape.

In analyzing circuits with delays, we usually do not introduce a new variable. We simply assume that there is no delay, find the circuit variables of interest, and then delay the answers by the amount of delay at the input. Note that the initial condition must be known at the time the sources turn on. For example, Equation 7.3, a formula for the step response of a variable in a first-order circuit, assumes that the input is applied at $t = 0$. If we delay the input by $t_0$ seconds—that is, input $= u(t - t_0)$—then we can rewrite this formula for any delayed first-order step response as follows:

$$x(t) = [x(\infty) + (x(t_0^+) - x(\infty))e^{-(t-t_0)/\tau}]u(t - t_0)$$     (7.8)

Example 7.9   *Delayed Response*

Find $v_C(t)$ in the circuit in Figure 7.24A.

We first assume that the input is $5u(t)$, and analyze. Since $v_C(t)$ is the step response of a first-order circuit, we can find it by inspection. The final value of $v_C$ is found by replacing the capacitor with an open circuit, and we find that $v_C(\infty) = 2.5$ V. The equivalent resistor seen by the capacitor is $70 \text{ k}\Omega$. Since the initial condition is given as 0 V,

$$v_C(t) = (2.5 - 2.5e^{-t/.7})u(t) \text{ V}$$

We now delay the response by .5 s. We replace $t$ *everywhere*, except in $v_C(t)$, by $(t - .5)$ and get the final answer as follows:

$$v_C(t) = (2.5 - 2.5e^{-(t-.5)/.7})u(t - .5) \text{ V}$$

*(continues)*

**Example 7.9**   *Continued*

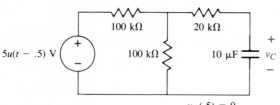

A. Original circuit

$v_C(.5) = 0$

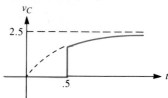

B. Undelayed step response

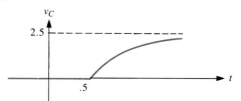

C. Delayed step response

D. Undelayed step response switched
in at $t - .5$, a common error

**FIGURE 7.24**

For comparison, we plot both the undelayed response and the delayed response, as shown in Figures 7.24B and 7.24C.

The common error that occurs in this informal technique is to replace just the $t$ in the unit step function:

$$v_C(t) = (2.5 - 2.5e^{-t/.7})u(t - .5) \text{ V}$$

The plot for this response is shown in Figure 7.24D. With this answer, we have not delayed the exponential. We merely switch the response in at $t = .5$.

The next step in the development of sequential switching analysis is the analysis of a circuit driven by a pulse. Such a circuit is shown in Figure 7.25. Let's first solve for $v(t)$ with analytical methods and then discuss solution by inspection.

We begin by writing the expression for $v_s(t)$. The pulse shown in Figure 7.25A can be constructed by subtracting a delayed step from the unit step:

$$v_s(t) = u(t) - u(t - .5) \tag{7.9}$$

We find $v(t)$ by superposition. The step response is

$$v_1(t) = (.5 - .5e^{-t/.4})u(t) \tag{7.10}$$

The response to the delayed step, $u(t - .5)$, is given by Equation 7.8. Since the

FIGURE 7.25   *Pulse Input*

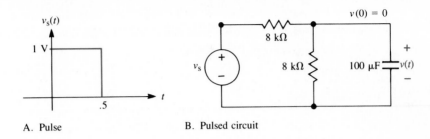

A. Pulse          B. Pulsed circuit

capacitor is initially uncharged in this example circuit, the initial condition for the delayed step response is zero:

$$v_2(t) = (.5 - .5e^{-(t-.5)/.4})u(t - .5)$$

The total answer is $v_1$ minus $v_2$:

$$v(t) = (.5 - .5e^{-t/.4})u(t) - (.5 - .5e^{-(t-.5)/.4})u(t - .5) \text{ V}$$

To gain some additional insight, let's find $v(t)$ for $t < .5$ and for $t > .5$:

$$v(t) = .5 - .5e^{-t/.4} \text{ V}, \qquad 0 < t < .5 \tag{7.11}$$

which is the answer that we already found in Equation 7.10, and

$$v(t) = .5 - .5e^{-t/.4} - .5 + .5(e^{-t/.4})(e^{.5/.4}) = 1.245e^{-t/.4} \text{ V}, \qquad t > .5 \tag{7.12}$$

Note that at $t = .5$, $v(t) = .357$ for both regions of time.

We can find $v(t)$ by inspection. Remember, the step response of a first-order circuit is completely determined by its initial condition, final value, and time constant. Finding the initial condition and time constant is fairly simple, but finding the final value requires more thought.

For $0 < t < .5$, the pulse looks like a step to the circuit. The circuit cannot anticipate that the step will be turned off at $t = .5$. Therefore, the final value of $v(t)$ for this range of time is found by assuming that the input is a step. We replace the capacitor with an open circuit and find that the final value is .5 V. Note that this is the value that $v(t)$ would reach if the input were left on.

The initial value is given as zero, and the time constant is found to be .4 s. Therefore, the response for $0 < t < .5$ is

$$v(t) = .5 - .5e^{-t/.4} \text{ V}, \qquad 0 < t < .5 \tag{7.13}$$

FIGURE 7.26    *Pulse Response*

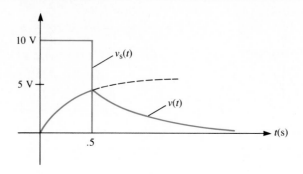

At $t = .5$, the input is turned off. Since $v(t)$ is a capacitor voltage, $v(.5^+) = v(.5^-)$. The initial value for the region $t > .5$ is found by evaluating Equation 7.13 at $t = .5^-$:

$$v(.5^-) = .5 - .5e^{-.5/.4} = .357$$

The final value is a function of the new input value. Since, in this case, the input is zero, the final value is also zero. The time constant is still .4 s. The response for $t > .5$ is

$$v(t) = .357e^{-(t-.5)/.4} \text{ V}, \qquad t > .5 \tag{7.14}$$

Remember, this part of the response is delayed by .5 s. Equation 7.13 is identical to Equation 7.11, and Equation 7.14 is the same as Equation 7.12. The input pulse and the response are shown in Figure 7.26. The dashed line indicates where the first part of the response would head if the step were not turned off.

Example 7.10    *Inductor Pulse Response*

Analyze the circuit shown in Figure 7.27A by inspection.
For $0 < t < 1\ \mu s$,

Initial value = 0      (given)
Final value = .67 mA      (inductor shorted)
Time constant = .67 $\mu s$      ($\tau = .01/15,000$)

Therefore,

$$i(t) = .67 - .67e^{-t/.67 \times 10^{-6}} \text{ mA}, \qquad 0 < t < 1\ \mu s$$

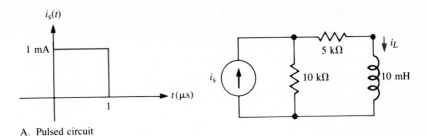

A. Pulsed circuit

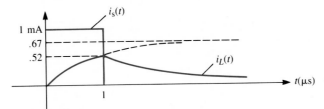

**FIGURE 7.27**    B. Pulse response

For $t > 1\ \mu s$,

Initial value $= .52$ mA     $(.67 - .67e^{-10^{-6}/.67 \times 10^{-6}})$

Final value $= 0$     (source turned off)

Time constant $= .67\ \mu s$     $(\tau_L = .01/15{,}000)$

Therefore,

$$i(t) = .52e^{-(t-10^{-6})/.67 \times 10^{-6}}\ \text{mA}, \qquad 1\ \mu s < t$$

The response is plotted in Figure 7.27B.

---

Now that we have considered delayed step and pulse responses, we are ready to deal with sequentially switched circuits. We will use inspection techniques to analyze these circuits. Remember, to find final values, assume that the switch will stay in its present position forever. Initial values are found from the previous region just before the switch is thrown.

In the sequentially switched circuit of Figure 7.28A, the switch is at position 1 for a long time. At $t = 0$, the switch is thrown to position 2. At $t = .25$ s, it is thrown to position 3, where it remains. We analyze this circuit as follows.

For position 1,     $v(0^-) = 5$ V     (initial condition for $0 < t < .25$ s)

FIGURE 7.28    *Sequentially Switched Response*

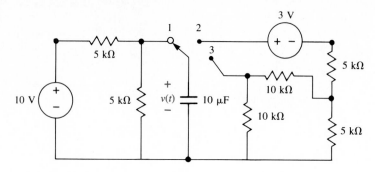

A. Original circuit

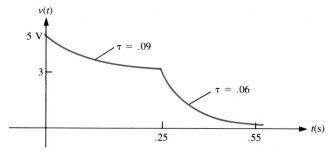

B. Capacitor voltage

For position 2,     Initial condition $= v(0^+) = v(0^-) = 5$
Final value $= 3$
Time constant $= .09$ s
$(R_{eq} = 5000 + (5000 \parallel 20,000) = 9 \text{ k}\Omega)$

Therefore,

$$v(t) = 3 + 2e^{-t/.09} \text{ V}, \qquad 0 < t < .25$$

For position 3,     Initial condition $= 3 + 2e^{-.25/.09} = 3.12$
Final value $= 0$
Time constant $= .06$ s
$(R_{eq} = 10,000 \parallel (10,000 + 5000) = 6 \text{ k}\Omega)$

Therefore,

$$v(t) = 3.12e^{-(t-.25)/.06} \text{ V}, \qquad t > .25$$

The total response is plotted in Figure 7.28B.

**Example 7.11**   *Sequential Switching*

Find $v(t)$ in the circuit shown in Figure 7.29A. The switch is at position 1 for a long time. At $t = 0$ s, the switch is moved to position 2. At $t = .1$ s, the switch is moved to position 3, where it remains.

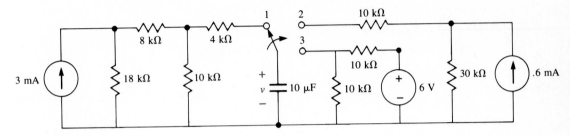

A. Original circuit

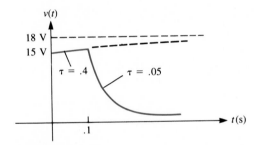

B. Capacitor voltage

**FIGURE 7.29**

For position 1,   $v(0^-) = 15$ V

For position 2,   Initial condition = 15
Final value = 18
Time constant = .4 s

Therefore,

$$v(t) = 18 - 3e^{-t/.4} \text{ V}, \quad 0 < t < .1$$

For position 3,   Initial condition = $18 - 3e^{-.1/.4} = 15.66$
Final value = 3
Time constant = .05 s

*(continues)*

**Example 7.11**   *Continued*

Therefore,

$$v(t) = 3 + 12.66e^{-(t-.1)/.05} \text{ V}, \qquad t > .1$$

The response is plotted in Figure 7.29B.

---

**Example 7.12**   *Voltage-Controlled Switch*

The switch in Figure 7.30A closes whenever $v_o < 2$ V and opens whenever $v_o > 5$ V. This type of switch can be built with semiconductor devices. Find $v_o(t)$.

This example presents us with a different type of problem than previous examples did. The timing of the switching is not given here but will be determined during the analysis. Since $v_o(0) = 0$, the switch is closed at $t = 0$. Therefore, the output voltage rises according to

$$v_o(t) = 10 - 10e^{-t/.375} \text{ V}$$

This response is plotted in Figure 7.30B.

When $v_o$ reaches 5 V, the switch opens. The time $t_1$ at which this occurs is found as follows:

$$5 = 10 - 10e^{-t_1/.375}$$

$$t_1 = .26 \text{ s}$$

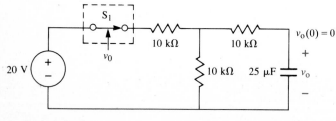

A. Original circuit

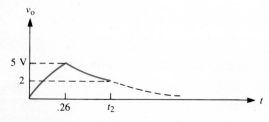

B. First sequence

C. Second sequence

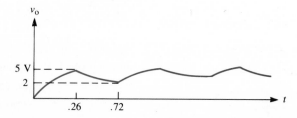

D. Complete response

**FIGURE 7.30**

With the switch open, the final value for $v_o$ is 0 V. The initial value is 5 V, and the time constant is now .5 s. Therefore, in this region,

$$v_o(t) = 5e^{-(t-.26)/.5} \text{ V}, \qquad .26 < t < t_2$$

The plot for this response is shown in Figure 7.30C.

At $t = t_2$, $v_o$ falls to 2 V, and the switch closes. The time $(t_2)$ at which this occurs is found as follows:

$$2 = 5e^{-(t_2-.26)/.5}$$

$$t_2 = .72 \text{ s}$$

With the switch closed, $v_o$ again rises toward 5 V with a time constant of .375 s. This process keeps repeating itself. The output voltage is plotted in Figure 7.30D.

## *7.7 THE SWITCHED CAPACITOR

As you probably know, the trend today is to build electronic devices with integrated circuits. These integrated circuits, which are built with silicon chips, have the advantage of being physically very small. Some circuit designs, however, require large resistor values, such as 100 kΩ. A resistor of this size requires a lot of valuable chip area. One solution to this problem is to replace the resistor with a switched capacitor. Small integrated capacitors can be made from semiconductor devices and have values in the pF range. Semiconductor devices also provide the necessary switching.

The general analysis of switched capacitors can be quite involved. However, given certain constraints, you know enough now to understand their behavior. A brief introduction to this important topic follows.

Figure 7.31A shows a capacitor that is switched from position 1 to position 2 and back again to position 1 every $T_s$ seconds. When the switch is at position 1, the capacitor voltage equals $v_1$. When the switch moves to position 2, the

**FIGURE 7.31**  *Switched Capacitor*

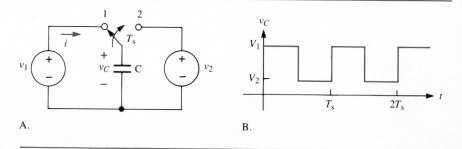

A.

B.

capacitor voltage equals $v_2$. We also assume, for now, that $v_1$ and $v_2$ are constant voltages. The capacitor voltage, $v_C$, is shown in Figure 7.31B.

Note that this case is an exception to the rule about capacitor voltage discontinuities. The jumps occur because there is no resistance in the circuit. The instantaneous change in capacitor voltage implies that the time constant is zero. This is, of course, just an approximation. It is valid as long as the actual time constant in a real circuit is small compared to the switching period $T_s$.

Let's find the amount of charge that is transferred from position 1 to position 2. Remember, for a capacitor,

$$q = Cv$$

When the switch is in position 1, the charge stored in the capacitor is

$$q_1 = Cv_1$$

At position 2, the charge is

$$q_2 = Cv_2$$

The difference in these two charges gives us the charge, $\Delta q$, transferred from position 1 to position 2:

$$\Delta q = q_1 - q_2 = C(v_1 - v_2)$$

This charge is delivered every $T_s$ seconds. Since current is equal to charge delivered per second, we find the current from position 1 to position 2 by dividing by $T_s$:

$$i = \frac{\Delta q}{T_s} = \frac{C}{T_s}(v_1 - v_2) \tag{7.15}$$

Now, examine the simple resistor circuit shown in Figure 7.32. The current in this circuit is given by

$$i = \frac{1}{R_C}(v_1 - v_2) \tag{7.16}$$

**FIGURE 7.32**   *Resistor Equivalent of Switched Capacitor*

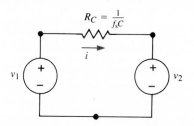

$$R_C = \frac{T_s}{C} = \frac{1}{f_s C} \qquad (7.17)$$

where the switching rate $f_s = 1/T_s$ and has units of hertz (Hz). In general, $v_1$ and $v_2$ are not constant but are time varying. In this case, Equation 7.17 is approximate and applies when $f_s$ is large compared to the time variations in $v_1$ and $v_2$.

---

**Example 7.13**    *Switched Capacitor as a Resistor*

A $100\,k\Omega$ resistor is required for a filter. If a switched capacitor of 1 pF is used, what is the required switching rate?

The answer is calculated from Equation 7.17 as follows:

$$f_s = \frac{1}{R_C C} = \frac{1}{10^5 \cdot 10^{-12}} = 10^7 \, \text{Hz} = 10 \, \text{MHz}$$

---

In most circuits, of course, $v_1$ and $v_2$ are not sources but are determined by the rest of the circuit. Analysis can then be quite difficult. However, certain op amp circuits lend themselves very well to the simplified analysis.

Consider the circuit shown in Figure 7.33A. You should recognize this circuit as an inverting integrator, where

$$v_o = -\frac{1}{RC_F} \int_{-\infty}^{t} v_{in}(t)\, dt$$

**FIGURE 7.33**    *Switched Capacitor Integrator*

---

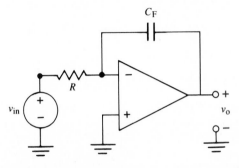

A. *RC* integrator

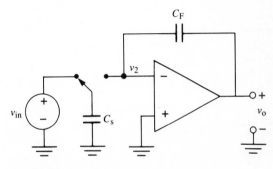

B. Switched capacitor integrator

In the switched capacitor circuit shown in Figure 7.33B, $v_2$ is approximately constant and equal to zero. Remember, in an ideal op amp, $v_- = v_+$.

If the capacitor, $C_s$, in this circuit is switched at a faster rate than the input voltage, $v_{in}$, is changing, then Equation 7.17 is valid and we get the following equivalence:

$$R = \frac{1}{C_s f_s}$$

The input–output relationship for the switched capacitor circuit is, therefore,

$$v_o = -\frac{C_s f_s}{C_F} \int_{-\infty}^{t} v_{in}(t)\, dt$$

**Example 7.14**    *Switched Capacitor Op Amp Circuit*

For the circuit shown in Figure 7.34A, draw the equivalent all-resistor circuit and find $v_o$.

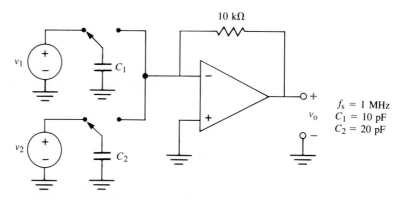

A. Switched capacitor summer

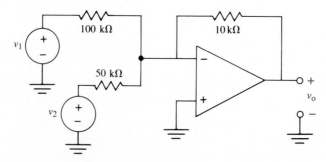

**FIGURE 7.34**    B. Resistor equivalent

For $C_1$,    $R_1 = \dfrac{1}{C_1 f_s} = 100\,\text{k}\Omega$

For $C_2$,    $R_2 = \dfrac{1}{C_2 f_s} = 50\,\text{k}\Omega$

Therefore, from the equivalent circuit of Figure 7.34B,

$$v_o = -.1v_1 - .2v_2$$

## 7.8  SUMMARY

This chapter showed how to analyze first- and some second-order circuits that contain switches and switch functions. General high-order circuits require some additional techniques, which will be considered in later chapters.

An important fact in this chapter is that capacitor voltages and inductor currents do not change instantaneously. Their values just before are the same as their values just after a switch has changed position. Understanding this fact is the key to finding all initial conditions in circuits.

The step function was introduced here to mathematically describe a discontinuity, or jump, in voltage or current. Then, the delayed step was described. You were shown how pulses can be formed from linear combinations of delayed steps.

The response of a circuit to switches that open or close at times other than $t = 0$ (or to delayed steps) was found. The key point to remember here is to replace every variable $t$ in the response with $t - t_1$, where $t_1$ is the delay time.

The results of the delayed response analysis were then used to find responses in circuits with sequential switching and constant inputs. Keep in mind that the circuit does not "know" that a switch is going to open or close at some later time. Therefore, when you are finding the final value of a switched DC circuit variable, still open all capacitors and short all inductors.

Finally, the switched capacitor was introduced. This device provides an equivalent high resistance without taking up as much valuable semiconductor chip area as a resistor does.

## ■ PROBLEMS

**7.1** A general first-order differential equation with a constant input can be written as

$$\frac{dy}{dt} + ay = F, \qquad y(0) = Y_0$$

Use this equation to prove Equation 7.3.

**7.2** The switch in the circuit shown in Figure P7.2 has been at position 1 for a long time. At $t = 0$, the switch is thrown to position 2. Find and sketch $v_C(t)$ for $t > 0$ if

   **a.**  $i_s = 0 \, \text{mA}$
   **b.**  $i_s = 1 \, \text{mA}$
   **c.**  $i_s = 2 \, \text{mA}$

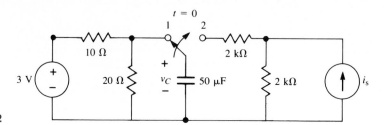

**FIGURE P7.2**

**7.3** The switch in the circuit shown in Figure P7.3 has been at position 1 for a long time. At $t = 0$, the switch is thrown to position 2. Find and sketch $v_C(t)$, $i_C(t)$, and $v_R(t)$ for $t > 0$.

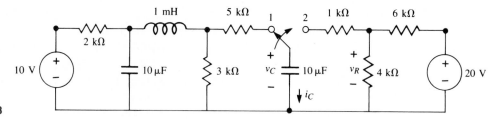

**FIGURE P7.3**

**7.4** The switch in the circuit shown in Figure P7.4 has been at position 1 for a long time. At $t = 0$, the switch is thrown to position 2. Find and sketch $v_C(t)$, $i_C(t)$, and $i_R(t)$ for $t > 0$.

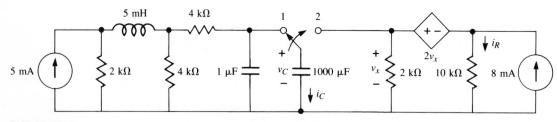

**FIGURE P7.4**

**7.5** Change the 20 V source in Figure P7.3 to $20e^{-50t}$ V. Find $v_C(t)$ and $v_R(t)$ for $t > 0$.

**7.6** The switch in the circuit shown in Figure P7.6 has been at position 1 for a long time. At $t = 0$, the switch is thrown to position 2. Find and sketch $i_L(t)$ for $t > 0$ if

   **a.**  $v_s = 0 \, \text{V}$
   **b.**  $v_s = 2 \, \text{V}$
   **c.**  $v_s = 10 \, \text{V}$

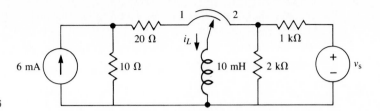

**FIGURE P7.6**

**7.7** The switch in the circuit shown in Figure P7.7 has been at position 1 for a long time. At $t = 0$, the switch is thrown to position 2. Find and sketch $i_L(t)$, $v_L(t)$, and $i_R(t)$ for $t > 0$.

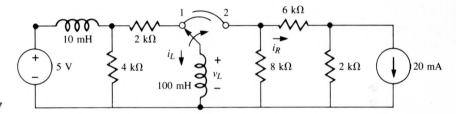

**FIGURE P7.7**

**7.8** The switch in the circuit shown in Figure P7.8 has been at position 1 for a long time. At $t = 0$, the switch is thrown to position 2. Find and sketch $i_L(t)$, $v_L(t)$, and $v_R(t)$ for $t > 0$.

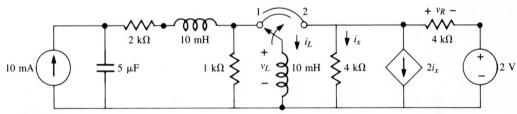

**FIGURE P7.8**

**7.9** Change the 20 mA source in Figure P7.7 to $5 \cos 10{,}000t$ mA. Find $i_L$ and $i_R$ for $t > 0$.

**7.10** Problem 6.36 of the preceding chapter described how to find initial conditions for any circuit variable in a first-order circuit. These initial conditions along with use of the differential operator allow us to directly solve for any circuit variable. Use these techniques to find and then sketch $v_R(t)$ for $t > 0$ from the circuit in Figure P7.3.

**7.11** Repeat Problem 7.10 to find and sketch $i_R(t)$ from the circuit in Figure P7.7.

**7.12** The switch in the circuit shown in Figure P7.12 has been at position 1 for a long time. At $t = 0$, the switch is thrown to position 2. Find and sketch $v_C(t)$, $i_L(t)$, and $v_R(t)$ for $t > 0$.

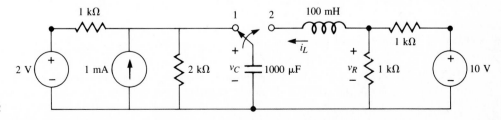

**FIGURE P7.12**

**7.13** The switch in the circuit shown in Figure P7.13 has been at position 1 for a long time. At $t = 0$, the switch is thrown to position 2. Find and sketch $v_C(t)$, $i_L(t)$, and $v_R(t)$ for $t > 0$.

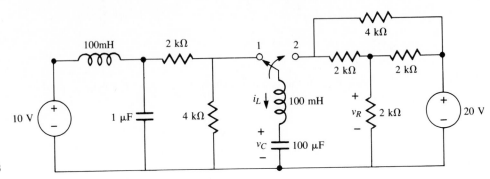

**FIGURE P7.13**

**7.14** The switch in the circuit shown in Figure P7.14 has been at position 1 for a long time. At $t = 0$, the switch is thrown to position 2. Find and sketch $v_C(t)$, $i_L(t)$, and $i_R(t)$ for $t > 0$.

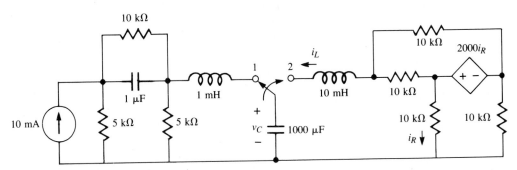

**FIGURE P7.14**

**7.15** Change the 20 V source in Figure P7.13 to $20e^{-10^4 t}$ V and find $v_C(t)$.

**7.16** The switch in the circuit shown in Figure P7.16 has been at position 1 for a long time. At $t = 0$, the switch is thrown to position 2. Find and sketch $i_L(t)$, $v_C(t)$, and $v_R(t)$ for $t > 0$.

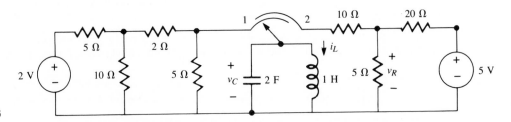

**FIGURE P7.16**

**7.17** The switch in the circuit shown in Figure P7.17 has been at position 1 for a long time. At $t = 0$, the switch is thrown to position 2. Find and sketch $v_C(t)$, $i_L(t)$, and $i_R(t)$ for $t > 0$.

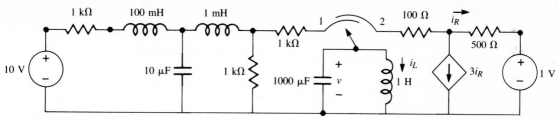

**FIGURE P7.17**

**7.18** Change the 5 V source in Figure P7.16 to $5e^{-2t}$ V and find $i_L(t)$ for $t>0$.

**7.19** For the circuit of Figure P7.13, find $v_R(0^+)$, $dv_R(0^+)/dt$, $dv_C(0^+)/dt$, and $di_L(0^+)/dt$.

**7.20** For the circuit of Figure P7.14, find $i_R(0^+)$, $di_R(0^+)/dt$, $dv_C(0^+)/dt$, and $di_L(0^+)/dt$.

**7.21** For the circuit of Figure P7.16, find $v_R(0^+)$, $dv_R(0^+)/dt$, $dv_C(0^+)/dt$, and $di_L(0^+)/dt$.

**7.22** For the circuit of Figure P7.17, find $i_R(0^+)$, $di_R(0^+)/dt$, $dv_C(0^+)/dt$, and $di_L(0^+)/dt$.

**7.23** Switch $S_1$ in the circuit of Figure P7.23 has been at position 1 for a long time, and switch $S_2$ has been open. At $t=0$, $S_1$ is thrown to position 2 and $S_2$ is closed. Find and sketch $v_1(t)$ and $v_2(t)$ for $t>0$.

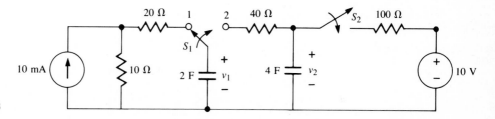

**FIGURE P7.23**

**7.24** The switch in the circuit of Figure P7.24 has been at position 1 for a long time. At $t=0$, the switch is thrown to position 2. Find and sketch $v_C(t)$ and $i_L(t)$ for $t>0$.

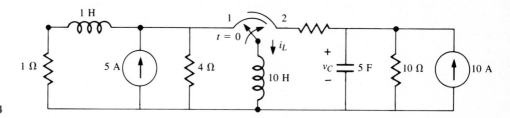

**FIGURE P7.24**

**7.25** Sketch the following functions:

   **a.**   $u(t) - u(t-10)$

   **b.**   $u(t) + .5u(t-5) + 2u(t-7) - 4u(t-12)$

   **c.**   $u(-t-5) + 5u(t+2) - 5u(t-2) + u(t-5)$

   **d.**   $[u(t-5) - u(t-10)] \cos 100t$

**7.26** Sketch the following functions:

    **a.**   $u(5t)$

    **b.**   $u(\cos 10t)$

    **c.**   $u(t^2)$

    **d.**   $u(t^2 - t - 6)$

**7.27** Write the mathematical expression for the graph shown in Figure P7.27.

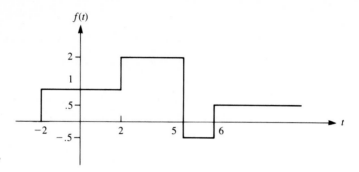

**FIGURE P7.27**

**7.28** Find and sketch $v_C(t)$ for $t > 0$ in the circuit shown in Figure P7.28.

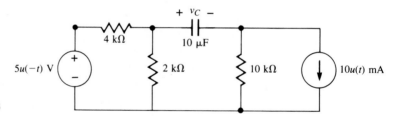

**FIGURE P7.28**

**7.29** Find and sketch $i_L(t)$ for $t > 0$ in the circuit shown in Figure P7.29.

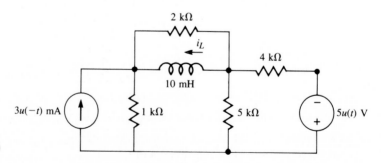

**FIGURE P7.29**

**7.30** Find $v_C(t)$ and $i_L(t)$ for $t > 0$ in the circuit shown in Figure P7.30.

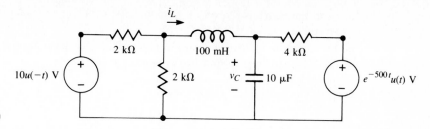

**FIGURE P7.30**

**7.31** Find and sketch $v_C(t)$ in the circuit shown in Figure P7.31 if

    **a.**   $\alpha = 0$

    **b.**   $\alpha = .001$

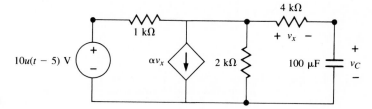

**FIGURE P7.31**

**7.32** Find and sketch $i_L(t)$ in the circuit shown in Figure P7.32 if

    **a.**   $\beta = 0$

    **b.**   $\beta = 10$

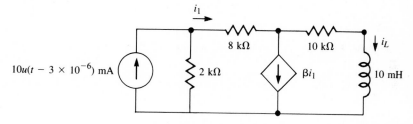

**FIGURE P7.32**

**7.33** Change the $10u(t)$ mA source in Figure P7.28 to $10u(t - .05)$ mA. Find $v_C(t)$ for $t > 0$.

**7.34** Change the $5u(t)$ V source in Figure P7.29 to $5u[t - (5 \times 10^{-6})]$ V. Find $i_L(t)$ for $t > 0$.

**7.35** The switch in the circuit shown in Figure P7.35 has been at position 1 for a long time. At

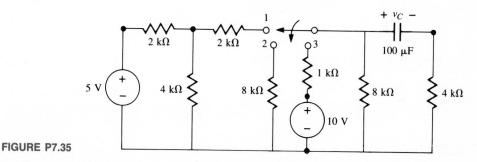

**FIGURE P7.35**

$t = 0$, the switch is thrown to position 2. At $t = 1$ s, the switch is thrown to position 3 and left there. Find and sketch $v_C(t)$ for $t > 0$.

**7.36** The switch in the circuit shown in Figure P7.36 has been at position 1 for a long time. At $t = 0$, the switch is thrown to position 2. At $t = .05$ ms, the switch is thrown to position 3 and left there. Find and sketch $i_L(t)$ for $t > 0$.

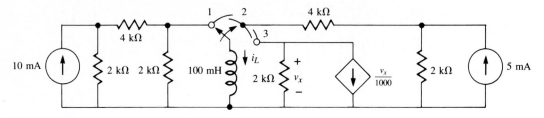

**FIGURE P7.36**

**7.37** The value of $R_D$ in Figure P7.37 is voltage dependent. For $v_D < 1$ V, $R_D = 500$ k$\Omega$. For $v_D > 2$ V, $R_D = 100$ $\Omega$. Find and sketch $v_D(t)$.

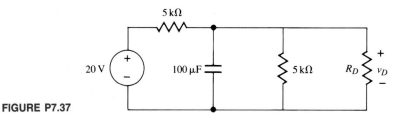

**FIGURE P7.37**

**7.38** Find $v_o$ for the switched capacitor op amp circuits shown in Figure P7.38. In all cases, $f_s = 10$ kHz.

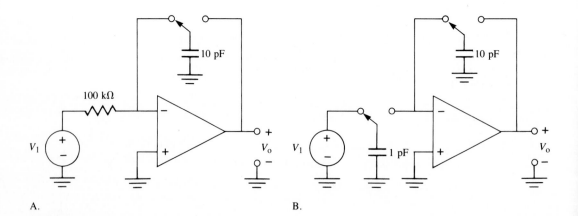

A.

B.

**FIGURE P7.38**

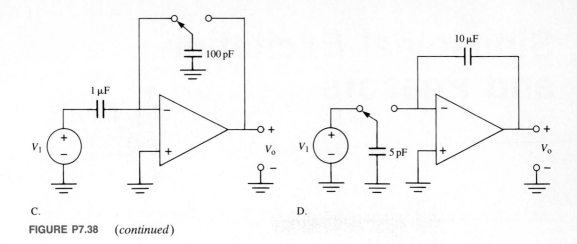

C.

D.

**FIGURE P7.38** (*continued*)

# Sinusoidal Excitation and Phasors

## 8.1  INTRODUCTION

In preceding chapters, we investigated the properties of several time functions that are commonly encountered forms for electrical circuit variables. These functions included step functions, exponentials, and sinusoids. Sinusoids are of such central importance to electrical engineers, however, that we will now devote an entire chapter to their study. There are several reasons why sinusoids are especially significant in electrical engineering. The most obvious reason is that virtually all electrical energy in the world is generated and transmitted in sinusoidal form. Also, sinusoids are used as the carrier signals in communications systems such as television and radio. There are other examples of the use of sinusoidal signals in practical applications, but probably the most important reason for modern electrical engineers to study sinusoidal excitation is a theoretical one. It can be demonstrated that most nonsinusoidal signals encountered in engineering can be treated mathematically as sums of sinusoidal components of various frequencies. This fact is the basis of Fourier analysis, which will be introduced in Chapter 12.

In this chapter, we will study the sinusoidal steady-state behavior of electrical circuits. A circuit is said to be in the sinusoidal steady state if its independent sources are purely sinusoidal as functions of time and if all parts of the natural response of the circuit have decayed to zero. This is not to say that the natural responses of circuits are of little interest to us. Transient responses, especially in circuits undergoing many switching operations, are extremely important to understand. However, a circuit designer or analyst is often more concerned with a circuit's forced response to a specific excitation because that is the signal that carries the intended information or energy. With this idea in mind, circuit analysis sometimes falls into two separate steps. First, a transient analysis is done for the circuit under expected start-up, switching, or failure conditions. Then, once the circuit passes this "test," attention can be turned to the forced response.

We will begin by reviewing the basic properties of sinusoidal signals and the classical methods with which sinusoidally excited circuits are analyzed. Such analyses are said to be carried out in the time domain because the sinusoidal functions are explicitly written as functions of time. Following this review, a new analytical technique will be introduced. This technique will change the description of sinusoids from real-valued functions of time to equivalent complex variable representations known as phasors. Circuit analysis is then performed in what is known as the frequency domain because the frequency of a sinusoid is its most central parameter when described in phasor form. Although this change from the time domain to the frequency domain might seem at first to be an unnecessary complication to a problem, we will see that it actually makes analysis of the sinusoidal steady state much easier.

## 8.2 TIME-DOMAIN ANALYSIS OF THE SINUSOIDAL STEADY STATE

### The General Sinusoid

Before we analyze circuits containing sinusoidal sources, let's briefly review the properties of sinusoidal time functions. Figure 8.1 shows a plot of the general sinusoid

$$x(t) = A \cos (\omega t + \theta) \tag{8.1}$$

where    $A$ = amplitude of the sinusoid

$\omega$ = radian frequency

$\theta$ = phase angle

When these parameters are known, the sinusoid is completely specified. Note that the peak-to-peak value from the most negative to the most positive point of the sinusoid is 2A.

From Figure 8.1, you can see that the sinusoid is a periodic function. It completes one period or cycle every $T$ seconds, where $T$ is the **period** of the sinusoid. The units of the argument of any sinusoid are radians, and, so, the *radian frequency* $\omega$ must have units of radians/second (rad/s). You may re-member from trigonometry that a sinusoid repeats itself every $2\pi$ radians. Thus, we can observe that

$$\omega \left( \frac{\text{radians}}{\text{seconds}} \right) \cdot T \left( \frac{\text{seconds}}{\text{cycle}} \right) = 2\pi \left( \frac{\text{radians}}{\text{cycle}} \right) \tag{8.2}$$

Another way of expressing the frequency of a sinusoid is with the parameter $f$, which has units of cycles/second and which is called, simply, the *frequency* of the sinusoid. More correctly, the units of $f$ are just $s^{-1}$ because cycles are dimension-less. By international agreement, the units of the frequency $f$ have been given the

**FIGURE 8.1**    *Plot of General Sinusoid $x(t) = A \cos (\omega t + \theta)$*

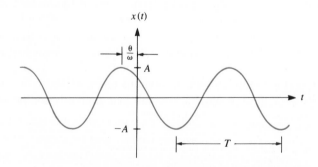

name hertz (Hz) in honor of the nineteenth century German physicist Heinrich Hertz. If there are $T$ seconds/cycle, it follows that

$$f(\text{Hz}) = \frac{1}{T} \left( \frac{\text{cycles}}{\text{second}} \right) \tag{8.3}$$

Combining Equations 8.2 and 8.3 leads to the well-known relationship between $f$ and $\omega$:

$$\omega \left( \frac{1}{f} \right) = 2\pi$$

or

$$\omega = 2\pi f \tag{8.4}$$

The remaining parameter from Equation 8.1, $\theta$, is known as the **phase angle** or, simply, the **phase** of $\cos(\omega t + \theta)$. Properly speaking, the units of $\theta$ should be in radians to be consistent with those of $\omega t$, but engineers often find it more convenient to express phase angles in degrees. Again, from trigonometry, you may remember that

$$1 \text{ radian} = \frac{180°}{\pi} \approx 57.3° \tag{8.5}$$

In this book, the degree symbol (°) is used whenever a phase shift is expressed in degrees. Otherwise, you should assume the units of radians. If expressed in degrees, the phase shift must be converted to radians before it is added to $\omega t$. Also, be careful when you use calculators or computers in such calculations. Make sure that the units you choose to work in are consistent with those of your calculator or programming language.

## Phase Shift

We have seen examples in earlier chapters of functions shifted along the time axis. A phase shift is just a special case of time shifting but with an extra consideration. A shift in phase by $\theta$ radians corresponds to a time shift of $\theta/\omega$ because $\cos(\omega t + \theta) = \cos[\omega(t + \theta/\omega)]$. In other words, the time shift depends on the frequency of the sinusoid. For instance, a 45° phase shift of a sinusoid of frequency $\omega = 100 \text{ rad/s}$ corresponds to a time shift of $(\pi/4)/100 = 7.854 \text{ ms}$. The same phase shift at $\omega = 10,000 \text{ rad/s}$, however, is a time shift of .07854 ms.

A cosine function and two phase-shifted versions are shown in Figure 8.2. $\phi$ is expressed in radians and is assumed to be positive. The asterisks mark corresponding points in the three functions. The function $\cos(\omega t + \phi)$ with its positive phase angle **leads** $\cos \omega t$ by $\phi$ because, for $\phi > 0$, points on the shifted

**FIGURE 8.2**  *Cosine and Two Phase-Shifted Versions*

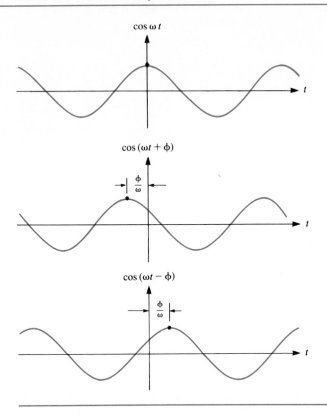

version (the location of a peak value, for instance) occur earlier in time than corresponding points on the unshifted cos $\omega t$. Similarly, cos $(\omega t - \phi)$ **lags** cos $\omega t$ by $\phi$ because corresponding points of the function cos $(\omega t - \phi)$ occur later in time when compared to the original cosine.

The idea of sinusoids leading or lagging one another is directly applicable to the signals found in electrical circuits. Consider the inductor shown in Figure 8.3A and assume its current to be cos $\omega t$. The voltage across the inductor is readily found to be

$$v(t) = L \, \frac{di}{dt}$$

$$= L \, \frac{d}{dt} \cos \omega t$$

$$= -\omega L \sin \omega t$$

$$= -\omega L \cos (\omega t - 90°)$$

$$= \omega L \cos (\omega t + 90°)$$

**FIGURE 8.3**    *Sinusoidal Steady-State v–i Relationships*

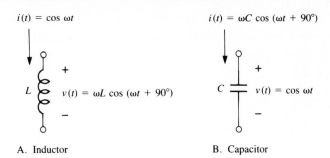

A. Inductor                                                                         B. Capacitor

where, in some of the steps, we have used trigonometric identities found in Appendix B. We see from the result that for sinusoidal excitation, voltage across an inductor leads the current by 90°. In the same way, we can show that capacitor current leads capacitor voltage by 90° (see Figure 8.3B). There is no phase shift between the voltage and current of a resistor. When there is a 0° phase shift between sinusoids, they are said to be **in phase.**

Finding the phase difference between two variables is not always quite as direct as in Figures 8.2 and 8.3. For instance, neither of the sinusoids in Figure 8.4 has a 0° phase angle. From the figure, however, we can see that if $x(t) = X \cos(\omega t + \theta)$ and $y(t) = Y \cos(\omega t + \phi)$, then $y(t)$ lags $x(t)$ by $\theta - \phi$.

It is recommended that certain steps first be performed if the phase difference between two signals is to be found. Phases can be directly compared only when both terms are sine functions or both are cosine functions. The

**FIGURE 8.4**    *Phase Difference between Sinusoids (Note: Functions are plotted with respect to ωt.)*

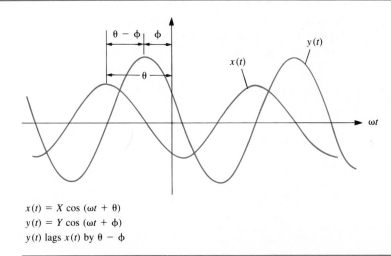

$x(t) = X \cos(\omega t + \theta)$
$y(t) = Y \cos(\omega t + \phi)$
$y(t)$ lags $x(t)$ by $\theta - \phi$

algebraic signs of the amplitudes must also be the same, although their magnitudes may be different. Finally, the phase difference between sinusoids is defined only for sinusoids of the same frequency. Example 8.1 shows the trigonometric manipulations that are sometimes needed before phases can be directly compared.

---

**Example 8.1**    *Phase Difference between Two Sinusoids*

Find the phase difference between the following two voltages:

$$v_1(t) = 5 \sin (5t - 145°)$$
$$v_2(t) = -20 \cos (5t + 220°)$$

We choose to write each voltage as a cosine function with positive amplitude. We use trigonometric identities found in Appendix B:

$$v_1(t) = 5 \cos (5t - 145° - 90°) = 5 \cos (5t - 235°)$$
$$v_2(t) = 20 \cos (5t + 220° - 180°) = 20 \cos (5t + 40°)$$

By using the rule developed with Figure 8.4, we can say that $v_2$ lags $v_1$ by

$$-235° - 40° = -275° = +85°$$

---

We could have expressed the answer to Example 8.1 in a different way by saying that $v_1$ leads $v_2$ by 85° without changing the meaning of the result. We could also say that $v_2$ leads $v_1$ by −85°. All of these statements are equivalent.

Recall from Chapter 6 that if a circuit is excited by a sinusoidal source of frequency $\omega$, then all of the steady-state voltages and currents in the circuit will be sinusoids of that same frequency. They can be expressed in either of the following two equivalent ways:

$$A \cos \omega t + B \sin \omega t = K \cos (\omega t + \theta) \qquad (8.6a)$$

where

$$K = \sqrt{A^2 + B^2} \qquad (8.6b)$$

and

$$\theta = -\tan^{-1} \frac{B}{A} \qquad (8.6c)$$

Equations 8.6b and 8.6c can be proven starting with Equation 8.6a and the known trigonometric expansion of its right-hand side:

$$A \cos \omega t + B \sin \omega t = K \cos \theta \cos \omega t - K \sin \theta \sin \omega t \qquad (8.7)$$

Equation 8.7 can be true only if the coefficients of the cosine terms on both sides of the equation are equal and if the coefficients of the sine terms are equal. That is,

$$A = K \cos \theta \qquad (8.8a)$$

and

$$B = - K \sin \theta \qquad (8.8b)$$

Taking the ratio of Equation 8.8a to 8.8b leads to the relationship of Equation 8.6c. Squaring and adding Equations 8.8a and 8.8b lead to the equation for the magnitude $K$. Keep in mind that in numerical calculations, $\theta$ can be either positive or negative.

For now, we will use the left side of Equation 8.6a as the assumed form for the solution to differential equations. Final answers are usually presented in the form of the right side of the equation because the sinusoid's magnitude and phase can be immediately seen.

With this general background, we can now solve some sinusoidal steady-state circuit problems.

---

**Example 8.2**   *Steady-State Circuit Analysis (Time Domain)*

Find the voltage $v(t)$ in the circuit of Figure 8.5.

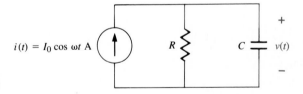

**FIGURE 8.5**

A KCL equation can be written as follows:

$$i(t) = \frac{v}{R} + C \frac{dv}{dt}$$

and it can be rearranged as

$$\frac{dv}{dt} + \frac{1}{RC} v = \frac{1}{C} I_0 \cos \omega t \qquad (8.9)$$

*(continues)*

**Example 8.2**    *Continued*

Because this circuit is in the sinusoidal steady state, the assumed form of the forced response is

$$v(t) = A \cos \omega t + B \sin \omega t \tag{8.10}$$

Substituting Equation 8.10 into Equation 8.9 leads to

$$\frac{d}{dt} (A \cos \omega t + B \sin \omega t) + \frac{1}{RC} (A \cos \omega t + B \sin \omega t) = \frac{I_0}{C} \cos \omega t$$

$$- A\omega \sin \omega t + B\omega \cos \omega t + \frac{A}{RC} \cos \omega t + \frac{B}{RC} \sin \omega t = \frac{I_0}{C} \cos \omega t \tag{8.11}$$

$$\left( B\omega + \frac{A}{RC} \right) \cos \omega t + \left( \frac{B}{RC} - A\omega \right) \sin \omega t = \frac{I_0}{C} \cos \omega t$$

Equating the coefficients of $\cos \omega t$ and of $\sin \omega t$ on opposite sides of Equation 8.11 leads to the following equations:

$$B\omega + \frac{A}{RC} = \frac{I_0}{C} \qquad \text{(coefficients of cosine terms)}$$

$$\frac{B}{RC} - A\omega = 0 \qquad \text{(coefficients of sine terms)}$$

These equations can be solved to find

$$A = \frac{RI_0}{1 + (\omega RC)^2}$$

$$B = \frac{\omega R^2 C I_0}{1 + (\omega RC)^2}$$

Therefore, the voltage is

$$v(t) = \frac{RI_0}{1 + (\omega RC)^2} \cos \omega t + \frac{\omega R^2 C I_0}{1 + (\omega RC)^2} \sin \omega t \text{ V}$$

By convention, we choose to express such results in the form of a cosine function with a phase shift:

$$v(t) = \frac{RI_0}{\sqrt{1 + (\omega RC)^2}} \cos (\omega t - \tan^{-1} \omega RC) \text{ V}$$

You should verify the algebraic and trigonometric steps used in arriving at the result. You should also find the expressions for the capacitor and resistor currents:

$$i_R(t) = \frac{I_0}{\sqrt{1 + (\omega RC)^2}} \cos(\omega t - \tan^{-1} \omega RC) \text{ A}$$

$$i_C(t) = \frac{\omega RCI_0}{\sqrt{1 + (\omega RC)^2}} \cos(\omega t - \tan^{-1} \omega RC + 90°) \text{ A}$$

**Example 8.3**   *Another Steady-State Circuit Analysis (Time Domain)*

Find the current through the 2 Ω resistor in the circuit of Figure 8.6. (*Note:* Mesh analysis seems a reasonable approach.)

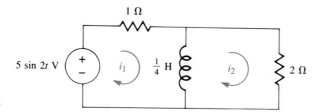

**FIGURE 8.6**

The KVL equations for mesh 1 and mesh 2 are

$$5 \sin 2t = i_1 + \frac{1}{4} \frac{d}{dt}(i_1 - i_2)$$

$$0 = \frac{1}{4} \frac{d}{dt}(i_2 - i_1) + 2i_2$$

We could simplify the problem by eliminating $i_1$, but for generality we will solve for both currents simultaneously. Both $i_1(t)$ and $i_2(t)$ are sinusoids of frequency $\omega = 2$, but different specific forms must be assumed:

$$i_1(t) = A \cos 2t + B \sin 2t$$

$$i_2(t) = C \cos 2t + D \sin 2t$$

Substituting these expressions into the differential equations and gathering terms lead to

$$5 \sin 2t = \left( A + \frac{B}{2} - \frac{D}{2} \right) \cos 2t + \left( B - \frac{A}{2} + \frac{C}{2} \right) \sin 2t$$

$$0 = \left( \frac{D}{2} - \frac{B}{2} + 2C \right) \cos 2t + \left( \frac{A}{2} - \frac{C}{2} + 2D \right) \sin 2t$$

(*continues*)

**Example 8.3** *Continued*

These give four equations and four unknowns:

$$A + \frac{B}{2} - \frac{D}{2} = 0$$

$$-\frac{A}{2} + B + \frac{C}{2} = 5$$

$$-\frac{B}{2} + 2C + \frac{D}{2} = 0$$

$$\frac{A}{2} - \frac{C}{2} + 2D = 0$$

You may wish to verify that $A = -8/5$, $B = 19/5$, $C = 4/5$, and $D = 3/5$. The current $i_2(t)$ is

$$i_2(t) = \frac{4}{5} \cos 2t + \frac{3}{5} \sin 2t = \cos (2t - 36.9°) \text{ A}$$

## 8.3  DEFINITION OF THE PHASOR

Examples 8.2 and 8.3 show the time-domain method for analyzing circuits operating in the sinusoidal steady state. We could go ahead and analyze any such circuit with this method, but Example 8.3 demonstrates how complicated this analysis can be for even a relatively simple circuit. A solution involving $N$ circuit variables requires the manipulation of $N$ integro-differential equations followed by the solution of $2N$ simultaneous algebraic equations. We would like to find an easier analytical approach to the solution of circuit problems in the sinusoidal steady state. The results of Example 8.2 give us a clue as to what that approach might be.

Example 8.2 is a good demonstration of the fact that when a circuit is excited by a sinusoidal source, all of the forced responses within the circuit are sinusoids of the same frequency. Although all variables share the same general form, each is unique in its magnitude and phase angle. This fact suggests a shorthand notation that can be used to express sinusoidal variables. If the frequency is known, we can write down just the magnitude and phase of a variable instead of a complete time-domain expression. In so doing, no information is lost.

Early in the twentieth century, the prominent American engineer Charles Proteus Steinmetz (1865–1923) was among the first to realize that complex variables can be used to advantage in describing sinusoidal time functions because, like sinusoids, they are characterized by magnitudes and angles. Consider a general complex number $z$. It can always be written in exponential or polar form as

$$z = |z|e^{j\theta}$$

where  $|z|$ = the magnitude of $z$

  $\theta$ = its angle

Again, $\theta$ should properly be expressed in radians, but engineers often prefer to use degrees.

We can now define a **phasor** to be a complex variable whose magnitude is the same as that of a corresponding sinusoidally varying voltage or current and whose angle is equal to the phase of the sinusoid. For example, consider the following four sinusoidal variables and their corresponding phasor representations:

$$v_a(t) = V_m \cos(\omega t + \phi) \rightarrow \mathbf{V}_a = V_m e^{j\phi} \tag{8.12}$$

$$v_1(t) = 25 \cos(50t + 45°) \rightarrow \mathbf{V}_1 = 25 e^{j45°}$$

$$v_2(t) = 8.13 \cos(4t - 134°) \rightarrow \mathbf{V}_2 = 8.13 e^{-j134°}$$

$$i_x(t) = .6 \cos(377t + 220°) \rightarrow \mathbf{I}_x = .6 e^{j220°}$$

As before, we have followed the convention of using lowercase letters as variable names for functions of time. Uppercase symbols are used for amplitudes and DC quantities. The standard notation for phasor variable names is to use bold uppercase letters. This convention of using bold print for phasor variable names is followed throughout the book. Bold print, however, is inconvenient to use with handwritten work. For working out problems by hand, we suggest an alternate notation of uppercase symbols with overbars. That is, $\mathbf{V}_a = \overline{V}_a$.

The use of arrows in Equation 8.12 and the following three relationships emphasizes that phasors are not in any way equal to the sinusoids to which they are related. In fact, phasors are not even functions of time because the magnitude and phase of a sinusoid do not change with time. As you can see from the examples just given, this definition of phasors is based on a presumed cosine form for time functions.

A compact mathematical relationship between a time function and its phasor representation can now be given. Consider the time function

$$v(t) = V_m \cos(\omega t + \theta)$$

and a related complex exponential shown here expanded by use of Euler's theorem:

$$V_m e^{j(\omega t + \theta)} = V_m \cos(\omega t + \theta) + jV_m \sin(\omega t + \theta)$$

Comparison shows that $v(t)$ is the real part of the complex exponential. That is,

$$v(t) = Re[V_m e^{j(\omega t + \theta)}] = Re[V_m e^{j\theta}e^{j\omega t}] = Re[\mathbf{V}e^{j\omega t}] \tag{8.13}$$

Equation 8.13 does not exactly say that $v(t)$ is the real part of its corresponding phasor. What it does say is that $v(t)$ is the real part of the phasor multiplied by a complex exponential to account for the sinusoidal time dependency.

Sometimes, to save a little effort, we write phasors in an abbreviated form. In this form, the complex exponential is suppressed and only the magnitude and angle of the phasor are written. For instance,

$$\mathbf{V} = 20e^{j170°} = 20\underline{/170°}$$

or

$$\mathbf{I} = 12e^{-j45°} = 12\underline{/-45°}$$

This more compact form is used primarily by power engineers. The change in form does not alter the fact that phasors are complex quantities.

The use of phasors in electrical circuit analysis moves the descriptions of problems from the time domain to the **frequency domain.** Real functions such as $\cos \omega t$ are called time-domain representations of signals because their independent variable or *domain* is time. As was pointed out earlier, phasors are not functions of time. They are frequency-domain representations of signals. Although the frequency $\omega$ is not explicitly included in the phasor, it must be implicitly understood. The transition from phasors back to the time domain can be performed only if the frequency is known. Remember that, by convention, phasor notation in this textbook is always based on an assumed cosine form in the time domain. Otherwise, the reverse process of writing the time function corresponding to a phasor would be ambiguous.

As they have been presented thus far, phasors are merely a shorthand method for writing sinusoidal signals. We will soon see, however, that the use of phasors can also be a powerful tool in the analysis of sinusoidal steady-state circuit problems. The use of phasors leads to a considerable reduction in effort and also provides insights to the sinusoidal problem that otherwise may not be obvious. In the next several sections, we will develop the use of phasors for circuit analysis. The development and use of phasor notation depend very much on an understanding of complex variables. Individuals feeling the need for a review in this area should read Appendix C before proceeding.

## 8.4 COMPLEX SIGNALS IN ELECTRICAL CIRCUITS

Consider a linear, time-invariant circuit that is otherwise unspecified. It is shown and labeled $N$ in Figure 8.7. Assume that the circuit's input function is $x(t) = A \cos(\omega t + \theta)$ and that the corresponding steady-state response or output is $y(t) = B \cos(\omega t + \phi)$. These signals may be either voltages or currents. Because the circuit is time-invariant, if the input is delayed by a time corresponding to a $90°$ phase shift, the response will be unchanged except for a delay by the same amount. Since $\cos(\omega t + \theta - 90°) = \sin(\omega t + \theta)$, an input to the circuit of $A \sin(\omega t + \theta)$ will give a response of $B \sin(\omega t + \phi)$. These input–output relationships are shown in Figures 8.7A and 8.7B.

Now we will make use of the circuit's assumed linearity. From the property of homogeneity, we can say that an input $jA \sin(\omega t + \theta)$ will cause a response of $jB \sin(\omega t + \phi)$, where, as before, $j = \sqrt{-1}$. This operation cannot be checked in

FIGURE 8.7    *Steps Demonstrating Validity of Use of Complex Exponentials To Represent Variables in Sinusoidal Steady State*

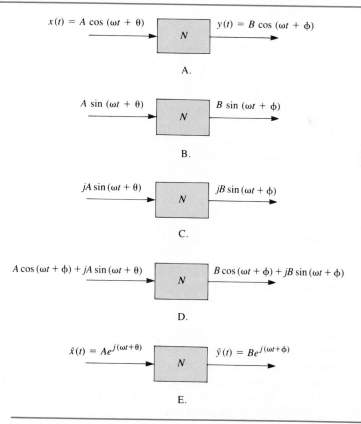

$x(t) = A \cos(\omega t + \theta)$     $N$     $y(t) = B \cos(\omega t + \phi)$

A.

$A \sin(\omega t + \theta)$     $N$     $B \sin(\omega t + \phi)$

B.

$jA \sin(\omega t + \theta)$     $N$     $jB \sin(\omega t + \phi)$

C.

$A \cos(\omega t + \phi) + jA \sin(\omega t + \theta)$     $N$     $B \cos(\omega t + \phi) + jB \sin(\omega t + \phi)$

D.

$\tilde{x}(t) = A e^{j(\omega t + \theta)}$     $N$     $\tilde{y}(t) = B e^{j(\omega t + \phi)}$

E.

the laboratory, but it is possible as a purely mathematical exercise. It is shown in Figure 8.7C.

Next, superposition can be used to show that an input of $A \cos(\omega t + \theta) + jA \sin(\omega t + \theta)$ will cause a response of $B \cos(\omega t + \phi) + jB \sin(\omega t + \phi)$ (Figure 8.7D).

We reach our final result by using Euler's theorem, which is repeated here in simplified form:

$$Ke^{j\alpha} = K \cos \alpha + jK \sin \alpha$$

In other words, Euler's relationship is a formula for transforming a complex number from polar to rectangular coordinates. Applying it to our present problem leads to the result that if

$$A \cos(\omega t + \theta) + jA \sin(\omega t + \theta) \rightarrow B \cos(\omega t + \phi) + jB \sin(\omega t + \phi)$$

then

$$\tilde{x}(t) = Ae^{j(\omega t + \theta)} \rightarrow \tilde{y}(t) = Be^{j(\omega t + \phi)}$$

is also a valid input–output pair for the circuit (Figure 8.7E). These complex exponentials are certainly not equal to the real-valued sinusoidal functions that we started with, but they are closely related. To distinguish between the two variable types, we add tilde marks ($\sim$) above the names given to variables in complex exponential form.

As a final step, we can rearrange our complex input–output pair into the form

$$Ae^{j\theta}e^{j\omega t} \rightarrow Be^{j\phi}e^{j\omega t}$$

or

$$\mathbf{X}e^{j\omega t} \rightarrow \mathbf{Y}e^{j\omega t}$$

where $\mathbf{X}$ and $\mathbf{Y}$ are the phasor representations of input and output, respectively.

What has all of this to do with our solutions of electrical circuit problems in the real world? Complex input and output variables might be perfectly valid mathematically, but it is not physically possible to apply a complex input to a real circuit. What must be kept in mind is that complex variables, like our earlier analytical techniques, are only mathematical tools for modeling the behavior of electrical circuits. If a mathematical model correctly describes the behavior of a physical system, then the analyst is free to use it, providing that the proper interpretation is applied to the results. In the present instance, if an input variable is of the form $x(t) = A \cos(\omega t + \theta)$, then it is mathematically allowable to substitute the complex quantity $\tilde{x}(t) = Ae^{j(\omega t + \theta)}$ if it is kept in mind that it is only the real (cosine) part of the resultant complex response that is of physical significance. If, in so doing, we can find the magnitude and phase of the response, then the problem is solved. To get a clearer idea of this concept, let's repeat the solution to Example 8.2.

**Example 8.4**  *Sinusoidal Circuit Analysis Using Complex Exponentials*

Repeat the solution of the $RC$ circuit problem of Example 8.2 using complex exponentials for the variables. The circuit diagram is repeated in Figure 8.8.

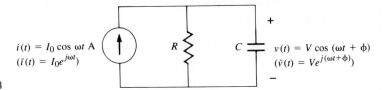

$i(t) = I_0 \cos \omega t$ A
$(\tilde{\imath}(t) = I_0 e^{j\omega t})$

$v(t) = V \cos (\omega t + \phi)$
$(\tilde{v}(t) = V e^{j(\omega t + \phi)})$

**FIGURE 8.8**

Although the input $i(t) = I_0 \cos \omega t$ is a purely real quantity, we can also recognize it to be the real part of the complex exponential $\tilde{\imath}(t) = I_0 e^{j\omega t}$. We mathematically apply this complex forcing function to the circuit. The corresponding assumed form for the response voltage is

$$\tilde{v}(t) = V e^{j(\omega t + \phi)}$$

where the actual response, $v(t)$, is the real part of $\tilde{v}(t)$. The complex exponential forms can be substituted into the differential equation:

$$\frac{\tilde{v}}{R} + C \frac{d\tilde{v}}{dt} = \tilde{\imath}(t)$$

$$\frac{1}{R} V e^{j(\omega t + \phi)} + j\omega C V e^{j(\omega t + \phi)} = I_0 e^{j\omega t}$$

The factor $e^{j\omega t}$ is common to all terms and, because it does not equal zero, can be divided out:

$$\frac{1}{R} V e^{j\phi} + j\omega C V e^{j\phi} = I_0 \tag{8.14}$$

$$V e^{j\phi} = \frac{R I_0}{1 + j\omega RC} \tag{8.15}$$

Two complex numbers can be equal only if both their magnitudes and phases are equal. That is,

$$V = \left| \frac{R I_0}{1 + j\omega RC} \right| = \frac{R I_0}{\sqrt{1 + (\omega RC)^2}}$$

*(continues)*

**Example 8.4**   *Continued*

and

$$\phi = \angle\left(\frac{RI_0}{1 + j\omega RC}\right) = -\tan^{-1} \omega RC$$

where the notation $\angle z$ denotes the angle of the complex number $z$. This notation should not be confused with the similar but distinctive notation for phasors that was introduced earlier. We can now write the response voltage as

$$v(t) = \frac{RI_0}{\sqrt{1 + (\omega RC)^2}} \cos{(\omega t - \tan^{-1} \omega RC)} \text{ V}$$

which is the same answer as in Example 8.2.

## 8.5   PHASORS AND IMPEDANCE

Thus far, the use of complex exponentials for representing sinusoidal signals does not seem to have saved us much effort. By making direct use of phasor notation, however, we can considerably streamline the calculations needed in the sinusoidal steady-state analysis of electrical circuits. Let's reconsider Equation 8.14 from Example 8.4 and repeat it in a slightly modified form:

$$\frac{1}{R}(Ve^{j\phi}) + j\omega C(Ve^{j\phi}) = (I_0 e^{j0°})   \tag{8.16}$$

The terms in parentheses are readily recognized to be phasors. $Ve^{j\phi}$ has a magnitude and a phase equal to those of $V\cos{(\omega t + \phi)}$, the assumed time-domain form of the response in Example 8.4. Similarly, $I_0 e^{j0°}$ has the magnitude, $I_0$, and phase, $0°$, of the forcing function. Once these relationships are recognized, Equation 8.16 can be rewritten as

$$\frac{1}{R}\mathbf{V} + j\omega C\mathbf{V} = \mathbf{I}   \tag{8.17}$$

Equation 8.17 is a KCL equation with the variables in phasor form. Because $\mathbf{I}$ is a phasor current, $(1/R)\mathbf{V}$ and $j\omega C\mathbf{V}$ must also be phasor currents. They are, in fact, phasor representations of the resistor and capacitor currents, respectively. We have seen the term $1/R$ before as the conductance of the resistor $R$. By analogy, $j\omega C$ must be the "conductance" of the capacitor.

The preceding observations seem to suggest that simple relationships exist between the phasor voltage and current for each of the three basic circuit elements. We will formally prove the **V–I** relationship for an inductor here and leave the proofs for resistive and capacitive elements as an independent exercise.

We begin in the time domain where the $v$–$i$ relationship for the inductor is well known. Figure 8.9 shows an inductor operating in the sinusoidal steady state with the most general possible assumed complex exponential forms for voltage and current. Substituting these assumed forms into the time-domain relationship between voltage and current gives

$$\tilde{v}(t) = L\,\frac{d\tilde{i}(t)}{dt}$$

$$V_0 e^{j(\omega t + \phi)} = L\,\frac{d}{dt}\left[I_0 e^{j(\omega t + \theta)}\right] = j\omega L I_0 e^{j(\omega t + \theta)}$$

$$V_0 e^{j\phi} = j\omega L I_0 e^{j\theta}$$

$$\mathbf{V}_L = j\omega L \mathbf{I}_L$$

The last equation tells us that the phasor voltage of an inductor can be found by multiplying the phasor current by $j\omega L$. The term $j\omega L$ is called the **impedance** of the inductor. Impedances are complex quantities that relate phasor voltages and currents. Impedance is given the general symbol $Z$. A subscript is often included to indicate the element or elements to which the impedance refers. The inverse of impedance is called **admittance** and is given the general symbol $Y$. For an inductor,

$$Z_L(j\omega) = \frac{\mathbf{V}}{\mathbf{I}} = j\omega L$$

and

$$Y_L(j\omega) = \frac{\mathbf{I}}{\mathbf{V}} = \frac{1}{j\omega L}$$

**FIGURE 8.9**   *Representations of an Inductor Operating in Sinusoidal Steady State*

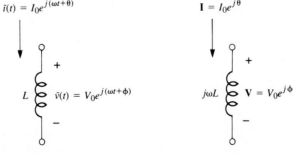

A. Time domain                    B. Frequency domain

The impedances and admittances of the three basic circuit elements are shown in Table 8.1. The impedances verify several of our earlier observations:

1. The impedance of a resistor is a real number. Thus, resistor voltage and current are in phase.
2. The impedance of an inductor has an angle of $+90°$. As shown earlier, inductor voltage leads current by $90°$.
3. The impedance of a capacitor has an angle of $-90°$. Thus, capacitor current leads voltage by $90°$.

It is clear from Table 8.1 that the magnitudes of $Z_C$ and $Z_L$ depend on frequency. This simple fact has very important implications in the design of electrical circuits. As an example, consider the tone controls of an audio amplifier. These are the bass and treble knobs of older or simpler amplifiers or the equalizer sections of more elaborate units. These controls can be used to emphasize some frequency ranges while suppressing others until a desired tonal quality is achieved. Such response characteristics are possible only because of the differing frequency dependencies of basic circuit elements. You will plot the magnitudes of $Z_L$ and $Z_C$ as functions of $\omega$ in one of the problems at the end of

**TABLE 8.1**  *Impedance and Admittance Relationships for Resistors, Capacitors, and Inductors*

|  | Resistor, R | Inductor, L | Capacitor, C |
|---|---|---|---|
| Impedance $Z = \dfrac{\mathbf{V}}{\mathbf{I}}$ | $Z_R(j\omega) = R$ | $Z_L(j\omega) = j\omega L$ | $Z_C(j\omega) = \dfrac{1}{j\omega C}$ |
| Admittance $Y = \dfrac{\mathbf{I}}{\mathbf{V}}$ | $Y_R(j\omega) = \dfrac{1}{R}$ | $Y_L(j\omega) = \dfrac{1}{j\omega L}$ | $Y_C(j\omega) = j\omega C$ |
| Magnitude of Impedance | $|Z_R| = R$ | $|Z_L| = \omega L$ | $|Z_C| = \dfrac{1}{\omega C}$ |
| as $\omega \to 0$ | (independent of frequency) | $|Z_L| = 0$ (short circuit) | $|Z_C| = \infty$ (open circuit) |
| as $\omega \to \infty$ | | $|Z_L| = \infty$ (open circuit) | $|Z_C| = 0$ (short circuit) |
| Angle of Impedance | $\angle Z_R = 0°$ (voltage and current in phase) | $\angle Z_L = +90°$ (voltage leads current by $90°$) | $\angle Z_C = -90°$ (current leads voltage by $90°$) |

the chapter. For now, we will consider only the special cases of $\omega = 0$ and $\omega \to \infty$. Table 8.1 shows the magnitudes of the three basic impedances for these extreme values for $\omega$.

The impedance concept is again consistent with results already arrived at in the time domain. As before, for DC ($\omega = 0$), an inductor is seen to act like a short circuit and a capacitor like an open circuit. At very high AC frequencies, these two elements reverse their behaviors. That is, as $\omega \to \infty$, inductors behave like open circuits and capacitors like short circuits. For any inductor–capacitor pair, there is one frequency between these extremes at which the magnitudes of their impedances are equal.

The impedances of the three basic circuit elements are purely real or purely imaginary. When several different elements are connected to form a circuit, however, the result is usually a complex impedance with both a real and an imaginary part. Impedances in a circuit can be mathematically combined by the same rules that apply to resistors in the time domain, the proof of which is derived in an end-of-chapter problem. The first step in such an analysis is to determine the circuit's frequency-domain description. The circuit is drawn, and beside each element is written its impedance. These separate impedances can then be reduced in number by the rules of series and parallel combination.

---

**Example 8.5**   *Calculating a Circuit's Input Impedance*

What is the equivalent input impedance of the circuit shown in Figure 8.10A? Its frequency-domain version (for $\omega = 2$) is shown in Figure 8.10B.

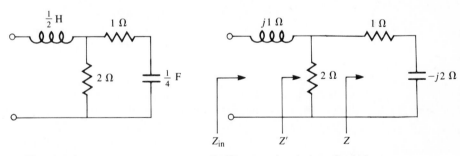

**FIGURE 8.10**   A. Time domain            B. Frequency domain ($\omega = 2$ rad/s)

The $1\,\Omega$ resistor and the capacitor are in series and can be added to give a combined equivalent impedance of

$$Z = 1 - j2\ \Omega$$

This impedance $Z$ is in parallel with the $2\,\Omega$ resistor. Their combined

*(continues)*

**Example 8.5**    *Continued*

equivalent can be calculated as

$$Z' = Z\|2 = \frac{(1 - j2)(2)}{(1 - j2) + 2} = \left(\frac{2 - j4}{3 - j2}\right)\left(\frac{3 + j2}{3 + j2}\right) = \frac{14}{13} - j\,\frac{8}{13}\,\Omega$$

Finally, the input impedance can be found by

$$Z_{\text{in}} = Z' + j1 = \frac{14}{13} - j\,\frac{8}{13} + j1 = \frac{14}{13} + j\,\frac{5}{13}\,\Omega$$

---

Before we move on to the next section, some points should be made about impedances and their calculation. Impedances enter into frequency-domain circuit analysis in ways that follow the same rules as for resistors in the time domain. They relate voltages and currents by a sort of Ohm's law, $\mathbf{V} = Z\mathbf{I}$. Combinations of impedances can be simplified by the same series and parallel rules that apply to resistors in the time domain. While the mechanics of manipulation for time-domain resistors and frequency-domain impedances are the same, the similarities end there. Ohm's law in the time domain is an instantaneous (valid at every instant of time) relationship between the voltage and current of a resistor. The voltage and current may be of any functional form and include both natural and forced components. The impedances that we have defined here are applicable to capacitors and inductors as well as to resistors. They are not, however, valid for all types of signals, and they do not relate instantaneous functions of time. As they are presented in this chapter, impedances are valid mathematical descriptions of passive elements only for circuits operating in the sinusoidal steady state. They relate complex-valued phasors and not instantaneous functions of time.

The confusion that sometimes exists between the concepts of impedance and resistance is often increased by the use of the term *ohms* and its symbol $\Omega$ for the units of impedance, even when the impedance is capacitive or inductive in origin. Choosing this terminology is entirely by convention.

In Example 8.5, we computed an input impedance that had real and imaginary parts, both of which were nonzero. This is usually the case. The real and imaginary parts of impedances are given special symbols:

$$Z(j\omega) = R(j\omega) + jX(j\omega) \tag{8.18}$$

The real part, $R$, of an impedance is called a **resistance** or a resistive impedance. The imaginary part, $X$, is called a **reactance** or a reactive impedance. For a totally passive circuit, $R$ cannot be negative, but $X$ may be of any value. $X$ is positive for a simple inductor and negative for a capacitor. The situation becomes less clear for elements in combination, especially in active circuits. However, if you were concerned with the effective behavior of a circuit, you would classify its equivalent reactance as inductive if $X > 0$ and capacitive if $X < 0$.

The inverse of an impedance is an admittance. Admittances also have real and imaginary parts:

$$Y(j\omega) = \frac{1}{Z(j\omega)} = G(j\omega) + jB(j\omega) \qquad (8.19)$$

$G$ is called the **conductance** or conductive admittance. $B$ is called the **susceptance** or susceptive admittance. $Y$, $G$, and $B$ are given the units of siemens (S), but again this terminology is entirely by convention and does not convey the same meaning as with a time-domain conductance. While it is true that $Y = 1/Z$, it does not follow that $G = 1/R$ or that $B = 1/X$. Finding the relationships between $G$, $B$, $R$, and $X$ will be dealt with in an end-of-chapter problem.

---

**Example 8.6**   *Input Impedance and Admittance*

Change the frequency of Example 8.5 to $\omega = 4$ and find the circuit's input impedance and admittance. Because the frequency has been changed, you cannot use the results of the earlier calculation. Begin the calculation by leaving the frequency unspecified and make the substitution later. The frequency-domain version of the circuit is shown in Figure 8.11.

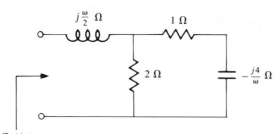

**FIGURE 8.11**   $Z_{in}(j\omega)$

Following the steps of Example 8.5, you should verify that

$$Z_{in}(j\omega) = \frac{(6 + 32/\omega^2) + j(9\omega/2 - 8/\omega)}{9 + 16/\omega^2} \; \Omega$$

Here is a good test of your ability to work with complex numbers. Notice that both the real and the imaginary parts of $Z_{in}$ are frequency dependent. Now, let $\omega = 4$:

$$Z_{in} = .8 + j1.6 \; \Omega$$

*(continues)*

**Example 8.6**    *Continued*

It follows that $R = .8\,\Omega$ and that $X = 1.6\,\Omega$:

$$Y_{in} = \frac{1}{Z_{in}} = \frac{1}{.8 + j1.6} = .25 - j.5\,\text{S}$$

Clearly, $G = .25\,\text{S}$ and $B = -.5\,\text{S}$.

## 8.6   PHASORS AND IMPEDANCE IN CIRCUIT ANALYSIS

It will not be proven here, but all of the basic circuit analysis techniques that we have been using in the time domain are also valid in the frequency domain where variables are in phasor form and passive elements are impedances. Mesh and node analysis, equivalent circuits, and all of the other techniques with which we are familiar in the time domain are still valid in the world of phasors and impedances.

**Example 8.7**    *Using Phasors to Find an Output Voltage*

Consider the ideal operational amplifier circuit shown in Figure 8.12. It is presented in its frequency-domain form. Find the output $\mathbf{V}_o$ as a function of the input $\mathbf{V}_i$.

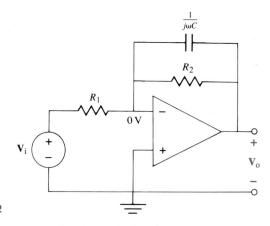

**FIGURE 8.12**

You can do so by writing a KCL equation at the inverting input of the amplifier:

$$\frac{\mathbf{V}_i}{R_1} + j\omega C\mathbf{V}_o + \frac{1}{R_2}\mathbf{V}_o = 0$$

This equation can be rearranged to find

$$\mathbf{V}_o = -\frac{R_2/R_1}{1 + j\omega R_2 C}\,\mathbf{V}_i$$

---

The result of Example 8.7 very clearly shows that we have done our calculations in the frequency domain. The independent variable in the expression for $\mathbf{V}_o$ is $\omega$, the radian frequency. Although the magnitude and phase of the output are independent of time, they are very much dependent on the frequency at which the circuit operates. Next, let's consider a numerical example.

---

**Example 8.8** *Node Analysis Using Phasors*

Calculate the value of $v_1(t)$ in the circuit of Figure 8.13.

Because the circuit is operating in the sinusoidal steady state, the phasor approach is appropriate. The first step is to find the impedance values of the passive elements and write the signal variables as phasors:

$$5 \sin 1000t = 5 \cos(1000t - 90°) \rightarrow 5e^{-j90°}$$

$$200\,\mu F \rightarrow \frac{1}{j\omega C} = -j5\,\Omega$$

$$20\,mH \rightarrow j\omega L = j20\,\Omega$$

These values are shown in Figure 8.14.

The two necessary node equations can be written as follows:

$$\frac{\mathbf{V}_1}{10} + \frac{\mathbf{V}_1}{-j5} + \frac{\mathbf{V}_1 - \mathbf{V}_2}{j20} + \frac{\mathbf{V}_1 - \mathbf{V}_2}{5} = 5e^{-j90°}$$

$$\frac{\mathbf{V}_2}{10} + \frac{\mathbf{V}_2 - \mathbf{V}_1}{j20} + \frac{\mathbf{V}_2 - \mathbf{V}_1}{5} = 0$$

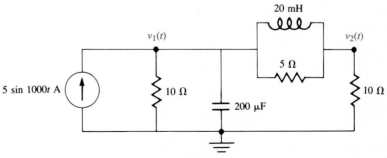

**FIGURE 8.13**

(*continues*)

**Example 8.8** *Continued*

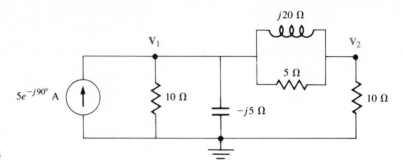

**FIGURE 8.14**

Each of these two equations can be multiplied by $j20$ and rearranged into

$$(-3 + j6)\mathbf{V}_1 + (-1 - j4)\mathbf{V}_2 = 100$$

$$(-1 - j4)\mathbf{V}_1 + (1 + j6)\mathbf{V}_2 = 0$$

Because we are solving only for $v_1(t)$, we use Cramer's rule:

$$\mathbf{V}_1 = \frac{\begin{vmatrix} 100 & (-1 - j4) \\ 0 & (1 + j6) \end{vmatrix}}{\begin{vmatrix} (-3 + j6) & (-1 - j4) \\ (-1 - j4) & (1 + j6) \end{vmatrix}} = \frac{100 + j600}{-24 - j20} = \frac{608.3e^{j80.5°}}{31.2e^{j219.8°}}$$

$$= 19.5e^{-j139.3°} \text{ V}$$

We can easily convert this phasor into the desired function of time:

$$v_1(t) = 19.5 \cos(1000t - 139.3°) \text{ V}$$

---

In order to perform calculations like those shown in Example 8.8, you must be adept at the use of complex numbers in both their rectangular and polar forms and in the conversion from one coordinate system to the other. This conversion is most easily done by using a hand calculator with direct polar to rectangular conversion. If this feature is not available, the basic conversion formulas can be used. They are as follows:

$$z = a + jb = Ke^{j\theta} = K \cos\theta + jK \sin\theta \qquad (8.20a)$$

$$K = \sqrt{a^2 + b^2} \qquad (8.20b)$$

$$\theta = \tan^{-1}\frac{b}{a} \qquad (8.20c)$$

FIGURE 8.15   *Finding Angle of a Complex Number*

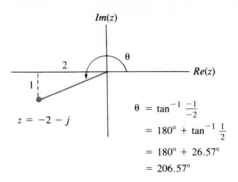

$$\theta = \tan^{-1}\frac{-1}{-2}$$
$$= 180° + \tan^{-1}\frac{1}{2}$$
$$= 180° + 26.57°$$
$$= 206.57°$$

You must be careful when you use the formula for finding the angle. The problem is that with a calculator, the argument $b/a$ must be calculated before the inverse tangent operation can be performed. Thus, the inverse tangent operation of virtually all calculators gives answers only in the first and fourth quadrants depending on whether $b/a$ has a positive or a negative algebraic sign. More properly applied, the inverse tangent operation requires knowledge of the individual algebraic signs of $a$ and $b$. If, for instance, we try to find $\theta = \tan^{-1}$ $(1/-1)$ by first simplifying the argument to $\theta = \tan^{-1}(-1)$, the result would seem to be $\theta = -45°$ rather than the correct answer of $+135°$. It is suggested that a quick sketch be made to locate $z = a + jb$ on the complex number plane. From it, the angle of $z$ can be reasonably estimated, thus avoiding errors. An example is shown in Figure 8.15.

**Example 8.9**   *Equivalent Circuits Using Phasors*

Find the Thévenin equivalent of the circuit in Figure 8.16. The change to the frequency domain has already been done. The source operates at $\omega_0 =$

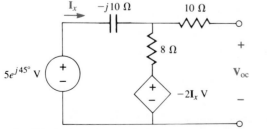

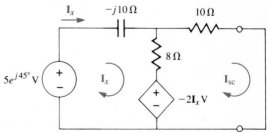

A. Open circuit case          B.  Short circuit case

FIGURE 8.16

*(continues)*

**Example 8.9**   *Continued*

100 rad/s. This value should be kept in mind when the results are interpreted.

Figure 8.16A shows the open circuit condition, and Figure 8.16B shows the short circuit. The mesh equations for these two load conditions are as follows:

Open circuit case:   $5e^{j45°} = -j10\mathbf{I}_x + 8\mathbf{I}_x - 2\mathbf{I}_x$

$\mathbf{V}_{oc} = 8\mathbf{I}_x - 2\mathbf{I}_x$

Short circuit case:   $5e^{j45°} = -j10\mathbf{I}_x + 8(\mathbf{I}_x - \mathbf{I}_{sc}) - 2\mathbf{I}_x$

$-2\mathbf{I}_x = 8(\mathbf{I}_{sc} - \mathbf{I}_x) + 10\mathbf{I}_{sc}$

You should fill in the steps necessary to demonstrate that

$$\mathbf{V}_{oc} = 2.57e^{j104°} \text{ V}$$

$$\mathbf{I}_{sc} = .158e^{j116.6°} \text{ A}$$

$$Z_{eq} = \mathbf{V}_{oc}/\mathbf{I}_{sc} = 15.87 - j3.55 \ \Omega$$

The equivalent circuit is shown in Figure 8.17. We have followed the convention of giving phasors in exponential form so that their magnitudes and phases are readily seen. The impedance is in rectangular form so that the values of its resistive and reactive parts are obvious.

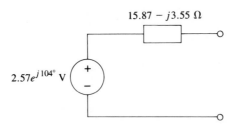

**FIGURE 8.17**

We usually think of the load for a one-port circuit such as in Example 8.9 as a simple passive element, but this is not always true. The circuit can just as easily be connected to devices that contain independent sources. When this is the case, the equivalent circuit shown in Figure 8.17 can be used only if those sources operate at $\omega_0 = 100$ rad/s.

## 8.7 PHASOR DIAGRAMS

Because phasors are complex numbers, they can be plotted in polar form as vectors on the complex number plane. The results of circuit analysis calculations are sometimes presented in this way. For instance, suppose a hypothetical single-source circuit has the following voltages and currents in phasor form:

$$\mathbf{V}_s = 2\underline{/0°}$$

$$\mathbf{I}_1 = 1\underline{/-55°}$$

$$\mathbf{I}_2 = 1.5\underline{/110°}$$

$$\mathbf{V}_1 = .7\underline{/170°}$$

They are plotted as vectors in Figure 8.18. Such plots are called **phasor diagrams.** Magnitudes of phasors are measured radially from the origin. Angles are measured in a counterclockwise direction starting from the positive real axis.

How would the phasor diagram for our hypothetical circuit change as a result of changes in the source? If $\mathbf{V}_s$ were increased in magnitude by a factor of, say, 10, the magnitude of every response variable and the length of every vector in the diagram would also be increased by 10. Similarly, if the source's phase were changed by 30°, the phasors of the diagram would all be rotated by that amount. These results follow from the circuit's linearity and time invariance. Although changing the source in magnitude or phase changes the phasor diagram in absolute terms, the *relative* lengths and positions of vectors in the phasor diagram do not change. The relative values of a circuit's response variables are functions of the circuit itself and the frequency at which it is excited but not of the magnitude or phase of the excitation.

Because vectors can be easily added graphically, a phasor diagram can also be used as the basis for actual circuit calculations. Such graphical calculations are

**FIGURE 8.18** *Example of a Phasor Diagram*

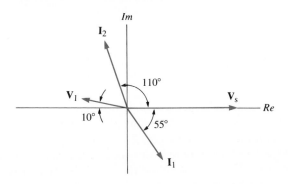

not done in a very formal way. Instead, a variable central to the problem is given an assumed value, and the remaining variables are found in a step-by-step process. This technique is best explained with an example.

---

**Example 8.10**    *Analysis Using a Phasor Diagram*

Consider the circuit shown in Figure 8.19. The source voltage is unspecified, but we do know that its frequency is $\omega = 500\,\text{rad/s}$. We want to find the relationship between all of the phasor variables in the circuit. As a way of getting started, let's assume some value for the voltage **V**. We could assume any value but will choose $\mathbf{V} = 1\underline{/0^\circ}$ to simplify the calculations.

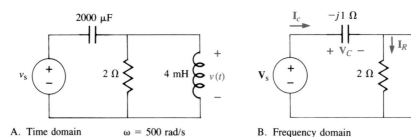

**FIGURE 8.19**    A. Time domain            $\omega = 500\,\text{rad/s}$            B. Frequency domain

If $\mathbf{V} = 1\underline{/0^\circ}$, then $\mathbf{I}_L$ and $\mathbf{I}_R$ must be

$$\mathbf{I}_L = \frac{\mathbf{V}}{j\omega L} = \frac{1\underline{/0^\circ}}{j2} = .5\underline{/-90^\circ}$$

$$\mathbf{I}_R = \frac{\mathbf{V}}{R} = \frac{1\underline{/0^\circ}}{2} = .5\underline{/0^\circ}$$

$\mathbf{V}$, $\mathbf{I}_L$, and $\mathbf{I}_R$ are plotted in Figure 8.20A. $\mathbf{I}_L$ and $\mathbf{I}_R$ can be added graphically. The result is $\mathbf{I}_C = .707\underline{/-45^\circ}$.

Next, we can find $\mathbf{V}_C$:

$$\mathbf{V}_C = \frac{\mathbf{I}_C}{j\omega C} = (.707\underline{/-45^\circ})(-j1) = .707\underline{/-135^\circ}$$

These additional values are shown in Figure 8.20B.

Finally, $\mathbf{V}_s$ can be found by the graphical addition of $\mathbf{V}_C$ and $\mathbf{V}$. The result can be measured to be $\mathbf{V}_s = .707\underline{/-45^\circ}$.

Now, it is unlikely that we would want the source to be $\mathbf{V}_s = .707\underline{/-45^\circ}$. Let's see how we would need to adjust the phasor diagram of Figure 8.20C if $\mathbf{V}_s = 1\underline{/0^\circ}$. A change to this value would involve a scaling in magnitude for

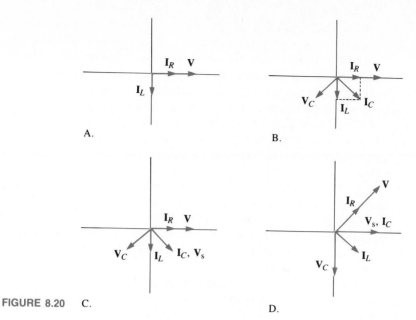

FIGURE 8.20

the source (and all other variables) of $1/.707 = 1.414$. Similarly, all phasors in the diagram should be rotated 45° in the counterclockwise direction. The end result is shown in Figure 8.20D.

The procedure of Example 8.10 is certainly a round-about way of arriving at the answer. Usually, we begin with knowledge of the source and determine the response analytically. Here, we have done just the opposite. The response was our starting point. We simply assumed a convenient value for it and worked our way backward through the circuit until we found the corresponding source value. A final adjustment was made to make the calculated source value agree with that which was specified. Although it seems somewhat involved, the approach actually requires much less work than our earlier methods because no formal sets of equations need to be solved. The approach's usefulness is limited to fairly simple circuits and depends somewhat on the ability to choose a good variable to start the calculation. For instance, in Example 8.10, making an assumption about $V_C$ would not have allowed us to complete the problem easily.

## 8.8 SUMMARY

When a circuit operates in the sinusoidal steady state, any variable consists entirely of a forced response because the natural response has decayed to zero.

All variables are sinusoids of the source frequency, and each can be uniquely described by a complex phasor that gives its magnitude and phase.

Although it is possible to solve such problems entirely in the time domain, the solutions can become very cumbersome and time consuming for large circuits. The use of complex-valued phasors simplifies the needed calculations. The use of phasors and impedances in the frequency domain changes differential equations of real-valued functions into algebraic equations of complex variables. This is one of the main advantages of the phasor approach because algebraic equations are easier to solve than differential equations even when complex variables are required.

Impedances in the frequency domain play much the same role as resistors do in the time domain. The impedances of each of the three basic circuit elements relate voltage to current by a sort of Ohm's law. In fact, all of the analytical techniques developed for resistive circuits in the time domain carry directly over to the frequency domain of phasors and impedances. The only change is that the variables become complex.

Finally, the phasor diagram has been shown as a convenient method for displaying the results of phasor calculations, and, to a certain extent, even a device for performing the calculations themselves. In an age of computers, phasor diagrams still remain useful because of the visual impact that they have for the viewer. At a glance, the interrelationships of a circuit's variables can be seen.

## ■ PROBLEMS

**8.1** Make sketches of the following as functions of time:

   **a.**   $x(t) = 2 \cos (t + 45°)$

   **b.**   $v(t) = 15 \sin (100t - 120°)$

   **c.**   $y(t) = -4 \cos (100t + 90°)$

   **d.**   $i(t) = -1.5 \cos (10{,}000t - \pi/4)$

**8.2** For each of the functions of Figure P8.2, express the variable as a function of time. Use both the sine and cosine forms.

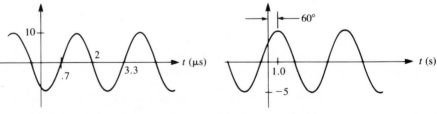

**FIGURE P8.2**   A.                                                    B.

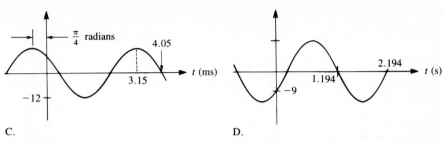

C.                                                          D.

**FIGURE P8.2** *(continued)*

**8.3 a.** For a frequency of $f = 100$ Hz, a time shift of 1.5 ms corresponds to what phase shift (in both degrees and radians)?

**b.** A phase shift of $-45°$ corresponds to what time shifts for the frequencies $\omega = 1$, 10, 100, and 1000 rad/s?

**c.** A sinusoid $x(t) = A \sin(\omega t + \theta)$ has an amplitude of 10 and a frequency of $\omega = 500$ rad/s. We also know that $x(0) = -3$. By how many milliseconds has the sine been shifted from the origin? (This question has an infinite number of answers. Give the smallest possible values, both positive and negative.)

**8.4** Determine the phase relationship between the following pairs of sinusoids. For each, state whether $y(t)$ is leading or lagging $x(t)$.

**a.** $x(t) = 5 \sin(200t + 45°)$,   $y(t) = 10 \sin(200t - 110°)$

**b.** $x(t) = -15 \cos(3t - 60°)$,   $y(t) = 7 \cos(3t + 135°)$

**c.** $x(t) = .3 \sin(10t + 170°)$,   $y(t) = 4 \cos(10t + 240°)$

**d.** $x(t) = -5 \sin(1000t - 80°)$,   $y(t) = \cos(1000t + 300°)$

**8.5** Use a time-domain solution to find the phase relationship between $v(t)$ and $i(t)$ in the circuit of Figure P8.5.

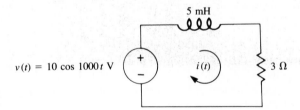

**FIGURE P8.5**

**8.6** Solve for the voltage $v_o(t)$ in the circuit of Figure P8.6 by using the classical differential equation approach.

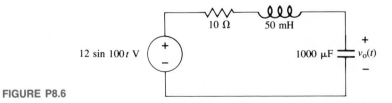

**FIGURE P8.6**

**8.7** Solve for the voltage $v_o(t)$ in the circuit of Figure P8.7 by using differential equations.

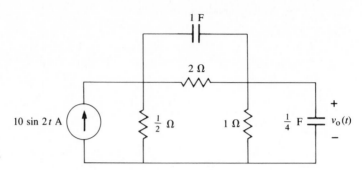

**FIGURE P8.7**

**8.8** Verify that the coefficients $A$ and $B$ found in Example 8.2 have the dimensions of voltage.

**8.9** Repeat Problem 8.5 using the complex exponential notation of Section 8.4.

**8.10** Repeat Problem 8.6 using the complex exponential notation of Section 8.4.

**8.11** Repeat Problem 8.7 using the complex exponential notation of Section 8.4.

**8.12** Solve for $i_L(t)$ in the circuit of Figure P8.12 by using the complex exponential notation of Section 8.4.

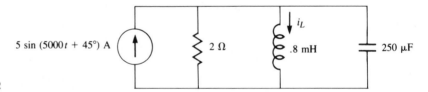

**FIGURE P8.12**

**8.13** Prove the impedance relationships given in Table 8.1 for

   **a.** The resistor

   **b.** The capacitor

**8.14 a.** Use a single-loop circuit as shown in Figure P8.14A to show that phasor voltages obey KVL. Begin with complex exponential notation.

   **b.** Use a single node pair circuit as shown in Figure P8.14B to show that phasor currents obey KCL.

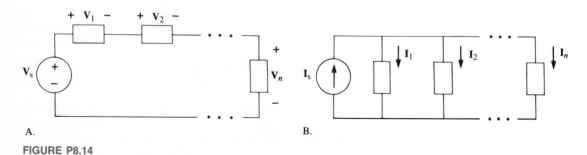

A.

B.

**FIGURE P8.14**

**8.15** Prove that, as with resistors in the time domain,

   **a.**   Impedances in series add.

   **b.**   Admittances in parallel add.

**8.16** Derive, in terms of phasors and impedance, rules for

   **a.**   Voltage division

   **b.**   Current division

   **c.**   Source transformation

**8.17** Plot, on a single graph, the magnitudes of impedance for capacitors of value $1 \, \mu F$, $2 \, \mu F$, and $3 \, \mu F$. The plots should be functions of $\omega$ for $\omega < 10$. Which capacitor has the largest impedance?

**8.18** Repeat Problem 8.17 for inductors of value $1 \, mH$, $2 \, mH$, and $3 \, mH$.

**8.19** Consider an inductor of value $5 \, mH$ and a capacitor of $.047 \, \mu F$. At what frequency will the magnitudes of their impedances be equal? Which will have the larger impedance at lower frequencies? Relate this latter answer to your knowledge of the behavior of inductors and capacitors for $\omega = 0$. At what frequencies will each have an impedance magnitude equal to that of a $100 \, \Omega$ resistor?

**8.20** In the circuit of Figure P8.20, over what frequency ranges will the magnitudes of each of the labeled voltages be greater than the other two?

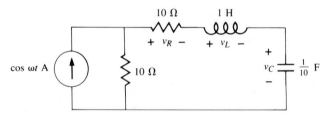

**FIGURE P8.20**

**8.21** In doing a quick analysis of electronic circuits, the following approximations are often made in order to simplify the calculations:

   i.   If two impedances in series differ in magnitude by tenfold or more, the smaller can be neglected.

   ii.  If two impedances in parallel differ in magnitude by tenfold or more, the larger can be neglected.

Over what ranges of frequencies can $L$, $C$, or both be ignored in the circuit of Figure P8.21? When must they be included? Do the same for the resistors. Redraw the circuit for each of the cases identified.

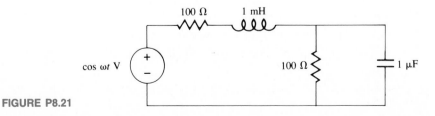

**FIGURE P8.21**

**8.22** Repeat Problem 8.21 for the circuit shown in Figure P8.22.

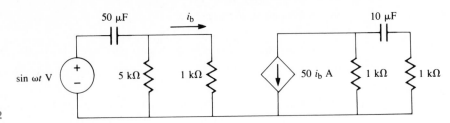

**FIGURE P8.22**

**8.23** Refer to Example 8.2. Plot the magnitude and phase of $v(t)$ as functions of $\omega$. Explain the behavior of the circuit at $\omega = 0$ and as $\omega \to \infty$ in light of what we know about the impedance behavior of the circuit's elements.

**8.24** If an impedance is $Z = R + jX$ and the corresponding admittance is $Y = G + jB$, find $G$ and $B$ as functions of $R$ and $X$. Find also the reverse—that is, $R$ and $X$ as functions of $G$ and $B$.

**8.25** Find the input impedance and admittance for each of the circuits in Figure P8.25. Use these results to draw the simplest possible series form and parallel form equivalent circuits for each.

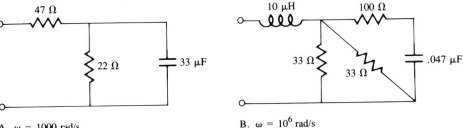

**FIGURE P8.25**    A. $\omega = 1000$ rad/s    B. $\omega = 10^6$ rad/s

**8.26** Use the answer to Example 8.7 to find the circuit's response when $R_1 = 10\text{ k}\Omega$, $R_2 = 20\text{ k}\Omega$, and $C = .1\ \mu\text{F}$ and for $f = 10$, 100, and 1000 Hz.

**8.27** Verify the results of Example 8.3 using phasor notation.

**8.28** Repeat the following problems using phasors and impedance:

   **a.**   Problem 8.5

   **b.**   Problem 8.6

   **c.**   Problem 8.7

   **d.**   Problem 8.12

**8.29** Find $i(t)$ in the circuit of Figure P8.29 using phasors.

**8.30** Use phasor notation to find the voltage $v_o(t)$ in the circuit of Figure P8.30.

**8.31** Use phasors and superposition to find $v_C(t)$ in the circuit of Figure P8.31.

**8.32** Use phasors to find the output voltage $v(t)$ in the circuit of Figure P8.32. Use the AC model for a BJT transistor given in Chapter 1. Assume parameter values of $\beta = 20$, $r_x = 50\ \Omega$, and $r_o = \infty$.

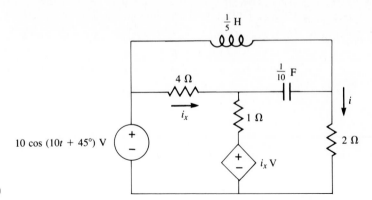

**FIGURE P8.29**

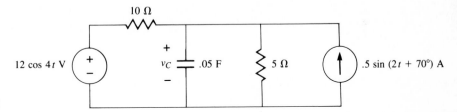

**FIGURE P8.30**

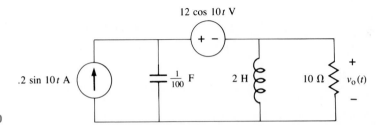

**FIGURE P8.31**

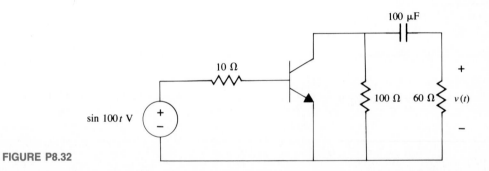

**FIGURE P8.32**

8.33 **a.** Find $V_o/V_i$ as a function of the two impedances $Z_1$ and $Z_2$ in Figure P8.33.

    **b.** Use the result of part a to find $v_o(t)$ when $v_i(t) = 5 \sin 2t$, $Z_1 = 10\,\Omega$, and $Z_2$ is the impedance of a .05 F capacitor.

    **c.** Repeat part b with the capacitor and resistor reversed in position.

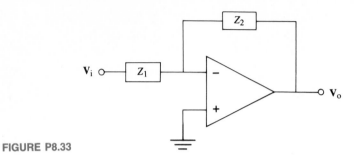

**FIGURE P8.33**

8.34 Write, but do not solve, the mesh impedance matrix equation for the circuit in Figure P8.34.

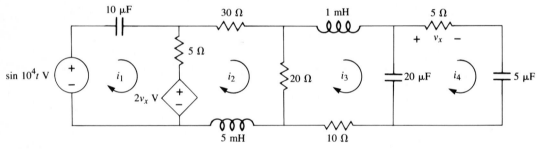

**FIGURE P8.34**

8.35 Write, but do not solve, the node admittance matrix equation for the circuit in Figure P8.35.

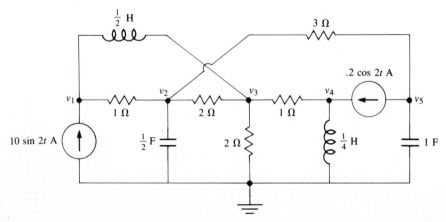

**FIGURE P8.35**

**8.36** Find the steady-state Norton equivalent of the circuit in Figure P8.36.

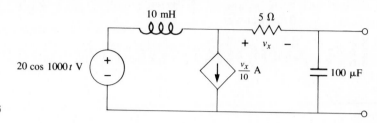

**FIGURE P8.36**

**8.37** Use a phasor diagram as the basis for calculating all of the voltages and currents in the parallel *RLC* circuit in Figure P8.37.

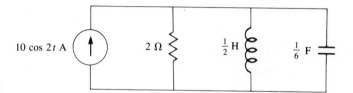

**FIGURE P8.37**

**8.38** Develop a phasor diagram for the circuit in Figure P8.38. It is known that the current in the 2 Ω resistor is 2 sin 100*t* A.

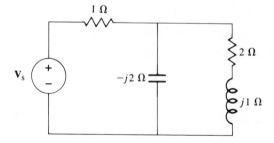

**FIGURE P8.38**

**8.39** Find the three voltages labeled in Figure P8.39. Draw a phasor diagram and comment on the relative magnitudes of these and of the source voltage. Explain the result.

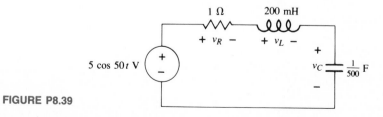

**FIGURE P8.39**

# The Laplace Transform and Its Applications

## 9.1  INTRODUCTION

The first seven chapters of this book described techniques for analyzing electrical circuits in the time domain. These techniques culminated in the analysis of dynamic circuits using differential equations. Applied with enough diligence and care, these classical methods can be used to analyze electrical circuits of any order.

In Chapter 8, a totally different approach was introduced. Phasors were used to move circuit analysis from the time domain to the frequency domain, thus changing differential equations into easier-to-solve algebraic equations. As useful as the phasor approach is, however, it has its limitations. First, as they have been so far described, phasors can be applied only to sinusoidal functions. Second, phasor techniques are limited to circuits in the steady state. Natural responses cannot be found, and there is no way to include knowledge of initial conditions in a problem solution.

Fortunately, phasors are just one of several related methods for moving problem descriptions from the time domain to the frequency domain. This chapter introduces one of the most powerful of these methods, the Laplace transform. The Laplace transform can be applied to the solution of linear, time-invariant, ordinary differential equations with suddenly applied excitations. Use of the transform does not change the final answer to a problem, but it does neatly sidestep our main difficulties with the solution of differential equations in the time domain. Like phasors, the Laplace transform changes differential equations into algebraic equations. Unlike phasors, the Laplace transform can be applied to a wide range of time functions. In addition, use of the Laplace transform in circuit analysis automatically includes initial conditions in such a way that the total response, both forced and natural, can be found in one operation. The Laplace transform is, indeed, a very useful circuit analysis tool!

What is a transform? A **transform** (or transformation) is a mathematical operation that restructures the description of a problem. The restructuring is intended to provide a form that is more easily manipulated or that lends an insight into the problem that otherwise would not be obvious. The phasor is a simple example of a transform. We did not refer to it as such in Chapter 8 only because it lacks the level of mathematical rigor usually associated with transforms.

A transformation with which you are familiar is the logarithm. It can be used to simplify arithmetic calculations involving multiplications and divisions of large numbers or to find fractional roots or powers of numbers. Although electronic calculators have made the teaching of logarithms for arithmetic calculations nearly obsolete, logarithms are still quite useful in the algebraic manipulation of variables in many scientific and engineering problems.

The steps used in applying the logarithm are very much like those of most other transformations. Suppose that we want to raise 18.6 to the power of 8.37. We cannot do this by the normal procedure of multiplying 18.6 by itself the necessary number of times, so we apply the logarithm:

$$\ln 18.6^{8.37} = 8.37 \ln 18.6 = 24.47$$

The transformation changes the problem into one of simple multiplication. To arrive at the desired form of the answer, we apply the antilogarithmic ($\ln^{-1}$) operation:

$$18.6^{8.37} = \ln^{-1} 24.47 = 4.22 \times 10^{10}$$

From this example, we see that if a transformation is to be of much use, a corresponding inverse transform must exist. Application of the Laplace transform follows this general pattern.

Use of the Laplace transform in engineering is attributed to two individuals. The British engineer Oliver Heaviside (1850–1925) popularized a technique very much like the Laplace transform method. He was less concerned with mathematical rigor than with developing the functional procedures that made his technique practical. Later, it was found that the earlier work of the French mathematician Pierre Laplace (1749–1825) substantiated Heaviside's techniques in all important respects and gave it credence among the theoreticians of the day.

This chapter also introduces the convolution integral. It is a compact time-domain formula for finding the response of a linear system to any forcing function if the system is initially at rest. Convolution can be developed from several points of view, but the Laplace transform offers a particularly convenient way of demonstrating its validity.

## 9.2  THE LAPLACE TRANSFORM

The Laplace transform is an operation that can be applied to functions of time. Not all time functions can be Laplace transformed, but when it exists, the **Laplace transform** of $f(t)$ is defined by the equation

$$F(s) = \int_{0^-}^{\infty} f(t)e^{-st}\, dt \tag{9.1}$$

where  $t = $ time

   $s = \sigma + j\omega$, a complex variable

Equation 9.1 is known as the **one-sided Laplace transform** because of its limits of integration. Note that the lower limit of integration is $t = 0^-$. Strictly speaking, a one-sided Laplace transform should have a lower limit of simply $t = 0$, and you will see it presented that way in many textbooks. Very often, however, the limit of $0^-$ is chosen so that the transform can properly be applied to the unit impulse, a function whose very interesting properties at $t = 0$ are discussed in Section 9.7. A two-sided transform can also be defined. Its limits of integration are $-\infty$ and $+\infty$. The one-sided transform is more commonly used in circuit analysis.

A rigorous understanding of Equation 9.1 requires a background in the theory of complex variables that is beyond the intended scope of this book. We can, however, make some basic observations.

By convention, lowercase letters are used to represent functions of time, and the corresponding uppercase letters are used for their transforms. For instance, the Laplace transform of the function $r(t)$ is given the symbol $R(s)$.

As already noted, the variable $s$ of the Laplace transform is complex. We will not have to make much use of this fact, but it will be helpful for our work in future chapters if we note that $s$ is a complex *frequency*. That is, it has the dimensions of $1/T$ and units of $s^{-1}$. This follows from its location in the exponent of Equation 9.1. Recall that a similar remark was made regarding the neper frequency introduced in Chapter 2.

Because the lower limit of integration in Equation 9.1 is $0^-$, the one-sided Laplace transform is not affected by the behavior of $f(t)$ for $t < 0$. The form of $f(t)$ for negative time does affect a circuit problem, however, by influencing the values of initial conditions. An additional advantage of selecting a lower limit of $0^-$ rather than $0$ is that it makes the initial conditions of switching problems unambiguous and generally easier to find.

Not all functions of time can be Laplace transformed. In order for a transform to exist for a function, it is sufficient that $f(t)$ be integrable in the sense that

$$\int_{0^-}^{\infty} |f(t)| e^{-\sigma t}\, dt < \infty \tag{9.2}$$

for some $\sigma > 0$, where $\sigma = Re(s)$. For instance, $e^{10t}$ is so integrable for $\sigma > 10$. However, there is no value of $\sigma$ for which the function $e^{t^2}$ meets the criterion. This requirement of absolute integrability (also called a convergence criterion) is not a serious restriction because Laplace transforms exist for virtually all functions of engineering interest.

Thus far, we have focused on how to find the Laplace transform of a function of time. What about the equally important inverse operation? If $F(s)$ is known, the corresponding function of time can be found through

$$f(t) = \frac{1}{2\pi j} \int_{\sigma - j\infty}^{\sigma + j\infty} e^{st} F(s)\, ds \tag{9.3}$$

Equation 9.3 defines the **inverse Laplace transform** operation. It has an intimidating appearance to anyone contemplating its use. Fortunately, we will be able to avoid using Equation 9.3 because a time function and its transform form a unique pair. Either one unambiguously implies the other. Symbolically,

$$f(t) \leftrightarrow F(s)$$

Our approach will be to develop a table of such transform pairs that includes all of the functions normally encountered in circuit analysis. A companion table showing how certain mathematical operations performed on $f(t)$ affect its transform will help us to tackle just about any circuit analysis problem that we might face.

The Laplace transform operation and its inverse are often indicated in an operational notation:

$$F(s) = \mathcal{L}\{f(t)\} \tag{9.4a}$$

$$f(t) = \mathcal{L}^{-1}\{F(s)\} \tag{9.4b}$$

Equation 9.4a is read as "$F(s)$ is the Laplace transform of $f(t)$." Equation 9.4b states that "$f(t)$ is the inverse Laplace transform of $F(s)$."

Just as $f(t)$ is a *time-domain* description of a function because its independent variable is $t$, $F(s)$ is a *frequency-domain* description of the same function. Its independent variable, or domain, is $s$, a complex frequency. In fact, the Laplace transform representation of a problem is often described as being in the $s$ **domain.** Regardless of terminology, keep in mind that the Laplace transform is just one of several related mathematical techniques for going from the time domain to the frequency domain. These techniques include phasors, which were discussed in Chapter 8, and Fourier analysis, a topic to be introduced in Chapter 12.

**FIGURE 9.1**   *Relationship Between a Problem's Solution in Time Domain and in s Domain*

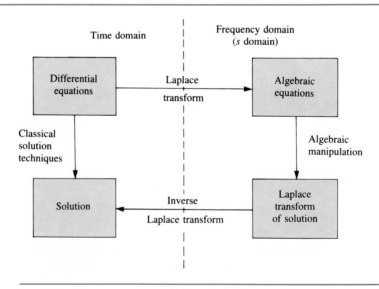

Our procedure in applying the Laplace transform is to take circuit problems from the time domain to the *s* domain. There, the answers will be more easily found. A final inverse transform operation returns us to the time domain and the form of the solution that we desire. These steps are outlined in Figure 9.1.

## 9.3 BASIC TRANSFORM PAIRS

Before we use the Laplace transform in the analysis of electrical circuits, we need to find the transforms of the time functions that we are likely to encounter. We will be helped toward this goal if we also determine what effects some common mathematical operations in the time domain have on the Laplace transforms of functions. Table 9.1 gives a useful selection of Laplace transforms of simple time functions. Table 9.2 lists some operational pairs of Laplace transforms. Some of the entries will be derived in this section; others are derived in later sections or in problems at the end of the chapter.

**TABLE 9.1** *Basic Laplace Transform Pairs*

| | $f(t)$ | $F(s)$ |
|---|---|---|
| 1. | $u(t)$ | $\dfrac{1}{s}$ |
| 2. | $\delta(t)$ | $1$ |
| 3. | $e^{-at}u(t)$ | $\dfrac{1}{s+a}$ |
| 4. | $tu(t)$ | $\dfrac{1}{s^2}$ |
| 5. | $\dfrac{t^n}{n!}u(t)$ | $\dfrac{1}{s^{n+1}}$ |
| 6. | $te^{-at}u(t)$ | $\dfrac{1}{(s+a)^2}$ |
| 7. | $\cos \omega t\, u(t)$ | $\dfrac{s}{s^2+\omega^2}$ |
| 8. | $\sin \omega t\, u(t)$ | $\dfrac{\omega}{s^2+\omega^2}$ |
| 9. | $e^{-at}\cos \omega t\, u(t)$ | $\dfrac{s+a}{(s+a)^2+\omega^2}$ |
| 10. | $e^{-at}\sin \omega t\, u(t)$ | $\dfrac{\omega}{(s+a)^2+\omega^2}$ |

**TABLE 9.2**

| Property | $f(t)$ | $F(s)$ |
|---|---|---|
| **1.** Linearity | $af_1(t) + bf_2(t)$ | $aF_1(s) + bF_2(s)$ |
| **2.** Differentiation | $\dfrac{df}{dt}$ | $sF(s) - f(0^-)$ |
| | $\dfrac{d^2f}{dt^2}$ | $s^2F(s) - sf(0^-) - f'(0^-)$ |
| **3.** Integration | $\displaystyle\int_{0^-}^{t} f(x)\,dx$ | $\dfrac{1}{s}F(s)$ |
| | $\displaystyle\int_{-\infty}^{t} f(x)\,dx$ | $\dfrac{1}{s}F(s) + \dfrac{1}{s}\displaystyle\int_{-\infty}^{0^-} f(x)\,dx$ |
| **4.** Time shift | $f(t - t_0)u(t - t_0)$ | $e^{-st_0}F(s)$ |
| **5.** Frequency shift | $e^{-at}f(t)$ | $F(s + a)$ |
| **6.** Scaling | $f(at)$ | $\dfrac{1}{a}F\left(\dfrac{s}{a}\right)$ |
| **7.** Multiplication by $t$ | $tf(t)$ | $-\dfrac{d}{ds}F(s)$ |
| **8.** Convolution | $\displaystyle\int_{0}^{t} h(\lambda)x(t - \lambda)\,d\lambda$ | $H(s)X(s)$ |
| | $\displaystyle\int_{0}^{t} h(t - \lambda)x(\lambda)\,d\lambda$ | |
| **9.** Periodicity | $f(t) = \displaystyle\sum_{n=0}^{+\infty} p(t - nt_0)$ | $F(s) = \dfrac{P(s)}{1 - e^{-st_0}}$ |
| **10.** Initial value theorem | $f(0^+) = \lim\limits_{s \to \infty} sF(s)$ | |
| **11.** Final value theorem | $\lim\limits_{t \to \infty} f(t) = \lim\limits_{s \to 0} sF(s)$ | |

## Unit Step Function

Let's begin with the simplest one-sided function of all—the unit step function, $u(t)$. In Chapter 7, we found this function to be useful for describing switching operations in circuits. Its transform can easily be found:

$$\mathcal{L}\{u(t)\} = \int_{0^-}^{\infty} e^{-st}u(t)\,dt = \int_{0^+}^{\infty} e^{-st}\,dt = -\frac{1}{s}e^{-st}\Big|_{0^+}^{\infty} = \frac{1}{s}$$

We have here our first transform pair. The Laplace transform of the unit step function is $1/s$, the first entry in Table 9.1.

## Simple Exponential

Now we turn our attention to an exponential function that begins at $t = 0$:

$$\mathcal{L}\{e^{-at}u(t)\} = \int_{0^-}^{\infty} e^{-st}e^{-at}u(t)\, dt = \int_{0^+}^{\infty} e^{-(s+a)t}\, dt = -\frac{1}{s+a}e^{-(s+a)t}\Big|_{0^+}^{\infty}$$

$$= \frac{1}{s+a}$$

The Laplace transforms of many functions of time are more easily shown using the properties of Table 9.2 rather than the defining integral itself. Two of the more important of these transforms are shown next.

## Time Differentiation

Since we intend to use the Laplace transform to solve differential equations, we should determine what effect time differentiation has on the Laplace transform of a function:

$$\mathcal{L}\left\{\frac{df}{dt}\right\} = \int_{0^-}^{\infty} \frac{df}{dt} e^{-st}\, dt$$

We can integrate by parts choosing

$$u = e^{-st} \qquad \text{and} \qquad dv = \frac{df}{dt}\, dt$$

Then,

$$du = -se^{-st}\, dt \qquad \text{and} \qquad v = f(t)$$

Thus,

$$\mathcal{L}\left\{\frac{df}{dt}\right\} = f(t)e^{-st}\Big|_{0^-}^{\infty} - \int_{0^-}^{\infty} f(t)(-se^{-st})\, dt$$

$$= 0 - f(0^-) + s\int_{0^-}^{\infty} f(t)e^{-st}\, dt \tag{9.5}$$

The term $f(t)e^{-st}$ goes to 0 as $t \to \infty$ because of the convergence condition (Equation 9.2) that we place on $f(t)$. If the remaining integral of Equation 9.5 is recognized as the defining expression for the Laplace transform of $f(t)$, we can rearrange the equation as

$$\mathcal{L}\left\{\frac{df}{dt}\right\} = sF(s) - f(0^-)$$

This result can easily be extended to higher-order derivatives:

$$\mathcal{L}\left\{\frac{d^2f}{dt^2}\right\} = \mathcal{L}\left\{\frac{d}{dt}\frac{df}{dt}\right\} = s\mathcal{L}\left\{\frac{df}{dt}\right\} - f'(0^-) = s^2F(s) - sf(0^-) - f'(0^-)$$

where $f'(t)$ is a commonly used notation that stands for the first derivative of $f(t)$. If the initial conditions are assumed to be zero, you can see that the Laplace transform has turned differentiation in the time domain into multiplication by the corresponding power of $s$ in the $s$ domain. An unexpected but highly desirable bonus is the fact that initial conditions are naturally included in the process. The Laplace transform of the first derivative of a function contains one initial condition, that of a second derivative has two initial conditions, and so forth.

## Time Integration

We now know how to treat derivatives and, by extension, differential equations. It seems natural to ask how integration in the time domain affects a Laplace transform. For example,

$$\mathcal{L}\left\{\int_{-\infty}^{t} f(x)\,dx\right\} = \mathcal{L}\left\{\int_{-\infty}^{0^-} f(x)\,dx\right\} + \mathcal{L}\left\{\int_{0^-}^{t} f(x)\,dx\right\} \tag{9.6}$$

Because the first integral on the right side of Equation 9.6 has definite limits of integration, it is simply a constant. The Laplace transform of a constant is the same as that of a step function. Therefore, we can use the known transform of $u(t)$ and the linearity property to say that

$$\mathcal{L}\left\{\int_{-\infty}^{0^-} f(x)\,dx\right\} = \frac{1}{s}\int_{-\infty}^{0^-} f(x)\,dx$$

The second term of Equation 9.6 must be evaluated directly:

$$\mathcal{L}\left\{\int_{0^-}^{t} f(x)\,dx\right\} = \int_{0^-}^{\infty} e^{-st}\left[\int_{0^-}^{t} f(x)\,dx\right]dt$$

We can integrate by parts choosing

$$u = \int_{0^-}^{t} f(x)\,dx \qquad \text{and} \qquad dv = e^{-st}\,dt$$

Then,

$$du = f(t)\, dt \qquad \text{and} \qquad v = -\frac{1}{s} e^{-st}$$

Thus,

$$\mathscr{L}\left\{ \int_{0^-}^{t} f(x)\, dx \right\} = -\frac{1}{s} e^{-st} \int_{0^-}^{t} f(x)\, dx \Big|_{0^-}^{\infty} + \frac{1}{s} \int_{0^-}^{\infty} e^{-st} f(t)\, dt$$

The first term is zero when evaluated at either of the limits. We conclude that

$$\mathscr{L}\left\{ \int_{0^-}^{t} f(x)\, dx \right\} = \frac{1}{s} F(s)$$

and that

$$\mathscr{L}\left\{ \int_{-\infty}^{t} f(x)\, dx \right\} = \frac{1}{s} F(s) + \frac{1}{s} \int_{-\infty}^{0^-} f(x)\, dx$$

Note that while differentiation in the time domain corresponds to multiplication by $s$ in the $s$ domain, integration leads to a division of the Laplace transform by $s$. In both cases, the behavior of the time function prior to $t = 0$ affects the operation.

## Unit Impulse Function

The derivative of the unit step function holds an important place in electrical circuit analysis. It is known as the unit impulse function and is given the symbol $\delta(t)$. Its Laplace transform is easily demonstrated:

$$\mathscr{L}\{\delta(t)\} = \mathscr{L}\left\{ \frac{d}{dt} u(t) \right\} = s\mathscr{L}\{u(t)\} - u(0^-) = \frac{1}{s} s - 0 = 1$$

The impulse function is discussed in more detail in Section 9.7.

## Sine and Cosine

Let's now find the transforms of $\sin \omega t\, u(t)$ and $\cos \omega t\, u(t)$. Notice that these are somewhat different from the steady-state sinusoids of Chapter 8 in that here the sinusoids are suddenly turned on at $t = 0$. Instead of starting with the defining integral to find the Laplace transforms, we will make use of the operation's linearity (the first entry of Table 9.2). Euler's theorem tells us that the cosine can be rewritten as

$$\cos \omega t = \frac{1}{2} [e^{j\omega t} + e^{-j\omega t}]$$

Then,

$$\mathcal{L}\{\cos \omega t\, u(t)\} = \mathcal{L}\left\{\frac{1}{2}\left[e^{j\omega t} + e^{-j\omega t}\right]u(t)\right\}$$

$$= \frac{1}{2}\,\mathcal{L}\{e^{j\omega t}u(t)\} + \frac{1}{2}\,\mathcal{L}\{e^{-j\omega t}u(t)\}$$

Nothing in the derivation of the Laplace transform of an exponential excludes complex exponents. We can, therefore, apply our earlier result and find that

$$\mathcal{L}\{\cos \omega t\, u(t)\} = \frac{1}{2}\,\frac{1}{s - j\omega} + \frac{1}{2}\,\frac{1}{s + j\omega} = \frac{s}{s^2 + \omega^2}$$

In a similar way, it can be shown that

$$\mathcal{L}\{\sin \omega t\, u(t)\} = \frac{\omega}{s^2 + \omega^2}$$

We conclude here by suggesting a way of deriving the Laplace transforms of exponentially damped sinusoids. This is most easily done if we first consider the fifth operational transform pair from Table 9.2.

## Frequency Shifting

Let's see what effect multiplication of a function by an exponential has on its Laplace transform:

$$\mathcal{L}\{e^{-at}f(t)\} = \int_{0^-}^{\infty} e^{-at}f(t)e^{-st}\, dt = \int_{0^-}^{\infty} f(t)e^{-(s+a)t}\, dt$$

The last expression can be recognized as the Laplace transform of $f(t)$, $s + a$ being the Laplace variable. That is,

$$\mathcal{L}\{e^{-at}f(t)\} = F(s + a)$$

where $F(s) = \mathcal{L}\{f(t)\}$. A problem at the end of the chapter asks you to show that applying this new rule to the already-known transforms of sine and cosine leads to the transforms for the exponentially damped sine and exponentially damped cosine that are given in Table 9.1.

## 9.4  THE LAPLACE TRANSFORM USED TO SOLVE EQUATIONS

With just what is known thus far, we can use the Laplace transform to simplify the solution of circuit problems and differential equations in general. Let's begin by applying the Laplace transform to the solution of the following differential equation (the two necessary initial conditions are also given):

$$\frac{d^2g}{dt^2} + 6\frac{dg}{dt} + 5g = 0$$

$$g(0^-) = 5$$

$$g'(0^-) = 2$$

The Laplace transform is a linear operation and can be applied to the equation as follows:

$$\mathscr{L}\left\{\frac{d^2g}{dt^2} + 6\frac{dg}{dt} + 5g\right\} = \mathscr{L}\{0\}$$

$$\mathscr{L}\left\{\frac{d^2g}{dg^2}\right\} + 6\mathscr{L}\left\{\frac{dg}{dg}\right\} + 5\mathscr{L}\{g(t)\} = \mathscr{L}\{0\}$$

$$[s^2G(s) - sg(0^-) - g'(0^-)] + 6[sG(s) - g(0^-)] + 5G(s) = 0$$

This algebraic equation can easily be solved for $G(s)$:

$$G(s) = \frac{5s + 32}{s^2 + 6s + 5} \tag{9.7}$$

At this point in any problem, we would like to be able to consult Table 9.1 to find the inverse transform of the solution. However, usually only the most basic transform pairs are shown in such tables. Fortunately, Equation 9.7 can be manipulated into a form which *is* recognizable from the table. Verify that

$$G(s) = \frac{5s + 32}{s^2 + 6s + 5} = \frac{-7/4}{s + 5} + \frac{27/4}{s + 1} \tag{9.8}$$

A technique for finding such equalities is described in Section 9.5. Consulting Table 9.1 now is more successful. Each of the two terms to the far right of Equation 9.8 is the Laplace transform of an exponential. The inverse is

$$g(t) = -\frac{7}{4}e^{-5t} + \frac{27}{4}e^{-t}, \qquad t > 0$$

Notice that we have not included a factor of $u(t)$ in the answer. To do so would imply that $g(t) = 0$ for $t < 0$, which would be in violation of the known initial condition of $g(0^-) = 5$. Instead of including the $u(t)$ that would seem to be suggested by the Laplace transform table, we simply state that the answer is valid only for $t > 0$.

Let's look at an elementary electrical circuit problem. The approach will be to first find a differential equation in the time domain and then apply the Laplace transform to it. Consider the circuit shown in Figure 9.2. A differential equation

FIGURE 9.2   *Differential Analysis of an RL Circuit*

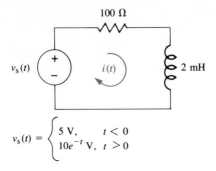

$$v_s(t) = \begin{cases} 5 \text{ V}, & t < 0 \\ 10e^{-t} \text{ V}, & t > 0 \end{cases}$$

and the necessary initial condition can be found:

$$v_s(t) = 100i(t) + .002 \frac{di(t)}{dt}^{\dagger}$$

Because there is a first-order derivative in the equation, we will need the initial condition of $i(0^-)$ in order to apply the Laplace transform. You should be able to show that $i(0^-) = .05$ A. Remember that, in applying the Laplace transform to the forcing function, it is a one-sided operation that is not affected by the behaviors of time functions for negative time. Consequently,

$$\frac{10}{s+1} = 100I(s) + .002[sI(s) - .05]$$

$$I(s) = \frac{.05s + 5000.05}{(s+1)(s+50,000)} = \frac{5000/49,999}{s+1} + \frac{-2500.05/49,999}{s+50,000}$$

The inverse transform can now be written:

$$i(t) = \frac{5000}{49,999} e^{-t} - \frac{2500.05}{49,999} e^{-50,000t} \text{ A}, \qquad t > 0$$

You should verify from this answer that $i(0) = .05$ A.

When Laplace transforms are used in the analysis of electrical circuits, it is usually to your advantage to begin with integro-differential equations rather than

---

$^{\dagger}$ In order to correctly apply the Laplace transform to electrical circuits undergoing discontinuities at $t = 0$, the written differential equation must be valid at least over the region of time $0^- < t < \infty$ to competely account for the effects of the discontinuity. This is in contrast to the classical time-domain approach where equations are sometimes written to be valid only for $t > 0$. See Section 9.8 for details.

differential equations in standard, classical form. The reason has to do with the way in which initial conditions enter into the problem. As an example, let's apply the Laplace transform to the analysis of the circuit shown in Figure 9.3. An integro-differential equation can easily be written:

$$i_s(t) = v_o(t) + \frac{20}{9} \int_{0^-}^{t} v_o(t)\, dt + i_L(0^-) + \frac{1}{9}\frac{dv_o(t)}{dt}$$

We apply the Laplace transform:

$$\frac{5}{s} = V_o(s) + \frac{20}{9s} V_o(s) + \frac{1}{s} i_L(0^-) + \frac{s}{9} V_o(s) - \frac{1}{9} v_C(0^-)$$

The last term of the equation has been written as $v_C(0^-)$ rather than $v_o(0^-)$ in order to emphasize that it is a capacitor voltage. The two initial conditions of the circuit can readily be found to be $v_C(0^-) = 0\ \text{V}$ and $i_L(0^-) = -5\ \text{A}$. The equation can now be solved for $V_o(s)$:

$$V_o(s) = \frac{90}{s^2 + 9s + 20} = \frac{-90}{s + 5} + \frac{90}{s + 4}$$

By now, we readily recognize the Laplace transforms of simple exponentials, and the inverse can be written:

$$v_o(t) = (-90e^{-5t} + 90e^{-4t})u(t)\ \text{V}$$

It is significant that the initial conditions needed in the previous example were the values of the inductor current and capacitor voltage that existed just prior to the discontinuity at $t = 0$. This is always true in the analysis of switched circuits when the Laplace transform is applied to integro-differential equations. No matter how complicated the circuit, the necessary initial conditions will always be the set of inductor currents and capacitor voltages just prior to the discontinuity.

FIGURE 9.3  *Integro-Differential Analysis of an RLC Circuit*

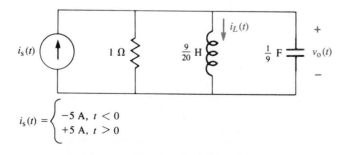

**Example 9.1**    *The Laplace Transform in a Simple Circuit Problem*

Analyze the circuit of Figure 9.4.

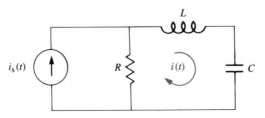

$R = 30 \ \Omega, \ L = 10 \text{ mH}, \ C = 50 \ \mu\text{F}$

$$i_s(t) = \begin{cases} -50 \text{ mA}, \ t < 0 \\ \ \ 50 \text{ mA}, \ t > 0 \end{cases}$$

**FIGURE 9.4**

Write a KVL equation around the nonsource mesh:

$$R(i - i_s) + L \frac{di}{dt} + \frac{1}{C} \int_{0^-}^{t} i \, dt + v_C(0^-) = 0$$

Apply the Laplace transform, making numerical substitutions at the same time. Note that $i_L(0^-) = 0$ A and $v_C(0^-) = -1.5$ V.

$$30\left[ I(s) - \frac{.05}{s} \right] + .01sI(s) + \frac{20,000}{s} I(s) - \frac{1.5}{s} = 0$$

$$I(s) = \frac{300}{(s + 1000)(s + 2000)} = \frac{.3}{s + 1000} - \frac{.3}{s + 2000}$$

$$= 300[e^{-1000t} - e^{-2000t}]u(t) \text{ mA}$$

## 9.5 PARTIAL FRACTION EXPANSION

You should now be fairly comfortable with using Laplace transforms to take problems from the time domain to the *s* domain. As noted earlier, however, transforms are not truly useful unless their inverses exist and can easily be found. The inverses allow us to present our answers in the time domain, which is usually the desired form. We have already seen the use of the inverse Laplace transform several times in Section 9.4, but many of the details were left out. In this section, a step-by-step procedure for finding the inverse Laplace transforms of circuit variables will be described.

In each of the analysis problems of Section 9.4, a magical step was performed in which the Laplace transform of the answer was put in a form whose inverse was recognizable. Of course, there was no magic to it. Oliver Heaviside developed the techniques of **partial fraction expansion** (PFE) that allow us to manipulate a wide class of complicated Laplace transforms into simpler forms whose corresponding time functions can be looked up in short tables of transform pairs.

You may have noticed that the Laplace transforms that have been introduced thus far are all in the form of ratios of polynomials in $s$. This is true of all the simple time functions that were discussed in Section 9.3. It is also true of the answers to the circuit problems of Section 9.4 and to all similar circuit problems. In contrast, electrical systems that involve time delays lead to Laplace transforms of a different kind. These transforms will be discussed later.

Assume that we want to find the inverse Laplace transform of $F(s)$ and that $F(s)$ is the ratio of two polynomials in $s$:

$$F(s) = \frac{N(s)}{D(s)} \tag{9.9}$$

The numerator polynomial, $N(s)$, is of degree $m$, and the denominator polynomial, $D(s)$, is of degree $n$. We proceed only if $F(s)$ is a proper rational function. For proper rational functions, $n > m$. Almost all Laplace transforms encountered in electrical engineering are proper, but when one is not, a long division must be done as a preliminary step:

$$F(s) = \frac{N(s)}{D(s)} = a_0 + a_1 s + \cdots + a_{m-n} s^{m-n} + \frac{N_1(s)}{D(s)} \tag{9.10}$$

The remaining ratio $N_1(s)/D(s)$ is then in proper form. The inverse transform of Equation 9.10 can be written as

$$f(t) = a_0 \delta(t) + a_1 \delta'(t) + \cdots + a_{m-n} \delta^{m-n}(t) + \mathscr{L}^{-1}\left\{\frac{N_1(s)}{D(s)}\right\} \tag{9.11}$$

The function $\delta(t)$ is the unit impulse function. The succeeding functions $\delta^k(t)$ are its various derivatives. Most often in circuit analysis, $F(s)$ is in proper form, and, so, the impulse function and its derivatives will not show up in the answer.

For the rest of this discussion, we will assume that $F(s)$ is a proper rational function and see how to put it into a more recognizable form. The partial fraction expansion of a proper function is based on the roots of its denominator polynomial. An example of such an expansion is shown in Equation 9.12 where, for simplicity, we assume that none of the roots are repeated:

$$F(s) = \frac{N(s)}{(s + p_1)(s + p_2)(s + p_3)} = \frac{K_1}{s + p_1} + \frac{K_2}{s + p_2} + \frac{K_3}{s + p_3} \tag{9.12}$$

The function of Equation 9.12 can be expanded into three relatively simple terms corresponding to each of the roots of the denominator polynomial. The constants

$K_1$, $K_2$, and $K_3$ are called **residues** of $F(s)$ at the roots of its denominator polynomial. We would like to be able to calculate the residues because then we could write the inverse transform of $F(s)$. Consider multiplying $F(s)$ by the factor $s + p_1$:

$$(s + p_1)F(s) = K_1 + (s + p_1)\frac{K_2}{s + p_2} + (s + p_1)\frac{K_3}{s + p_3} \qquad (9.13)$$

Now, let $s = -p_1$, the value of the root corresponding to the multiplying factor that we have chosen. When we do this, both the second and third terms on the right side of Equation 9.13 become zero. We are left with

$$K_1 = (s + p_1)F(s)|_{s=-p_1}$$

To generalize, we can say that if the denominator polynomial of $F(s)$ has a simple (nonrepeated) root at $s = -p_j$, then the residue of the corresponding term of the partial fraction expansion can be found using the formula

$$\boxed{K_j = (s + p_j)F(s)|_{s=-p_j}} \qquad (9.14)$$

---

**Example 9.2**    *PFE with Simple Real Roots*

Verify the expansion used in Equation 9.8.

$$G(s) = \frac{5s + 32}{s^2 + 6s + 5} = \frac{5s + 32}{(s + 5)(s + 1)} = \frac{K_1}{s + 5} + \frac{K_2}{s + 1}$$

$$K_1 = \frac{(s + 5)(5s + 32)}{(s + 5)(s + 1)}\bigg|_{s=-5} = \frac{5(-5) + 32}{-5 + 1} = -\frac{7}{4}$$

$$K_2 = \frac{(s + 1)(5s + 32)}{(s + 5)(s + 1)}\bigg|_{s=-1} = \frac{5(-1) + 32}{-1 + 5} = \frac{27}{4}$$

A check with Equation 9.8 shows these to be the residues that were claimed earlier without proof.

---

**Example 9.3**    *Another PFE with Simple Real Roots*

Expand $F(s)$ and invert to find $f(t)$.

$$F(s) = \frac{2s^2 + 12s + 16}{(s + 1)(s + 2)(s + 3)} = \frac{K_1}{s + 1} + \frac{K_2}{s + 2} + \frac{K_3}{s + 3}$$

$$K_1 = \frac{2s^2 + 12s + 16}{(s + 2)(s + 3)}\bigg|_{s=-1} = 3$$

$$K_2 = \frac{2s^2 + 12s + 16}{(s + 1)(s + 3)}\bigg|_{s=-2} = 0$$

$$K_3 = \frac{2s^2 + 12s + 16}{(s + 1)(s + 2)}\bigg|_{s=-3} = -1$$

Therefore,

$$F(s) = \frac{3}{s + 1} - \frac{1}{s + 3}$$

Table 9.1 can now be consulted to find the inverse:

$$f(t) = [3e^{-t} - e^{-3t}]u(t)$$

---

To this point, we have considered only the case of simple, real roots. $D(s)$ can also have multiple roots at a location, and it can have complex conjugate roots. The possibility of multiple complex conjugate roots will not be considered because it virtually never occurs in practice. For purposes of discussion, we will look at a hypothetical function that contains all of these usual possibilities:

$$F(s) = \frac{N(s)}{(s + p_1)(s + p_2)^r(s^2 + bs + c)}$$

$$= \frac{K_1}{s + p_1} + \frac{K_{21}}{s + p_2} + \frac{K_{22}}{(s + p_2)^2} + \cdots + \frac{K_{2r}}{(s + p_2)^r} + \frac{As + B}{s^2 + bs + c}$$

(9.15)

The correctness of Equation 9.15 can be demonstrated mathematically. We will accept it as true and devote ourselves to finding ways to calculate the constants. Notice that we choose to combine pairs of complex conjugate roots into single quadratic terms. Thus, we avoid any complex number arithmetic. Problem 9.14 at the end of the chapter investigates the approach of treating such roots separately. $K_1$ and $K_{21}$ are the residues of $F(s)$ at the roots $s = -p_1$ and $s = -p_2$, respectively. The other constants are not residues. $K_1$ can be found by Equation 9.14. The constant $K_{2r}$ can be found by a similar formula:

$$K_{2r} = (s + p_2)^r F(s)\big|_{s=-p_2}$$

(9.16)

More generally, the constants $K_{j1}$, $K_{j2}$, ..., $K_{jr}$ can be found either by the formula

$$K_{jk} = \frac{1}{(r-k)!} \frac{d^{r-k}}{ds^{r-k}} [F(s)(s+p_j)^r]\Big|_{s=-p_j} \qquad (9.17)$$

or by another approach that will be demonstrated by example.

---

**Example 9.4**    *PFE Including Multiple Real Roots*

Expand $F(s)$ and invert to find $f(t)$.

$$F(s) = \frac{s}{(s+1)(s+2)^2} = \frac{K_1}{s+1} + \frac{K_{21}}{s+2} + \frac{K_{22}}{(s+2)^2}$$

$$K_1 = (s+1)F(s)\Big|_{s=-1} = \frac{s}{(s+2)^2}\Big|_{s=-1} = -1$$

$$K_{22} = (s+2)^2 F(s)\Big|_{s=-2} = \frac{s}{s+1}\Big|_{s=-2} = 2$$

Substitution of the known constants into the original expression gives

$$\frac{s}{(s+1)(s+2)^2} = \frac{-1}{s+1} + \frac{K_{21}}{s+2} + \frac{2}{(s+2)^2} \qquad (9.18)$$

Equation 9.18 is an identity that must be true for all values of $s$. Cross-multiplying to clear the fractions and rearranging yield

$$s = -1(s+2)^2 + K_{21}(s+1)(s+2) + 2(s+1)$$

$$(-1+K_{21})s^2 + (-3+3K_{21})s + (-2+2K_{21}) = 0$$

If this polynomial is to be true for all values of $s$, each of its coefficients must equal zero. Any of the three leads to the conclusion that $K_{21} = 1$. Then,

$$F(s) = \frac{-1}{s+1} + \frac{1}{s+2} + \frac{2}{(s+2)^2}$$

Consulting Table 9.1 shows the inverse to be

$$f(t) = [-e^{-t} + e^{-2t} + 2te^{-2t}]u(t)$$

Example 9.4 demonstrates our general approach to the PFE process. First, we calculate the most easily found residues and constants through Equations 9.14 and 9.16. Second, we make use of the fact that any PFE equation is an identity, and we find the remaining constants by cross-multiplying and setting the coefficients to zero. This second step is called the **method of undetermined coefficients.** The next example shows how this approach works for an expansion involving a quadratic term in the denominator.

**Example 9.5**  *PFE Including Quadratic Terms*

Expand and invert $F(s)$.

$$F(s) = \frac{5s^2 + 9s + 31}{(s + 1)(s^2 + 2s + 10)} = \frac{K_1}{s + 1} + \frac{As + B}{s^2 + 2s + 10}$$

Note that the numerator of the second term is not a simple constant. It is the polynomial $As + B$. The residue, $K_1$, is found in the usual way:

$$K_1 = \left.\frac{5s^2 + 9s + 31}{s^2 + 2s + 10}\right|_{s=-1} = 3$$

Now, substituting the numerical value of $K_1$ into the equation and cross-multiplying give

$$5s^2 + 9s + 31 = 3(s^2 + 2s + 10) + (As + B)(s + 1)$$
$$= (3 + A)s^2 + (A + B + 6)s + (30 + B)$$

Rather than gather all terms to one side, we simply equate coefficients of like powers of $s$:

$3 + A = 5$

$A + B + 6 = 9$

$30 + B = 31$

Any two of these equations lead to the result that $A = 2$ and $B = 1$. We can now write

$$F(s) = \frac{3}{s + 1} + \frac{2s + 1}{s^2 + 2s + 10}$$

We know the inverse of the first term, but what about the second?

*(continues)*

**Example 9.5**    *Continued*

Consulting Table 9.1 shows that the only likely candidates are the exponentially damped sinusoids, but we must manipulate the expression before its inverse can be found. We begin by adding a number to the first two terms of the denominator such that the total will be a perfect square. This added number will be borrowed from the constant of 10:

$$\frac{2s+1}{s^2+2s+10} = \frac{2s+1}{(s^2+2s+1)+9} = \frac{2s+1}{(s+1)^2+3^2}$$

Now that the denominator is in the proper form for the transforms of exponentially damped sines or cosines, we can turn our attention to the numerator:

$$\frac{2s+1}{s^2+2s+10} = \frac{2(s+1)-1}{(s+1)^2+3^2} = 2\frac{s+1}{(s+1)^2+3^2} - \frac{1}{3}\frac{3}{(s+1)^2+3^2}$$

Therefore,

$$F(s) = \frac{3}{s+1} + 2\frac{s+1}{(s+1)^2+3^2} - \frac{1}{3}\frac{3}{(s+1)^2+3^2}$$

and

$$f(t) = \left[3e^{-t} + 2e^{-t}\cos 3t - \frac{1}{3}e^{-t}\sin 3t\right]u(t)$$

---

We now know how to move back and forth between the time domain and the $s$ domain. The partial fraction expansion techniques that have been introduced here will work for almost any circuit problem that we are likely to encounter. The methods shown are convenient but, by no means, unique. Several of the problems at the end of the chapter lead you through other methods of computing the constants associated with the expansion.

When paper and pencil methods are used, the hardest part of a partial fraction expansion calculation can be at the very beginning. We must determine the roots of the denominator polynomial $D(s)$ before we can proceed. This can be a difficult task for high-order systems. Computers or calculators are usually employed. So, why not write a computer program that performs the entire expansion calculation? In fact, many such programs exist.

Before you can make intelligent use of such programs, however, you should be able to perform the calculations yourself. It is risky to make extensive use of a computer program if you have little understanding of the principles upon which it is based. Almost inevitably, you will encounter a special case and be unable to explain why the program gives strange results. This same comment applies to nearly all of the analytical techniques introduced in this book. The computer is an immensely important aid in the modern practice of engineering, but it can never substitute for basic understanding.

## 9.6 THE INITIAL AND FINAL VALUE THEOREMS

There are still several entries in Tables 9.1 and 9.2 that have not yet been discussed. In this section, we will prove and then demonstrate the usefulness of the initial value theorem and the final value theorem. These two theorems provide links between a time function and its Laplace transform that allow us to check our calculations.

As with several of our earlier proofs, we will proceed indirectly. Consider the Laplace transform of the derivative of a function of time. We assume that Laplace transforms exist for both $f(t)$ and $f'(t)$:

$$\int_{0^-}^{\infty} e^{-st} \frac{df}{dt} \, dt = sF(s) - f(0^-) \tag{9.19}$$

We take the limit of both sides of Equation 9.19 as $s \to \infty$. At the same time, we rewrite the integral:

$$\lim_{s \to \infty} \left[ \int_{0^-}^{0^+} e^{-st} \frac{df}{dt} \, dt + \int_{0^+}^{\infty} e^{-st} \frac{df}{dt} \, dt \right] = \lim_{s \to \infty} [sF(s) - f(0^-)]$$

For the first integral, $t \approx 0$ and the exponential is equal to 1. When the limit is taken, the exponential of the second integral becomes zero. Therefore,

$$\lim_{s \to \infty} \left[ \int_{0^-}^{0^+} \frac{df}{dt} \, dt \right] = \lim_{s \to \infty} [sF(s) - f(0^-)]$$

$$\lim_{s \to \infty} [f(0^+) - f(0^-)] = \lim_{s \to \infty} [sF(s) - f(0^-)] \tag{9.20}$$

The two terms on the left-hand side of Equation 9.20 are constants that are independent of $s$. We conclude that

$$f(0^+) = \lim_{s \to \infty} sF(s) \tag{9.21}$$

Equation 9.21 is the **initial value theorem.**

The final value theorem is proven in a similar manner. We begin with the Laplace transform of a derivative and let $s$ go to 0:

$$\lim_{s \to 0} [sF(s) - f(0^-)] = \lim_{s \to 0} \int_{0^-}^{\infty} e^{-st} \frac{df}{dt} \, dt = \int_{0^-}^{\infty} \frac{df}{dt} \, dt = \lim_{t \to \infty} \int_{f(0^-)}^{f(t)} df$$

$$= \lim_{t \to \infty} [f(t) - f(0^-)]$$

Therefore,

$$\lim_{t\to\infty} f(t) = \lim_{s\to 0} sF(s) \tag{9.22}$$

Equation 9.22 is the **final value theorem.** Together with Equation 9.21, it can be used as a check on the correctness of either a transform or an inverse transform calculation. Be careful, however: Successfully "passing" the initial and final value tests does not mean that a transform pair is necessarily correct!

---

**Example 9.6**    *Applying the Initial and Final Value Theorems*

Find the initial and final values of $10te^{-5t}u(t)$.

$$10te^{-5t}u(t) \leftrightarrow \frac{10}{(s+5)^2}$$

$$f(0^+) = \lim_{s\to\infty} \frac{10s}{(s+5)^2} = 0$$

$$\lim_{t\to\infty} f(t) = \lim_{s\to 0} \frac{10s}{(s+5)^2} = 0$$

The initial value is readily verified by looking at the function of time. The second relationship is less easily proven in the time domain.

---

You have to be careful in applying the final value theorem because many functions are of indeterminate value as $t\to\infty$. For instance, $\sin \omega t$ does not approach a specific value for large $t$. Even so, application of the final value theorem indicates a value of 0, an obviously meaningless result. There is a simple way of knowing ahead of time whether the final value theorem can be meaningfully applied to the Laplace transform $F(s) = N(s)/D(s)$. The real parts of the roots of $D(s)$ must be strictly less than zero.

---

**Example 9.7**    *Validity of the Final Value Theorem*

1.   $F(s) = \dfrac{1}{s^2 + 2s + 2} \leftrightarrow f(t) = e^{-t} \sin t\, u(t)$

The denominator has roots of $-1 \pm j$. The real parts are negative, so the final value theorem is meaningful:

$$\lim_{t\to\infty} f(t) = \lim_{s\to 0} sF(s) = 0$$

2. $F(s) = \dfrac{10s}{s^2 + 4} \leftrightarrow f(t) = 10 \cos 2t\, u(t)$

The denominator has roots of $\pm j2$. The real parts $(0)$ are not negative. It is meaningless to apply the final value theorem.

3. $F(s) = \dfrac{5}{s - 10} \leftrightarrow f(t) = 5e^{+10t}u(t)$

$D(s)$ has a single root at $s = +10$. The theorem cannot be applied.

---

In examples like those in this chapter, the initial and final value theorems are used as checks on partial fraction expansion calculations. The final value theorem, however, has a much more important application in the area of automated control theory and design. There, the final value theorem is used to determine the difference or error between the desired and actual behavior of systems as time goes to infinity.

## 9.7 SINGULARITY FUNCTIONS

The unit step function $u(t)$ has proven to be very useful to us, especially in the description of switched circuits. It is just one of a larger class of functions, called **singularity functions,** that consist of $u(t)$ and all of its successive derivatives and integrals. Although there are an infinite number of functions in the set, only a few are commonly used in circuit analysis. These are $u(t)$ and its lowest-order derivatives and integrals.

The integral of $u(t)$ is shown in Figure 9.5. For obvious reasons, it is called the **unit ramp function** and is given the symbol $r(t)$. Its Laplace transform can be

**FIGURE 9.5**  *Ramp Function r(t) as Integral of Unit Step Function*

---

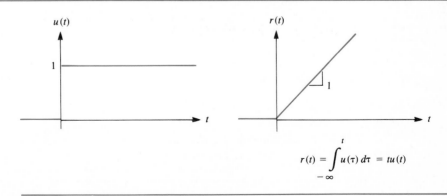

$$r(t) = \int_{-\infty}^{t} u(\tau)\, d\tau = tu(t)$$

found by using the fact that $r(t)$ is the integral of $u(t)$:

$$R(s) = \mathcal{L}\{r(t)\} = \mathcal{L}\left\{\int_{-\infty}^{t} u(\tau)\, d\tau\right\} = \frac{1}{s}\,\mathcal{L}\{u(t)\} = \frac{1}{s^2}$$

The derivative of the unit step function was briefly introduced earlier in this chapter. It is known as the **unit impulse function** and has the symbol $\delta(t)$. Because it is an important function, its properties should be discussed carefully.

The derivative of $u(t)$ does not exist in the usual mathematical sense because it is undefined at $t = 0$. However, $u(t)$ can be thought of as the limiting case of a continuum of ramp-like functions such as are shown in Figure 9.6. The derivative of this ramp-like function is everywhere defined. From the figure, we see that it is a rectangular pulse. When the ramp-like function approaches $u(t)$, its derivative is zero everywhere except at the origin. There, the derivative is infinite. You can see from the limiting process that the area under the derivative function is always 1. The properties of the unit impulse function can be summarized as follows:

$$\delta(t) = \begin{cases} 0, & t \neq 0 \\ \infty, & t = 0 \end{cases} \tag{9.23}$$

and

$$\int_{-\infty}^{\infty} \delta(t)\, dt = 1 \tag{9.24}$$

FIGURE 9.6   *Relationship Between Step and Impulse Functions*

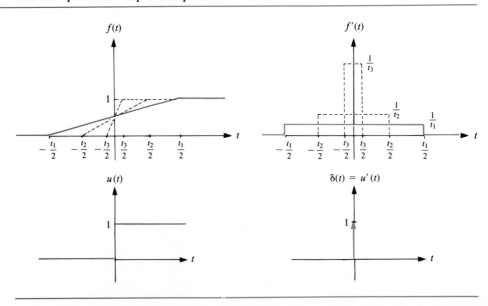

Because we cannot sketch a function whose only nonzero value is infinity, we plot $\delta(t)$ by showing a vertical arrow whose height equals the total area under the function. Although it is physically impossible to generate a voltage or current that is an impulse function, one can be reasonably approximated in an electrical circuit by a brief pulse of energy if the duration of the pulse is short compared to the time constants of the circuit.

The Laplace transform of $\delta(t)$ was first found in Section 9.3. We will rederive it here based on one of the impulse function's more interesting characteristics, its **sifting property.** If a function $f(t)$ is defined at $t = t_0$, then

$$\int_{-\infty}^{\infty} f(t)\delta(t - t_0)\, dt = \int_{-\infty}^{\infty} f(t_0)\delta(t - t_0)\, dt = f(t_0) \int_{-\infty}^{\infty} \delta(t - t_0)\, dt = f(t_0)$$

$$(9.25)$$

The first line follows from the fact that $\delta(t - t_0)$ is zero everywhere except at $t = t_0$. Using Equation 9.25, the Laplace transform of the impulse function can be written directly without calculation:

$$\mathcal{L}\{\delta(t)\} = \int_{0^-}^{\infty} \delta(t)e^{-st}\, dt = \int_{0^-}^{\infty} \delta(t)(1)\, dt = 1$$

(using the sifting property)

This result, of course, is the same as the one that was arrived at earlier using a different approach.

Although individual singularity functions are quite simple, more complicated functions can be created by combining singularity functions with well-chosen amplitudes and locations. For instance, the rectangular pulse of Figure 9.7 can be written as the sum of two step functions: $p(t) = u(t) - u(t - t_0)$.

Figure 9.8 shows examples of waveforms of engineering or scientific interest that can be written as sums of singularity functions. When the individual segments

**FIGURE 9.7**   *Rectangular Pulse as Sum of Two Step Functions*

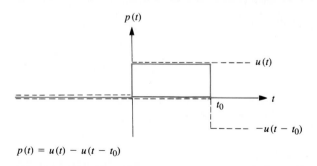

FIGURE 9.8    *Functions That Can Be Expressed as Sums of Singularity Functions*

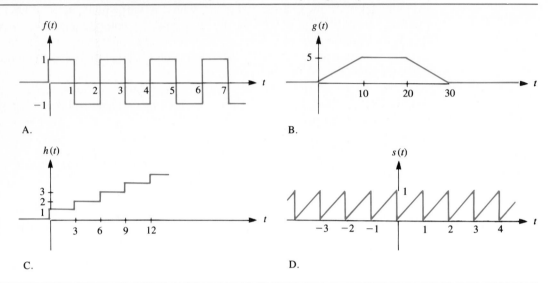

A.

B.

C.

D.

of a function can be expressed as polynomials, it is straightforward to write the entire function as a sum of singularity functions. As a simple example of this, consider the square wave function of Figure 9.8A. Starting from the left, it can be "built up" by the successive addition of step functions that account for all of the rising and falling edges of the square wave. Thus, $f(t)$ can be analytically written as

$$f(t) = u(t) - 2u(t-1) + 2u(t-2) - 2u(t-3) + \cdots$$

**Example 9.8**    *Functions That Are Sums of Singularity Functions*

Write analytical expressions for $g(t)$, $h(t)$, and $s(t)$ shown in Figure 9.8.

$$g(t) = \frac{1}{2}\, r(t) - \frac{1}{2}\, r(t-10) - \frac{1}{2}\, r(t-20) + \frac{1}{2}\, r(t-30)$$

$$h(t) = u(t) + u(t-3) + u(t-6) + u(t-12) + \cdots$$

$$s(t) = \cdots + p(t+2) + p(t+1) + p(t) + p(t-1) + \cdots$$

where $p(t) = r(t) - r(t-1) - u(t-1)$

Let's now return our attention to the square wave function $f(t)$ in Figure 9.8A. We can use the transform of the unit step function and the known effect of time shifting (see Table 9.2) to write the Laplace transform of $f(t)$:

$$F(s) = \frac{1}{s} - 2\frac{e^{-s}}{s} + 2\frac{e^{-2s}}{s} - 2\frac{e^{-3s}}{s} + 2\frac{e^{-4s}}{s} - \cdots$$

This is the correct Laplace transform of the square wave, but it is not necessarily in as compact a form as possible. Functions that are periodic for $t > 0$ can be treated in a special way. Consider a function that is created by repeating a pulse, $p(t)$, every $T$ seconds. We can write it as

$$f(t) = p(t) + p(t - T) + p(t - 2T) + \cdots$$

The Laplace transform of $f(t)$ can be written:

$$F(s) = P(s) + e^{-sT}P(s) + e^{-2sT}P(s) + \cdots = P(s)\sum_{n=0}^{\infty} e^{-nsT}$$

$$= \frac{P(s)}{1 - e^{-sT}}$$

---

**Example 9.9**    *The Transform of a One-Sided Periodic Function*

Find the Laplace transform of the triangle wave function in Figure 9.9.

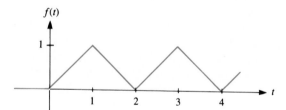

**FIGURE 9.9**

Its first pulse can be written as follows:

$$p(t) = r(t) - 2r(t - 1) + r(t - 2)$$

$$P(s) = \frac{1}{s^2} - \frac{2}{s^2}e^{-s} + \frac{1}{s^2}e^{-2s} = \frac{1 - 2e^{-s} + e^{-2s}}{s^2} = \frac{(1 - e^{-s})^2}{s^2}$$

The function $f(t)$ is written

$$f(t) = p(t) + p(t - 2) + p(t - 4) + \cdots$$

$$F(s) = \frac{P(s)}{1 - e^{-2s}} = \frac{1}{s^2}\left(\frac{1 - e^{-s}}{1 + e^{-s}}\right)$$

The Laplace transforms of functions like those of Example 9.9 are not in our earlier form of $F(s) = N(s)/D(s)$. The methods of partial fraction expansion cannot be used to find their inverses. However, in Problem 9.30 at the end of the chapter, you will demonstrate how a simple long division of $F(s)$ will put the Laplace transform in a more recognizable form.

## 9.8    A COMPARISON OF METHODS

We are now in a position to compare three different mathematical methods that can be used to analyze the responses of dynamic circuits to suddenly applied excitations. One of these methods is the classical differential equation approach that was discussed in Chapters 6 and 7. The other two methods are variations on the application of the Laplace transform. The Laplace transform can be applied either to differential equations of standard form or to integro-differential equations. There are differences among the three methods. Some are obvious, and some are not so obvious. To help distinguish these differences, let's review the three methods and then apply each to the analysis of an electrical circuit.

The classical analysis of an electrical circuit in the time domain often begins with finding a differential equation that describes the response variable for $t > 0$. This approach avoids the circuit's discontinuity at $t = 0$.* (For purposes of discussion in this section, we will always assume that a discontinuity occurs at $t = 0$.) Because the solutions of such equations are only valid for positive time, initial conditions must be found at $t = 0^+$. For an $n$th-order circuit, these initial conditions are the response variable and its first $n - 1$ derivatives. Finding these can be a difficult job and requires an analysis of the circuit's behavior just before and just after the discontinuity.

In order to correctly apply the Laplace transform of Equation 9.1 to a differential equation, the equation must be written to include any discontinuity at $t = 0$. Therefore, the equation must be valid over at least the range of time $0^- < t < \infty$. As we will see, such equations often contain impulse functions, but these are easily handled with the Laplace transform. An advantage of this approach is that while the same number of initial conditions is required, the conditions are evaluated at $t = 0^-$. They are easily found if the circuit happens to be in a DC steady state prior to the discontinuity.

In a similar way, if the Laplace transform is to be applied to an integro-differential equation, that equation must be valid for $0^- < t < \infty$. The initial conditions needed for an $n$th-order circuit are the $n$ inductor currents and capacitor voltages that exist in the circuit at $t = 0^-$. These are usually easy to find and have a degree of physical significance that is lacking in the more mathematical initial conditions of the other two methods.

---

* There are other approaches to the use of differential equations, but they have their own difficulties.

FIGURE 9.10 *Series RLC Circuit Used in Examples 9.10, 9.11, and 9.12*

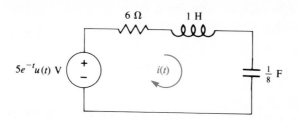

Now, let's consider the series $RLC$ circuit shown in Figure 9.10 and analyze it by each of the three methods.

**Example 9.10** *The Differential Equation in Circuit Analysis*

First analyze the circuit in Figure 9.10 entirely in the time domain. Begin by writing an equation for $t > 0$:

$$5e^{-t} = \frac{di}{dt} + 6i + 8 \int_0^t i \, dt + v_C(0), \qquad t > 0$$

Differentiating leads to

$$-5e^{-t} = \frac{d^2i}{dt^2} + 6\frac{di}{dt} + 8i, \qquad t > 0$$

You should verify that the initial conditions are

$$i(0^+) = 0 \, \text{A} \qquad \text{and} \qquad i'(0^+) = 5 \, \text{A/s}$$

and that $i(t)$ is

$$i(t) = \left[ -\frac{5}{3} e^{-t} + 5e^{-2t} - \frac{10}{3} e^{-4t} \right] u(t) \, \text{A}$$

**Example 9.11** *The Laplace Transform Applied to Differential Equations*

Now, repeat the analysis and apply the Laplace transform to the differential equation.

Begin with

$$5e^{-t}u(t) = \frac{di}{dt} + 6i + 8 \int_{-\infty}^t i \, dt \qquad (9.26)$$

*(continues)*

**Example 9.11**    *Continued*

Notice that $u(t)$ has been retained within the expression for the forcing function and that the equation is valid for *all* time. Differentiating and reorganizing lead to

$$-5e^{-t}u(t) + 5e^{-t}\delta(t) = \frac{d^2i}{dt^2} + 6\frac{di}{dt} + 8i$$

The coefficient of the impulse on the left-hand side is $5e^{-t}$ evaluated at $t = 0$. Therefore,

$$-5e^{-t}u(t) + 5\delta(t) = \frac{d^2i}{dt^2} + 6\frac{di}{dt} + 8i$$

In applying the Laplace transform, the appropriate initial conditions are evaluated at $t = 0^-$:

$$i(0^-) = 0\,\text{A} \qquad \text{and} \qquad i'(0^-) = 0\,\text{A/s}$$

Both conditions are zero because the circuit is not excited for negative time. Using these, you can write

$$\frac{-5}{s+1} + 5 = s^2I(s) + 6sI(s) + 8I(s)$$

$$I(s) = \frac{5}{s^2 + 6s + 8} - \frac{5}{(s+1)(s^2 + 6s + 8)}$$

By partial fraction expansion, you can verify that the result is as in Example 9.10:

$$i(t) = \left[ -\frac{5}{3}e^{-t} + 5e^{-2t} - \frac{10}{3}e^{-4t} \right]u(t)\,\text{A}$$

---

**Example 9.12**    *The Laplace Transform Applied to Integro-Differential Equations*

Finally, apply the Laplace transform to the integro-differential equation (Equation 9.26) written in a somewhat different form:

$$5e^{-t}u(t) = \frac{di}{dt} + 6i + 8\int_{0^-}^{t} i\, dt + v_C(0^-)$$

The necessary initial conditions are

$$i_L(0^-) = 0\,\text{A} \qquad \text{and} \qquad v_C(0^-) = 0\,\text{V}$$

The transformed equation is

$$\frac{5}{s+1} = sI(s) + 6I(s) + \frac{8}{s} I(s)$$

$$I(s) = \frac{5s}{(s+1)(s^2 + 6s + 8)}$$

This last expression can be inverted to arrive at the same answer as was just found in Examples 9.10 and 9.11.

---

The method selected to analyze any particular electrical circuit problem is largely a matter of personal preference. However, applying the Laplace transform to integro-differential equations is generally the method of choice because of the ease with which initial conditions are treated and the particular physical importance that capacitor voltages and inductor currents have in circuit analysis. For multivariable problems, any application of the Laplace transform is preferable to time-domain techniques because the transform changes differential equations into algebraic equations, thus simplifying the separation of variables.

We seem to have built a strong argument against the use of differential equations and in favor of the Laplace transform. In doing so, we have overlooked one very important advantage of the time-domain approach: Differential equations can be used to model the behavior of any electrical circuit. The Laplace transform can be used only on systems that are linear and time-invariant. Many, but not all, electrical circuits can be modeled in this way. When nonlinear circuits such as those containing diodes are encountered, however, they cannot be analyzed with Laplace transform techniques. It is valid, however, to use differential equations in the modeling of such circuits.

## 9.9 THE TRANSFORMED CIRCUIT

We have examined two ways in which the Laplace transform can be used in the analysis of electrical circuits. In this section, we will carry the process one step further. Instead of applying the transform to the time-domain equations that describe a circuit, we will find a transformed version of the circuit itself. This procedure eliminates entirely the need to write integro-differential equations. Instead, we write algebraic equations in terms of the transformed circuit variables. To see how this procedure might be done, let's consider separately the *V–I* relationships of the basic circuit elements.

Figure 9.11A shows a resistor with its assumed voltage and current. Its *v–i* relationship is

$$v(t) = Ri(t)$$

FIGURE 9.11  *Representations of a Resistor*

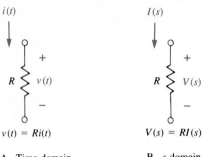

$v(t) = Ri(t)$                    $V(s) = RI(s)$

A. Time domain                 B. *s* domain

As shown in Figure 9.11B, application of the Laplace transform yields

$$V(s) = RI(s) \tag{9.27}$$

Equation 9.27 can be rewritten as

$$R = \frac{V(s)}{I(s)}$$

Equation 9.28, which appears next, is the ratio of a frequency-domain representation of a voltage to that of a current. Such ratios are called **impedances.** Impedances will be given the general symbol $Z$, just as when they were introduced in Chapter 8 as ratios of phasors. Thus, for a resistor

$$Z_R(s) = \frac{V(s)}{I(s)} = R \tag{9.28}$$

You can see that the impedance of a resistor is identical to its resistance. The same is not true for inductors or capacitors.

A capacitor is shown in Figure 9.12A. Its behavior is described by

$$i(t) = C \frac{dv(t)}{dt}$$

which, in the *s* domain, becomes

$$I(s) = sCV(s) - Cv_C(0^-) \tag{9.29}$$

FIGURE 9.12    *Representations of a Capacitor*

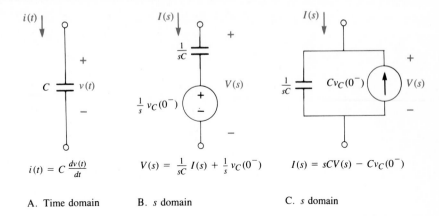

A. Time domain

$i(t) = C\dfrac{dv(t)}{dt}$

B. *s* domain

$V(s) = \dfrac{1}{sC} I(s) + \dfrac{1}{s} v_C(0^-)$

C. *s* domain

$I(s) = sCV(s) - Cv_C(0^-)$

or, in an alternate form,

$$V(s) = \frac{1}{sC} I(s) + \frac{1}{s} v_C(0^-) \tag{9.30}$$

Equation 9.29 tells us that the current through a capacitor is the sum of two parts. One is due to the voltage across the capacitor for $t > 0$, and the other is due to the capacitor's initial condition. A circuit model with elements to account for these two current components is shown in Figure 9.12C. The independent source that accounts for $v_C(0^-)$ is called an **initial condition generator.** Figure 9.12B shows how the capacitor can be modeled based on the voltage equation of Equation 9.30. When $v_C(0^-) = 0$ V, either equation can be used to show that the impedance of a capacitor is

$$Z_C(s) = \frac{V(s)}{I(s)} = \frac{1}{sC}, \qquad v_C(0^-) = 0 \tag{9.31}$$

Figure 9.13 shows the time- and *s*-domain models for the inductor. Their proofs are left to Problem 9.31 at the end of the chapter. From the models, you can see that the impedance of an inductor is

$$Z_L(s) = \frac{V(s)}{I(s)} = sL, \qquad i_L(0^-) = 0 \tag{9.32}$$

When Laplace-transformed variables are used, there are two ways in which both the inductor and the capacitor can be modeled. The particular models chosen will depend on whether node or mesh analysis is used. Let's look at some examples.

FIGURE 9.13    *Representations of an Inductor*

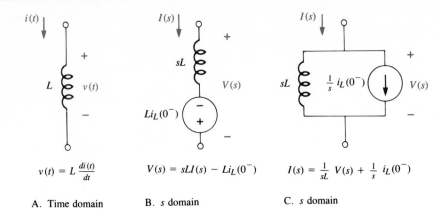

$$v(t) = L \frac{di(t)}{dt}$$

A.  Time domain

$$V(s) = sLI(s) - Li_L(0^-)$$

B.  *s* domain

$$I(s) = \frac{1}{sL} V(s) + \frac{1}{s} i_L(0^-)$$

C.  *s* domain

Example 9.13    *Transforming a Simple Circuit*

Use the transformed version of the circuit in Figure 9.14A to find $v_C(t)$.
Choose the parallel model for the capacitor in order to facilitate a node
analysis approach.

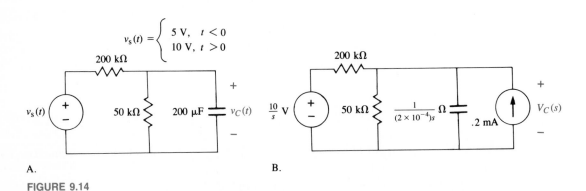

A.

B.

FIGURE 9.14

The initial value of the capacitor voltage is easily seen to be

$$v_C(0^-) = \frac{50,000}{200,000 + 50,000} \cdot 5 = 1 \text{ V}$$

Figure 9.14B shows the transformed circuit. From it is written a KCL equation at the top node of the capacitor:

$$\frac{V_C}{50,000} + (2 \times 10^{-4})sV_C + \frac{1}{200,000}\left(V_C - \frac{10}{s}\right) = 2 \times 10^{-4}$$

$$V_C(s) = \frac{1}{s + (1/8)} + \frac{1/4}{s[s + (1/8)]} = \frac{2}{s} - \frac{1}{s + (1/8)}$$

The inverse is

$$v_C(t) = 2 - e^{-t/8}\,V, \qquad t > 0$$

**Example 9.14**    *Transforming a Multi-Mesh Circuit*

A circuit that requires more than one equation is shown in Figure 9.15A. Using mesh analysis will make the series models for the inductor and capacitor good choices.

You should verify the *s*-domain circuit model shown in Figure 9.15B. From it are written two mesh equations:

$$\frac{4}{s + 1} = 2I_1 + \frac{2}{s} + \frac{1}{s}(I_1 - I_2)$$

$$-\frac{2}{s} + \frac{1}{s}(I_2 - I_1) + (s + 2)I_2 - 1 = 0$$

Since $V_o(s)$ is $2I_2(s)$, the equations can be solved to find

$$V_o(s) = \frac{2s^2 + 7s + 9}{(s + 1)[s^2 + (5s/2) + 2]}$$

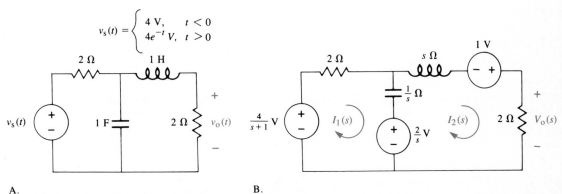

A.

B.

FIGURE 9.15

*(continues)*

**Example 9.14**   *Continued*

Partial fraction expansion techniques are used to manipulate $V_o(s)$ into a more recognizable form:

$$V_o(s) = \frac{8}{s+1} + \frac{-6s-7}{s^2+5s/2+2}$$

$$= \frac{8}{s+1} - 6\frac{s+5/4}{(s+5/4)^2+7/16} + \frac{2}{\sqrt{7}}\frac{\sqrt{7}/4}{(s+5/4)^2+7/16}$$

The inverse is

$$v_o(t) = 8e^{-t} - 6e^{-5t/4}\cos\frac{\sqrt{7}}{4}t + \frac{2}{\sqrt{7}}e^{-5t/4}\sin\frac{\sqrt{7}}{4}t \text{ V}, \qquad t>0$$

## 9.10   SYSTEM FUNCTIONS

As a way of introducing the topic of this section, let's consider the circuit of Figure 9.16. We wish to find a relationship between the Laplace transform of the current $i(t)$ and that of the source voltage. First, we write an integro-differential equation:

$$v_s(t) = Ri + L\frac{di}{dt} + \frac{1}{C}\int_{0^-}^{t} i(t)\, dt + v_C(0^-)$$

Then, we apply the Laplace transform:

$$V_s(s) = RI(s) + L[sI(s) - i_L(0^-)] + \frac{1}{sC}I(s) + \frac{1}{s}v_C(0^-)$$

$$= \left(R + sL + \frac{1}{sC}\right)I(s) - Li_L(0^-) + \frac{1}{s}v_C(0^-)$$

$$I(s) = \underbrace{\frac{V_s(s)}{R+sL+(1/sC)}}_{\text{zero-state response}} + \underbrace{\frac{Li_L(0^-)-[v_C(0^-)/s]}{R+sL+(1/sC)}}_{\text{zero-input response}} \qquad (9.33)$$

**FIGURE 9.16**   *Circuit Whose Transformed Response Is Given by Equation 9.33*

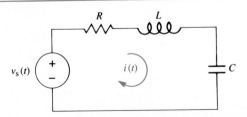

Equation 9.33 has been arranged in a particular way. From it, we see that $I(s)$ (or any other response variable) is composed of two parts. One part depends on the input function and the properties of the circuit. The other part depends on the circuit's initial conditions as well as the properties of the circuit. The first part of the response is called the **zero-state response.** The **state** of a system is defined as the minimum amount of information at an instant of time with which you can predict the future behavior of the source-free system.* The state of an electrical circuit is known when the energies stored in all of its inductors and capacitors are given. If no energy is stored in a circuit, it is said to be in the **zero state.** When $v_C(0^-)$ and $i_L(0^-)$ are both zero, the response of Equation 9.33 reduces to just the first term. Hence, the name zero-state response.

It is sometimes overlooked that circuits can respond to initial energy storage in addition to externally applied voltages or currents. When the independent source of a single-input circuit is set equal to zero, the remaining response is called the **zero-input response.** Problem 9.37 at the end of the chapter investigates the relationships among the forced, natural, zero-input, and zero-state responses.

Let's consider in more detail the zero-state response. When the two initial conditions of our example circuit are set to zero, Equation 9.33 reduces to

$$I(s) = \frac{1}{R + sL + (1/sC)} \, V_s(s)$$

or

$$\frac{I(s)}{V_s(s)} = \frac{1}{R + sL + (1/sC)} \tag{9.34}$$

Equation 9.34 is a ratio of the Laplace transform of the response to that of the input function. Because we have done nothing to specify or limit the input in any way, the ratio must be a property of the circuit that is valid for any input–output pair. Such ratios are called **system functions.** They are often given special symbols. For instance,

$$H(s) = \frac{I(s)}{V_s(s)} \tag{9.35a}$$

where

$$H(s) = \frac{1}{R + sL + (1/sC)} \tag{9.35b}$$

The function $H(s)$ shown in Equation 9.35 is so important that we will repeat it in more general terms. Refer to Figure 9.17, where a single-input/single-

---

* In this and the next section, the general term *system* is used frequently because the concepts covered can be applied to any type of system, not just electrical circuits.

**FIGURE 9.17**    *Block Diagram Representation of a Single-Input/Single-Output System*

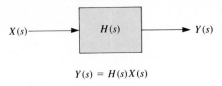

$$Y(s) = H(s)X(s)$$

output system is shown. $X(s)$ is the Laplace transform of the input, and $Y(s)$ is the transform of the output. If the system is linear, time invariant, initially at rest, and contains no independent sources (dependent sources are allowed), then its input–output relationship can be written as

$$Y(s) = H(s)X(s) \tag{9.36a}$$

or

$$H(s) = \frac{Y(s)}{X(s)} \tag{9.36b}$$

where $H(s)$ is the system function.

The system representation of Figure 9.17 is known as a **block diagram.** It is clearly very abstract and gives little insight into the physical workings of the system upon which it is based. It has the advantage of being extremely compact in notation and also in being applicable to systems of any kind. Whenever they are used, block diagrams imply that only the zero-state behavior of the system is being modeled. The system is assumed to be initially at rest.

**Example 9.15**    *Finding and Using a System Function*

Imagine that an experiment has been performed on the system of Figure 9.18. A unit step function has been applied and the response measured.

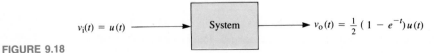

**FIGURE 9.18**

The result of the experiment is shown in the figure. We can find the system function $H(s)$ if the Laplace transforms of the input and output are known. These are as follows:

$$V_i(s) = \frac{1}{s}$$

$$V_o(s) = \frac{.5}{s} - \frac{.5}{s+1} = \frac{.5}{s(s+1)}$$

$$H(s) = \frac{V_o(s)}{V_i(s)} = \frac{.5}{s+1}$$

The response of the system to any forcing function can now be found. For instance,

$$v_i(t) = e^{-2t}u(t) \rightarrow V_i(s) = \frac{1}{s+2}$$

$$V_o(s) = H(s)V_i(s) = \frac{.5}{s+1} \cdot \frac{1}{s+2} = \frac{.5}{s+1} - \frac{.5}{s+2}$$

which, in the time domain, is

$$v_o(t) = .5(e^{-t} - e^{-2t})u(t)$$

---

We can gain some additional insight into system functions and the relationship of Equation 9.36 by considering the special case of $x(t) = \delta(t)$. Because $\mathcal{L}\{\delta(t)\} = 1$, the Laplace transform of the system's response is

$$Y(s) = H(s) \tag{9.37}$$

The response of a system when all initial conditions are zero and when the input is $\delta(t)$ is known as the system's **impulse response.** It is very often given the symbol $h(t)$, although other lowercase letters are sometimes used. Reflection upon Equation 9.37 and the assumptions that led to it reveal that

$$H(s) = \mathcal{L}\{h(t)\} \tag{9.38}$$

That is, the system function relating a pair of input and output variables is equal to the Laplace transform of the corresponding zero-state impulse response.

The response of a system to any input will, in general, contain within it the natural frequencies of the system. The impulse response is no exception to this rule. In fact, the impulse response is made up entirely of a natural response because $\delta(t) = 0$ for $t > 0$. Since $h(t)$ can be found from $H(s)$, the system function must also contain the system's natural frequencies.* To see where they might be

---

* For electrical circuits with multiple inputs, the natural frequencies observed may vary somewhat depending on where the circuit is excited. They may also vary depending on whether the excitation is with a current or a voltage source. The system's set of natural frequencies in such a case is the combination of these individually elicited natural frequencies.

hidden, let's reconsider the system function already derived for the series *RLC* circuit:

$$H(s) = \frac{1}{R + sL + (1/sC)}$$

Now, let's consider the denominator polynomial of $H(s)$ and set it equal to zero:

$$R + sL + \frac{1}{sC} = 0$$

We can put this equation into a more standard form:

$$s^2 + \frac{R}{L}s + \frac{1}{LC} = 0$$

In Chapter 6, this equation was shown to be the characteristic equation of a series *RLC* circuit. We have arrived at a general result not limited to our specific example: The denominator polynomial of a system function when set equal to zero is identical to the characteristic equation of the system. The roots of each are the natural frequencies of the response variable.

## 9.11   THE CONVOLUTION INTEGRAL

We saw in Section 9.10 that the system function $H(s)$ is the basis for a very simple $s$-domain relationship between the input and the response of a linear system when it is initially at rest. In this section, we will see that a simple relationship between input and output also exists in the time domain.

To prove this new relationship, we choose the known starting point of

$$Y(s) = H(s)X(s)$$

where $x(t)$ is the input to a system, $y(t)$ is the response, and $H(s)$ is the system function that relates their Laplace transforms. We assume that both $x(t)$ and $h(t)$ are zero for $t < 0$. We want to find $y(t)$, which is

$$y(t) = \mathcal{L}^{-1}\{H(s)X(s)\} = \frac{1}{2\pi j} \int_{\sigma - j\infty}^{\sigma + j\infty} H(s)X(s)e^{st}\, ds$$

We replace $X(s)$ with the defining integral for the Laplace transform, using $\tau$ as a dummy variable of integration:

$$y(t) = \frac{1}{2\pi j} \int_{\sigma - j\infty}^{\sigma + j\infty} H(s)\left[\int_{0^-}^{\infty} x(\tau)u(\tau)e^{-s\tau}\, d\tau\right]e^{st}\, ds$$

Because $x(\tau)u(\tau) = 0$ for $\tau < 0$, we can change the lower limit of integration with respect to $\tau$ from $0^-$ to $0$. At the same time, we change the order of integration:

$$y(t) = \int_0^\infty x(\tau)\left[\frac{1}{2\pi j}\int_{\sigma-j\infty}^{\sigma+j\infty}[H(s)e^{-s\tau}]e^{st}\,ds\right]d\tau$$

The term $H(s)e^{-s\tau}$ can be recognized as the Laplace transform of the impulse response delayed by $\tau$ seconds. Using the time-shift property from Table 9.2, we can write

$$y(t) = \int_0^\infty x(\tau)h(t-\tau)u(t-\tau)\,d\tau$$

But, $h(t-\tau)u(t-\tau) = 0$ whenever $\tau > t$. Therefore,

$$y(t) = \int_0^t x(\tau)h(t-\tau)\,d\tau \tag{9.39}$$

Equation 9.39 is the **convolution integral.** It can be used to calculate a linear, time-invariant system's response to a general input $x(t)$ if the system's impulse response is known. Keep in mind the conditions under which Equation 9.39 is valid:

1. The excitation, $x(t)$, is suddenly applied at $t = 0$. That is, $x(t) = x(t)u(t)$.
2. The system is *causal.* Responses in causal systems cannot begin before the input is applied. That is, $h(t) = h(t)u(t)$.
3. The system is initially at rest.

When these conditions are met, the convolution integral can equivalently be written as

$$y(t) = \int_0^t h(\tau)x(t-\tau)\,d\tau \tag{9.40}$$

Choosing between the alternate forms of the convolution integral is largely a matter of convenience. We will use Equation 9.39 to graphically interpret, in a series of steps, the meaning of the integral. Refer to Figure 9.19.

The integral of Equation 9.39 contains within it the product of two functions, $x(\tau)$ and $h(t-\tau)$. We begin by finding these two functions. Using $\tau$ as the independent variable for $x(\cdot)$ and $h(\cdot)$ is merely a bookkeeping procedure that does not change the shapes or locations of the functions (Figure 9.19A). The next step is to consider $h(-\tau)$. This function is found by folding $h(\tau)$ about the vertical axis. To arrive at $h(t-\tau)$, we add $t$ to the argument, which causes a shift along

FIGURE 9.19    *Graphical Interpretation of Convolution Integral*

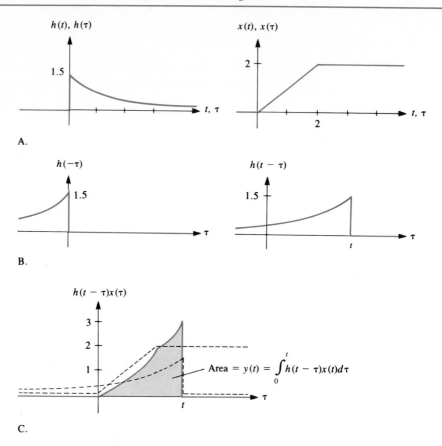

A.

B.

C.

the $\tau$ axis as shown in Figure 9.19B for an assumed positive value for $t$. Finally, the two functions $x(\tau)$ and $h(t - \tau)$ are overlayed and their product computed. The convolution integral tells us that $y(t)$ is the area under this product curve between the limits $\tau = 0$ and $\tau = t$ (Figure 9.19C). This area changes as a function of time.

**Example 9.16**    *Convolution in Circuit Analysis*

Use convolution to determine the step response of the simple *RC* circuit shown in Figure 9.20.

First, the impulse response must be found. Although this could be done using time-domain methods, use the transformed circuit. Note that the

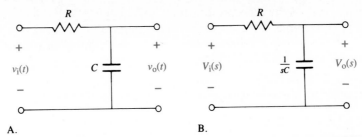

FIGURE 9.20   A.                                            B.

circuit is initially at rest. Voltage division easily shows that

$$\frac{V_o(s)}{V_i(s)} = \frac{1/sC}{R + (1/sC)} = \frac{1/RC}{s + (1/RC)}$$

The impulse response is

$$h(t) = \mathcal{L}^{-1}\left\{\frac{1/RC}{s + (1/RC)}\right\} = \frac{1}{RC}\, e^{-t/RC} u(t)$$

To simplify the calculations, assume $RC = 1\,\text{s}$:

$$h(t) = e^{-t}u(t)$$

The convolution integral can now be applied; use Figure 9.21 as an aid:

$$v_o(t) = \int_0^t v_i(\tau) h(t - \tau)\, d\tau = \int_0^t u(\tau) e^{-(t-\tau)} u(t - \tau)\, d\tau = e^{-t}\int_0^t e^{\tau}\, d\tau$$

$$= e^{-t}[e^t - 1] = 1 - e^{-t}\,\text{V}, \qquad t > 0$$

You should verify this result by using Laplace transforms.

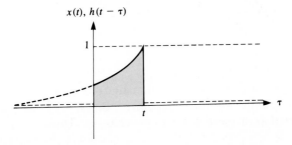

$x(t),\, h(t - \tau)$

FIGURE 9.21

When the two functions to be convolved are more complicated than those in Example 9.16, a sketch that graphically interprets the calculation is almost a necessity in order to keep the limits of integration clear in your mind.

---

**Example 9.17**    *Convolution for a Finite Duration Response*

Figure 9.22A shows the impulse response of a hypothetical system and an input function $x(t)$. Both are rectangular pulses, but their widths are different.

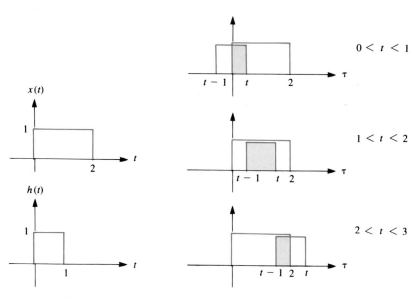

A. Functions of Example 9.17          B. Graphical interpretation of functions' convolution

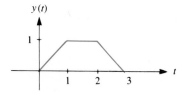

**FIGURE 9.22**    C. Resultant output

Figure 9.22B shows that the convolution integral must be interpreted differently for three separate ranges of time because the nature of the overlap of the two functions changes. Thus,

for $0 < t < 1,$     $y(t) = \int_0^t (1 \cdot 1) \, d\tau = t$

for $1 < t < 2$,     $y(t) = \int_{t-1}^{t} (1 \cdot 1) \, d\tau = 1$

and for $2 < t < 3$,     $y(t) = \int_{t-1}^{2} (1 \cdot 1) \, d\tau = 3 - t$

Outside of the range $0 < t < 3$, $y(t) = 0$ because there is no overlap of the functions where they are nonzero. The output $y(t)$ is shown in Figure 9.22C. When two functions of finite duration are convolved, the duration of the result is equal to the sum of their individual durations.

## 9.12 SUMMARY

This chapter showed how the Laplace transform can be used to simplify the analysis of dynamic systems such as *RLC* circuits. Through it, the system's complete response can be found in one operation. Because initial conditions are handled with relative ease when compared with classical differential equation approaches, the Laplace transform is well suited to the analysis of linear, time-invariant switched circuits. Multivariable problems are also made much easier by the use of the Laplace transform.

The Laplace transform can be applied at several stages of a circuit analysis problem, ranging from finding the transformed version of the circuit to transforming the formal differential equation that relates input to output. For several reasons, it is usually best either to transform the circuit itself before writing any equations or, as a second choice, to apply the transform to the integro-differential equations.

Do not assume from this chapter that the Laplace transform is uniquely suited to the solution of switched circuit problems. The time-domain technique of state-space analysis, for instance, simplifies the solution of dynamic electrical circuits by choosing inductor currents and capacitor voltages as the problem variables. State-space analysis has some advantages over frequency-domain techniques in that it can be applied to nonlinear problems and is well suited to digital computation.

Systems with single inputs and single outputs lend themselves to particularly simple mathematical representations in the *s* domain. If the system is linear, time invariant, and initially at rest, then the ratio of the Laplace transform of the response to that of the input is a constant property of the system. This ratio is called the system function. It is the Laplace transform of the system's impulse response. A single-input system may have several system functions associated with it depending on the number of defined output variables.

The impulse response is also the basis of a relationship between input and output in the time domain. The convolution integral that has been defined is not of the most general form. Its limits of integration are 0 and $t$ because the system is assumed to be causal and its forcing function is suddenly applied at $t = 0$. With these and the additional restrictions of linearity and no initial energy storage, the convolution integral can be used to calculate the response of a system to any arbitrary input if its impulse response is known. Convolution can be difficult to perform by hand calculation. However, it is straightforward to implement using numerical techniques on a digital computer.

When a system is not initially at rest, its response cannot be found from just the system function. Instead, the response consists of both zero-state and zero-input components. There are no particular shortcuts to finding these components. Formal application of the Laplace transform is required.

## ■ PROBLEMS

**9.1** Use Tables 9.1 and 9.2 to find the Laplace transforms of the following functions:
   a.  $5t$
   b.  $-10$
   c.  $-10u(t)$
   d.  $5 \sin (t/2 - \pi/6)u(t/2 - \pi/6)$
   e.  $20e^{-10t} \cos 200t\, u(t)$

**9.2** Sketch the following functions and find their Laplace transforms:
   a.  $A \cos \omega t\, u(t)$
   b.  $A \cos (\omega t - \theta)u(t)$
   c.  $A \cos (\omega t - \theta)u(\omega t - \theta)$

**9.3** Use entries 7 and 8 from Table 9.1 and entry 5 from Table 9.2 to find the Laplace transforms of exponentially damped sines and exponentially damped cosines.

**9.4** Prove the following operational transform pairs from Table 9.2:
   a.  Linearity
   b.  Time shifting
   c.  Time scaling
   d.  Multiplication by $t$ (*Hint*: Start with $F(s)$ and take its derivative with respect to $s$.)

**9.5** The transform of $\cos t$ is $s/(s^2 + 1)$. Use this fact and the scaling theorem (entry 6, Table 9.2) to derive the transform of the more general $\cos \omega t$.

**9.6** Use entry 7 of Table 9.2 and entry 4 of Table 9.1 to find the transform of $t\, r(t)$.

**9.7** Prove that $\mathcal{L}\{te^{-at}u(t)\} = 1/(s + a)^2$.

**9.8** Find the transforms of the third- and fourth-order derivatives of $f(t)$.

**9.9** Use Tables 9.1 and 9.2 to find the transforms of the following:

   a.  The integral of $r(t)$, $\dfrac{t^2}{2}\, u(t)$

   b.  The derivative of the unit impulse
   c.  $t^{n-1}u(t)$
   d.  The polynomial $(t^3 + 2t^2 - t)u(t)$
   e.  $200te^{-10t} \cos 2t\, u(t)$

**9.10** Find the Laplace transforms of $v(t)$ for each of the following differential equations (initial conditions are given):

a. $25 \dfrac{dv}{dt} + 50v = 100u(t);$ $\qquad v(0^-) = -5$

b. $\dfrac{d^2v}{dt^2} + 3 \dfrac{dv}{dt} + 2v = 0;$ $\qquad v(0^-) = 5, \ v'(0^-) = 0$

c. $\dfrac{d^2v}{dt^2} + 2 \dfrac{dv}{dt} + v = 10u(t);$ $\qquad v(0^-) = 1, \ v'(0^-) = -1$

d. $\dfrac{d^2v}{dt^2} + 2 \dfrac{dv}{dt} + 101v = -5e^{-t}u(t);$ $\qquad v(0^-) = 0, \ v'(0^-) = 0$

e. $\dfrac{d^3v}{dt^3} + 8 \dfrac{d^2v}{dt^2} + 19 \dfrac{dv}{dt} + 12v = 0;$ $\qquad v(0^-) = 0, \ v'(0^-) = 1, \ v''(0^-) = 2$

**9.11** Find the Laplace transforms of $i(t)$ for each of the following integro-differential equations:

a. $\dfrac{di}{dt} + 8 \displaystyle\int_{0^-}^{t} i \, dt + 6i = 0;$ $\qquad i(0^-) = 3$

b. $6 \displaystyle\int_{-\infty}^{t} i \, dt + 4i = y(t);$ $\qquad y(t) = 5u(t), \ \displaystyle\int_{-\infty}^{0^-} i \, dt = -2$

c. $\dfrac{di}{dt} + 5i + 4 \displaystyle\int_{0^-}^{t} i \, dt = e^{-t}u(t);$ $\qquad i(0^-) = 2$

d. $\dfrac{di}{dt} + 4 \displaystyle\int_{-\infty}^{t} i \, dt + 4i = 3u(t);$ $\qquad i(0^-) = 3, \ 4 \displaystyle\int_{-\infty}^{0^-} i \, dt = -6$

**9.12** Write time-domain equations and from them find the Laplace transforms of the labeled output variables in

a. Figure P9.12A
b. Figure P9.12B

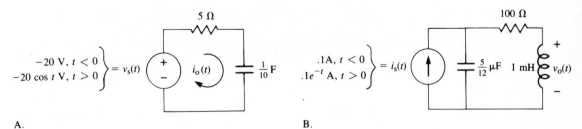

A.

B.

FIGURE P9.12

**9.13** Use partial fraction expansion methods to find the inverses of each of the following transforms:

a. $F(s) = \dfrac{10}{s(s+1)(s+2)}$

**b.** $F(s) = \dfrac{4(s+1)}{s^2(s+2)}$

**c.** $F(s) = \dfrac{s^2 + 4}{s(s^2 + 5s + 4)}$

**d.** $F(s) = \dfrac{1}{(s+5)(s+10)^2}$

**e.** $F(s) = 10\,\dfrac{s+2}{s(s^2 + 4s + 20)}$

**9.14** Prove from Equation 9.14 that residues related to complex conjugate roots are themselves complex conjugates. That is, $K_1 = K_2^*$ in the following equation:

$$F(s) = \frac{1}{s^2 + bs + c} = \frac{K_1}{s + \sigma - j\omega} + \frac{K_2}{s + \sigma + j\omega}$$

Although the case considered here is not general, the result is.

**9.15** Find the inverse transform of $F(s)$:

$$F(s) = \frac{K}{s + \sigma - j\omega} + \frac{K^*}{s + \sigma + j\omega} \qquad \text{where} \qquad K = |K|e^{j\theta}$$

**9.16** Find the inverse transform of $F(s)$ by each of two methods:

**a.** $F(s) = \dfrac{s}{(s+1)(s^2 + 2s + 5)} = \dfrac{K_1}{s+1} + \dfrac{As + B}{s^2 + 2s + 5}$

**b.** $F(s) = \dfrac{s}{(s+1)(s^2 + 2s + 5)} = \dfrac{K_1}{s+1} + \dfrac{K}{s+1-j2} + \dfrac{K^*}{s+1+j2}$

**9.17** Find the inverse transform for each of the answers in Problem 9.10.

**9.18** Find the inverse transform for each of the answers in Problem 9.11.

**9.19** Find the inverse of the two answers found in Problem 9.12.

**9.20** The method of undetermined coefficients can be used as the sole basis for finding the constants needed in a partial fraction expansion. Expand each of the following Laplace transforms by two different methods. First, use the appropriate formulas and techniques of Section 9.5. Then, use only the method of undetermined coefficients to find all constants in one operation.

**a.** $G(s) = \dfrac{s+10}{(s+5)(s+15)}$

**b.** $F(s) = \dfrac{s^2 + 2s + 5}{s(s+3)(s+8)}$

**c.** $H(s) = \dfrac{s}{(s+10)(s^2 + 4s + 29)}$

**9.21** Any PFE equation is an identity valid for all choices of $s$. One method for finding the residues and other constants is to evaluate the equation for specific values of $s$. For instance,

$$\frac{2}{(s+1)(s+2)} = \frac{K_1}{s+1} + \frac{K_2}{s+2}$$

could be evaluated at $s = 0$ and $s = 1$ to get two equations for finding $K_1$ and $K_2$. Use this technique for finding the expansions of Problem 9.20.

9.22 Verify the correctness of Equation 9.17 for the following specific example. First, use the formula; then, repeat the partial fraction expansion by the method of undetermined coefficients. Repeat once again using the technique described in Problem 9.21.

$$F(s) = \frac{s^2 + 1}{(s+1)(s+2)^3}$$

9.23 Completing the square is often a necessary step in the partial fraction expansion process when exponentially damped sinusoids are involved. Find a general rule for completing the square of the quadratic polynomial $s^2 + as + b$ to get it in the form $(s + \alpha)^2 + \omega^2$ and apply the rule to the following numerical examples:

a.  $s^2 + 4s + 13$
b.  $s^2 + 2000s + 2,000,000$
c.  $s^2 + 200s + 12,500$

9.24 Use the initial and final value theorems to determine whether any of the following transform pairs is incorrect. If it is inappropriate to do so, state why.

a.  $(5e^{-t} - 2e^{-2t})u(t) \leftrightarrow \dfrac{3s + 8}{s^2 + 3s + 2}$

b.  $10e^{-t} \cos 2t\, u(t) \leftrightarrow 10\,\dfrac{s + 1}{s^2 + 2s + 5}$

c.  $(3te^{-t} + 10e^{-2t})u(t) \leftrightarrow \dfrac{10s^2 + 20s + 26}{s^3 + 4s^2 + 5s + 2}$

d.  $5 \cos 10t\, u(t) \leftrightarrow \dfrac{5s}{s^2 + 100}$

9.25 Verify the result of applying the final value theorem in Example 9.6 by applying L'Hôpital's rule directly to the time function.

9.26 Explain why the final value theorem is valid for $F(s) = N(s)/D(s)$ only when all the roots of $D(s)$ have negative real parts.

9.27 Sketch the following functions of time:

a.  $u(t - 3)$
b.  $u(t + 3)$
c.  $u(-t + 3)$
d.  $u(-t - 3)$
e.  $r(t - 6)$
f.  $(t - 6)u(t)$
g.  $t\, r(t)$
h.  $t\, r(t + 2)$

**9.28** Find Laplace transforms for the one-sided periodic functions shown in Figure P9.28.

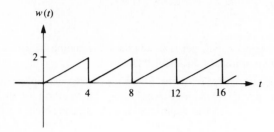

A.

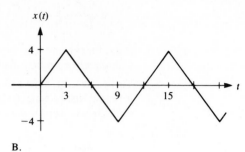

B.

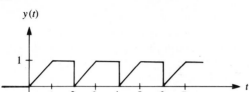

C.

**FIGURE P9.28**

**9.29** Use singularity functions to write analytical expressions for the functions shown in Figure P9.29.

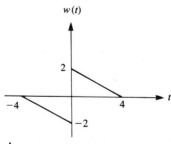

A.

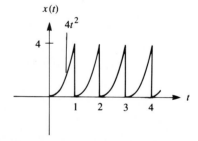

B.

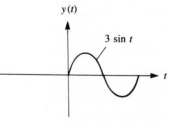

**FIGURE P9.29**    C.

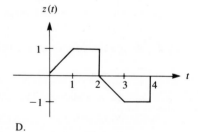

D.

**9.30** Find the inverse Laplace transform of $F(s)$ in Example 9.9 by first performing a long division of $(1 - e^{-s})/(1 + e^{-s})$.

**9.31** Prove the correctness of the *s*-domain inductor models shown in Figure 9.13.

**9.32** Transform the circuit of Figure P9.32 and use it to find $i(t)$.

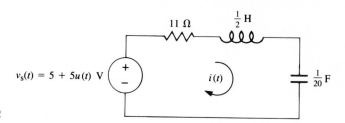

**FIGURE P9.32**

**9.33** Transform the circuit of Figure P9.33 and use it to find $v(t)$.

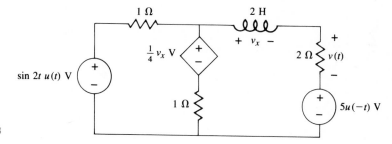

**FIGURE P9.33**

**9.34** Transform the circuit of Figure P9.34 and use it to find $i_o(t)$.

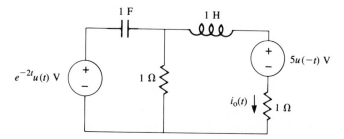

**FIGURE P9.34**

**9.35** Transform the circuit of Figure P9.35 and, by inspection, write its mesh impedance equation. Do not solve.

**9.36** Transform the circuit of Figure P9.36 and, by inspection, write its node admittance equation. Do not solve.

**9.37** Use the Laplace transform to solve the following differential equation. In the process, identify the zero-state, zero-input, forced, and natural responses.

$$\frac{d^2v}{dt^2} + 4\frac{dv}{dt} + 4v = 20e^{-10t}u(t)$$

$$v(0^-) = 1 \quad \text{and} \quad v'(0^-) = -1$$

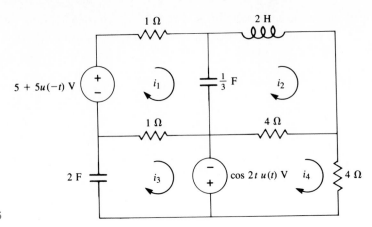

FIGURE P9.35

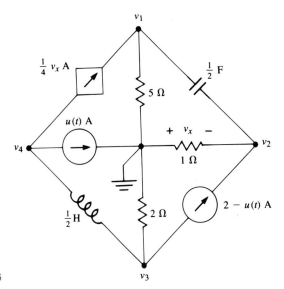

FIGURE P9.36

**9.38** Choose values for the initial conditions of the circuit shown in Figure P9.38 that will cause the natural response to be zero.

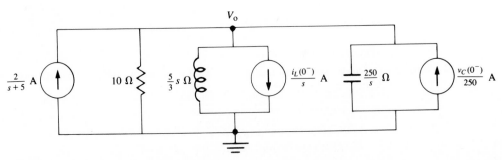

FIGURE P9.38

**9.39** Find system functions relating input to output for the circuits shown in Figure P9.39.

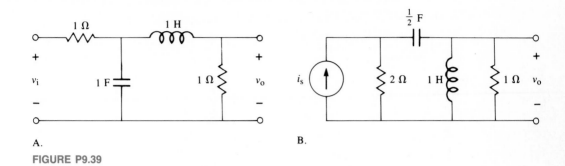

A.

B.

**FIGURE P9.39**

**9.40** Four output variables are defined for the circuit shown in Figure P9.40. Find the four corresponding system functions. Are they similar in any way?

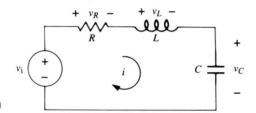

**FIGURE P9.40**

**9.41** Find the system function $V_o(s)/V_i(s)$ for the op amp circuit shown in Figure P9.41 when

   **a.** $Z_1(s) = R_1$, $Z_2(s) = R_2$

   **b.** $Z_1(s) = R_1 + (1/sC_1)$, $Z_2(s) = R_2$

   **c.** $Z_1(s) = R_1$, $Z_2(s) = R_2 \| 1/sC_2$

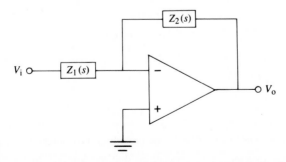

**FIGURE P9.41**

**9.42** Find $V_o(s)/V_i(s)$ for the circuit shown in Figure P9.42.

**9.43** Convolve the pairs of functions given in Figure P9.43.

**9.44** Use convolution to find the step response of the circuit shown in Figure P9.44.

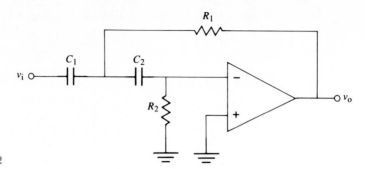

**FIGURE P9.42**

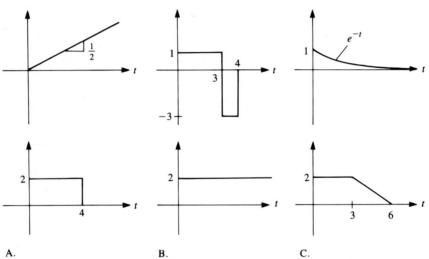

**FIGURE P9.43**     A.                              B.                              C.

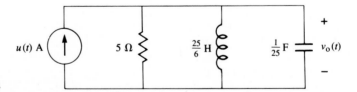

**FIGURE P9.44**

9.45 Redo Example 9.16 using the resistor voltage as the output variable. The lower limit of the convolution integral should be $0^-$ to unambiguously include the impulse.

9.46 Prove, by use of a variable substitution, that the form of the convolution integral shown in Equation 9.40 follows from that of Equation 9.39.

# System Functions and the $s$ Plane

10

## 10.1  INTRODUCTION

Chapters 8 and 9 described two methods for transforming circuit problems from the time domain to the frequency domain—phasors and the Laplace transform. The phasor approach discussed in Chapter 8 is quite useful but limited in its application (so far) to the sinusoidal steady state. The Laplace transform introduced in Chapter 9 is mathematically more rigorous than the phasor. In addition, it can be applied to a broader class of functions that includes, among others, ramps and exponentials in addition to sinusoids. Phasors and Laplace transforms seem very different from each other, and, in some ways, they are. However, it seems reasonable to assume that they are related to each other in some way because, after all, both are mathematical transformations from the time domain to the frequency domain.

This chapter unifies and broadens our understanding of the frequency domain. The approach taken will extend the use of phasors beyond pure sinusoids to include exponentially damped sinusoids. This larger class of functions can be characterized by the complex frequency $s = \sigma + j\omega$. As we will see, the impedances associated with these more general phasors will be functions of the frequency variable $s$. As with the earlier discussion of phasors, the focus here will be on finding the forced responses of circuits.

The main topics of this chapter are clearly a direct generalization of our work with phasors in the sinusoidal steady state. It will become evident, however, that the approach taken here can also be considered as a special case of the application of the Laplace transform. For instance, the impedance forms developed in this chapter will look exactly as they do when Laplace transforms are used and initial conditions are assumed to be zero. This seeming duplication of effort is justified because the approach in this chapter will give us additional insight into the description of electrical circuit behavior and prepare us for the topic of Chapter 11.

A large part of the discussion is devoted to the system function, a concept introduced in Chapter 9. A system function is a ratio of polynomials of the complex frequency variable $s$. It relates the response variable of a circuit to its input. In Chapter 9, the two variables were in the form of Laplace transforms. Here, the variables will be in phasor form. In either case, the system function will look the same. The two interpretations of system functions will be compared and contrasted in Section 10.8.

Many of the techniques introduced in this chapter can be conveniently visualized by considering the *s* plane. The *s* plane is a two-dimensional mapping of all possible values of the complex frequency variable *s*. The frequencies that characterize many types of forcing functions can be located on the *s* plane. More importantly, frequencies that characterize the natural behavior of circuits can also be mapped on the *s* plane. Then, the *s* plane becomes a convenient tool for the interpretation of linear electrical circuit behavior.

## 10.2   COMPLEX FREQUENCIES AND THE s PLANE

Consider a linear electrical circuit excited by an exponentially damped sinusoidal source. That is, $x(t) = Ae^{\sigma t} \cos(\omega t + \phi)$. Before proceeding, note that although exponentially damped sinusoids are often seen in the natural responses of underdamped circuits, they are not of much practical importance as forcing functions. Nonetheless, in studying them, we can arrive at some important results about electrical circuits. Chapter 6 demonstrated that the general form of the forced response for any variable in a circuit is a linear combination of the forcing function and its derivatives. That is, voltages and currents in the circuit will be in the form

$$y(t) = K_0 x(t) + K_1 \frac{dx}{dt} = K_0 Ae^{\sigma t} \cos(\omega t + \phi) + K_1 \frac{d}{dt} Ae^{\sigma t} \cos(\omega t + \phi)$$

$$= Be^{\sigma t} \cos(\omega t + \theta)$$

Thus, when the input to a linear circuit is an exponentially damped sinusoid, the forced part of any response will be of the same general exponentially damped form. It will be different only in amplitude and phase.

There is one important exception, however. If the circuit being analyzed has a pair of natural frequencies equal to $\sigma \pm j\omega$, then the form that the response takes on is $(Ct + D)e^{\sigma t} \cos(\omega t + \theta)$ and the general results of this chapter do not apply. This special case will be considered in more detail in Section 10.9.

Before we continue with the mathematical development, we should consider in some detail the nature of the function with which we are working:

$$f(t) = Ae^{\sigma t} \cos(\omega t + \phi) \tag{10.1}$$

Several special forms of this general function depend on the values of $\sigma$ and $\omega$:

$$f(t) = A \cos(\omega t + \phi), \quad \sigma = 0 \tag{10.2a}$$

$$f(t) = Ae^{\sigma t} \cos\phi = Ke^{\sigma t}, \quad \omega = 0 \tag{10.2b}$$

$$f(t) = A \cos\phi = K, \quad \sigma = 0, \ \omega = 0 \tag{10.2c}$$

As you can see, the exponentially damped sinusoid contains within it the cases of the pure sinusoid, the simple exponential, and the DC signal. Specific examples are shown in Figure 10.1.

One thing that the functions of Equation 10.2 have in common is that they all have frequencies associated with them. The function $A \cos(\omega t + \phi)$, for instance, is characterized by the radian frequency $\omega$. In a similar way, the exponential $Ke^{\sigma t}$ is characterized by the neper frequency $\sigma$. The constant $K$ has a frequency (radian and neper) of zero. These definitions of frequency should be familiar to you from earlier chapters in this book.

There is another way in which to describe the frequencies of time functions. We first encountered it when the natural responses of second-order circuits were

FIGURE 10.1    *The Function* $f(t) = e^{\sigma t}\cos \omega t$ *for Several Values of* $\sigma$ *and* $\omega$

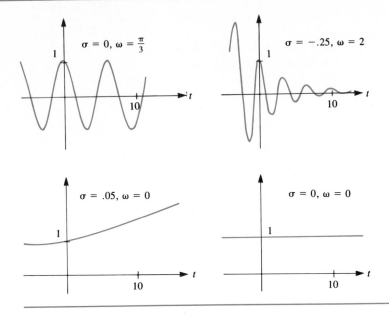

studied. As you will remember, if a circuit has two natural frequencies at, say, $s_1 = -10 - j200$ and $s_2 = -10 + j200$, then the natural response will be of the form

$$y_N(t) = K_1 e^{(-10-j200)t} + K_2 e^{(-10+j200)t} = A e^{-10t}\cos 200t + B e^{-10t}\sin 200t$$

$$= K e^{-10t}\cos (200t + \theta)$$

This same relationship applies to any exponentially damped sinusoid, not just to those that might be present in natural responses. We will generalize by saying that any time function that can be written as a sum of complex exponentials

$$f(t) = K_1 e^{s_1 t} + K_2 e^{s_2 t} + K_3 e^{s_3 t} + \cdots \tag{10.3}$$

is said to be characterized by the **complex frequencies** $s_1$, $s_2$, $s_3$, and so on. The constant coefficients $K_i$ are also complex. If $f(t)$ is a real function, then the individual terms of the summation must be present in complex conjugate pairs. Euler's relationship provides a simple example:

$$\sin \omega t = \left(\frac{1}{j2}\right)e^{j\omega t} + \left(\frac{-1}{j2}\right)e^{-j\omega t} \tag{10.4}$$

The real sinusoidal function on the left side of Equation 10.4 is characterized by the single real radian frequency $\omega$. The equivalent exponential representation of the function, however, contains within it *two* complex frequencies. They are the conjugates $+j\omega$ and $-j\omega$. Note that the coefficients of the two exponentials on the right side of Equation 10.4 are also complex conjugates of each other. This is necessary if the two terms are to sum to a real quantity.

It is a simple matter to generalize and show that any exponentially damped sinusoid can also be written as a sum of two complex exponentials:

$$Ae^{\sigma t}\cos{(\omega t + \phi)} = \frac{A}{2}\, e^{j\phi}e^{(\sigma + j\omega)t} + \frac{A}{2}\, e^{-j\phi}e^{(\sigma - j\omega)t} = K_1 e^{s_1 t} + K_2 e^{s_2 t}$$

$$= K_1 e^{s_1 t} + K_1^* e^{s_1^* t} = K_1 e^{s_1 t} + (K_1 e^{s_1 t})^* \tag{10.5}$$

where the asterisk symbol * denotes the complex conjugate operation. The exponentially damped sinusoid, then, is also characterized by a conjugate pair of complex frequencies, $\sigma + j\omega$ and $\sigma - j\omega$, and a conjugate pair of constant coefficients, $(A/2)e^{j\phi}$ and $(A/2)e^{-j\phi}$.

The frequencies of a time function such as in Equations 10.3, 10.4, or 10.5 are given the general symbol $s$. They can be plotted on a complex number plane called the $s$ **plane.** There is a unique position on the $s$ plane for every possible complex frequency. The relationship between complex frequency locations and the corresponding time functions is shown in Figure 10.2. With the exception of frequencies appearing on the real $(\sigma)$ axis, all frequencies must be present in complex conjugate pairs.

FIGURE 10.2    *Time Functions and Their Complex Frequencies*

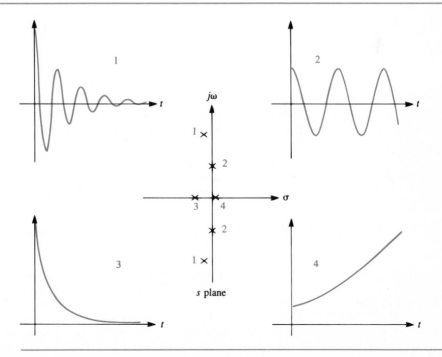

## 10.3   THE GENERALIZED PHASOR AND IMPEDANCE

As with the pure sinusoid, an exponentially damped sinusoid can be represented in the frequency domain by a complex-valued phasor. The phasor notation is based on the fact that

$$Ae^{\sigma t}\cos(\omega t + \phi) = Re[Ae^{\sigma t}e^{j(\omega t+\phi)}] = Re[Ae^{j\phi}e^{(\sigma+j\omega)t}]$$
$$= Re[Ae^{j\phi}e^{st}] \tag{10.6}$$

where $s = \sigma + j\omega$ is a complex frequency. We define a generalized phasor to be a complex number whose magnitude and phase are equal to those of a corresponding exponentially damped sinusoid. Notice that this is the same definition that was used for the phasors that described purely sinusoidal variables in Chapter 8. We now have the correspondence

$$f(t) = Ae^{\sigma t}\cos(\omega t + \phi) \tag{10.7a}$$

$$\mathbf{F} = Ae^{j\phi} \tag{10.7b}$$

$$f(t) = Re[\mathbf{F}e^{st}] \tag{10.7c}$$

where $\mathbf{F}$ is the phasor representation of the time function $f(t)$. Like the more restricted sinusoidal phasor, the generalized phasor is a frequency-domain representation of a signal. The same notation has been used as in Chapter 8 because the sinusoidal steady-state phasor is really just a special case of Equation 10.7. You can see the correspondence by setting $\sigma$ to zero.

Consider, once again, a system with an exponentially damped sinusoidal input. The response will be another exponentially damped sinusoid of the same complex frequency but with different magnitude and phase. This situation is displayed in Figure 10.3. It can be proven that if by some means the complex input $Ae^{j\phi}e^{st}$ could be applied to the system, the response would be $Be^{j\theta}e^{st}$. Although we cannot actually apply a complex input function in the laboratory, it is quite possible to do so mathematically. As in Chapter 8, we will simply remember that, by convention, it is the real parts of the complex forcing function and its response that are the input–output pair in which we are interested. We can distinguish between a real-valued time function $x(t)$ and the corresponding complex exponential by giving the latter the symbol $\tilde{x}(t)$.

**FIGURE 10.3**   *Exponentially Damped Sinusoidal Input and Output Signals*

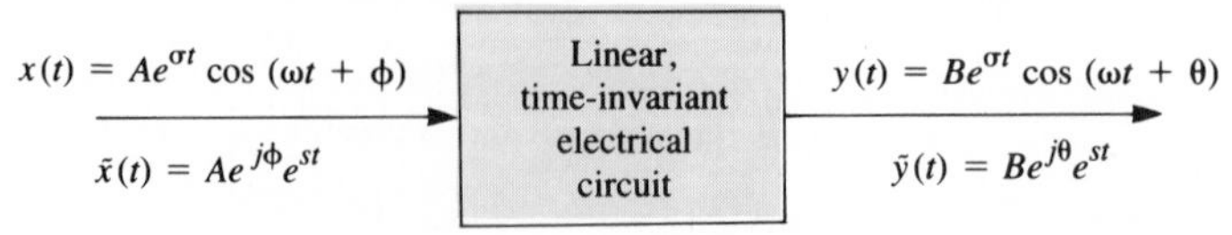

$x(t) = Ae^{\sigma t}\cos(\omega t + \phi)$ → | Linear, time-invariant electrical circuit | → $y(t) = Be^{\sigma t}\cos(\omega t + \theta)$

$\tilde{x}(t) = Ae^{j\phi}e^{st}$ → | | → $\tilde{y}(t) = Be^{j\theta}e^{st}$

**Example 10.1** *Solving a Differential Equation Using Phasors*

Consider the differential equation that relates $x(t)$ and $y(t)$:

$$\frac{d^3y}{dt^3} + 2\frac{d^2y}{dt^2} - \frac{dy}{dt} + 3y = 2\frac{dx}{dt} + 5x$$

Let $x(t) = e^{-t}\cos t$. Rather than solving the equation by using real-valued functions, use a complex exponential form for the input and a corresponding form for $y(t)$ (note that the magnitude of $x(t)$ is 1 and that its phase is $0°$):

$$x(t) = e^{-t}\cos t \quad \rightarrow \quad \tilde{x}(t) = e^{st}$$

and

$$\tilde{y}(t) = Ye^{j\theta}e^{st}$$

where $Y$ and $\theta$ are the assumed magnitude and phase of the function $y(t)$ and $s = -1 + j$. Substitute these two expressions into the differential equation:

$$\frac{d^3}{dt^3}Ye^{j\theta}e^{st} + 2\frac{d^2}{dt^2}Ye^{j\theta}e^{st} - \frac{d}{dt}Ye^{j\theta}e^{st} + 3Ye^{j\theta}e^{st} = 2\frac{d}{dt}e^{st} + 5e^{st}$$

This equation becomes, after some manipulation,

$$Ye^{j\theta} = \frac{2s + 5}{s^3 + 2s^2 - s + 3}\bigg|_{s=-1+j} = \frac{3 + j2}{6 - j3} = .54e^{j60.3°}$$

The phasor **Y** has been found. In the time domain, it becomes

$$y(t) = .54e^{-t}\cos(t + 60.3°)$$

Next, we will establish the generalized impedance relationships for the three basic passive electrical circuit elements. Consider the voltage across and the current in a capacitor when they are in the form of exponentially damped sinusoids. The current and voltage will be represented by complex exponentials. Keep in mind that for each, it is the real part of the complex quantity that is of interest. Figure 10.4 gives the assumed forms for these quantities. The time-domain relationship between a capacitor's voltage and current will be our starting point. If we use the complex representations defined in Figure 10.4, then

$$\tilde{i}(t) = C\frac{d\tilde{v}(t)}{dt}$$

**FIGURE 10.4    *Capacitor with Exponentially Damped Sinusoidal Excitation***

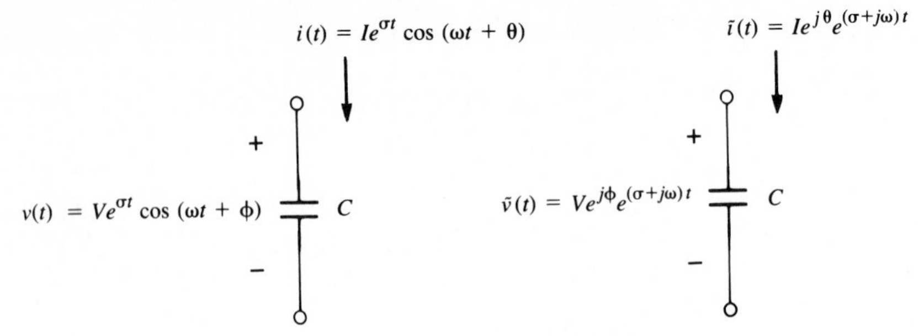

$i(t) = Ie^{\sigma t} \cos(\omega t + \theta)$                      $\tilde{\imath}(t) = Ie^{j\theta}e^{(\sigma + j\omega)t}$

$v(t) = Ve^{\sigma t} \cos(\omega t + \phi)$                $\tilde{v}(t) = Ve^{j\phi}e^{(\sigma + j\omega)t}$

A. Real-valued functions                    B. Corresponding complex exponentials

$$Ie^{\sigma t}e^{j(\omega t + \theta)} = \frac{d}{dt}\, Ve^{\sigma t}e^{j(\omega t + \phi)}$$

$$Ie^{j\theta}e^{st} = C\,\frac{d}{dt}\,(Ve^{j\phi}e^{st}) = sCVe^{j\phi}e^{st}$$

$$Ie^{j\theta} = sCVe^{j\phi}$$

We observe that the phasor representations for voltage and current are

$$\mathbf{I}_C = Ie^{j\theta} \qquad \text{and} \qquad \mathbf{V}_C = Ve^{j\phi}$$

The relationship between these two phasors is

$$\frac{\mathbf{V}_C}{\mathbf{I}_C} = \frac{1}{sC} = Z_C(s) \tag{10.8a}$$

or

$$\frac{\mathbf{I}_C}{\mathbf{V}_C} = sC = Y_C(s) \tag{10.8b}$$

The ratio of $\mathbf{V}_C$ to $\mathbf{I}_C$ is the generalized impedance, $Z_C(s)$, of a capacitor; the inverse is the admittance, $Y_C(s)$. It is left as an exercise (Problem 10.10 at the end of the chapter) to show that the impedance relationships for inductors and resistors are as shown in Table 10.1.

The impedances defined in Table 10.1 are, of course, more general in their application than are the sinusoidal steady-state impedance relationship found in Chapter 8 and given in Table 8.1. To move from one impedance form to the other is quite easy, however. Simply replace the complex frequency *s* with *jω* or vice versa. Even though the mathematical transition from the sinusoidal to the more general phasor is straightforward, not all of the rules that apply to the sinusoidal

**TABLE 10.1**   *Generalized Impedances and Admittances*

|  | *Resistor, R* | *Inductor, L* | *Capacitor, C* |
|---|---|---|---|
| *Impedance,* $Z = \dfrac{\mathbf{V}}{\mathbf{I}}$ | $Z_R = R$ | $Z_L = sL$ | $Z_C = \dfrac{1}{sC}$ |
| *Admittance,* $Y = \dfrac{\mathbf{I}}{\mathbf{V}}$ | $Y_R = \dfrac{1}{R}$ | $Y_L = \dfrac{1}{sL}$ | $Y_C = sC$ |

case are valid here. For example, in the sinusoidal steady state, capacitor current always leads capacitor voltage by 90°. With the more general exponentially damped sinusoid, however, any phase relationship can exist between the voltages and currents of capacitors and inductors.

As in the sinusoidal steady state, we can still observe that for *DC* frequencies ($s = 0$), capacitors act as open circuits and inductors as short circuits. For infinite frequency ($|s| = \infty$), the opposite is true: Capacitors act as short circuits and inductors as open circuits.

If you have studied the Laplace transform, you will recognize the impedances and admittances of Table 10.1 to be identical to those derived through the Laplace transform when initial conditions are zero. Because, in this chapter, we are concerned with finding the forced responses of systems, initial conditions are irrelevant to our present work.

## 10.4   GENERALIZED PHASORS IN CIRCUIT ANALYSIS

Just as with the sinusoidal steady-state phasor, all of the standard analytical techniques developed in the early chapters of this book are valid when signal variables are represented by generalized phasors and passive elements are represented by impedances. KVL, KCL, mesh and node analysis, and equivalent circuits still apply when we use the generalized phasor and impedance notation.

---

**Example 10.2**   *Mesh Analysis with Generalized Phasors*

Consider the circuit shown in Figure 10.5A. Find $i_2(t)$ using mesh analysis.

The first step is to transform the circuit into its frequency-domain representation, as shown in Figure 10.5B. Signal variables become phasors, and passive elements become impedances (or admittances). We could, at this point, find the numerical value of each impedance at the specified complex frequency, but it is generally much easier to carry $s$ along as a variable to be evaluated as the last step to the solution. Thus, we can write

*(continues)*

**Example 10.2**    *Continued*

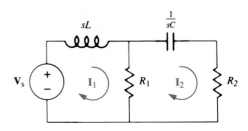

$v_s(t) = 5e^{-t} \cos 2t$ V

$R_1 = 3\ \Omega$    $R_2 = 4\ \Omega$

$L = 2$ H    $C = \frac{1}{5}$ F

FIGURE 10.5    A. Time-domain representation          B. Frequency-domain representation

the two mesh equations for the circuit as follows:

$$\mathbf{V}_s = sL\mathbf{I}_1 + R_1(\mathbf{I}_1 - \mathbf{I}_2)$$

$$0 = \frac{1}{sC}\,\mathbf{I}_2 + R_2\mathbf{I}_2 + R_1(\mathbf{I}_2 - \mathbf{I}_1)$$

Substituting in the source and element values and rearranging lead to

$$5e^{j0°} = (3 + 2s)\mathbf{I}_1 - 3\mathbf{I}_2 \tag{10.9a}$$

$$0 = -3\mathbf{I}_1 + \left(3 + 4 + \frac{5}{s}\right)\mathbf{I}_2 \tag{10.9b}$$

We then solve Equations 10.9a and 10.9b for the phasor current $\mathbf{I}_2$:

$$\mathbf{I}_2 = \frac{15s}{14s^2 + 22s + 15}$$

All that is left is to evaluate the expression for $\mathbf{I}_2$ at the frequency of the forcing function:

$$\mathbf{I}_2 = \frac{15s}{14s^2 + 22s + 15}\Big|_{s=-1+j2}$$

$$= \frac{15(-1 + j2)}{14(-1 + j2)^2 + 22(-1 + j2) + 15}$$

$$= .665e^{-j77.2°} \text{ A}$$

As a function of time, the current is written as

$$i_2(t) = .665e^{-t} \cos(2t - 77.2°) \text{ A}$$

It is left as an exercise for you to demonstrate that

$$i_1(t) = 1.402e^{-t} \cos{(2t - 95.63°)} \text{ A}$$

The forcing function in Example 10.2 is characterized by the two complex frequencies $s = -1 + j2$ and $s = -1 - j2$. Notice, however, that only the frequency $s = -1 + j2$ entered numerically into the calculations. More will be said about this seeming oversight in Section 10.7.

### Example 10.3    Node Analysis with Generalized Phasors

Consider the circuit shown in Figure 10.6A. Its two sources are at the same frequency, $s = -4$, so we can immediately proceed to a solution in the usual way. If the two sources were at different frequencies, the principle of superposition would have to be used because the passive elements in the circuit would present different impedances to each of the two sources.

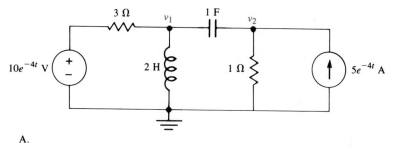

A.

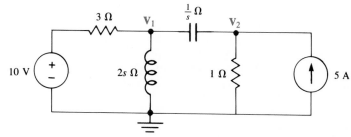

FIGURE 10.6    B.

Node equations can be written from the transformed circuit of Figure 10.6B.

$$\frac{1}{3}(V_1 - 10) + \frac{1}{2s}V_1 + s(V_1 - V_2) = 0$$

$$s(V_2 - V_1) + V_2 = 5$$

(continues)

**Example 10.3**   *Continued*

Rearranging leads to

$$\left(s + \frac{1}{3} + \frac{1}{2s}\right)\mathbf{V}_1 - s\mathbf{V}_2 = \frac{10}{3}$$

$$-s\mathbf{V}_1 + (s + 1)\mathbf{V}_2 = 5$$

The phasor $\mathbf{V}_1$ is found to be

$$\mathbf{V}_1 = \frac{50s^2 + 20s}{8s^2 + 5s + 3}\bigg|_{s=-4} = \frac{800 - 80}{128 - 20 + 3} = 6.49 \text{ V}$$

The phasor $\mathbf{V}_1$ is a real number because it corresponds to a simple exponential and, therefore, does not have a phase angle:

$$v_1(t) = 6.49e^{-4t} \text{ V}$$

## 10.5   SYSTEM FUNCTIONS

In Section 10.4, circuit problems were solved using notation based on the complex frequency $s = \sigma + j\omega$. The values of output variables were calculated in response to specific input forcing functions. This method is, in many ways, even more useful when the forcing function is not immediately specified. For example, consider the parallel *RLC* circuit of Figure 10.7. The relationship between the voltage $\mathbf{V}_o$ and the current $\mathbf{I}_s$ is found to be

$$\mathbf{I}_s = \left(\frac{1}{R} + \frac{1}{sL} + sC\right)\mathbf{V}_o$$

$$\mathbf{V}_o = \frac{1}{C} \cdot \frac{s}{s^2 + (1/RC)s + 1/LC} \mathbf{I}_s \qquad\qquad (10.10)$$

**FIGURE 10.7**   *Parallel RLC Circuit*

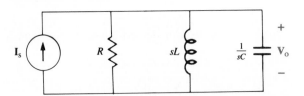

From Equation 10.10, we see that there is a term of proportionality between the phasor input and output; we will give it the symbol $H(s)$:

$$\mathbf{V}_o = H(s)\mathbf{I}_s$$

where

$$H(s) = \frac{1}{C} \cdot \frac{s}{s^2 + (1/RC)s + 1/LC}$$

Once the forcing function has been specified, we can find the response by evaluating $H(s)$ at the appropriate frequency. Let's assume that the input current is $i_s(t) = 2e^{-t}\sin(4t + 45°)$ A. As was mentioned earlier, although the forcing function is characterized by the conjugate frequencies $s = -1 \pm j4$, our calculations need consider only the frequency that is in the upper half of the $s$ plane. That is, $s = -1 + j4$. If we assume in addition that $R = L = C = 1$, then we can calculate $H(-1 + j4)$:

$$H(s) = \frac{s}{s^2 + s + 1}$$

$$H(-1 + j4) = \frac{-1 + j4}{(-1 + j4)^2 + (-1 + j4) + 1} = .266e^{-j90.9°}$$

The phasor voltage in response to the given current is, then,

$$\mathbf{V}_o = H(-1 + j4)\mathbf{I}_s = (.266e^{-j90.9°})(2e^{j45°})$$

which corresponds to a time-domain response of

$$v_o(t) = (2)(.266)e^{-t}\sin(4t + 45° - 90.9°)$$
$$= .532e^{-t}\sin(4t - 45.9°) \text{ V}$$

With $H(-1 + j4)$ already calculated, the system's response to any forcing function of that frequency can be found. For example, an input of

$$i_s(t) = -4e^{-t}\cos(4t + 220°) \text{ A}$$

causes the response

$$v_o(t) = (-4)(.266)e^{-t}\cos(4t + 220° - 90.9°)$$
$$= -1.064e^{-t}\cos(4t + 129.1°) \text{ V}$$

and

$$i_s(t) = 12e^{-t}\sin(4t - 30°) \text{ A}$$

leads to

$$v_o(t) = 3.192e^{-t} \sin (4t - 120.9°) \text{ V}$$

The responses to forcing functions of other frequencies can also be determined, but $H(s)$ must be recalculated at each new frequency. You should verify that the circuit described by Equation 10.10 has the following input–output pairs:

$$i_s(t) = 3e^{-.5t} \text{ A}$$

$$v_o(t) = -2e^{-.5t} \text{ V}$$

$$i_s(t) = -2 \cos t \text{ A}$$

$$v_o(t) = -2 \cos t \text{ V}$$

$$i_s(t) = 20e^t \cos t \text{ A}$$

$$v_o(t) = 7.84e^t \cos (t - 11.3°) \text{ V}$$

The term $H(s)$ defined in the preceding discussion is known as a **system function** or **network function** of the electrical circuit. With it, we can rewrite Equation 10.10 as either

$$\mathbf{V}_o = H(s)\mathbf{I}_s \tag{10.11a}$$

or

$$H(s) = \frac{\mathbf{V}_o}{\mathbf{I}_s} \tag{10.11b}$$

We see, in these equations, that system functions can be used to define the mathematical relationship between input–output pairs in a circuit when the variables are in phasor form. The term *network function* is usually limited to electrical networks, but the term *system function* can be applied to the problems of many different disciplines.

Because the system function of Equation 10.11b is the ratio of a phasor voltage to a phasor current, it is an impedance. Not all system functions are impedances. They may also have the dimensions of admittance, or they may be dimensionless. For instance, consider again the circuit in Figure 10.7. We could have found the system function that relates the current in the resistor to that of the source. We could also have done the same if the inductor or capacitor current were defined to be the response variable. All three resulting system functions would be ratios of currents and, therefore, dimensionless.

For any circuit, as many different system functions can be found as there are definable input–output pairs. If the input and output variables are located at different places within the circuit, the system function is called a **transfer function**

FIGURE 10.8 *Several Types of System Functions*

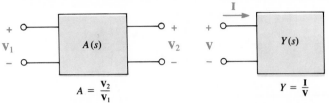

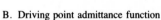

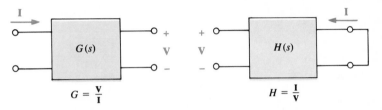

A. Voltage transfer function

B. Driving point admittance function

C. Impedance transfer function (transimpedance)

D. Admittance transfer function (transadmittance)

because information or energy is transferred from one location to another. Often, the input and output variables are defined at a single location in the circuit, as, for example, when the input is an applied voltage and the response is the current from the source into the circuit. These system functions are called **driving point functions.** Driving point functions are either impedances or admittances. Additional terminology is given in Figure 10.8, but for now, the general term *system function* will serve our needs. The symbols used for system functions are uppercase letters such as $H$, $G$, $Y$, or $Z$.

**Example 10.4** *Finding a Transfer Function*

Find the transfer function that relates $\mathbf{V}_o$ to $\mathbf{V}_s$ in the circuit shown in Figure 10.9. The elements are already given in impedance form.

Using node analysis, the following equations can be written:

$$\frac{\mathbf{V}-\mathbf{V}_s}{s/2} + \frac{\mathbf{V}}{2} + \frac{\mathbf{V}-\mathbf{V}_o}{4/s} = 0$$

$$\frac{\mathbf{V}_o-\mathbf{V}}{4/s} + \frac{\mathbf{V}_o}{4} + \frac{\mathbf{V}_x}{2} = 0$$

$$\mathbf{V}_x = -\mathbf{V}$$

*(continues)*

**Example 10.4**   *Continued*

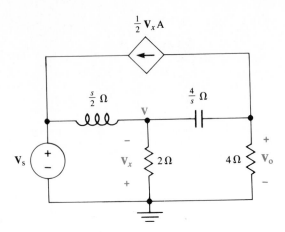

**FIGURE 10.9**

Eliminating $V_x$ and $V$, $V_o$ is found as a function of $V_s$:

$$V_o = 8 \frac{s+2}{s^2 + 10s + 8} V_s$$

To find the circuit's response to, say, $v_s(t) = .5 \sin (t + 10°)$ V, evaluate $H(s)$ at $s = j$:

$$H(j) = 8 \frac{j+2}{j^2 + 10j + 8} = 1.465 e^{-28.4°}$$

The response is

$$v_o(t) = (1.465)(.5) \sin (t + 10° - 28.4°)$$
$$= .733 \sin (t - 18.4°) \text{ V}$$

---

**Example 10.5**   *Voltage Transfer Function of an Op Amp Circuit*

Find the voltage transfer function of the ideal operational amplifier circuit shown in Figure 10.10.
    A KCL equation written at the inverting input of the amplifier yields

$$\frac{V_i}{(1/sC_1) + R_1} + \frac{V_o}{\dfrac{R_2/sC_2}{R_2 + (1/sC_2)}} = 0$$

which leads to

$$\frac{V_o}{V_i} = - \frac{sR_2C_1}{(1 + sR_1C_1)(1 + sR_2C_2)}$$

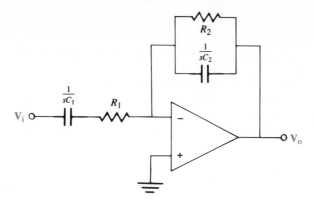

**FIGURE 10.10**

## 10.6 POLES, ZEROS, AND THE s PLANE

Once a system function is known, it can be used to calculate the forced response to any particular excitation (as long as it can be characterized by complex frequencies). As we have seen, the calculations can be performed analytically. They can also be interpreted graphically using the s plane. Before examples of graphical calculations can be shown, however, some additional concepts relating to system functions must be introduced.

Consider a general system function (Equation 10.12a) shown first as a ratio of two polynomials in s (Equation 10.12b) and then in an equivalent factored form (Equation 10.12c). It is assumed that there are no factors common to both the numerator polynomial, $N(s)$, and the denominator polynomial, $D(s)$. $K$ is a numerical constant. As with all real polynomials, the roots of $N(s)$ and $D(s)$ may be either real-valued or complex. If complex, they must occur in conjugate pairs. Finally, they may be either simple roots (only one occurring at a location) or multiple. The equations are as follows:

$$H(s) = K \frac{N(s)}{D(s)} \tag{10.12a}$$

$$H(s) = K \frac{s^m + a_{m-1}s^{m-1} + \cdots + a_1 s + a_0}{s^n + b_{n-1}s^{n-1} + \cdots + b_1 s + b_0} \tag{10.12b}$$

$$H(s) = K \frac{(s - z_1)(s - z_2) \cdots (s - z_m)}{(s - p_1)(s - p_2) \cdots (s - p_n)} \tag{10.12c}$$

System functions can be evaluated for any frequency on the s plane, but some values of s are more interesting than others. For instance, for some values of s, the magnitude of a system function will be zero. Such frequencies are called **zeros** of the system function. These can be values of s for which the numerator polynomial, $N(s)$ of Equation 10.12, is equal to zero. Thus, $s = z_1, z_2, \ldots z_m$ are all zeros of the system. A **pole** of the system is defined as any frequency $s$ at which the magnitude of the system function becomes infinite. Roots of the denominator polynomial satisfy this requirement. The frequencies $s = p_1, p_2, \ldots p_n$ are all poles of the system function of Equation 10.12.

The poles and zeros found by setting $D(s)$ and $N(s)$ to zero are known as the **critical frequencies** of the system. These frequencies are not the only poles and zeros that a system can have, however. If $N(s)$ and $D(s)$ are of different degree ($m \neq n$), then either poles or zeros will exist at $|s| = \infty$. Consider, for example, Equation 10.13:

$$G(s) = \frac{(s+1)^2}{s(s^2 + 2s + 10)} \qquad (10.13)$$

$G(s)$ has two zeros at $s = -1$. It has one pole at $s = 0$ and two others at $s = -1 \pm j3$ (the roots of $s^2 + 2s + 10$). Because the order of the denominator ($n = 3$) exceeds that of the numerator ($m = 2$), the system function also has a zero at $|s| = \infty$. That is,

$$G(\infty) = \frac{(\infty + 1)^2}{\infty(\infty^2 + 2\infty + 10)} = 0$$

Because the basic definition of a zero is any frequency $s$ for which $G(s) = 0$, we conclude that $|s| = \infty$ must be a zero of $G(s)$. In fact, all system functions have equal numbers of poles and zeros. Whenever $n > m$, there are $n - m$ zeros at infinity. Similarly, if $m > n$, there are $m - n$ poles at infinity.

While poles and zeros at infinity are of interest to specialists in systems theory, we will be concerned only with the finite poles and zeros of a system. It is useful to mark their locations on the $s$ plane. The locations of finite poles are marked with X's and those of zeros with O's. The result is called a **pole–zero plot.** For example, the system function shown plotted in Figure 10.11 has four finite poles and one finite zero. Its three zeros at infinity are not shown. The constant multiplier is given in the box.

A system function can be uniquely reconstructed from its pole–zero plot only if the constant multiplier $K$ is known. The constant will be shown in this book as in Figure 10.11, although you may find other books that do so in other ways.

A complete pole–zero plot, including the constant $K$, is a graphical representation of a system function. It contains no more or no less information than does the analytical expression of the function. It seems reasonable, then, that calculations performed analytically with $H(s)$ can also be performed graphically on the $s$ plane. Section 10.7 will describe how graphical calculations are done. In the past, graphical calculations of this sort were an important analytical tool for the engineer. Although today the calculations are done by computers, you should still learn the graphical process in order to more fully understand the results of formal calculations and their relation to the $s$ plane.

FIGURE 10.11    *A Pole–Zero Plot*

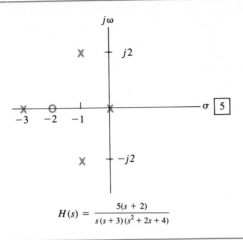

$$H(s) = \frac{5(s + 2)}{s(s + 3)(s^2 + 2s + 4)}$$

## *10.7    A GRAPHICAL INTERPRETATION OF SYSTEM FUNCTIONS

An inspection of Equation 10.12c (repeated below) shows that except for the constant $K$, each factor of the system function $H(s)$ has the same form:

$$H(s) = K \frac{(s - z_1)(s - z_2) \cdots (s - z_m)}{(s - p_1)(s - p_2) \cdots (s - p_n)}$$

Each is a complex number that is the difference between the frequency at which the function is evaluated and one of the poles or zeros. Complex numbers and their sums or differences can be represented by vectors on the $s$ plane. Figure 10.12 shows how the difference $s - s_k$ can be graphically calculated using the vector representations of $\mathbf{s}$ and $\mathbf{s}_k$. Vector names are given in boldface type. Stated most simply, $\mathbf{s} - \mathbf{s}_k$ is found by drawing a vector from the tip of vector $\mathbf{s}_k$ to the tip of vector $\mathbf{s}$. The magnitude of the complex number $s - s_k$ is the length of the difference vector, and its angle is the angle that the vector makes when it is measured counterclockwise from the positive real axis of the $s$ plane. The magnitude and angle of a vector do not change as it is moved without rotation on the $s$ plane.

Once it is understood how to represent the individual factors of Equation 10.12c with vectors on the $s$ plane, it is a simple matter to graphically calculate the system function's magnitude and angle. The first step is to draw the pole–zero plot to scale. The next step is to draw a vector from each pole and zero to the frequency at which the system function is to be evaluated. The length and angle of each vector are then measured. The final step is to combine the individual

FIGURE 10.12    *Graphical Construction of* $s - s_k$

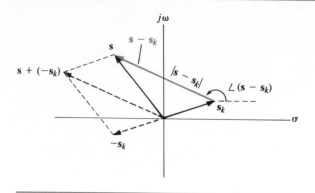

magnitudes and angles so as to calculate the system function itself. This last step is most easily seen if Equation 10.12c is rewritten with each term in polar form:

$$H(s) = K \frac{M_{z_1}(s)e^{j\theta_{z_1}(s)}M_{z_2}(s)e^{j\theta_{z_2}(s)} \cdots M_{z_m}(s)e^{j\theta_{z_m}(s)}}{M_{p_1}(s)e^{j\theta_{p_1}(s)}M_{p_2}(s)e^{j\theta_{p_2}(s)} \cdots M_{p_n}(s)e^{j\theta_{p_n}(s)}} \tag{10.14}$$

where    $M_{z_k}(s) = |s - z_k|$
           $\theta_{z_k} = \angle(s - z_k)$

The operator $\angle$ denotes the angle of a complex quantity. The magnitude and

FIGURE 10.13    *General Example of Graphically Calculating Value of a System Function*

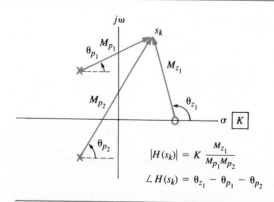

angle of $H(s)$ can be written as follows:

$$|H(s)| = K \frac{M_{z_1} M_{z_2} \cdots M_{z_m}}{M_{p_1} M_{p_2} \cdots M_{p_n}} \tag{10.15a}$$

$$\angle H(s) = \theta_{z_1} + \theta_{z_2} + \cdots + \theta_{z_m} - \theta_{p_1} - \theta_{p_2} - \cdots - \theta_{p_n} \tag{10.15b}$$

Figure 10.13 shows a general example of the graphical approach.

Let's see how this graphical calculation method works with a specific numerical example.

---

**Example 10.6** *Graphical Calculation of a Forced Response*

A system function is given as

$$H(s) = \frac{\mathbf{V}_o}{\mathbf{V}_i} = 3\,\frac{s^2 + 2s}{s^2 + 4s + 8}$$

Find the system's forced response to $v_i(t) = 5 \sin t$.

To do this calculation graphically, first determine and plot the finite poles and zeros of the system function:

$$H(s) = 3\,\frac{s(s+2)}{(s+2-j2)(s+2+j2)}$$

The two zeros are at $s = 0, -2$; the poles are at $s = -2 \pm j2$. These values are plotted in Figure 10.14 along with the vectors needed to evaluate $H(j1)$.

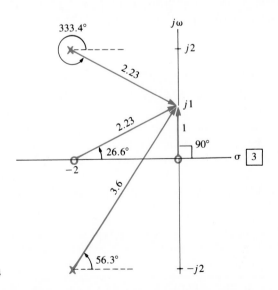

**FIGURE 10.14**

*(continues)*

**Example 10.6**  *Continued*

The magnitudes and angles of the individual vectors have been measured and are shown in the figure. The system function can now be evaluated:

$$|H(j1)| = \frac{(3)(1)(2.23)}{(2.23)(3.6)} = .83$$

$$\angle H(j1) = 90° + 26.6° - 333.4° - 56.3° = 86.9°$$

The forced response of the system is

$$v_o(t) = (.83)(5)\sin(t + 86.9°) = 4.15\sin(t + 86.9°)\ V$$

You should verify this result analytically.

---

Computers and calculators have made the graphical computation of $H(s)$ obsolete. Still, the graphical technique and the pole–zero plot upon which it is based are often used to quickly determine the general characteristics of systems. Controls engineers, in particular, make effective use of this approach. For them, the plot–zero plot can be an invaluable tool for approximating system behavior. We will return to this idea in Chapter 11.

Before we move on to the next topic, we should address a point that is often confusing to those new to the use of the generalized phasor approach to system functions. Section 10.2 showed that any exponentially damped sinusoid can be characterized by two complex frequencies that are conjugates of each other. That is, $x(t) = Ae^{\sigma t}\cos \omega t$ can be equivalently written as

$$x(t) = \frac{A}{2}\left(e^{(\sigma + j\omega)t} + e^{(\sigma - j\omega)t}\right) \tag{10.16}$$

In several previous examples, however, only the one frequency $\sigma + j\omega$ has been used in developing answers. What happened to the conjugate frequency $\sigma - j\omega$? The easiest way out of this apparent dilemma is to say that the procedure of considering the actual forcing function to be the real part of a complex quantity requires only that we be aware of the characterizing frequency in the upper half of the $s$ plane. You may verify this fact by rereading the pertinent parts of Sections 10.2 and 10.3. With more mathematical formalism, however, we can say that the response of a linear system to the forcing function of Equation 10.16 is

$$y(t) = \frac{A}{2}H(\sigma + j\omega)e^{(\sigma + j\omega)t} + \frac{A}{2}H(\sigma - j\omega)e^{(\sigma - j\omega)t}$$

It follows from the general properties of functions of complex variables that

$H(\sigma + j\omega)$ and $H(\sigma - j\omega)$ are complex conjugates of each other. That is, if

$$H(\sigma + j\omega) = |H|e^{j\theta}$$

then

$$H(\sigma - j\omega) = |H|e^{-j\theta}$$

It is left as an exercise (Problem 10.30 at the end of the chapter) to show that these facts imply that if $x(t) = Ae^{\sigma t} \cos \omega t$, then

$$y(t) = |H|Ae^{\sigma t} \cos (\omega t + \theta)$$

In other words, the magnitude and angle of $H(\sigma + j\omega)$ alone are enough to determine the system's response.

## 10.8 PHASORS, LAPLACE TRANSFORMS, AND COMPLETE RESPONSES

We have encountered system functions twice in this book, once briefly in Chapter 9 on the Laplace transform and now in this chapter as we study generalized phasors and the $s$ plane. For any given circuit problem, system functions arrived at from either of these two approaches will be identical in appearance and cannot be distinguished from one another. System functions, then, must contain the same information regardless of the way in which they are derived. This statement might be somewhat surprising to you because the procedures used in these approaches seem to be very different. In this section, we will see that these differences are in matters of interpretation only. That is, the Laplace and the phasor techniques are two approaches to the same general frequency-domain method.

When a system function is treated as a ratio of two phasors, it is generally used to find only the forced response of a system such as an electrical circuit to a particular class of input functions (those that can be characterized by complex frequencies). Initial conditions do not enter into the problem because the natural response is assumed to have died out. Under these conditions, the total response is found by evaluating the system function at the frequency of the forcing function and scaling the input with the result.

When the system function relates two Laplace transforms, the procedure is very different. The Laplace transform of the forcing function is multiplied by the system function, and the inverse transform of the product is found by the standard techniques discussed in Chapter 9. In contrast to the phasor approach, the inverse transform of the response contains both the forced *and* the natural responses of the system to an excitation suddenly applied at $t = 0$ assuming that the system is initially at rest. Another difference between these two approaches is that the Laplace transform can be applied to a wider class of time functions than can the generalized phasor notation.

The fact that the Laplace transform approach includes the natural response of the system suggests that any system function must contain the natural frequencies of the system it describes. This suggestion is, in fact, correct. The relationship between the system function and the natural frequencies can be shown quite easily by considering a general differential equation that might describe the relationship between input and response in a linear system:

$$b_n \frac{d^n y}{dt^n} + b_{n-1} \frac{d^{n-1} y}{dt^{n-1}} + \cdots + b_0 y = a_m \frac{d^m x}{dt^m} + a_{m-1} \frac{d^{m-1} x}{dt^{m-1}} + \cdots + a_0 x$$

If complex exponential forms are assumed such that $\tilde{x}(t) = \mathbf{X} e^{st}$ and $\tilde{y}(t) = \mathbf{Y} e^{st}$, then it can be shown that

$$(b_n s^n + b_{n-1} s^{n-1} + \cdots + b_0) \mathbf{Y} = (a_m s^m + a_{m-1} s^{m-1} + \cdots + a_0) \mathbf{X}$$

This equation can be arranged into system function form as a ratio of two phasors:

$$\frac{\mathbf{Y}}{\mathbf{X}} = \frac{a_m s^m + a_{m-1} s^{m-1} + \cdots + a_0}{b_n s^n + b_{n-1} s^{n-1} + \cdots + b_0}$$

It can now be recognized that the denominator polynomial of a system function, when set equal to zero, is the same as the characteristic equation of the system's differential equation. We know that the roots of the characteristic equation are the natural frequencies of the system and that the roots of the denominator polynomial are, by definition, the system's finite poles. We conclude that the poles of a system are identical to its natural frequencies. We again assume that there are no roots common to the numerator and denominator polynomials (that is, that there can be no cancellation between poles and zeros). In practice, it is unusual for poles and zeros to cancel each other by having identical locations. This caution having been stated, the relationship between poles and natural frequencies allows us to use knowledge of the system function to write down the *form* of the natural part of the response. We already know that the system function can be used to easily find a response to a forcing function characterized by s-plane frequencies. With additional knowledge of the response variable's initial conditions, we have enough information to use the phasor approach for finding complete responses, even when the excitation is suddenly applied.

---

**Example 10.7**    *Finding a Complete Response from a Pole–Zero Plot*

Consider the circuit shown in Figure 10.15A. The input signal is $v_s(t)$, and the response variable is $i_C(t)$. Initial conditions for the response variable are given. The pole–zero plot of the system function $H(s) = \mathbf{I}_C / \mathbf{V}_s$ is shown in Figure 10.15B. The forced part of the response for $t > 0$ can be found by evaluating the system function at $s = -1$, the frequency of excitation.

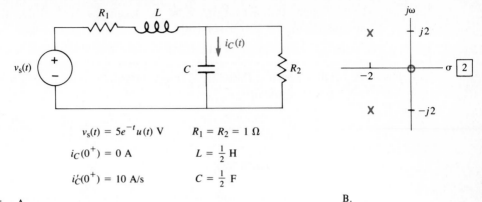

$$v_s(t) = 5e^{-t}u(t) \text{ V} \qquad R_1 = R_2 = 1 \, \Omega$$

$$i_C(0^+) = 0 \text{ A} \qquad L = \tfrac{1}{2} \text{ H}$$

$$i_C'(0^+) = 10 \text{ A/s} \qquad C = \tfrac{1}{2} \text{ F}$$

**FIGURE 10.15**  A.
B.

The system function is found to be

$$\frac{\mathbf{I}_C}{\mathbf{V}_s} = 2 \frac{s - 0}{[s - (-2 - j2)][s - (-2 + j2)]} = 2 \frac{s}{s^2 + 4s + 8}$$

Evaluated at $s = -1$, it becomes

$$\frac{\mathbf{I}_C}{\mathbf{V}_s} = 2 \frac{-1}{+1 - 4 + 8} = -\frac{2}{5}$$

The forced response is, therefore, $i_{C_F}(t) = -2e^{-t}$, $t > 0$. The form of the natural response can be found from the pole locations of the system function; that is, $s = -2 \pm j2$:

$$i_{C_N}(t) = Ae^{-2t} \cos 2t + Be^{-2t} \sin 2t$$

The complete response, then, is

$$i_C(t) = i_{C_F}(t) + i_{C_N}(t) = -2e^{-t} + Ae^{-2t} \cos 2t + Be^{-2t} \sin 2t \text{ A}, \qquad t > 0$$

In order to find the constants $A$ and $B$, we must apply the two initial conditions

$$i_C(0^+) = 0 = -2 + A$$
$$i_C'(0^+) = 10 = 2 - 2A + 2B$$

from which we see that $A = 2$ and $B = 6$. The complete response is, therefore,

$$i_C(t) = -2e^{-t} + 2e^{-2t} \cos 2t + 6e^{-2t} \sin 2t \text{ A}, \qquad t > 0$$

Let's repeat the circuit problem of Example 10.7 using Laplace transform methods rather than phasors.

---

**Example 10.8***    *The Complete Response Using Laplace Transforms and the System Function*

Again, consider the circuit shown in Figure 10.15. Note that the circuit is initially at rest. It is, therefore, correct to say that

$$I_C(s) = H(s)V_s(s)$$

where the input and output variables are in Laplace transform form. By using the known transform of $v_s(t)$, we can write

$$I_C(s) = \frac{2s}{s^2 + 4s + 8} \cdot \frac{5}{s + 1}$$

Partial fraction expansion is needed to change $I_C(s)$ to a more recognizable form:

$$\frac{2s}{s^2 + 4s + 8} \cdot \frac{5}{s + 1} = \frac{K}{s + 1} + \frac{Cs + D}{s^2 + 4s + 8}$$

$$K = \left( \frac{2s}{s^2 + 4s + 8} \bigg|_{s=-1} \right)(5) = -2 \tag{10.17}$$

You should verify that, by the method of undetermined coefficients, $C = 2$ and $D = 16$. Also, show that

$$i_C(t) = -2e^{-t} + 2e^{-2t} \cos 2t + 6e^{-2t} \sin 2t \text{ A}, \qquad t > 0$$

---

We see that the phasor and Laplace transform approaches yield the same answers—just as they should—despite the fact that our use of the system function was very different in each example. Or was it? To see the similarities of these two interpretations of system functions, we should separately consider the natural and forced parts of the response.

The form of the natural response of any system is governed by the location of its natural frequencies, which are the roots of the denominator polynomial of $H(s)$. These natural frequencies can be taken from the pole locations on the s plane, as in Example 10.7. They can also be the basis for a partial fraction

---

*If you have not previously studied the Laplace transform, you should skip over this example and the remainder of Section 10.8.

expansion, as in Example 10.8. These observations are straightforward and as expected. It is when the forced part of the response is considered that the connection between the two approaches seems, at first, less obvious. The critical link is to look once again at Equation 10.17, the formula for finding the residue of the expansion term corresponding to the forced response:

$$K = \left( \frac{2s}{s^2 + 4s + 8} \bigg|_{s=-1} \right)(5) = H(-1)\mathbf{I}_C$$

In other words, we find the residue corresponding to the forced response in exactly the same way as we find the response phasor; the system function is evaluated at the frequency of the excitation and then multiplied by the forcing function in phasor form.

Thus, it is true that the use of phasors and the use of Laplace transforms to represent variables in circuit analysis are not distinctly different approaches. They are two different aspects of the same general frequency-domain method. System functions arrived at by these two approaches contain identical information about the circuit they describe. It is only our interpretation and use of the system function under these two conditions that seem to vary.

If your goal in a problem is to find only the forced response to an excitation that can be characterized by a frequency $s$, then it is better to treat $H(s)$ as a ratio of phasors and to simply compute its value at the frequency of excitation. If, however, you wish to find the complete response to a suddenly applied excitation, then the Laplace transform technique will be easier. It yields both the natural and forced responses in one operation and includes the circuit's initial conditions in a particularly easy way.

## 10.9  CIRCUIT STABILITY

Great variety occurs in the patterns of the pole–zero plots that describe physically realizable systems. (A physically realizable system is one that could actually be constructed.) All pole–zero plots conform to certain basic rules, however. First, the pole–zero plot of any physical system must be symmetrical about the real axis of the $s$ plane. In other words, if a complex pole exists, the conjugate pole must also be present. The same is true for zeros. A second check that can be applied is based on the concept of system stability.

There are several ways in which to define system stability. One way is to say that a system such as an electrical circuit is **stable** if its natural response is transient. Otherwise, if the system were excited, its responses might increase in size without limit until some component was damaged causing the system to fail! This requirement cannot be met by any system that has poles with positive real parts. Such poles, which are in the right half of the $s$ plane, would lead to natural responses that became larger without bound as time increased. Left-half-plane (LHP) poles correspond to natural responses that are transient (die out). A system needs to have only one right-half-plane (RHP) pole in order to be

**unstable.** If it has no RHP poles but has simple (nonrepeated) poles on the $j\omega$ axis, it is **marginally stable** because, although the natural response is not transient, at least it does not increase with time. Figure 10.16 shows pole–zero plots for systems that are stable, unstable, and marginally stable.

Now, let's turn our attention specifically to the stability of electrical circuits. All electrical circuits that contain only passive elements are stable and will have only LHP poles. An ideal resistanceless circuit could have poles on the $j\omega$ axis and be marginally stable, but this situation can never happen in a physical system. In contrast, a circuit that includes a dependent source may or may not be stable depending on the nature of the source. For example, consider the circuit shown in Figure 10.17. The voltage transfer function of the circuit is given in Equation 10.18 (you should derive it yourself to find out how involved such calculations can be):

$$\frac{V_o}{V_i} = \frac{sR_1R_2C}{s^2[LC(R_1 + R_2) - KLR_1R_2C] + s(L + R_1R_2C) + R_1} \tag{10.18}$$

**FIGURE 10.16**   *Relationship of Stability to Pole–Zero Plots*

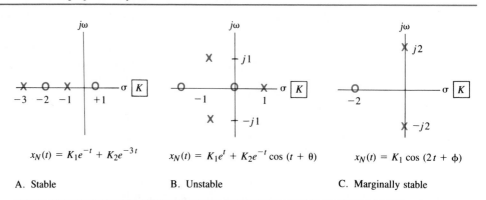

$$x_N(t) = K_1e^{-t} + K_2e^{-3t}$$

A. Stable

$$x_N(t) = K_1e^{t} + K_2e^{-t}\cos(t + \theta)$$

B. Unstable

$$x_N(t) = K_1\cos(2t + \phi)$$

C. Marginally stable

**FIGURE 10.17**   *Circuit Whose Stability Depends on Value of K*

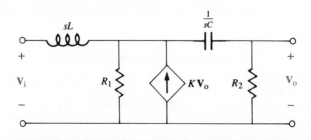

For ease of computation, let's assume element values of $R_1 = R_2 = L = C = 1$. Equation 10.18 then becomes

$$\frac{V_o}{V_i} = \frac{s}{s^2(2-K) + 2s + 1} \tag{10.19}$$

The single finite zero of the circuit is not affected by the choice of the constant $K$. The locations of the two poles, however, can change dramatically for different $K$. The circuit can be under, over, or critically damped depending on the numerical value for $K$. It can also be demonstrated that the circuit is actually unstable whenever $K > 2$. Figure 10.18 displays the pole and zero locations of the circuit for several different values of $K$.

Thus far, the discussion of stability has been focused on the natural behavior of circuits. Our definition has been one of **asymptotic stability.** A closely related definition of circuit stability states that a circuit is stable if, whenever it is excited by a forcing function that is bounded in magnitude, its response is also bounded. This is the so-called **bounded-input–bounded-output** (BIBO) definition of stability. For completeness, therefore, we should also consider forced responses and the very interesting case of exciting a circuit at one of its natural frequencies. This condition is difficult, if not impossible, to achieve in practice, but we can get close enough to it so that we should understand its theoretical and practical significance.

What happens when a circuit is excited at one of its natural frequencies? Let's assume that we have a stable electrical circuit with a pole at $s = p_k$. If the circuit is excited by a source of that same frequency, we cannot find the forced response through an evaluation of $H(p_k)$. Such a calculation would lead to the conclusion that the forced response was infinite. For instance, if

$$H(s) = \frac{(s+1)(s+3)}{s(s+2)(s+4)}$$

FIGURE 10.18   *Pole and Zero Locations for Circuit of Figure 10.17*

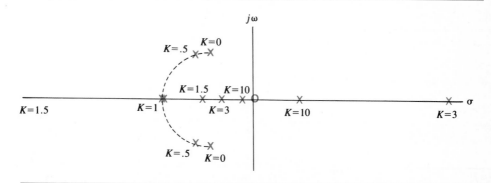

then

$$H(-2) = \frac{(-2+1)(-2+3)}{-2(-2+2)(-2+3)} = \frac{(-1)(1)}{(-2)(0)(1)} = \infty$$

This conclusion would seem to violate our assumption that the circuit is stable. The calculation on which this puzzling conclusion is based, however, is in violation of one of the basic assumptions of this chapter. A system function can be used to relate phasors as in Equation 10.11 only if none of the frequencies of the forcing function are equal to any of the natural frequencies of the circuit.

When a circuit is excited at one of its natural frequencies, Laplace transforms or classical differential equation techniques must be used to find the response. Consider, for example, the simple $RL$ circuit shown in Figure 10.19A. This first-order $RL$ circuit has only one natural frequency, and the circuit is driven by an exponential forcing function. As long as $a$, the frequency of the forcing function, is not equal to $-R/L$, the natural frequency of the circuit, the forced response is easily found to be $i_F(t) = H(a)e^{at}$, where

$$H(s) = \frac{1}{R+sL} \qquad \text{and} \qquad H(a) = \frac{1}{R+aL}$$

The total response is $i(t) = H(a)e^{at} + K_1 e^{-(R/L)t}$, where $K_1$ is a constant dependent upon the initial condition of the circuit. Two such responses (1 and 3) are shown in Figure 10.19B.

If $a = -R/L$, then $H(a) = \infty$ and the system function approach can be used only in its Laplace transform interpretation. By using either the Laplace trans-

FIGURE 10.19    *Simple RL Circuit Excited by an Exponential Source*

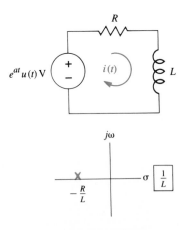

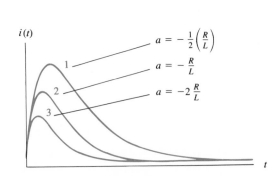

A. Example circuit with natural
   frequency of $-R/L$

B. Examples of response current

form or differential equations, the solution can be found to be

$$i(t) = \frac{1}{L} \, t e^{-(R/L)t} + K_2 e^{-(R/L)t}$$

where $K_2$ is a constant that depends on the initial condition of the circuit. This response is bounded for all $t > 0$ (see Figure 10.19B). So, although the system function "blows up" at its natural frequency, its response does not. The circuit, therefore, satisfies BIBO stability.

     This example, although restricted to a real natural frequency in the left half of the *s* plane, is entirely reflective of the stability analysis of any circuit that is excited at a natural frequency. It is not necessary to consider an excitation coinciding with an RHP pole of a circuit because we already know that any such circuit is unstable. Poles on the $j\omega$ axis, however, provide an interesting special situation that is covered in Problem 10.35 at the end of the chapter.

## 10.10   THE FORCED RESPONSE AS A FUNCTION OF *s*

Now, let's exercise our imaginations just a bit. Thus far, our task has been to compute the forced responses of circuits to specific individual excitations. For example, we might find the response of a circuit to $.02 \cos(1200t - 125°)$ or to $-20e^{-t}$. What we wish to consider now is not the response of a circuit to one particular excitation, but rather the behavior of the circuit as a function of the complex frequency *s*. The behavior of a circuit (or any type of system) is mathematically described by its system function. For instance, an input **X** and response **Y** are related by

$$\mathbf{Y} = H(s)\mathbf{X} \tag{10.20}$$

     Equation 10.20 is really two equations in one because it relates complex quantities. Separate equations can be written for magnitude and angle:

$$|\mathbf{Y}| = |H(s)| \, |\mathbf{X}| \tag{10.21a}$$

and

$$\angle \mathbf{Y} = \angle \mathbf{X} + \angle H(s) \tag{10.21b}$$

The behavior of a circuit is described analytically by $H(s)$ and Equations 10.21a and 10.21b. We would like, in addition, to develop a graphical representation or visualization of how $H(s)$ varies as a function of *s*. We do so not because the result will be used quantitatively. Instead, the exercise will give us a qualitative appreciation of the *s*-domain description of circuits. In particular, we will have a better understanding of how poles and zeros affect circuit behavior.

     The complete graphical description of $H(s)$ is a four-dimensional problem. The frequency variable *s* has a real part, $\sigma$, and an imaginary part, $\omega$. $H(s)$ also has a real and an imaginary part, but, as Equation 10.21 shows, it is more useful for us to consider its magnitude and angle. Ideally, we would like to sketch the variations in the magnitude and angle of $H(s)$ as functions of $s$. In practice, plotting the angle of $H(s)$ in this way is quite difficult because it can have both positive and negative values. We will be satisfied with plotting $|H(s)|$, which is always positive.

Think of the $s$ plane as being laid out on a horizontal surface such as the floor or a table top and with its two axes drawn at right angles to each other. This plane will be our zero reference or datum for the purposes of plotting magnitude. The magnitude of a system function can be represented as a contoured surface that rises above the reference plane. At each point $s$, the magnitude of $H(s)$ is equal to the vertical distance from the reference plane to the contoured surface. This idea may be easier to visualize if we look at a specific example.

Consider the circuit shown in Figure 10.20A. Its voltage transfer function is, in general terms,

$$H(s) = \frac{V_o}{V_i} = \frac{sR/L}{s^2 + (sR/L) + (1/LC)}$$

For the specified element values, this equation becomes

$$H(s) = 2\,\frac{s}{s^2 + 2s + 26}$$

$H(s)$ has one finite zero at $s = 0$ and poles at $s = -1 \pm j5$.

**FIGURE 10.20**   $|V_o/V_i|$ *as a Function of Complex Frequency s*

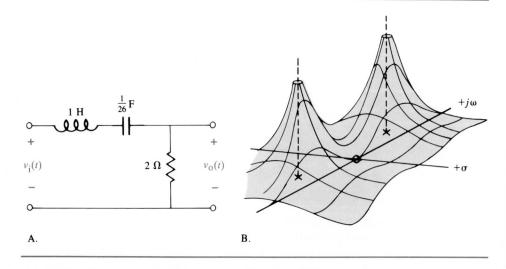

A.                                                                    B.

$|H(s)|$ is shown plotted over the $s$ plane in Figure 10.20B. Because the figure is a three-dimensional graph drawn on a two-dimensional surface, a little imagination is required to interpret it. Even more imagination is needed to sketch it! Because this process of imagination is rather abstract, you may benefit from a physical analogy or model.

Think of the contoured surface that makes up the graph as being a thin, perfectly flexible sheet. We start by laying the sheet out flat on the $s$ plane. At locations of $s$ corresponding to zeros of the system function, the sheet is pinned down to the $s$ plane. At locations of poles, the sheet is propped up by long rods, much like a tent being supported by its poles. This physical model cannot be a perfect analog of the magnitude function because the rods supporting the sheet must be finite in length even though the magnitude is infinite at the location of a pole. Nonetheless, the model is useful in helping us to visualize the terrain of peaks and valleys that make up such a plot.

Thus, we have a visual and intuitively appealing description of the behavior of a system. We see that a system responds selectively to excitations at different frequencies. The forced response will be large for excitations near a natural frequency (pole) and small for excitations near a zero of the system. In between these two extreme conditions, there will be a continuous gradation of response size.

Before we proceed, we should note that the plot of Figure 10.20B accounts for all of the poles and zeros of the circuit, not just for those whose locations are finite. Certainly, the two complex conjugate poles are very dramatically displayed. The zero at the origin can also be seen. Finally, even the zero that exists at infinity is accounted for by the fact that the plot asymptotically declines toward the reference plane as $|s|$ approaches infinity in any direction.

Be cautious in viewing plots such as Figure 10.20B. They show, quite correctly, that the magnitude of the circuit's forced response becomes arbitrarily large as $s$ approaches the location of a pole. As noted in Section 10.9, however, our interpretation of system functions as ratios of phasors breaks down when systems are excited at their natural frequencies (poles). For this case, the natural and forced responses cannot be obviously separated. The total response of any stable circuit excited at one of its natural frequencies will always be finite.

## 10.11  SUMMARY

This chapter has expanded our understanding of phasors beyond the sinusoidal steady state to include the entire class of exponentially damped sinusoidal functions. These functions can be conveniently characterized by the complex frequency $s = \sigma + j\omega$. This more general class of phasors has the same appearance as in the purely sinusoidal case and is manipulated and interpreted in the same way. Impedances, which are ratios of phasor voltages to phasor currents, are functions of the complex frequency variable $s$.

The generalized phasor approach can be used to find the forced response of a circuit to any exponentially damped sinusoid. Its greatest usefulness, however, lies in the clear way in which it leads to the concepts of system functions and the $s$ plane. System functions are concise descriptions of circuit behavior. For circuits such as those in this book, system functions are ratios of polynomials in $s$ that relate response variables to forcing functions. Of particular interest are those values of $s$ for which the system function has an infinite magnitude. They are called poles of the system function. Its zeros are those values of $s$ for which the system function becomes zero.

Both the poles and zeros of a system function can be plotted on the $s$ plane. The poles of a system function are identical to the system's natural frequencies. Poles in the right half of the $s$ plane correspond to circuit instability. A stable circuit must have all of its poles in the left half of the $s$ plane. In addition to the pole and zero frequencies of a circuit, the frequencies of forcing functions can also be located on the $s$ plane. Thus, just as the system function is a compact analytical description of circuit behavior, the $s$ plane can be used as a convenient graphical summary of the information known about a circuit and the signal that excites it. Either the system function or its pole–zero plot can be used to characterize the behavior of the circuit with respect to both its natural and its forced responses.

It was not just by coincidence that the symbol $s$ is chosen to represent complex frequencies. The same symbol was used in Chapter 9 for the independent variable of the Laplace transform. These two seemingly different methods are actually just different ways of looking at the same general frequency-domain approach. The Laplace transform is the more mathematically rigorous and general method of the two. However, the phasor approach, when appropriate, can more quickly lead to a solution and, arguably, give a more intuitive insight into the behavior of a circuit. The phasor approach is perhaps the more appropriate when we are interested in only the forced response of a linear circuit. Laplace transforms are better tailored to problems that include initial conditions and natural responses. As with other analytical methods introduced in this book, choosing between the phasor and Laplace approaches is a matter of personal preference and the requirements of particular circuit problems.

## ■ PROBLEMS

10.1 Fill in the steps necessary to complete the proof given in Section 10.2 that states that if a linear, time-invariant circuit has an input of $Ae^{\sigma t} \cos (\omega t + \phi)$, all forced responses in the circuit will be of the form $Be^{\sigma t} \cos (\omega t + \theta)$.

10.2 Determine $\sigma$ and $\omega$ for each of the functions shown in Figure P10.2. Write the analytical expression for each function.

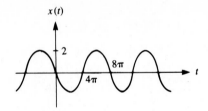

A.

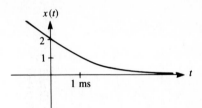

B.

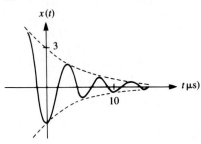

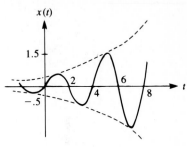

C.

D.

**10.3** Sketch the functions of time corresponding to the following given values of $\sigma$ and $\omega$ and of $f(0)$. Assume cosine forms with zero phase shifts. The amplitudes may be positive or negative.

**a.**  $\sigma = 0$,  $\omega = 2000$,  $f(0) = -2$

**b.**  $\sigma = .5$,  $\omega = 0$,  $f(0) = 10$

**c.**  $\sigma = -1$,  $\omega = 10$,  $f(0) = 1$

**d.**  $\sigma = -2$,  $\omega = 10\pi$,  $f(0) = -1$

**e.**  $\sigma = -1$,  $\omega = 20$,  $f(0) = 5$

**10.4** Demonstrate that an exponentially damped sinusoid can be written in the complex exponential form of Equation 10.5.

**10.5** Write general analytical expressions for the time functions whose complex frequencies are shown plotted on the $s$ plane in Figure P10.5.

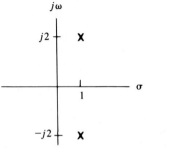

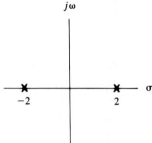

**FIGURE P10.5**   A.

B.

*(continues)*

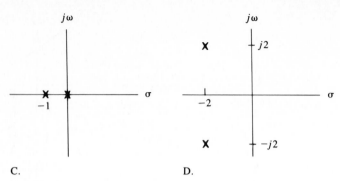

C.                           D.

**FIGURE P10.5** (*continued*)

**10.6** Refer to Figure 10.3. Prove that the complex exponential input–output pair is valid.

**10.7** Use complex exponential variable representations to find the forced responses for each of the following differential equations:

**a.** $\dfrac{dy}{dt} + 2y = a(t)$, $\quad a(t) = 2e^{-t}\sin t$

**b.** $\dfrac{d^2g}{dt^2} + 5\dfrac{dg}{dt} + 4g = 3h(t)$, $\quad h(t) = e^{-t}\cos 2t$

**c.** $\dfrac{d^2x}{dt^2} + 2\dfrac{dx}{dt} + 10x = \dfrac{dy}{dt} + 4y$, $\quad y(t) = 4e^t\sin(3t + 30°)$

**d.** $\dfrac{d^2v}{dt^2} - 2v = \dfrac{d^3i}{dt^3} + \dfrac{di}{dt} - 2i$, $\quad i(t) = -5\cos 4t$

**10.8** Use complex exponential notation to find the forced response $i_R(t)$ in the circuit of Figure P10.8.

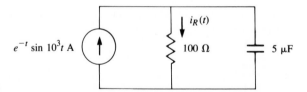

**FIGURE P10.8**

**10.9** Use complex exponential notation to find the forced response $v_o(t)$ in the circuit of Figure P10.9.

**10.10** Prove the generalized impedance relationships $Z_R$ and $Z_L$ of Table 10.1.

**10.11** Find the corresponding time functions given the following values:

**a.** $F = 1.5e^{j75°}$, $\quad s = 10 + j20$

**b.** $X = -8$, $\quad s = 5$

**c.** $Y = .3e^{-j130°}$, $\quad s = -1 + j100$

**d.** $V = 120e^{-j45°}$, $\quad s = j100$

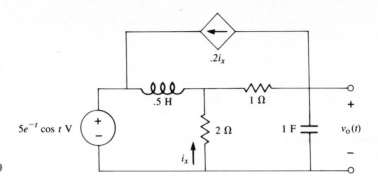

**FIGURE P10.9**

**10.12** Find the corresponding phasors given the following:
    **a.** $x(t) = 10e^{2t} \cos(5t + 37°)$
    **b.** $g(t) = 13e^{-6t}$
    **c.** $v(t) = .05e^{-10t} \sin(20t - 60°)$
    **d.** $y(t) = .05 \cos(100t - 150°)$

**10.13** Use generalized phasor and impedance techniques to find the current $i_2(t)$ in the circuit of Figure P10.13.

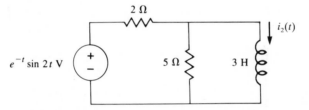

**FIGURE P10.13**

**10.14** Repeat the following problems by making immediate use of phasors:
    **a.** Problem 10.8
    **b.** Problem 10.9

**10.15** Use phasors to find the voltage $v_o(t)$ in the circuit of Figure P10.15.

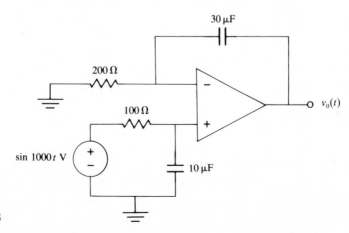

**FIGURE P10.15**

**10.16** Use phasors and the principle of superposition to determine the voltage $v_o(t)$ in the circuit of Figure P10.16.

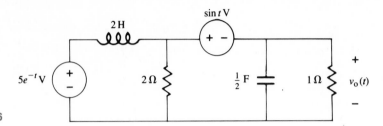

**FIGURE P10.16**

**10.17** For the circuit shown in Figure P10.17, find the system functions $\mathbf{I}_1/\mathbf{V}_s$, $\mathbf{I}_2/\mathbf{V}_s$, and $\mathbf{V}_o/\mathbf{V}_s$. What are the similarities in their forms?

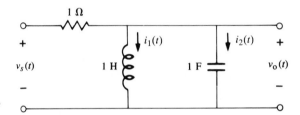

**FIGURE P10.17**

**10.18** Find the transfer function $\mathbf{V}_o/\mathbf{V}_s$ in the circuit of Figure P10.18.

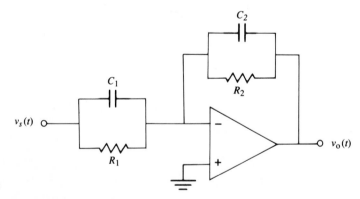

**FIGURE P10.18**

**10.19** Find the response of the circuit of Problem 10.18 if $C_1 = 10\ \mu\mathrm{F}$, $R_1 = 1\ \mathrm{k}\Omega$, $C_2 = 20\ \mu\mathrm{F}$, $R_2 = 250\ \Omega$, and $v_s(t) = \sin 100t$ V.

**10.20** Find the system function $\mathbf{V}_2/\mathbf{V}_1$ in the circuit of Figure P10.20.

**10.21** Find the driving point impedances of the circuits shown in Figure P10.21.

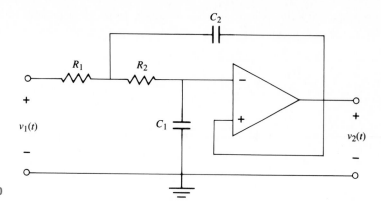

**FIGURE P10.20**

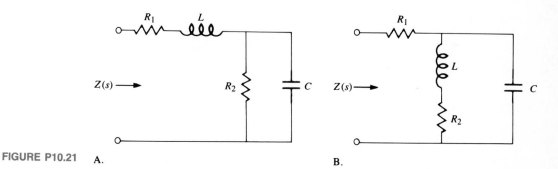

**FIGURE P10.21**    A.                                                    B.

**10.22** Draw pole–zero plots for the following system functions:

a.   $H(s) = 5 \dfrac{s+2}{s^2 + 2s + 10}$

b.   $G(s) = \dfrac{2s - 10}{(s+1)(s^2 + 4s + 3)}$

c.   $A(s) = -3 \dfrac{s^3 + 9s}{(s+1)(s+2)(s-3)}$

d.   $Y(s) = \dfrac{s^2}{s^2 + 2s + 101}$

**10.23** a.   Find the system function $\mathbf{I}_C/\mathbf{V}_s$ for the circuit shown in Figure 10.15A. Your answer should be in terms of the element variable names $R_1$, $R_2$, $L$, and $C$.

b.   For the specified element values, verify that the system function has a pole–zero plot as shown in Figure 10.15B.

**10.24** Evaluate each of the system functions of Problem 10.22 for $s = -1 + j$. Do so both analytically and graphically.

**10.25** Find the system functions whose pole–zero plots are as shown in Figure P10.25. Discuss the stability of each.

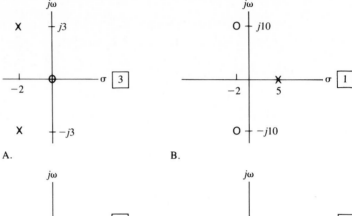

A.                                      B.

C.                                      D.

**FIGURE P10.25**

**10.26** Determine the natural frequencies of the following:
  **a.** Circuit of Problem 10.13
  **b.** Circuit of Problem 10.15
  **c.** Circuit of Problem 10.17
  **d.** System functions of Problem 10.22

**10.27** A system's pole–zero plot is shown in Figure P10.27. Redraw it to a larger scale and use it to graphically determine the system's forced responses to the following input functions:
  **a.** $v_i(t) = 10 \sin (2t - 27°)$
  **b.** $v_i(t) = 32e^{-3t} \cos (4t + 45°)$
  **c.** $v_i(t) = 2.3e^{5t}$

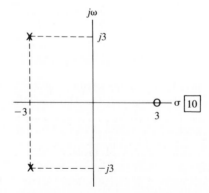

**FIGURE P10.27**

**10.28** Find the total response of the circuit whose system function is

$$H(s) = \frac{\mathbf{V}_o}{\mathbf{V}_s} = \frac{s+1}{(s+2)(s+3)}$$

when $v_s(t) = 5e^{-t}u(t)$, $v_o(0^+) = 0$, and $v'_o(0^+) = 5$. What general conclusion do you draw?

**10.29** Find the system function $\mathbf{I}_o/\mathbf{I}_s$ in the circuit of Figure P10.29, draw the corresponding pole–zero plot, and, for an input of $i_s(t) = e^{-3t}$ A, determine the forced response both analytically and graphically.

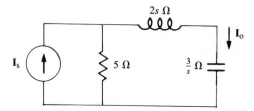

**FIGURE P10.29**

**10.30** As outlined in Section 10.7, prove that if a forcing function to a system is characterized by a conjugate pair of complex frequencies, the system function need be evaluated only at the frequency with a positive imaginary part in order to find the response. That is, if $H(\sigma + j\omega) = |H|e^{j\theta}$ and the input is $x(t) = Ae^{\sigma t}\cos \omega t$, then the response will be $y(t) = |H|Ae^{\sigma t}\cos(\omega t + \theta)$.

**10.31** An incomplete pole–zero plot is shown in Figure P10.31. It is also known that $H(-1) = 10$. Complete the plot and write the corresponding system function. Is the system stable?

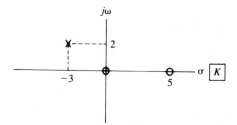

**FIGURE P10.31**

**10.32** Stable systems must have no RHP poles, but the same is not necessarily true of zeros.

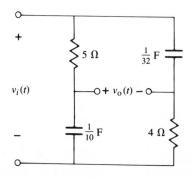

**FIGURE P10.32**

Determine the system function for the bridge circuit shown in Figure P10.32 and draw its pole–zero plot.

**10.33** Determine by analytical calculation the complete response of the circuit shown in Figure P10.33.

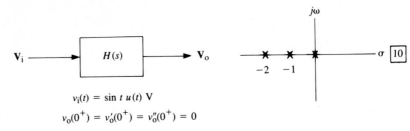

$$v_i(t) = \sin t\, u(t) \text{ V}$$

**FIGURE P10.33**         $$v_o(0^+) = v_o'(0^+) = v_o''(0^+) = 0$$

**10.34 a.** Determine the system function for the *RC* circuit shown in Figure P10.34.

  **b.** Using the techniques of this chapter, find the total response to $e^{-990t}u(t)$ when the circuit is initially at rest.

  **c.** Use the techniques of earlier chapters (either differential equations or Laplace transforms) to find the response to $e^{-1000t}u(t)$.

  **d.** On one graph, sketch the responses from parts b and c.

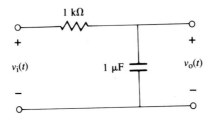

**FIGURE P10.34**

**10.35** Demonstrate that for the system whose pole–zero plot is shown in Figure P10.35, the response to $\sin \omega t$ is bounded if $\omega \neq \omega_k$ and unbounded if $\omega = \omega_k$. You may use Laplace transform or differential equation techniques.

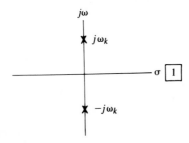

**FIGURE P10.35**

**10.36** A system function is $10/(s + 10)$. Find and plot the system's forced responses when the input is of the form $e^{st}$ and $s = -1, -5, -8, -9,$ and $-9.9$.

**10.37** Find the total response of the system in Problem 10.36 for each of the separate forcing functions. Assume that the response variable has an initial value of zero. Sketch the results.

**10.38** Verify the system function of Equation 10.19.

**10.39** For what ranges of $K$ will the system described by Equation 10.19 be stable and overdamped, underdamped, and critically damped?

**10.40** The circuit shown in Figure P10.40 is a Wien bridge oscillator. Notice that it has an output, $v_o$, but *no input*. Find the characteristic equation for $v_o$ and from it plot the circuit's natural frequencies on the $s$ plane. Based on the results, comment on the fact that the output oscillates (is sinusoidal) in the absence of an input.

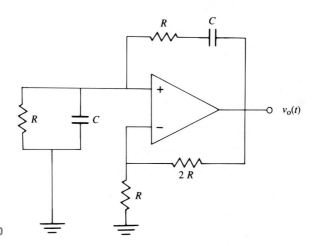

**FIGURE P10.40**

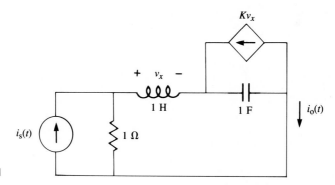

**FIGURE P10.41**

**10.41** Find the system function $\mathbf{I}_o/\mathbf{I}_s$ for the circuit shown in Figure P10.41. Plot its poles for various values of $K$. Find a value of $K$ for which the system is

   **a.** Overdamped

   **b.** Underdamped

   **c.** Critically damped

   **d.** Unstable

# Frequency Response

## 11.1   INTRODUCTION

This chapter introduces one of the most interesting and useful concepts in electrical engineering—the frequency response characteristics of systems. The application of frequency response techniques is not limited to electrical engineering problems. These techniques are equally useful in fields as diverse as acoustics and geophysics. Because the topic of this book is electrical circuit analysis, however, the examples will naturally emphasize circuit applications.

The frequency response characteristics of a circuit are a complete description (either analytical or graphical) of its sinusoidal steady-state behavior as a function of frequency. Remember from Chapter 8 that when a circuit is sinusoidally excited, every voltage and current in the circuit is a sinusoid of the same frequency as the excitation. Each signal variable is uniquely described by its magnitude and phase angle. The magnitudes and phases are functions not only of the elements and topology of the circuit, but also of the frequency of the sinusoidal excitation.

Once the general frequency response characteristics of a circuit are known, the particular steady-state response to any specific sinusoidal input can be found by evaluating the characteristics at the frequency of excitation. The importance of knowing the frequency response behavior of a circuit, however, goes well beyond merely being able to find the responses to particular sinusoidal excitations. It can be shown that the responses of a stable circuit to many forcing functions can be calculated mathematically from the circuit's frequency response characteristics. That is, a circuit's frequency response characteristics contain enough information to completely describe its general behavior.

The frequency response characteristics of circuits are often used to broadly categorize their functional usefulness. The operations of many electronic circuits depend upon their selective responses to sinusoids of different frequencies. This filtering capability is what allows, for instance, the bass/treble knob of a simple radio to affect the tonal quality of the music produced. Other filters have been designed to block stray 60 Hz noise from power and lighting sources to prevent the noise from contaminating other electrical signals.

Finding the frequency response curves of a circuit is often the main goal of an analysis problem. In a complementary way, the design of electrical circuits often begins with a set of desired frequency response curves. That is, an engineer may be given a set of response curves and asked to design a circuit that meets the specifications. Thus, an understanding of frequency response characteristics is important in both the analysis and the design of electrical circuits.

In this chapter, several methods for determining and displaying the frequency response characteristics of electrical circuits will be developed. The first of these methods will be an entirely analytical computation, and the results will be displayed graphically. Following will be a graphical calculation technique that allows the practiced engineer to inspect the pole–zero plot of a system and to estimate from it the corresponding frequency response. Finally, Bode diagrams will be introduced. Bode techniques allow us to quickly, and with reasonable accuracy, display a system's frequency response characteristics over a wide range of frequencies without having to perform any involved calculations.

## 11.2 FREQUENCY RESPONSE BASICS

Usually, we find a circuit's frequency response characteristics by first finding its system function. System functions relate response variables to input variables and can be interpreted either as ratios of Laplace transforms or as ratios of phasors. The latter interpretation will be used here because our interest for the moment is limited to steady-state behavior. We must assume that the circuits under study are stable so that the steady state actually exists. Under this assumption, if the input forcing function can be characterized by a complex frequency $s$, then the response $\mathbf{Y}$ can be found as follows:

$$\mathbf{Y} = H(s)\mathbf{X} \qquad \text{(11.1)}$$

where $\mathbf{Y}$, $H(s)$, and $\mathbf{X}$ are all complex numbers. $\mathbf{X}$ and $\mathbf{Y}$ are phasor representations of the input and output variables, and $H(s)$ is the system function evaluated at the frequency of excitation. By the rules of complex number arithmetic, the magnitude and phase of $\mathbf{Y}$ are

$$|\mathbf{Y}| = |H(s)|\,|\mathbf{X}| \qquad \text{(11.2a)}$$
$$\angle\mathbf{Y} = \angle H(s) + \angle\mathbf{X} \qquad \text{(11.2b)}$$

where, as in earlier chapters, $\angle$ is a symbol to denote the angle of a complex term.

You may wish to reread the beginning of Section 10.4 as a review of the use of Equations 11.1 and 11.2. That discussion illustrates two points that will motivate our work in this chapter. First, the value of a system function depends on the frequency of the input signal but not on its magnitude or phase. It is true that to find the response to a specific input, we need to know the input's magnitude and phase, but the behavior of the circuit itself depends only on the frequency at which it is excited. Thus, $H(s)$ alone is enough to completely describe the circuit's behavior. Second, unless a given circuit is to be operated at only one frequency, it is necessary to evaluate $H(s)$ for all possible values of input frequency including, apparently, the entire $s$ plane. Fortunately, most circuits that an engineer is likely to encounter can be described more simply. It can be demonstrated mathematically that a stable circuit (no RHP poles) is completely described if its steady-state behavior is known for all sinusoidal frequencies. This fact considerably narrows the scope of the problem because it limits our attention to the $j\omega$ axis of the $s$ plane.

The sinusoidal steady state is a special-case use of the system function. Mathematically, we substitute $s = j\omega$ in Equations 11.1 and 11.2 and arrive at

$$\mathbf{Y} = H(j\omega)\mathbf{X} \tag{11.3}$$

where

$$|\mathbf{Y}| = |H(j\omega)|\,|\mathbf{X}| \tag{11.4a}$$

and

$$\angle\mathbf{Y} = \angle H(j\omega) + \angle\mathbf{X} \tag{11.4b}$$

Equations 11.3 and 11.4 describe the frequency response behavior of a circuit. By officially accepted definition, the frequency response characteristics of a single-input/single-output circuit are a complete description of the frequency-dependent magnitude and phase differences between input and output when the circuit is operating in the sinusoidal steady state.*

Taken together, $|H(j\omega)|$ and $\angle H(j\omega)$ are an analytical description of the frequency response behavior of a circuit. Very often, they are plotted as functions of $\omega$ in order to have a graphical representation of the information. The main purpose of this chapter is to present various ways in which this graphing can be done.

Because $H(j\omega)$ is a special case of $H(s)$, frequency response plots can be related to the three-dimensional plots of $|H(s)|$ that were discussed qualitatively in Section 10.10. To find a plot of $|H(j\omega)|$ versus $\omega$, we can simply "slice" the $|H(s)|$ graph vertically along the $j\omega$ axis (i.e., $\sigma = 0$) and look at the resultant cross section. This cross section is a plot of $|H(j\omega)|$. Figure 11.1 shows the graphical relationship between $|H(s)|$ and $|H(j\omega)|$ for the circuit defined in Figure 10.20. The cross section is shown as the shaded area of Figure 11.1A, and the corresponding $|H(j\omega)|$ curve is plotted in the normal manner in Figure 11.1B. A plot of $\angle H(j\omega)$ for the same circuit is shown in Figure 11.1C. The three-dimensional plot of $H(s)$ is shown, not for its analytical use, but instead, to emphasize again that the frequency response characteristics of a system are a limited but special part of its more general description.

The fact that the sinusoidal steady-state frequency response curves in Figures 11.1B and 11.1C are plotted for negative as well as positive values of $\omega$ is often puzzling. After all, who ever heard of a real sinusoid with a negative radian frequency? While it is true that $+\omega$ is the frequency of the real transcendental function $\cos \omega t$, you should remember from Chapter 10 that the same function expressed as a sum of complex exponentials requires that the two frequencies $+j\omega$ and $-j\omega$ be used. Negative frequencies, then, are a necessary mathematical result

---

* *IEEE Standard Dictionary of Electrical and Electronics Terms*, 2d ed., IEEE, Inc., 1978.

FIGURE 11.1    *Frequency Response Curves as a Subset of the More General s Plane Description*

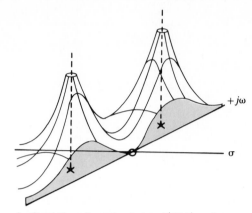

A. Quasi three-dimensional plot of $|H(s)|$ sectioned
vertically along $j\omega$ axis

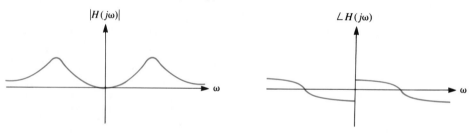

B. Corresponding magnitude function          C. Corresponding phase function

of using the complex frequency $s$ plane in our analysis. At this point, we need not be overly concerned with these mathematical formalities. It is only the positive values of $\omega$ that have physical significance for us, so we will plot frequency response curves as functions of positive $\omega$ only. As can be seen in the curves of Figures 11.1B and 11.1C, there is no loss of information because the magnitude is an even function of $\omega$ and the phase is an odd function of $\omega$. That is, $|H(-j\omega)| = |H(j\omega)|$ and $\angle H(-j\omega) = -\angle H(j\omega)$.

In the remainder of this chapter, several ways in which to calculate and display frequency response information will be developed.

## 11.3  LINEAR FREQUENCY RESPONSE PLOTS

In this section, we will look at several examples of frequency response characteristics found by direct analytical calculation. The results will be plotted on linear scales as functions of $\omega$. Frequency response data are often plotted as functions of $f$ (Hz) rather than $\omega$ (rad/s). This choice involves only a scaling of the frequency axis and does not change the shape of the curves. We will plot the curves as functions of $\omega$ because this frequency variable is more directly related to the $s$ plane.

**Example 11.1**   *Low-Pass RC Filter*

Shown in Figure 11.2 is a simple *RC* circuit along with its voltage transfer function and the corresponding pole–zero plot. The system function is evaluated for $s = j\omega$, and its magnitude and phase angle are calculated:

$$H(j\omega) = \frac{1/RC}{j\omega + (1/RC)} \tag{11.5}$$

$$|H(j\omega)| = \frac{1/RC}{\sqrt{\omega^2 + (1/RC)^2}} \tag{11.6a}$$

$$\angle H(j\omega) = -\tan^{-1}\omega RC \tag{11.6b}$$

The magnitude and phase curves are shown in Figure 11.3. These curves are the plotted results of numerical calculation based on Equations 11.6a and 11.6b, but their basic features can be determined by a less complete inspection of the analytical expressions. $|H(j\omega)|$ is maximum with respect to $\omega$ when $\omega = 0$. It monotonically decreases as $\omega$ increases toward $+\infty$, asymptotically approaching zero for very high frequencies.

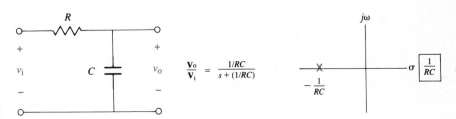

**FIGURE 11.2**

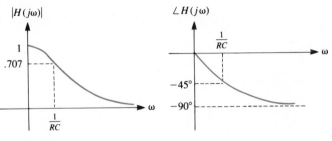

**FIGURE 11.3**   A. Magnitude       B. Phase

*(continues)*

**Example 11.1**   *Continued*

The phase, $\angle H(j\omega)$, is also easy to plot for this simple *RC* circuit. At $\omega = 0$, $\angle H(j0) = -\tan^{-1} 0 = 0°$. As $\omega$ approaches $+\infty$, the angle asymptotically approaches $-90°$.

The magnitude plot for a circuit is often used to characterize its electrical filtering properties. In general terminology, a **filter** is a device that selectively operates upon its inputs. Some inputs are allowed to pass through the filter, while others are blocked. An air filter, for instance, should pass gases of various kinds while trapping dust and other particulates. The word *filter* has a more specific meaning in electrical engineering applications. To an electrical engineer, a filter is a circuit that acts selectively upon its input signals on the basis of frequency. Signals at some frequencies pass through the filter with little change, while others may be greatly reduced in amplitude, or **attenuated.**

From Figure 11.3, we see that for the simple *RC* transfer function of Example 11.1, output signal amplitudes are nearly as large as input amplitudes when the frequency is low. For high frequencies, however, the output amplitude is small compared to the input. Thus, we say that this *RC* circuit is a **low-pass filter.** The boundary between the frequency range that is "passed through" and the range that is attenuated is not sharp. However, by convention and for mathematical reasons that will be discussed later, the frequency at which the magnitude falls to $1/\sqrt{2}$ or .707 of its maximum value is considered to be a significant landmark in magnitude curves like the one in Figure 11.3. For this circuit, the frequency that corresponds to a magnitude of $1/\sqrt{2}$ is easily seen to be $\omega = 1/RC$ and is known as the **cutoff frequency** for the low-pass filter. It is sometimes called the **half-power point.** Power delivered to a resistive load is $V^2/R$. If $V$ is decreased to .707 of its maximum value or $V_{max}/\sqrt{2}$, then the power delivered will be $P = (V_{max}/\sqrt{2})^2/R = P_{max}/2$.

**Example 11.2**   *High-Pass RC Filter*

Consider the circuit shown in Figure 11.4. This circuit is the same simple *RC* circuit as in Example 11.1, but now the output voltage is taken across the resistor. The magnitude and phase angle functions are as follows:

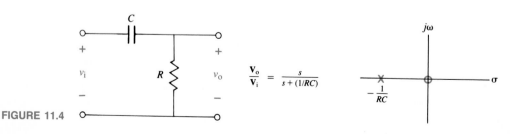

$$\frac{\mathbf{V_o}}{\mathbf{V_i}} = \frac{s}{s + (1/RC)}$$

**FIGURE 11.4**

$$|H(j\omega)| = \frac{\omega}{\sqrt{\omega^2 + (1/RC)^2}} \ , \qquad \omega > 0 \qquad\qquad (11.7\text{a})$$

$$\angle H(j\omega) = 90° - \tan^{-1}\omega RC \ , \qquad \omega > 0 \qquad\qquad (11.7\text{b})$$

The magnitude and phase angle functions are plotted as functions of $\omega$ in Figure 11.5. From the analytical expression for magnitude, it is easy to see that $|H(j0)| = 0$ and that $|H(j\omega)|$ asymptotically approaches a value of 1 for high frequencies. Less obvious but also true is the fact that this function monotonically increases as $\omega$ approaches infinity. A few intermediate points, easily determined by calculator or computer, indicate the exact shape of this magnitude function. The most interesting intermediate frequency is that of $\omega = 1/RC$. For this frequency, $|H(j\omega)|$ is $1/\sqrt{2}$ of its maximum (high-frequency) value.

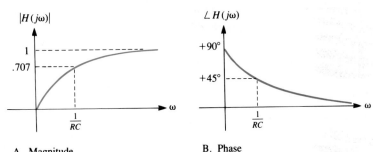

**FIGURE 11.5**    A. Magnitude          B. Phase

An electrical circuit with a magnitude function such as the one shown in Figure 11.5 is called a **high-pass filter.** We call $\omega = 1/RC$ the low-frequency cutoff of the filter.

Examples 11.1 and 11.2 show how very different the frequency response characteristics can be for system functions within one circuit. For this simple $RC$ circuit, the response characteristics can be either high-pass or low-pass depending on which variable is chosen as the output. Still, because the two examples are from the same circuit, there are similarities in their response curves because the denominator polynomials (and, therefore, the poles) of the two system functions are identical. Thus, for instance, the cutoff frequencies are identical in the two circuits, although we attach very different significances to them.

**Example 11.3**    *Bandpass RLC Filter*

As a final example of the purely analytical approach to frequency response determination, consider the input impedance of the parallel $RLC$ circuit

(continues)

**Example 11.3**  *Continued*

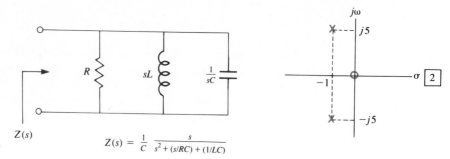

$$Z(s) = \frac{1}{C} \frac{s}{s^2 + (s/RC) + (1/LC)}$$

**FIGURE 11.6**

shown in Figure 11.6. The locations of the poles are for $C = 1/2\,\mathrm{F}$, $R = 1\,\Omega$, and $L = 1/13\,\mathrm{H}$. The magnitude and phase functions, plotted in Figure 11.7, are as follows:

$$|Z(j\omega)| = \frac{\omega/C}{\sqrt{[(1/LC) - \omega^2]^2 + (\omega/RC)^2}}, \qquad \omega > 0 \tag{11.8a}$$

$$\angle Z(j\omega) = +90° - \tan^{-1} \frac{\omega/RC}{(1/LC) - \omega^2}, \qquad \omega > 0 \tag{11.8b}$$

Plotting the magnitude and phase of this system function is more complicated than for Examples 11.1 and 11.2. Even so, a calculator can be used to obtain accurate plots for given values of $R$, $L$, and $C$. The shapes of the response curves can vary quite a bit depending on the particular element values chosen. For the curves in Figure 11.7, we selected values of $R$, $L$, and $C$ that cause the system function to have complex conjugate poles.

Note that the magnitude function equals zero for $\omega = 0$ and that it approaches zero for high frequencies. Elsewhere, it must be positive. One way to find the frequency at which the magnitude of the impedance is maximum is to look for the minimum magnitude of the corresponding

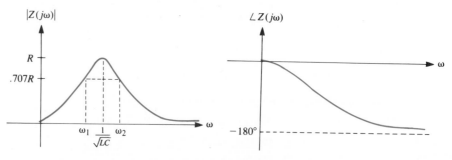

**FIGURE 11.7**   A. Magnitude                                        B. Phase

admittance:

$$Y(j\omega) = \frac{1}{Z(j\omega)} = \frac{1}{R} + \frac{1}{j\omega L} + j\omega C = \frac{1}{R} + j\left(\omega C - \frac{1}{\omega L}\right)$$

Because the real part of $Y(j\omega)$ is frequency independent, we can minimize $|Y(j\omega)|$ by minimizing the imaginary part. The imaginary part actually vanishes for $\omega = 1/\sqrt{LC}$. At this frequency, $Y(j\omega) = 1/R$ and $Z(j\omega) = R$. It is desirable for you to develop insight of the sort that led to this conclusion, but it is not necessary in order to plot frequency response curves. Curves like those in Figure 11.7 are actually determined and plotted by direct analytical calculation.

---

The nature of the frequency response characteristics of the second-order circuit in Example 11.3 is quite different from that of the first-order circuit in Examples 11.1 and 11.2. In the second-order circuit, signals of both extremely high and extremely low frequencies are greatly attenuated. Signals over a range or band of frequencies between these extremes are passed through with little loss. Circuits with frequency characteristics like these are known as **bandpass filters.** In other examples, you will see that second-order filters are not always bandpass in nature. It is true, however, that higher-order filters are more capable of complicated frequency response behavior than are lower-order circuits.

We can define two cutoff frequencies for the circuit of Example 11.3. That is, there are two frequencies at which the magnitude function is $1/\sqrt{2}$ of its maximum value. Determining their exact locations is left as an exercise for you in Problem 11.23 at the end of the chapter. For now, we will simply refer to them as $\omega_1$ and $\omega_2$, as shown in Figure 11.7. We define the **bandwidth** (BW) for the circuit to be $\omega_2 - \omega_1$. For a bandpass filter, the frequency at which the magnitude function is maximum is called the **center frequency** $(\omega_0)$. The center frequency for this circuit has already been found to be $\omega_0 = 1/\sqrt{LC}$. Although its name might cause you to think otherwise, the center frequency is not halfway between the upper and lower cutoff frequencies.

A wide variety of frequency response characteristics can be found in electrical filters. Presented here are basic examples of simple low-pass, high-pass, and bandpass filters. These three types, along with a few others, make up the vast majority of commonly used filters. Once the frequency response characteristics of a circuit have been found, they can be used to determine the steady-state response to any sinusoidal forcing function.

---

**Example 11.4**   *Determining Steady-State Response from Frequency Response Curves*

Consider the hypothetical frequency response curves shown in Figure 11.8.

*(continues)*

**Example 11.4**   *Continued*

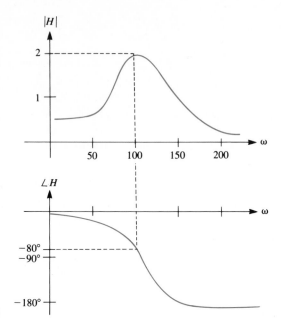

FIGURE 11.8

If the input is $x(t) = 15 \sin (100t - 45°)$, we need to evaluate the system function at $\omega = 100$ rad/s. From the curves, you can see that $|H(j100)| = 2$ and that $\angle H(j100) = -80°$. Thus, the response will be

$$y(t) = (2)(15) \sin (100t - 45° - 80°) = 30 \sin (100t - 125°)$$

You should verify the following input–output pairs:

$$x(t) = -20 \cos (50t + 60°)$$
$$y(t) = -12 \cos (50t + 40°)$$

$$x(t) = .5 \sin (150t + 45°)$$
$$y(t) = .5 \sin (150t - 125°)$$

$$x(t) = 8 \cos (200t - 80°)$$
$$y(t) = 2 \cos (200t + 100°)$$

In addition to introducing several basic filter types and terminology, the preceding examples suggest the complexity of analytical computations required for a complete frequency response plot for any but the simplest of electrical circuits. Of course, these calculations are easily done with the aid of a digital computer. However, even when a computer is available, an engineer is often in need of techniques that allow him or her to rapidly, if with somewhat less precision, arrive at frequency response plots for systems under study. These techniques provide methods for quickly gaining a general sense of the frequency-dependent characteristics of a system and for giving a convenient check on the analytical calculations that must ultimately be done in any complete analysis or design.

## *11.4  A GRAPHICAL APPROACH TO FREQUENCY RESPONSE PLOTS

Frequency response calculations for a circuit usually begin with the system function. We have already seen several standard mathematical forms for system functions. The form chosen will depend on the nature of the specific calculation to be done. When first determined from a circuit, $H(s)$ is most often in the form of a ratio of polynomials in $s$:

$$H(s) = K \frac{s^m + a_{m-1}s^{m-1} + a_{m-2}s^{m-2} + \cdots + a_1 s + a_0}{s^n + b_{n-1}s^{n-1} + b_{n-2}s^{n-2} + \cdots + b_1 s + b_0} \tag{11.9}$$

This form of $H(s)$ is also most often chosen when a circuit's frequency response is determined by analytical calculation. It was the form chosen in Section 11.3. $H(s)$ is first rewritten as a function of $j\omega$:

$$H(j\omega) = K \frac{(j\omega)^m + a_{m-1}(j\omega)^{m-1} + \cdots + a_1(j\omega) + a_0}{(j\omega)^n + b_{n-1}(j\omega)^{n-1} + \cdots + b_1(j\omega) + b_0} \tag{11.10}$$

Equation 11.10 can then be evaluated at any frequency of interest.

If the precision of an analytical calculation is not required, a graphical technique can be used to find the frequency response of a circuit. This approach depends on knowing the locations of the finite poles and zeros of the system function. These locations are seen by rewriting Equation 11.9 as follows:

$$H(s) = K \frac{(s - z_1)(s - z_2) \cdots (s - z_m)}{(s - p_1)(s - p_2) \cdots (s - p_n)} \tag{11.11}$$

where $z_1, z_2, \ldots, z_m$ and $p_1, p_2, \ldots, p_n$ are the locations of the zeros and poles, respectively. When evaluated at $s = j\omega$, Equation 11.11 becomes

$$H(j\omega) = K \frac{(j\omega - z_1)(j\omega - z_2) \cdots (j\omega - z_m)}{(j\omega - p_1)(j\omega - p_2) \cdots (j\omega - p_n)} \tag{11.12}$$

In Section 10.7, we saw how expressions such as Equation 11.12 can be interpreted graphically. Each term in parentheses is a complex number that can be treated as a vector on the $s$ plane. Each vector is drawn from a pole or zero of $H(s)$ to the sinusoidal frequency of interest on the $j\omega$ axis. Each vector has a magnitude and an angle that are dependent on $\omega$. This relationship is more clearly seen when the individual terms are expressed in polar form:

$$H(j\omega) = K \frac{M_{z_1}(\omega)e^{j\theta_{z_1}(\omega)}M_{z_2}(\omega)e^{j\theta_{z_2}(\omega)}\cdots M_{z_m}(\omega)e^{j\theta_{z_m}(\omega)}}{M_{p_1}(\omega)e^{j\theta_{p_1}(\omega)}M_{p_2}(\omega)e^{j\theta_{p_2}(\omega)}\cdots M_{p_n}(\omega)e^{j\theta_{p_n}(\omega)}} \quad (11.13)$$

$M_{z_k}(\omega)$ is the magnitude of $j\omega - z_k$, and $\theta_{z_k}(\omega)$ is its angle. In terms of these newly defined quantities, the magnitude and phase of the system function are

$$|H(j\omega)| = K \frac{M_{z_1}(\omega)M_{z_2}(\omega)\cdots M_{z_m}(\omega)}{M_{p_1}(\omega)M_{p_2}(\omega)\cdots M_{p_n}(\omega)} \quad (11.14a)$$

$$\angle H(j\omega) = \theta_{z_1}(\omega) + \theta_{z_2}(\omega) + \cdots + \theta_{z_m}(\omega) - \theta_{p_1}(\omega) - \theta_{p_2}(\omega)$$
$$- \cdots - \theta_{p_n}(\omega) \quad (11.14b)$$

A graphical interpretation of Equations 11.14a and 11.14b is shown in Figure 11.9 for a relatively simple system function.

FIGURE 11.9   *Graphical Interpretation of $H(j\omega)$*

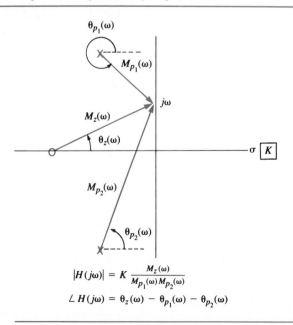

$$|H(j\omega)| = K \frac{M_z(\omega)}{M_{p_1}(\omega)M_{p_2}(\omega)}$$
$$\angle H(j\omega) = \theta_z(\omega) - \theta_{p_1}(\omega) - \theta_{p_2}(\omega)$$

In Chapter 10, this graphical method was used to evaluate $H(s)$ at individual frequencies. What we will do now is to consider what happens to the lengths and angles of the vectors as $\omega$ varies from zero to infinity. One possible approach to this problem would be to consider a set of specific test frequencies and do a graphical computation of $H(j\omega)$ for each. This approach could be as accurate as our patience and care would allow, but it would also be more time consuming than a straightforward analytical calculation. Instead, we will use the approach for *estimating* frequency response curves.

Let's exercise our imaginations in order to visualize what happens to the lengths and angles of the vectors as $\omega$ varies. With practice, this inspection technique can be used to estimate the general nature of the circuit's frequency response characteristics. The results will be imprecise when compared to a direct analytical calculation, but they will be arrived at very quickly.

To develop an understanding of the graphical inspection technique, let's consider three simple system functions similar to those of the circuits in Examples 11.1, 11.2, and 11.3.

---

**Example 11.5**   *Frequency Response by Inspection—a Single Pole*

Consider a system whose single pole is shown in Figure 11.10. A constant multiplier of $K = a$ normalizes the DC magnitude to 1. Graphically, the system function consists of this multiplier and a vector corresponding to the single pole. This vector is shown for several different values of $\omega$. Note that the length of the vector $M_p(\omega)$ monotonically increases for increasing magnitudes of $\omega$. Because $M_p(\omega)$ is in the denominator of the magnitude

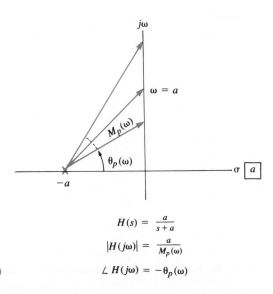

$$H(s) = \frac{a}{s+a}$$

$$|H(j\omega)| = \frac{a}{M_p(\omega)}$$

**FIGURE 11.10**   $\angle H(j\omega) = -\theta_p(\omega)$

*(continues)*

**Example 11.5**    *Continued*

function, however, the magnitude of the system function must decrease monotonically with increasing $\omega$. At $\omega = 0$, $M_p = a$ and $|H(j0)| = 1$. At $\omega = a$, we see that $M_p = \sqrt{2}a$ and that $|H(ja)| = 1/\sqrt{2}$. This information is really all that is needed to sketch a reasonably accurate magnitude plot. The resulting magnitude plot is shown in Figure 11.11A.

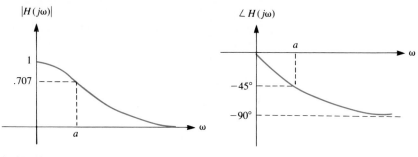

FIGURE 11.11    A. Magnitude                                   B. Phase

Similar reasoning allows us to estimate the phase plot. Focusing our attention on $\theta_p(\omega)$, we see that it equals zero for $\omega = 0$ and monotonically increases to $+90°$ as $\omega$ approaches infinity. The phase of the system function thus starts at $0°$ and approaches $-90°$ at high frequencies. Note that at $\omega = a$, $\theta_p = +45°$ and the phase of the system function is $-45°$.

An inspection of the magnitude plot of the single-pole system of Example 11.5 reveals that it acts as a low-pass filter. With this in mind, you should re-examine the pole–zero plot of the low-pass *RC* circuit of Figure 11.2.

**Example 11.6**    *Frequency Response by Inspection—a Pole and a Zero*

Now, consider a system with a pole at $s = -a$ and a zero at the origin. The constant multiplier $K$ will be assumed equal to 1. The pole–zero plot and related equations are shown in Figure 11.12.

Because $H(s)$ has a zero at the origin, $M_z(0) = 0$ and $|H(j0)| = 0$. Both vectors monotonically increase in length as $\omega$ increases, with $M_p$ always the longer of the two. Therefore, $|H(j\omega)|$ is always less than 1. For very high frequencies, the vector lengths become nearly identical and the system function's magnitude approaches 1. The angle $\theta_z$ is not defined for $\omega = 0$. However, for all positive frequencies, no matter how small, $\theta_z = +90°$. Thus, it is easily seen through an inspection of the vector diagram, that $\angle H(j\omega) = +90°$ for vanishingly small $\omega$ and that it approaches zero for high

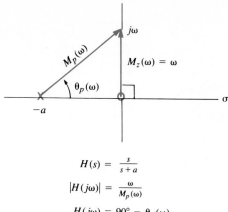

$$H(s) = \frac{s}{s+a}$$

$$|H(j\omega)| = \frac{\omega}{M_p(\omega)}$$

**FIGURE 11.12**          $$H(j\omega) = 90° - \theta_p(\omega)$$

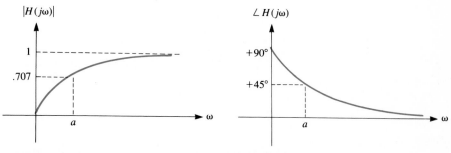

**FIGURE 11.13**    A. Magnitude                          B. Phase

frequencies. Clearly, this system is a high-pass filter, as is seen in the plot of the magnitude function in Figure 11.13.

The graphical inspection technique is somewhat more difficult to apply to systems with poles or zeros not on the real axis. The simplest such case to consider is that of one complex conjugate pair of poles. The frequency response characteristics for such a pair is very much dependent on the exact locations of the poles.

**Example 11.7**    *Frequency Response by Inspection—Conjugate Poles*

The pole–zero plot of Figure 11.14 consists of a conjugate pair of poles that we will assume to be much closer to the $j\omega$ axis than they are to the real axis. In terms of the quantities defined in Figure 11.14, $\alpha << \omega_d$.

*(continues)*

Example 11.7    *Continued*

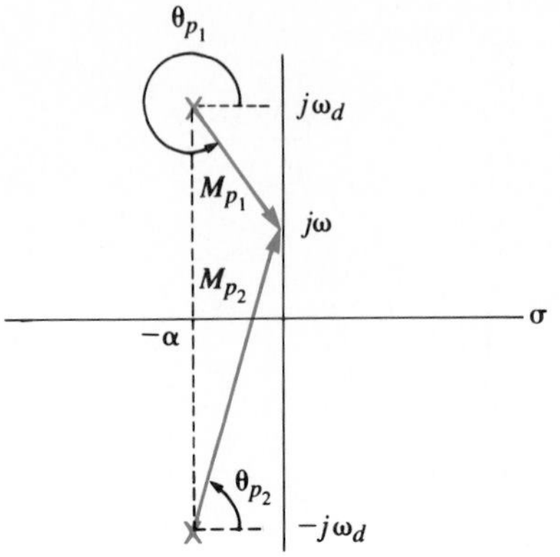

$$H(s) = \frac{1}{s^2 + 2\alpha s + (\alpha^2 + \omega_d^2)}$$

$$|H(j\omega)| = \frac{1}{M_{p_1} M_{p_2}}$$

FIGURE 11.14          $\angle H(j\omega) = -\theta_{p_1} - \theta_{p_2}$

One reason that the graphical inspection technique is a bit more difficult here is that $M_{p_1}$ is not monotonic as $\omega$ increases from zero to $+\infty$. Also, for $0 < \omega < \omega_d$, one of the vector lengths is increasing and the other is decreasing as $\omega$ changes. We can sketch reasonably accurate approximations for the frequency response curves if we consider separately three distinct frequency ranges: $\omega << \omega_d$, $\omega >> \omega_d$, and $\omega \approx \omega_d$. The first of these cases is shown in Figure 11.14. Vector interpretations of the latter two cases are shown in Figure 11.15.

**Case 1:** $\omega << \omega_d$ $(\omega \approx 0)$.    When $\omega \approx 0$ and is increasing, $M_{p_2}$ is increasing at about the same rate that $M_{p_1}$ is decreasing. Their product, then, stays approximately constant. Therefore, $|H(j\omega)|$ is nearly constant and equal to $1/(\alpha^2 + \omega_d^2)$ for frequencies near 0. Also for low frequencies, the angles $\theta_{p_1}$ and $\theta_{p_2}$ are opposite in sign and nearly equal in size. Therefore, $\angle H(j\omega) \approx 0°$.

**Case 2:** $\omega >> \omega_d$.    The vector representation for this case is shown in Figure 11.15A. Both $M_{p_1}$ and $M_{p_2}$ asymptotically approach $\omega$ for large $\omega$. With little error, then, we can say that $|H(j\omega)| \approx 1/\omega^2$, a function that approaches zero at high frequencies. It is easy to show that the phase function approaches $-180°$.

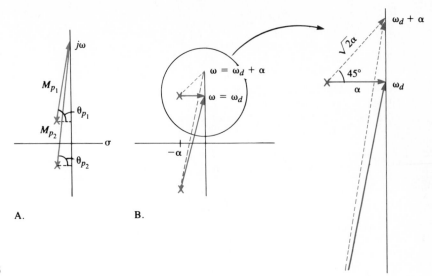

**FIGURE 11.15**

**Case 3:** $\omega \approx \omega_d$. Consider $\omega = \omega_d$ (solid lines of Figure 11.15B). The length and angle of the vector from the nearby pole are $M_{p_1}(\omega_d) = \alpha$ and $\theta_{p_1}(\omega_d) = 0°$. When $\omega$ increases to $\omega_d + \alpha$, the length of this vector increases by 41% to $\sqrt{2}\alpha$. The angle increases to $+45°$. A decrease in frequency to $\omega = \omega_d - \alpha$ causes the same change in $M_{p_1}(\omega)$ and an equal but opposite change in $\theta_{p_1}(\omega)$.

For the same changes in $\omega$, the magnitude and angle of the vector from the more distant pole undergo only small changes. The values of these changes depend on the relative sizes of $\alpha$ and $\omega_d$. For instance, if $\omega_d = 10$ and $\alpha = 1$, $M_{p_2}(\omega)$ and $\theta_{p_2}(\omega)$ change by approximately 10% and .3°, respectively, as $\omega$ increases from $\omega_d - \alpha$ to $\omega_d + \alpha$. With little error, therefore, we can treat $M_{p_2}(\omega)$ and $\theta_{p_2}(\omega)$ as constants when $\omega$ is in the near vicinity of the pole $p_1$. Thus, we conclude that for frequencies in the near vicinity of one of the poles, changes in the magnitude and phase of the system function are dominated by changes in the length and angle of the vector from the nearest pole. For the system function of Figure 11.14, the approximation is best when the distance from the poles to the $j\omega$ axis is much less than their distance from the real axis. That is, the approximation is best when $\alpha << \omega_d$.

The complete magnitude and phase curves ($\omega > 0$) are shown in Figure 11.16. The peak of the magnitude occurs at a frequency near, but not exactly equal to, $\omega_d$.

(*continues*)

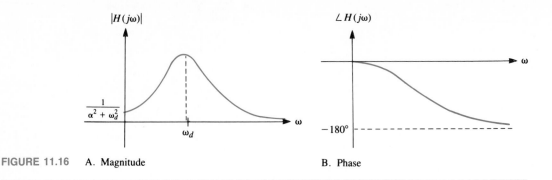

FIGURE 11.16    A. Magnitude                                                    B. Phase

The magnitude curve of Example 11.7 has a nonzero DC value, and it asymptotically approaches zero at high frequencies. It thus describes a low-pass filter. For some choices of $\omega_d$ and $\alpha$, however, the filter can have a very sharp peak near $\omega = \omega_d$. In such cases, you might wish to define a localized bandwidth for the peak even though the filter is not truly bandpass in nature. From the discussion of Example 11.7, it is evident that the lower cutoff frequency would be approximately $\omega_d - \alpha$ and that the upper cutoff frequency would be approximately $\omega_d + \alpha$. The localized bandwidth, then, would be approximately $2\alpha$.

Now that we have seen a few examples of determining frequency response curves through graphical inspection, we are in a position to state some general rules that describe the relationship between a circuit's poles and zeros and its frequency response curves. Consideration of Example 11.7 leads to the conclusion that if poles and zeros are relatively isolated from one another when compared to their distance to the $j\omega$ axis, then frequency response functions are dominated by the pole or zero nearest the frequency under consideration. In Equation 11.14, all terms of the magnitude and phase functions are considered to be virtually constant except for those from the nearest pole or zero. This approximation introduces a degree of error but allows us to easily sketch the magnitude curves for certain types of pole–zero plots. If we imagine a test frequency as it changes location on the $j\omega$ axis, we can make the following general observations about the magnitude function:

1.  The magnitude function will increase as $\omega$ approaches the vicinity of a pole or leaves the vicinity of a zero.
2.  The magnitude function will decrease as $\omega$ approaches a zero or leaves a pole.
3.  A localized maximum will occur near the value of $\omega$ closest to a pole.
4.  A localized minimum will occur near the value of $\omega$ closest to a zero.
5.  If the number of finite poles, $n$, exceeds the number of finite zeros, $m$, ($n > m$), then the magnitude function will decrease as $K/\omega^{n-m}$ for large $\omega$.
6.  If $n < m$, then the magnitude function will asymptotically approach $K\omega^{m-n}$ for large $\omega$.
7.  If $m = n$, then the magnitude function asymptotically approaches the constant $K$ for large $\omega$.

       Similar rules can be developed for phase functions, but consideration here will be limited to magnitudes. The technique is best understood by simply looking at some examples. Figure 11.17 shows three pole–zero plots and their corresponding magnitude functions estimated through graphical inspection. In viewing them, notice that the locations of maxima and minima in the response curves correspond to pole and zero locations on the *s* plane.

       The visual estimation of frequency response curves works well for simple systems whose poles and zeros are relatively distant from one another compared to their distances to the *jω* axis. What if this is not the case? Consider the example of Figure 11.18. Several poles and zeros are shown for a hypothetical system, and, in contrast to the earlier examples, they are clustered in one area of the *s* plane. Shown also are the vectors connecting each pole and zero with the test frequency *jω*. With such a pole–zero plot, it is a hopelessly complicated task to mentally keep track of all of these vectors and their sometimes contradictory effects as *ω* varies. Try it yourself and you will soon be convinced! In cases like this, we would have to return to an analytical approach.

       Despite its shortcomings, the inspection technique is extremely valuable as a

**FIGURE 11.17**   *Magnitude Curves Estimated from Pole-Zero Plots*

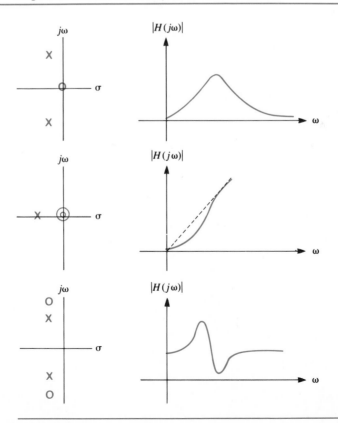

FIGURE 11.18    *A System Too Complicated for Graphical Estimation*

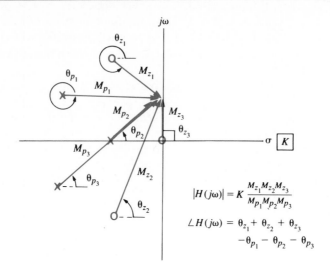

$$|H(j\omega)| = K \frac{M_{z_1}M_{z_2}M_{z_3}}{M_{p_1}M_{p_2}M_{p_3}}$$

$$\angle H(j\omega) = \theta_{z_1} + \theta_{z_2} + \theta_{z_3}$$
$$\qquad\qquad -\theta_{p_1} - \theta_{p_2} - \theta_{p_3}$$

means of quickly estimating the frequency response characteristics of simple systems. It is also worthwhile as an exercise in developing a feeling for what poles and zeros are and how they relate to a system's behavior. Even so, we must keep in mind its relatively limited applicability. Section 11.6 will introduce the Bode diagram, a method that allows us to quickly sketch the frequency response curves for any system, no matter how complicated. The resultant response curves are remarkably accurate. Bode techniques, then, can be thought of as a compromise between the purely analytical approach of Section 11.3 and the inspection methods described here.

## 11.5   RESONANCE

The system function of Example 11.7 as well as the circuit described in Example 11.3 have localized peaks in their magnitude curves that are the result of a phenomenon known as **resonance.** The principle of resonance is applicable to many areas of natural science and engineering. As a result, resonance, even as it applies to electrical engineering, can be defined in several different ways and from several different points of view. You may come across definitions for resonance that seem different from the discussion here. Note, however, that all such definitions are consistent with one another. The engineer's approach to resonance will be emphasized here, but the discussion will still be as general as possible.

Every resonant system, regardless of the physical type, involves an oscillation of stored energy between one form and another. The stored energy of a resonant mechanical system, for instance, oscillates back and forth between potential and kinetic energy. In resonant electrical circuits, energy is alternately stored in the magnetic field of an inductance and the electric field of a capacitance. Because both inductors and capacitors must be present, any resonant electrical circuit must be of order two or greater. However, not all second- and higher-order circuits necessarily display resonance.

The oscillations of energy just mentioned are part of a system's natural response. Electrical engineers, however, usually recognize the presence of resonance in a system in terms of its frequency response behavior. Resonance in an electrical circuit will be defined as an enhancement of its forced response due to sinusoidal excitation when the circuit is excited near one of its natural frequencies.* The existence of resonance can be most easily recognized if the magnitude curve of the circuit's frequency response has a sharp localized peak.

Resonance exists in any system that has a complex conjugate pair of poles. This is a necessary as well as sufficient condition for resonance because, without complex natural frequencies, oscillations of stored energy could not exist in the unforced system.** The degree to which the phenomenon of resonance is present, however, depends upon the exact positions of the poles on the $s$ plane. The simplest system that can exhibit resonance is a second-order system with no zeros. For purposes of discussion, let's consider the system function of Equation 11.15. Its denominator polynomial is in a standard form that has been found to be useful.[†]

$$H(s) = \frac{\omega_n^2}{s^2 + 2\zeta\omega_n s + \omega_n^2}$$

(11.15)

You may remember having seen this polynomial earlier in Chapter 6 as one form of the characteristic equation of a second-order circuit. The locations of the poles of $H(s)$ obviously depend on the values of $\omega_n$ and $\zeta$, the system's **damping ratio.** For any fixed value of $\omega_n$, the pole locations are uniquely described by $\zeta$. The two poles either are on the real axis or are complex conjugates of each other on a circle of radius $\omega_n$. $\zeta < 0$ corresponds to poles in the right half of the $s$ plane, a situation that will not be considered here because a basic assumption of this

---

* *IEEE Standard Dictionary of Electrical and Electronics Terms,* 2d ed., IEEE, Inc., 1978.

** As the discussion implies, if a passive circuit has poles located off the real axis of the $s$ plane, the circuit must contain both inductors and capacitors.

[†] The denominator of Equation 11.15 is not the only possible standard form for quadratic terms in system functions. Another form, widely used in the areas of electronics and communications, is $s^2 + (\omega_n s/Q) + \omega_n^2$. $\omega_n$ has the same interpretation as in Equation 11.15. $Q$ is called the system's **quality factor.** Details for this form are developed in the problems at the end of the chapter.

chapter is that the system is stable. $\zeta > 1$ leads to unequal poles on the negative real axis, a nonresonant condition. $\zeta = 1$ corresponds to a double pole at $s = -\omega_n$. When $0 < \zeta < 1$, the poles will be complex conjugates of each other. $\zeta = 0$ (no damping) leads to poles at $\pm j\omega_n$. Thus, $\omega_n$ is known as the **undamped natural frequency** of the system. These details are shown in Figure 11.19A.

Figure 11.19B shows the dependence of the magnitude function on $\zeta$ when $\omega_n$ is fixed. Resonance peaking (a localized maximum in the magnitude function that is due to resonance) occurs only for $0 < \zeta < .707$. That is, resonance peaking occurs only when the poles are nearer the $j\omega$ axis than they are to the real axis. Although the most strict definition of resonance includes any situation in which

**FIGURE 11.19**    *Resonance as a Function of $\zeta$*

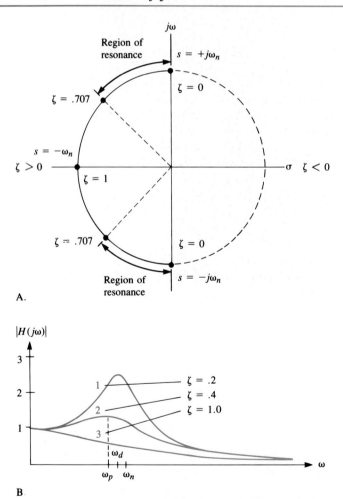

A.

B

poles are complex, most engineers recognize resonance only in lightly damped systems (small $\zeta$) with truly pronounced peaking in their magnitude curves. In such systems, the poles are quite near the $j\omega$ axis.

Figure 11.19B shows three specific frequencies that are related to resonance effects. We have already touched upon each, but now they will be defined more carefully. Two of the frequencies are related to the system's natural response. The imaginary parts of the pole locations are $\pm\omega_d$, the frequency of sinusoidal oscillation of the corresponding part of the natural response. The undamped natural frequency is $\omega_n$. It is the frequency of oscillation that would exist if there were no losses (damping) in the system. The third frequency, $\omega_p$, is related to the system's forced response. It is the sinusoidal frequency of excitation at which the magnitude function is maximum. Thus, $\omega_p$ is known as the **resonance frequency** of the system's response variable. These three frequencies are shown for the case of $\zeta = .4$ in Figure 11.19B. As the figure shows, for a two-pole system with no zeros, $\omega_p < \omega_d < \omega_n$.

In electronics applications, it is often desirable to design circuits with very pronounced resonance peaking because of the good frequency selection qualities that such circuits have. For instance, the earliest radio tuners were lightly damped, second-order circuits with variable resonance frequencies. In lightly damped circuits, the three frequencies just defined are very close in numerical value. For $\zeta = 0$, they are all actually equal. Therefore, electronics engineers seldom distinguish between these three frequencies. Controls engineers, however, most often wish to avoid any pronounced resonance effects. In the types of systems they work with, it is important to understand the difference in meanings of the frequencies $\omega_d$, $\omega_n$, and $\omega_p$.

Many different electrical circuits exhibit resonance effects. The only properties that they must have in common are that they be of at least second order and that they contain both inductors and capacitors.* Higher-order circuits can exhibit several different resonances, not unlike the several resonances that can be heard when someone is singing in the shower. Despite the richness of potential examples, electrical engineers nearly always discuss resonance in terms of two standard circuits. They are the series and the parallel $RLC$ circuits. The development of some of the details related to their resonances is left as an exercise in the problems at the end of the chapter.

Before we move on to the next topic, one point of the discussion should be clarified. Our definition of resonance has been based on the frequency response characteristics of a circuit. It has been said that circuits with resonance are often recognized by the localized maxima that can be seen in their magnitude functions. This is certainly true, but the reverse is not. That is, magnitude functions with localized maxima in them do not necessarily imply a resonance effect. Examples

---

*The requirement that both inductors and capacitors be present is true only for passive circuits. Active filters containing only operational amplifiers, resistors, and capacitors can also display resonance-like effects.

of bandpass *RC* or *RL* circuits whose magnitude functions have maxima at frequencies above DC are easy to find. Do not jump to any hasty conclusions based on a brief inspection of a magnitude curve!

## 11.6  BODE DIAGRAMS

Section 11.4 developed a technique for estimating the frequency response characteristics of a circuit based on an inspection of its pole–zero plot. The approach allows us to very quickly sketch the response curves, but the results are obviously not very accurate. Another shortcoming of the approach is that it is difficult to apply to any but the simplest of circuits. This difficulty is due primarily to the multiplicative interactions between the individual terms of the magnitude expression repeated here:

$$|H(j\omega)| = K \frac{M_{z_1}(\omega)M_{z_2}(\omega)\cdots M_{z_m}(\omega)}{M_{p_1}(\omega)M_{p_2}(\omega)\cdots M_{p_n}(\omega)} \tag{11.16}$$

This section introduces the **Bode diagram.** This method of displaying frequency response characteristics was popularized by the American engineer H. W. Bode. Bode diagrams are frequency response curves plotted as functions of logarithmically scaled frequency axes. In addition, the magnitude functions themselves are also logarithmically scaled. As will be shown, this method of presenting frequency response curves is, on the one hand, much quicker than the purely analytical approach of Section 11.3 and, on the other hand, much more accurate than our estimations of Section 11.4. The Bode approach has come to be one of the most common methods of presenting frequency response characteristics.

Bode diagrams are based on a mathematical transformation of Equation 11.16 into a form that changes its multiplicative interactions into additions. The mathematical operation that accomplishes this task is a logarithmic transformation:

$$\boxed{|H(j\omega)|_{\text{dB}} = 20\log_{10}|H(j\omega)|} \tag{11.17}$$

We say that the magnitude of $H(j\omega)$ in **decibels** (dB) is 20 times $\log_{10}$ of the linear magnitude. Any scale factor and any log base would serve the desired purpose, but the decibel definition is universally accepted. One of the decibel's early applications was to the physics of sound. The unit was named to honor Alexander Graham Bell for his work in that field.

Before we proceed, we should become more comfortable with the change in scale caused by this decibel transformation. Table 11.1 lists some representative linear magnitudes and their corresponding decibel values. Negative decibel values correspond to system function magnitudes that are less than 1. Positive decibel

**TABLE 11.1**   *Correspondence between Selected Linear and Decibel Magnitudes*

| $|H|$ | $|H|_{dB}$ |
|---|---|
| 0 | $-\infty$ dB |
| 1/1000 | $-60$ dB |
| 1/100 | $-40$ dB |
| 1/10 | $-20$ dB |
| 1 | 0 dB |
| 10 | $+20$ dB |
| 100 | $+40$ dB |
| 1000 | $+60$ dB |
| $\infty$ | $+\infty$ dB |

values correspond to magnitudes greater than 1. We need not worry about the fact that the logarithm is not defined for negative arguments; the magnitude of a system function is, by definition, always positive! Try to keep in mind that 0 dB does not mean that a circuit's output is zero. Instead, it corresponds to a system function whose magnitude equals 1 and, therefore, an output of equal amplitude to the input. Zero output corresponds to a system function magnitude of zero, which, in decibels, is $-\infty$ dB.

Because curves will be plotted as functions of logarithmic frequency, we should also familiarize ourselves with that scale. Shown here is a representative logarithmically scaled axis:

Generally, only those frequencies that are even powers of 10 are marked. They are spaced evenly along the axis. Each tenfold change in frequency is called a **decade.** The scale is obviously nonlinear. For instance, $\omega = 0$ rad/s (or $f = 0$ Hz) can never be shown because it is always infinitely far to the left! In addition, note that two times any decade frequency is at a point approximately one third of the way to the next higher power of 10. A threefold increase (or decrease) covers about one half of a decade, and a fivefold change covers a little more than two thirds of a decade.

Let's satisfy ourselves that the decibel transformation has the desired effect. We need to apply the transformation of Equation 11.17 to the magnitude function of Equation 11.16:

$$|H(j\omega)|_{dB} = 20 \log |H(j\omega)| = 20(\log K + \log M_{z_1} + \cdots + \log M_{z_m}$$
$$- \log M_{p_1} - \cdots - \log M_{p_n}) \tag{11.18}$$

Clearly, the desired effect has been achieved. The contributions from the individual pole and zero terms are separated in Equation 11.18. If we could determine the form of the frequency-dependent contribution due to each of the individual terms, it would be a simple matter to add them all together to find the magnitude of the frequency response of the system. The individual terms of Equation 11.18 are, with the exception of the constant, dependent on frequency. Our next step will be to investigate the nature of these dependencies.

The frequency response of $H(s)$ depends on its pattern of poles and zeros. Our examples should include all of the situations that we are likely to find in real, stable, physical systems. These situations include poles or zeros that are

—At the origin
—Simple and on the negative real axis
—Real and multiple
—Complex conjugate pairs

A prototype system function that contains these possible critical frequency locations is

$$H(s) = \frac{K}{s(s + \omega_1)(s + \omega_2)^2(s^2 + 2\zeta\omega_n s + \omega_n^2)} \qquad (11.19)$$

As a basic example, a system function that has no finite zeros has been chosen. This does not limit us in any way because, as you can see from Equation 11.18, the effect that a zero has on the Bode diagrams for a system is just the negative of the effect of a pole at the same location.

Equation 11.19 introduces some changes in notation for $H(s)$. First, the use of the symbol $\omega$ has been chosen to represent the locations of real-valued poles (or zeros, if there are any). This symbol does not imply that such poles or zeros are located on the $j\omega$ axis. They are, of course, on the real axis of the $s$ plane. In Equation 11.19, for instance, we see that $H(s)$ has a simple pole at $s = -\omega_1$ and a double pole at $s = -\omega_2$. As we will see in later examples, however, significant features of frequency response curves occur at radian frequencies equal in magnitude to the distance of such poles or zeros from the origin. In Bode analysis, complex conjugate poles or zeros are always considered in their quadratic forms. We are using the quadratic form that is given in Equation 11.15.

To make the following analysis easier, we change the form of the system function to

$$H(j\omega) = \frac{K'}{j\omega[1 + (j\omega/\omega_1)][1 + (j\omega/\omega_2)]^2[(j\omega/\omega_n)^2 + 2\zeta(j\omega/\omega_n) + 1]} \qquad (11.20)$$

Because Bode analysis is concerned only with the sinusoidal steady state, $j\omega$ has been substituted for $s$. In addition, each of the real-root terms has been refactored to the form $1 + (j\omega/\omega_k)$. The quadratic term was similarly factored. These factorizations cause the constant term $K$ to change to $K' = K/(\omega_1 \omega_2^2 \omega_n^2)$.

The individual terms of Equation 11.20 can be thought of as building blocks that contribute, each in its own way, to the frequency response characteristics of the overall system. We will consider the behavior of each of these building blocks separately, beginning with the simplest and working toward the more complicated.

**Case 1:** *The Constant.* Suppose that a system function consists of just a constant term; that is, $H(s) = K$. For $K > 0$, $\angle H(j\omega) = 0°$; for $K < 0$, $\angle H(j\omega) = \pm 180°$. To find the decibel magnitude of the function, we apply Equation 11.17:

$$|H|_{dB} = 20 \log |K|$$

This simple constant is easily plotted as a function of $\omega$.

**Case 2:** *A Pole at the Origin.* Consider a system function with a simple pole at the origin of the $s$ plane:

$$H(j\omega) = \frac{1}{j\omega}$$

For $\omega > 0$ (negative frequencies are meaningless in Bode analysis), $\angle H(j\omega) = -90°$. The decibel magnitude is

$$|H(j\omega)|_{dB} = 20 \log \frac{1}{\omega} = -20 \log \omega \tag{11.21}$$

Equation 11.21 is a logarithmic curve when plotted against a linear frequency scale. Such a curve is not easy to draw accurately. As was said earlier, however, Bode frequency response curves are plotted as functions of logarithmically scaled frequency axes. Thus, Equation 11.21 can be drawn as a straight line, as is shown in Figure 11.20A. Figure 11.20B shows the phase curve. Notice that for every tenfold increase in frequency, there is a 20 dB decrease in $|H|_{dB}$. Therefore, we say that the decibel magnitude curve of Figure 11.20A falls off (decreases) at a rate of 20 dB/decade. Notice also that it passes through 0 dB at $\omega = 1$ rad/s.

Because $\omega = 0$ can never be seen in a Bode diagram, we draw the vertical and horizontal axes so that they do not cross. To do otherwise might suggest the existence of an origin with respect to the frequency axis.

**Case 3:** *A Simple Pole on the Negative Real Axis.* Consider the next most complicated building block, a system function with a simple pole at $s = -\omega_k$ (we have normalized the system function for a DC magnitude of 1):

$$H(j\omega) = \frac{1}{1 + (j\omega/\omega_k)}$$

FIGURE 11.20    *Bode Frequency Response Curves for a Simple Pole at Origin*

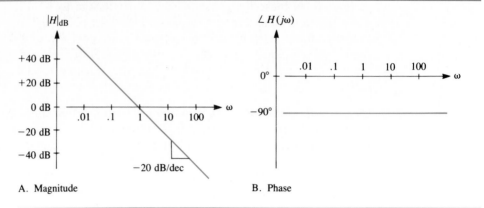

A. Magnitude

B. Phase

$$|H(j\omega)| = \frac{1}{\sqrt{1 + (\omega/\omega_k)^2}}$$

$$|H(j\omega)|_{dB} = 20\log|H(j\omega)| = -20\log\left[1 + \left(\frac{\omega}{\omega_k}\right)^2\right]^{1/2}$$

$$= -10\log\left[1 + \left(\frac{\omega}{\omega_k}\right)^2\right]$$

To see how this magnitude can be plotted, consider the extreme values for frequency (remember that only positive frequencies are allowed).

For $\omega << \omega_k$:   $|H|_{dB} \approx -10\log(1 + 0) = 0\,dB$

For $\omega >> \omega_k$:   $|H|_{dB} \approx -10\log\left(\frac{\omega}{\omega_k}\right)^2 = -20\log\left(\frac{\omega}{\omega_k}\right)$

$$= -20\log\omega + 20\log\omega_k$$

For very low frequencies ($\omega << \omega_k$), we see that the decibel magnitude contribution of a simple pole is approximately a constant of 0 dB and can be plotted as a horizontal straight line. At high frequencies ($\omega >> \omega_k$), the decibel magnitude is also a straight line. It has a slope of $-20\,dB/decade$ and passes through 0 dB at $\omega = \omega_k$. Figure 11.21A shows the results of the analysis. When the low-frequency and high-frequency approximations are extended to intersect, the result is a **straight-line approximation** of the magnitude function. For obvious reasons, $\omega_k$ is called the **break point** or corner frequency of the diagram.

The straight-line approximation is just that—an approximation. It is exact

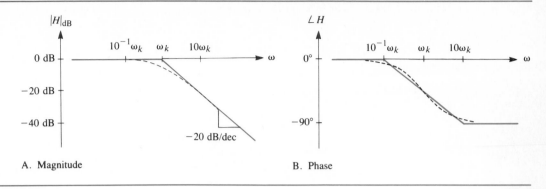

A. Magnitude

B. Phase

only at $\omega = 0$ and at $\omega = +\infty$, two points not even appearing on the graph. It can be shown that the maximum error between the straight-line approximation and an exact plot of decibel magnitude occurs at the break point. There, the exact decibel magnitude is

$$-10 \log \left[ 1 + \left( \frac{\omega_k}{\omega_k} \right)^2 \right] = -10 \log 2 \approx -3 \text{ dB}$$

When this information is used to improve upon the straight-line approximation, a remarkably accurate plot can be drawn, as is shown in Figure 11.21A with a dashed line. Note that $-3$ dB corresponds to a linear magnitude of $1/\sqrt{2}$. This point is often called the **3 dB point** and, for a single-pole system, is synonymous with the term *half-power point* that was defined in Section 11.3.

The logarithmic transformation of Bode analysis is not really needed in order to find the total phase curve from the individual pole and zero contributions. For ease of interpretation, however, these curves are also plotted against logarithmic scales for frequency. The phase of a system function composed of one simple pole is given by the following equation:

$$\angle H(j\omega) = \angle \left( \frac{1}{1 + (j\omega/\omega_k)} \right) = -\tan^{-1} \frac{\omega}{\omega_k}$$

This function is plotted in Figure 11.21B. Both the straight-line approximation and the exact form are shown. The straight-line approximation is $0°$ at very low frequencies, $-45°$ at the break point, and $-90°$ at very high frequencies. There is a fall-off of $45°$/decade for one decade on either side of the break point. The greatest difference between the actual phase curve and the straight-line approximation is $5.7°$. It occurs at both one decade above and one decade below the break point.

**Case 4:** *Multiple Real Poles.* Consider a system function with $n$ poles, all located at $s = -\omega_k$. The normalized system function is

$$H(j\omega) = \frac{1}{[1 + (j\omega/\omega_k)]^n}$$

The magnitude and phase expressions are

$$|H|_{dB} = -10n \log\left[1 + \left(\frac{\omega}{\omega_k}\right)^2\right]$$

$$\angle H(j\omega) = -n \tan^{-1}\frac{\omega}{\omega_k}$$

Each is equal to the contribution from a simple pole at $s = -\omega_k$ multiplied by $n$, the number of poles at $-\omega_k$. Thus, the magnitude plot is still $0\,dB$ at low frequencies but falls off at $20n\,dB/decade$ at high frequencies. At the break point, the exact magnitude in decibels is $-3n\,dB$. These details will be shown more explicitly in later examples.

**Case 5:** *Complex Conjugate Poles.* Consider one last situation, the case of complex conjugate poles. As was said earlier, these poles are always considered in their quadratic form. Assume that

$$H(s) = \frac{1}{(s/\omega_n)^2 + 2\zeta(s/\omega_n) + 1}$$

Then,

$$H(j\omega) = \frac{1}{(j\omega/\omega_n)^2 + 2\zeta(j\omega/\omega_n) + 1}$$

$$|H(j\omega)| = \frac{1}{\sqrt{[1 - (\omega/\omega_n)^2]^2 + [2\zeta(\omega/\omega_n)]^2}}$$

For $\omega \ll \omega_n$: $\quad |H(j\omega)|_{dB} = 0\,dB$

For $\omega \gg \omega_n$: $\quad |H(j\omega)|_{dB} = -10 \log\left(\frac{\omega}{\omega_n}\right)^4 = -40 \log\frac{\omega}{\omega_n}$

$$= -40 \log \omega + 40 \log \omega_n$$

$$\angle H(j\omega) = -\tan^{-1}\frac{2\zeta(\omega/\omega_n)}{1 - (\omega/\omega_n)^2}$$

Straight-line approximations for the magnitude and phase of this quadratic term are shown by the bold lines in Figure 11.22.

While the straight-line approximations for second-order terms are simple to

**FIGURE 11.22** *Second-Order Bode Frequency Response Curves for Several Values of $\zeta$*

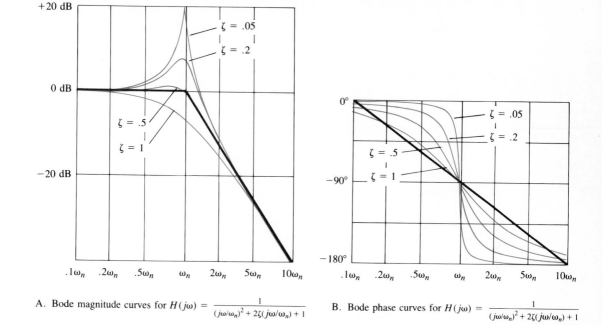

A. Bode magnitude curves for $H(j\omega) = \dfrac{1}{(j\omega/\omega_n)^2 + 2\zeta(j\omega/\omega_n) + 1}$

B. Bode phase curves for $H(j\omega) = \dfrac{1}{(j\omega/\omega_n)^2 + 2\zeta(j\omega/\omega_n) + 1}$

sketch, more exact plots are not nearly as easy as for simple first-order poles. To avoid the necessity of referring to the rather cumbersome analytical expression in order to improve on the second-order straight-line approximations, families of curves have been developed for both the decibel magnitude and the phase of normalized second-order terms. These are the curved lines of Figure 11.22.

Remember that for complex conjugate poles in stable systems, $\zeta$ varies between 0 and +1. $\zeta = 1$ corresponds to multiple poles on the negative real axis. Note that for $\zeta = 1$, the magnitude and phase plots of Figure 11.22 agree with those already derived for two real equal poles at $s = -\omega_n$. An undamped system ($\zeta = 0$) has poles on the imaginary axis. The curves do not include this case, but it can be seen that as $\zeta$ becomes smaller, the peak in magnitude becomes larger. For $\zeta = 0$, the peak has a value of infinity located at $\omega = \omega_n$. Also in the limit, the phase curve has a sudden decrease of 180° at $\omega = \omega_n$, just as you would expect as the frequency of evaluation passes through the location of a pole.

Before we turn to some specific examples of Bode diagrams, we should discuss a few points of interpretation. A Bode magnitude curve is based on a log/log transformation. The decibel magnitude is computed as $20 \log |H|$. The result, however, is plotted on a linear scale. No such mathematical transformation is applied to the frequency variable, but in order to have plots made up of straight lines, frequency is scaled logarithmically. Thus, Bode diagrams are drawn on *semilog* graph paper.

For the purposes of our studies and, in fact, for many applications, a straight-line approximation will give us the necessary level of accuracy in our Bode diagrams. A common exception to this rule is when significant peaking is associated with a second-order term. This peaking is added to the straight-line approximation by determining $\zeta$ and then referring to the family of curves of Figure 11.22. For simplicity, we will always use straight-line approximations for phase, even though their use can lead to large errors for small values of $\zeta$.

Finally, it should be emphasized that both linear and Bode frequency response curves contain the same information. The logarithmic transformation is one of scale only. The advantage of Bode analysis lies in the ease with which frequency response diagrams can be drawn, particularly when the curves cover wide ranges of magnitude or frequency. This ease is a result of the fact that each individual contribution can be drawn as just two straight lines for magnitude and three for phase. The results are quite accurate unless significant resonance is present. Because they are composed of straight lines, the individual contributions are easily added. The sum of any number of straight lines is itself a straight line. This point and other points will be illustrated in the examples that follow. In each example, the individual contributions as well as the complete Bode curves are shown. You should fill in some of the intermediate details in order to verify the correctness of the results.

---

**Example 11.8**   *Bode Diagram I*

Consider the system function $H(s)$.

$$H(s) = \frac{10,000}{(s + 10)(s + 100)}$$

Make the substitution $s = j\omega$ and normalize the individual terms.

$$H(j\omega) = \frac{10,000}{(j\omega + 10)(j\omega + 100)} = \frac{10,000}{(10)(100)[(j\omega/10) + 1][(j\omega/100) + 1]}$$

Three terms contribute to the magnitude:

1. $K' = 10,000/(10)(100) = 10$, which gives $20 \log 10 = +20$ dB
2. Simple pole with a break point at $\omega = 10$
3. Simple pole with a break point at $\omega = 100$

Because $K'$ is positive, it does not affect the phase. The Bode magnitude and phase curves for $H(s)$ are shown in Figure 11.23.

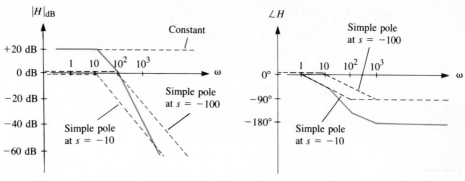

FIGURE 11.23    A. Magnitude                                    B. Phase

Once you have gained some experience in drawing Bode diagrams from system functions, you will be able to work directly from $H(s)$ rather than from $H(j\omega)$. The individual pole and zero terms must still be refactored, of course, as is done in the next example.

**Example 11.9**    *Bode Diagram II*

Normalize the individual terms of $H(s)$, keeping $s$ as the variable of the system function.

$$H(s) = \frac{9000s}{(s + .1)(s + 30)^2} = \frac{100s}{[(s/.1) + 1][(s/30) + 1]^2}$$

Notice that there is a double pole at $s = -30$. On a logarithmic scale, $\omega = 30$

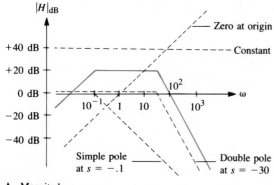

A. Magnitude

FIGURE 11.24

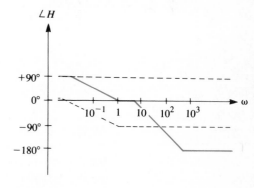

B. Phase

*(continues)*

**Example 11.9**  *Continued*

is approximately halfway between 10 and 100. The magnitude and phase curves for this example are shown in Figure 11.24.

**Example 11.10**  *Bode Diagram III*

Consider a system function containing a quadratic term.

$$H(s) = \frac{1000(s + .1)}{s^2 + 4s + 100}$$

Comparing the quadratic term to the standard form of $s^2 + 2\zeta\omega_n s + \omega_n$ leads to

$$\omega_n^2 = 100 \qquad \text{so} \qquad \omega_n = 10$$
$$2\zeta\omega_n = 4 \qquad \text{so} \qquad \zeta = .2$$

$$H(j\omega) = \frac{(j\omega/.1) + 1}{(j\omega/10)^2 + 2(.2)(j\omega/10) + 1}$$

The magnitude curves of Figure 11.22A show that with $\zeta = .2$, there is a resonance peak near $\omega = 10$ that is about 8 dB greater than the straight-line approximation. The response curves are shown in Figure 11.25.

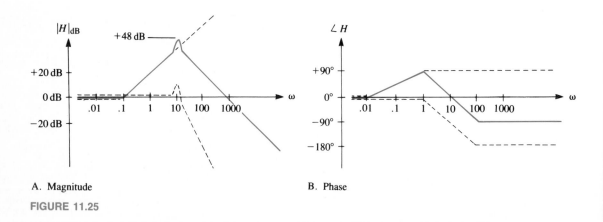

A. Magnitude                                B. Phase

**FIGURE 11.25**

We have seen several examples of how Bode diagrams can be drawn to describe the frequency response characteristics of a system. Now, let's use Figure 11.25 to see how they can be used to find the output of a system excited at a specific sinusoidal frequency. The approach is almost as easy as with the linear response curves of Section 11.3. All that is needed is to find the value of the system function at the frequency of interest. This value can be found by reading the magnitude and phase of the system function from the plotted response curves. For instance, look at Figure 11.25 and assume a frequency of excitation of $\omega = 1$ rad/s. We see that $|H(j1)|_{dB} \approx +20$ dB, which corresponds to a magnitude of $|H(j1)| = 10$. Similarly, the phase curve shows that $\angle H(j1) \approx +90°$. Armed with this information, we can now find the system's response to any excitation at $\omega = 1$ rad/s. Assume that $H(s) = V_o/V_i$. Then, the following input–output pairs are correct:

$$v_i(t) = 15 \sin (t - 30°)$$
$$v_o(t) = (10)(15) \sin (t - 30° + 90°) = 150 \sin (t + 60°)$$

$$v_i(t) = 1.2 \cos (t + 80°)$$
$$v_o(t) = (10)(1.2) \cos (t + 80° + 90°) = 12 \cos (t + 170°)$$

Similarly,

$$v_i(t) = -25 \cos (10^4 t + 45°)$$
$$v_o(t) = \left(\frac{1}{10}\right)(-25) \cos (10^4 t + 45° - 90°) = -2.5 \cos (10^4 t - 45°)$$

$$v_i(t) = .01 \sin (30t + 25°)$$
$$v_o(t) = (31.6)(.01) \sin (30t + 25° - 45°) = .316 \sin (30t - 20°)$$

In the last case, the use of a hand calculator was needed to find that the decibel value of $+30$ dB corresponds to a system function magnitude of 31.6.

## 11.7 SUMMARY

This chapter has described three techniques for determining and plotting frequency response curves. The technique chosen will depend on your immediate needs. Each has its advantages and disadvantages. If all that is needed is a quick idea of the frequency response behavior of a fairly simple circuit, then the graphical inspection technique is a good choice. For more accuracy, Bode diagrams can be easily drawn while still avoiding detailed analytical calculations. The straightforward analytical calculation is the ultimate in precision. Even so, the results of analytical calculations done with computers are often presented in Bode diagram form rather than as linear plots.

Basic examples of low-pass, high-pass, and bandpass filters have been shown. The terms *bandwidth*, *cutoff frequency*, and *resonance* have been defined. The latter term (resonance) applies only to circuits of order two or greater, and then, in passive circuits, only when both inductors and capacitors are present. In modern active circuit design, however, resonance-like effects can be present in circuits containing only resistors and capacitors along with operational amplifiers.

Regardless of the technique used, all of the frequency response curves have been in the form of magnitude and phase curves, each plotted separately as functions of $\omega$. Do not be left with the impression that this is the only form used in plotting frequency response information. Nyquist plots, for instance, simultaneously plot magnitude and phase onto one linear polar plot using $\omega$ as a parameter of the resultant curve. The Nichols chart uses a similar approach but on a decibel-scaled Cartesian coordinate system. These forms are useful in certain specialized applications. The forms used in the examples in this chapter are common to all areas of electrical engineering.

You may not, at this time, fully appreciate why the goal of finding a description of a circuit's response to sinusoids of all frequencies is important enough to have had an entire chapter devoted to it. The earlier steady-state examples, after all, were of circuits with only one sinusoidal frequency present at a time. It is true that in many important applications such as power systems, sinusoids appear in their pure form. Sinusoids, in fact, are probably the most common of signal waveforms encountered by electrical engineers. Even when a signal is not a pure sinusoid, however, it is often possible to treat it as the sum of an infinite number of sinusoids of different frequencies. This is the topic of the next chapter. For such a case, knowledge of a circuit's frequency response characteristics and an application of the principle of superposition allow us to determine the circuit's response to the nonsinusoidal input signal. We are, however, getting a little ahead of ourselves. For now, we can consider frequency response techniques to be quick and efficient methods for determining a circuit's response to a sinusoid of any frequency.

## ■ PROBLEMS

**11.1** $H(s)$ is a voltage transfer function:

$$H(s) = \frac{10(s + 2)}{s^2 + 2s + 5}$$

Interpret $H(s)$ as a ratio of phasors and use it to find, if possible, the complete responses to the following input functions. If it is not possible to do so with the information provided, explain why not.

a. $v_i(t) = 10 \cos t$ V

b. $v_i(t) = 10 \cos 2t$ V

c. $v_i(t) = 10 \cos 3t$ V

d. $v_i(t) = -2 \sin (2t + 45°)$ V

e. $v_i(t) = 15$ V

f. $v_i(t) = 20 \cos (5t + 60°)u(t)$ V

g. $v_i(t) = -.5t + 3$ V

h. $v_i(t) = 1.5e^{2t}$ V

i. $v_i(t) = 10e^{3t} \cos t$ V

j. $v_i(t) = \cos .1t + \cos t + \cos 10t + \cos 100t + \cos 1000t$ V

11.2 For the pole–zero plots shown in Figure P11.2, analytically find the systems' steady-state responses to inputs of $x(t) = \cos 500t$ and $x(t) = \cos 1000t$.

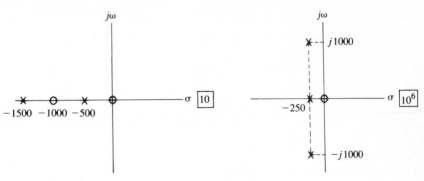

**FIGURE P11.2**  A.  B.

11.3 The magnitude and phase of a driving-point impedance are plotted in Figure P11.3. Use the curves to determine the complete voltage responses to the following input currents. If it is not possible to do so, explain why not.

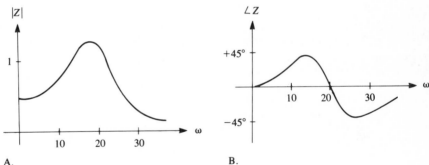

**FIGURE P11.3**  A.  B.

  **a.**  $i_s(t) = 6 \sin(10t - 135°)$ A

  **b.**  $i_s(t) = -10 \cos(10t + 30°)$ A

  **c.**  $i_s(t) = 6e^{-2t} \sin(10t - 135°)$ A

  **d.**  $i_s(t) = 10 + 16 \sin 10t + 8 \sin 20t - 50 \cos(30t - 45°)$ A

  **e.**  $i_s(t) = 20t + 10$ A

11.4 Show that the magnitude of a complex system function $H(j\omega)$ is an even function of $\omega$ and that the phase is an odd function. That is, $|H(j\omega)| = |H(-j\omega)|$ and $\angle H(j\omega) = -\angle H(-j\omega)$. *Hint:* Begin with a general expression for $H(s)$ and pay attention to the even and odd powers of $s$.

11.5 Determine and plot linear frequency response curves for the following system functions. Use a programmable calculator or computer, if necessary. For each, characterize the filter type.

  **a.**  $H(s) = \dfrac{8}{s^2 + 6s + 8}$

**b.** $G(s) = \dfrac{s}{s+10}$

**c.** $A(s) = \dfrac{s+12}{s+6}$

**11.6** Analytically determine and plot the frequency response curves $|H|$ and $\angle H$ of the circuit in Figure P11.6 for each of the following four sets of element values:

**a.** $L = 1\,\text{mH}, \quad R = 1\,\text{k}\Omega$

**b.** $L = 1\,\text{mH}, \quad R = 100\,\Omega$

**c.** $L = 1\,\text{mH}, \quad R = 10\,\text{k}\Omega$

**d.** $L = 10\,\text{mH}, \quad R = 10\,\text{k}\Omega$

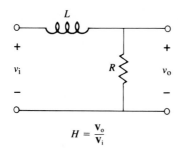

$$H = \frac{V_o}{V_i}$$

**FIGURE P11.6**

**11.7** Repeat Problem 11.6 for the circuit shown in Figure P11.7.

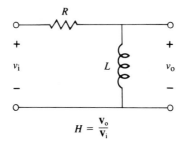

$$H = \frac{V_o}{V_i}$$

**FIGURE P11.7**

**11.8 a.** Analytically determine and plot the frequency response curves for $I_C/V_s$ in Figure P11.8.

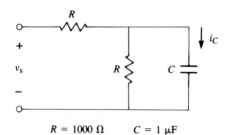

**FIGURE P11.8**        $R = 1000\,\Omega$        $C = 1\,\mu\text{F}$

**b.** Rescale the frequency axis of part a to $f$ (Hz).

**c.** Characterize the signal filtering properties of the electrical circuit.

11.9 For the circuit in Problem 11.8, select values for $R$ and $C$ in order to have a cutoff frequency of 10 kHz.

11.10 Consider the circuit of Example 11.3 (Figure 11.6). Plot the magnitude and phase curves of its input impedance for $L = 1/13$ H and $C = 1/2$ F and for $R = 1\,\Omega$, $2\,\Omega$, $4\,\Omega$, and $10\,\Omega$. Use a computer for the calculations if you wish. In each case, estimate the center frequency and bandwidth.

11.11 Plot the magnitude and phase frequency response curves for $\mathbf{V}_R/\mathbf{V}_s$, $\mathbf{V}_L/\mathbf{V}_s$, and $\mathbf{V}_C/\mathbf{V}_s$ of the circuit in Figure P11.11. For each case, characterize the filtering properties of the system function. From the curves, estimate the center and cutoff frequencies, when they exist.

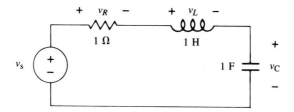

**FIGURE P11.11**

11.12 Refer back to Problem 11.11. Mathematically find the exact values of center and cutoff frequencies for each of the three system functions.

11.13 Consider the system function $H(s)$.

$$H(s) = \frac{200}{s^2 + 2s + 100}$$

**a.** Plot the magnitude and phase curves.

**b.** Estimate the cutoff frequency of the low-pass filter. It is the frequency at which the magnitude is .707 of its DC value.

11.14 Determine and plot the frequency response curves for the operational amplifier circuit shown in Figure P11.14 when:

**a.** $Z_1$ consists of a 100 µF capacitor and $Z_2$ is a 1000 $\Omega$ resistor.

**b.** $Z_1$ consists of a 1000 $\Omega$ resistor and $Z_2$ is a 100 µF capacitor.

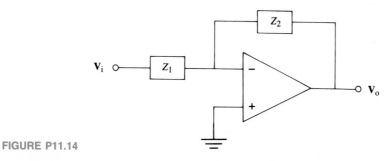

**FIGURE P11.14**

c. $Z_1$ is a 200 Ω resistor in series with a 100 µF capacitor and $Z_2$ is a parallel combination of a 10 kΩ resistor and a 1 µF capacitor.

11.15 Sketch the linear frequency response characteristics of the system function $\mathbf{I}_L / \mathbf{V}_s$ in Figure P11.15.

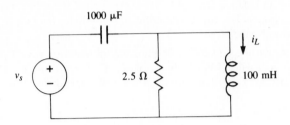

**FIGURE P11.15**

11.16 Use the graphical inspection technique to sketch the general shape of the magnitude functions corresponding to the pole–zero plots of Figure P11.16.

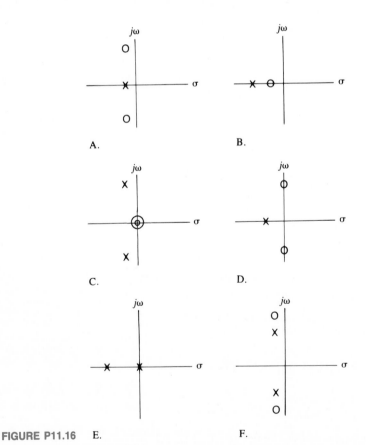

**FIGURE P11.16**  E.

**11.17** Repeat Problem 11.16 for the following system functions:

a. $H(s) = \dfrac{10s}{s+5}$

b. $H(s) = \dfrac{s+1}{s+10}$

c. $H(s) = \dfrac{s^2 + 2s + 100}{s^2 + 2s + 10{,}000}$

d. $H(s) = \dfrac{s}{s^2 + 6s + 5}$

**11.18** Estimate the locations of poles and zeros for system functions having the magnitude functions shown in Figure P11.18.

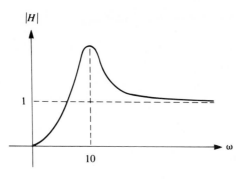
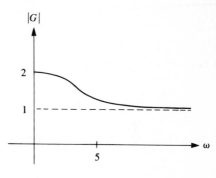

**FIGURE P11.18**   A.                                                              B.

**11.19** The standard form for a quadratic term in a system function has been $s^2 + 2\zeta\omega_n s + \omega_n^2 = 0$. Plot the loci of points on the $s$ plane corresponding to the following:

a. $\omega_n = 10$ with $\zeta$ as a variable parameter

b. $\omega_n = 30$ with $\zeta$ as a variable parameter

c. $\zeta = .4$ with $\omega_n$ as a variable parameter

d. $\zeta = .8$ with $\omega_n$ as a variable parameter

e. $\omega = 5$ and $\zeta = .707$

**11.20** An alternate standard form to the one developed in Section 11.4 for quadratic terms is $s^2 + (\omega_n s/Q) + \omega_n^2 = 0$.

a. Plot, with $Q$ as a parameter, the locus of points on the $s$ plane corresponding to locations of the roots of the polynomial for $\omega_n = 1$.

b. Repeat part a for $Q = 1.25$ and $\omega_n$ as a variable parameter.

c. For what range of positive values of $Q$ are the roots complex conjugates?

d. For what range of positive $Q$ will resonant peaking exist if this quadratic form describes the locations of a pair of poles?

**11.21** Standard forms for second-order low-pass, bandpass, and high-pass filters are, respectively,

a. $H(s) = \dfrac{H_0 \omega_n^2}{s^2 + (\omega_n s/Q) + \omega_n^2}$

**b.** $H(s) = \dfrac{H_0\omega_n s/Q}{s^2 + (\omega_n s/Q) + \omega_n^2}$

**c.** $H(s) = \dfrac{H_0 s^2}{s^2 + (\omega_n s/Q) + \omega_n^2}$

By hand or with the aid of a computer, plot the magnitudes of these three system functions for $\omega_n = 1$, $Q = 2$, and $H_0 = 1$. Also, for each of the three cases, relate the locations of the zeros to the shape of the magnitude curve.

**11.22** For the bandpass system of Problem 11.21b, analytically determine and plot the magnitude for $H_0 = 1$, $\omega_n = 100$, and $Q = 1, 2$, and 10.

**11.23** Determine, as functions of $\omega_n$ and $Q$, the upper and lower cutoff frequencies of the system of Problem 11.21b.

**11.24** Find and plot the frequency response characteristics of the $RC$ circuit shown in Figure P11.24. Is resonance present?

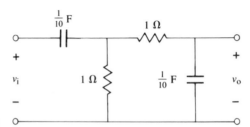

**FIGURE P11.24**

**11.25** Consider the series $RLC$ circuit shown in Figure P11.25.

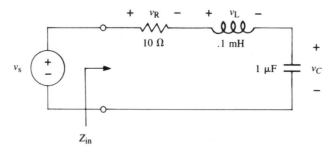

**FIGURE P11.25**            $Z_{in}$

**a.** Find the resonance frequency. It is the frequency at which $Z_{in}$ is a real number.
**b.** Find the system functions $V_R/V_s$, $V_L/V_s$, and $V_C/V_s$.
**c.** Find $\zeta$, $Q$, and $\omega_n$ for the circuit.
**d.** Plot (as in Section 11.3) the frequency response curves for $V_R/V_s$, $V_L/V_s$, and $V_C/V_s$.
**e.** Use the frequency response curves to verify KCL at the resonance frequency.

**11.26** Answer the following questions relating to the parallel resonance circuit shown in Figure P11.26.

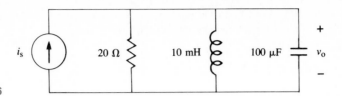

**FIGURE P11.26**

    **a.** Find the system function $V_o/I_s$.
    **b.** Find $\zeta$, $Q$, and $\omega_n$.
    **c.** Draw the frequency response curves.
    **d.** Select different values of $R$ that will give $Q$'s of 10 and 20.

**11.27** Draw Bode magnitude and phase curves for the circuit of Problem 11.6 under the four different conditions given.

**11.28** Draw Bode magnitude and phase diagrams for the system functions given. For each, use the curves to find the response to inputs of 1, $\cos t$, and $\cos 1000t$. Also, characterize the filter behaviors of the systems as reflected by their magnitude functions.

    **a.** $\quad H(s) = \dfrac{100}{(s+1)(s+10)}$      **b.** $\quad H(s) = \dfrac{1000s}{(s+1)(s+10)}$

    **c.** $\quad H(s) = \dfrac{10s}{(s+1)^2(s+10)}$      **d.** $\quad H(s) = \dfrac{s}{(s+1)(s^2+4s+100)}$

**11.29** Draw Bode diagrams for the system functions of Problem 11.17.

**11.30** Draw Bode frequency response curves for the circuit of Problem 11.14c.

**11.31** Draw Bode magnitude and phase curves for the following system functions:

    **a.** $\quad H(s) = \dfrac{20s^2}{(s+1)(s+10)}$      **b.** $\quad G(s) = \dfrac{s^2+2s+100}{s^2+10s+100}$

    **c.** $\quad A(s) = \dfrac{s^2+10s+400}{s^2+2s+400}$

**11.32** Draw Bode magnitude and phase curves for the circuit of Problem 11.15.

**11.33** A parallel $RLC$ circuit is shown in Figure P11.33. Draw a pole–zero plot and the Bode frequency response curves for the system function $V/I_s$ for each of the following sets of element values:

    **a.** $\quad L = 1/2\,\text{H}, \quad C = 1/50\,\text{F}, \quad R = 10\,\Omega$
    **b.** $\quad L = 1/2\,\text{H}, \quad C = 1/50\,\text{F}, \quad R = 2.5\,\Omega$
    **c.** $\quad L = 1/2\,\text{H}, \quad C = 1/50\,\text{F}, \quad R = .5\,\Omega$

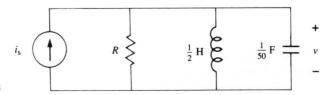

**FIGURE P11.33**

11.34 Shown in Figure P11.34 are the Bode magnitude and phase curves for a system. Use them to determine the corresponding system function.

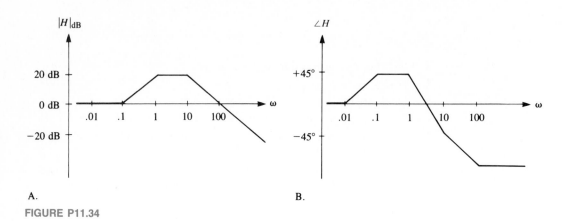

A.                                                                           B.

**FIGURE P11.34**

11.35 An experiment has been performed on an electrical filter. In it, the system function was experimentally measured at various frequencies. Use the data of Figure P11.35 to draw an estimated pole–zero plot for the filter.

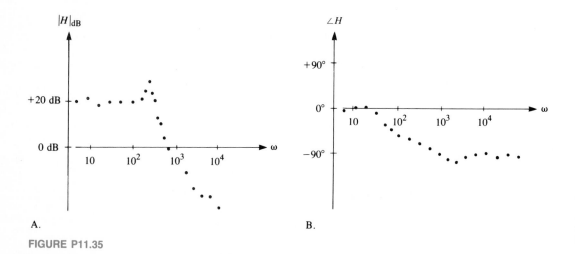

A.                                                                           B.

**FIGURE P11.35**

11.36 Draw Bode frequency response curves for the transistor circuit shown in Figure P11.36. Use the AC model for the bipolar junction transistor introduced in Chapter 1. Parameter values are $r_x = 1000\,\Omega$, $\beta = 100$, and $r_o = 10\,\text{k}\Omega$.

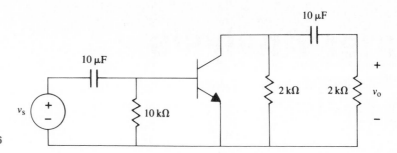

**FIGURE P11.36**

**11.37** The magnitude function for an $n$th-order Butterworth filter is

$$|H(\omega)| = \frac{1}{\sqrt{1 + \omega^{2n}}}$$

Draw Bode and linear magnitude curves for $n = 1, 2,$ and $3$. The function does not fit any of the standard forms discussed in Section 11.6. Your answer will have to be developed numerically either by hand or with the aid of a computer.

**11.38** Investigate the reason for normalizing the factored terms of a system function before you draw Bode diagrams. Determine and plot magnitude and phase curves for a normalized and non-normalized system function:

a. $H(s) = \dfrac{1}{(s/a) + 1}$

b. $G(s) = \dfrac{1}{s + a}$

c. Use the two rules suggested by parts a and b to draw Bode magnitude curves for the following system function (that is, draw the curve twice, first by one rule and then by the other):

$$A(s) = \frac{10s}{(s + .1)(s + 10)}$$

**11.39** For equation 11.15,

a. Find $|H(j\omega)|$

b. Find $\omega_p$, the frequency for maximum $|H(j\omega)|$. Hint: Maximize $|H(j\omega)|$ by minimizing its denominator.

c. For what range of $\zeta$ does $\omega_p$ exist (is real)?

# Fourier Analysis

12

## 12.1   INTRODUCTION

The title of this book, *Introduction to Circuit Analysis*, suggests that it focuses solely on the properties of electrical components and the systems that can be formed from their interconnections. By now, you should appreciate the fact, however, that electrical circuits cannot be studied separately from the signals that excite and exist within them. The two broad areas of analysis, the time domain and the frequency domain, are defined in terms of the mathematical descriptions of these signals. Until now, the emphasis has been on describing circuits and their response behavior. Signal properties have been investigated, but only to support the study of the circuits themselves. In this chapter, our focus will shift somewhat to the development of mathematical descriptions of a very special kind for signals. The method that will be introduced is known as Fourier analysis.

Fourier analysis, named for the French mathematician Joseph Fourier (1768–1830), is a technique for computing the frequency content of signals. The meaning of the term *frequency content* will become clear as we proceed. We have already used to our advantage the fact that many signals can be characterized by complex frequencies. This fact was the basis for the phasor notation developed in Chapters 8 and 10. The signals studied there could be characterized by just one or two frequencies. Many signals of engineering or scientific interest cannot be characterized by a small number of frequencies. Instead, they are composed of an infinite number of sinusoidal components, each with its own magnitude and phase.

Fourier analysis is a body of mathematics through which we can calculate the sinusoidal frequency content of time signals. That is, we can find the frequencies, amplitudes, and phases of all the individual components that go into making up the signal. Fourier analysis is mathematically sophisticated, but, in this book, we will not attempt to prove its validity with the utmost rigor. Instead, we will focus on its main functional techniques and emphasize an intuitive understanding of the approach and its results. All of the important aspects of Fourier analysis, however, will be covered. If more detail is desired, you may wish to consult a book that specializes in the topic.

## 12.2   THE FREQUENCY CONTENT OF SIGNALS

Everyone has experienced the phenomenon of a complicated signal that is composed of a number of individual components. The music produced by an orchestra, for example, is the sum of the "signals" generated by the separate instruments. Another example, also from the world of music, is closer to the point of this chapter. From physics, you are probably aware of the fact that instruments such as violins do not produce pure musical tones. A pure tone is one that varies sinusoidally at a given frequency. For instance, the note middle A consists of sinusoidal pressure waves at $f = 440$ Hz. The sounds produced by musical instruments consist not only of such pure tones but also of their harmonic overtones.

The harmonics are sinusoids whose frequencies are multiples of that of the main, or fundamental, tone. The amplitudes of the harmonics are related to the construction of the instrument.

Before we begin to formalize the techniques of Fourier analysis in the next section, let's more convincingly justify the central point that many types of time-varying signals can be written as sums of sinusoids of different frequencies. Consider the square wave shown in Figure 12.1A. If we tried to approximate it with a single sine wave, it would make sense to choose one that repeats itself at the same rate as the square wave. As shown in Figure 12.1B, the amplitude should be chosen to be somewhat larger than that of the square wave in order to minimize the overall error in the approximation. The approximation can be improved by adding another sinusoid at three times the frequency of the first. Notice from Figure 12.1C that this second contribution makes additions to the first approximation where it is too small and subtractions where it is too large. Figure 12.1D shows how an additional sinusoidal contribution at five times the frequency of the first can make further improvements to the overall approximation. In the limit, our approximation can become an exact replica of the original signal; an infinite number of well-chosen sinusoidal components, each with a particular frequency, amplitude, and phase, are required.

This intuitive demonstration certainly is not a rigorous proof that signals can be thought of as sums of related sinusoids. It does, however, give us some confidence before we proceed to the more abstract mathematics of Fourier analysis.

**FIGURE 12.1**   *A Periodic Square Wave and Successive Approximations*

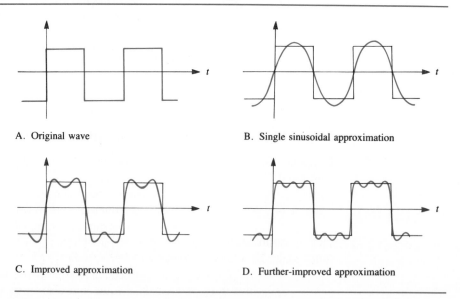

A. Original wave

B. Single sinusoidal approximation

C. Improved approximation

D. Further-improved approximation

## 12.3 THE FOURIER SERIES IN TRIGONOMETRIC FORM

The area of Fourier analysis that will be described in this section applies only to periodic functions of time. A **periodic function** is one that repeats itself at fixed intervals $T$, as described mathematically by the following equation:

$$f(t) = f(t \pm T) \tag{12.1}$$

for any value of $t$. The smallest value of $T$ for which Equation 12.1 is true is the fundamental **period** of the function. Figure 12.2 shows some examples of periodic functions.

Fourier demonstrated that a periodic function satisfying certain conditions can be written as a summation of harmonically related sinusoids:

$$f(t) = a_0 + \sum_{n=1}^{\infty} (a_n \cos n\omega_0 t + b_n \sin n\omega_0 t) \tag{12.2}$$

**FIGURE 12.2** *Periodic Functions*

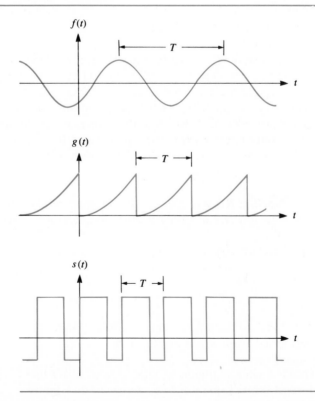

where $\omega_0 = 2\pi/T$ rad/s is the radian frequency of the sinusoid whose period $T$ is identical to the period of $f(t)$. Equation 12.2 defines the **Fourier series** representation of $f(t)$. The constants $a_n$ and $b_n$ are known as **Fourier coefficients.** The constant $a_0$ is the average (DC) value of $f(t)$.

Equation 12.2 can be written in another, equally useful, form that combines sines and cosines of equal frequency:

$$f(t) = a_0 + \sum_{n=1}^{\infty} A_n \cos(n\omega_0 t + \phi_n) \tag{12.3}$$

where    $A_n = \sqrt{a_n^2 + b_n^2}$

$\phi_n = -\tan^{-1}\left(\dfrac{b_n}{a_n}\right)$

The individual terms of Equation 12.3 are the **harmonics** of $f(t)$. The sinusoid with the lowest frequency ($\omega_0$) is the **fundamental** or first harmonic; that with the frequency $2\omega_0$ is the second harmonic; and that with the frequency of $n\omega_0$ is the $n$th harmonic. Each of the harmonics has a unique amplitude and phase. It is important to emphasize that a Fourier series does not contain sinusoids of any arbitrary frequency. Only terms that are harmonically related to $f(t)$ are present in its Fourier series.

Our goal here is to calculate the Fourier coefficients of periodic functions. These quantities are defined based on the series form of Equation 12.2. When it comes to interpreting the results, however, Equation 12.3 is often more useful. You should be comfortable with both equations. Before we discover how the coefficients are calculated, we state the sufficient conditions that, if met, guarantee that a Fourier series exists for the function $f(t)$.

A Fourier series for $f(t)$ exists if $f(t)$ satisfies the following properties, which are known as the **Dirichlet conditions:**

1. The function $f(t)$ has only a finite number of discontinuities in one period.
2. The function $f(t)$ has only a finite number of maxima and minima in one period.
3. The function $f(t)$ converges in the sense that

$$\int_{t_0}^{t_0+T} |f(t)|\, dt < \infty \tag{12.4}$$

The time $t_0$ is any arbitrary starting point for the integration, which is performed over one complete period of $f(t)$.

All periodic signals common to engineering satisfy these conditions.

Inspection of Equation 12.2 shows us that Fourier series are composed of three kinds of terms: a constant ($a_0$) and harmonically related sines and cosines.

It is possible to treat the constant as a cosine of frequency zero, but we are keeping it separate for clarity. Together, these terms comprise a set of **basis functions** for periodic signals. Harmonic sinusoids are not the only possible basis set for periodic functions, but they are among the most useful for engineers, especially because of all we know about describing the sinusoidal frequency response behavior of systems.

Our chosen basis set possesses a property that lets us easily calculate the Fourier coefficients of a periodic function. This important property is that the basis set is **orthogonal** over the period $T$. When $f_j(t)$ and $f_k(t)$ are both functions in a basis set, the set is orthogonal if

$$\int_{t_0}^{t_0+T} f_j(t) f_k(t)\, dt = 0, \qquad j \neq k \tag{12.5}$$

Again, $t_0$ is any arbitrary starting point, and the integration is performed over exactly one period. The term $f_j(t) f_k(t)$ is a cross product if $j \neq k$. All such cross-product terms must integrate to zero over a period $T$ if the basis set is orthogonal. Thus, for the sinusoidal basis set,

$$\int_{t_0}^{t_0+T} a_0 \cos n\omega_0 t\, dt = 0, \qquad \text{for all } n \tag{12.6a}$$

$$\int_{t_0}^{t_0+T} a_0 \sin n\omega_0 t\, dt = 0, \qquad \text{for all } n \tag{12.6b}$$

$$\int_{t_0}^{t_0+T} \sin n\omega_0 t \cos m\omega_0 t\, dt = 0, \qquad \text{for all } n, m \tag{12.6c}$$

$$\int_{t_0}^{t_0+T} \sin n\omega_0 t \sin m\omega_0 t\, dt = 0, \qquad \text{for all } n \neq m \tag{12.6d}$$

$$\int_{t_0}^{t_0+T} \cos n\omega_0 t \cos m\omega_0 t\, dt = 0, \qquad \text{for all } n \neq m \tag{12.6e}$$

The correctness of Equation 12.6c will be demonstrated here; verification of the other equations is left to you in Problem 12.1 at the end of the chapter. Appendix B can be consulted to find that

$$\int_{t_0}^{t_0+T} \sin n\omega_0 t \cos m\omega_0 t\, dt = \frac{1}{2} \int_{t_0}^{t_0+T} [\sin (n-m)\omega_0 t + \sin (m+n)\omega_0 t]\, dt$$

The two sine waves on the right side of this equation are being integrated over integer numbers of their respective periods. Because a sine wave has an average value of zero over its period, the total result is zero and Equation 12.6c is seen to be correct.

Now, let's see how the property of orthogonality allows us to calculate the Fourier coefficients of a periodic function. Consider the following integration:

$$\int_{t_0}^{t_0+T} \cos n\omega_0 t\, f(t)\, dt = \int_{t_0}^{t_0+T} \cos n\omega_0 t$$

$$\times \left[ a_0 + \sum_{k=1}^{\infty} (a_k \cos k\omega_0 t + b_k \sin k\omega_0 t) \right] dt$$

There is an infinite number of product terms within the integral sign on the right side of the equation. All but one are cross products and will integrate to zero according to Equation 12.6. For $k = n$, we have the only term that is not a cross product, $(\cos n\omega_0 t)(\cos n\omega_0 t)$. We are left with

$$\int_{t_0}^{t_0+T} \cos n\omega_0 t\, f(t)\, dt = \int_{t_0}^{t_0+T} a_n \cos^2 n\omega_0 t\, dt$$

$$= \frac{1}{2} \int_{t_0}^{t_0+T} a_n(1 + \cos 2n\omega_0 t)\, dt = a_n\left(\frac{T}{2}\right)$$

We have again made use of the fact that when a sinusoid is integrated over an integer number of its periods, the result will be zero. The last equation can now be rewritten to give an expression for $a_n$:

$$a_n = \frac{2}{T} \int_{t_0}^{t_0+T} \cos n\omega_0 t\, f(t)\, dt$$

In a similar way, the coefficients of the sine terms can be calculated:

$$b_n = \frac{2}{T} \int_{t_0}^{t_0+T} \sin n\omega_0 t\, f(t)\, dt$$

Because $a_0$ is the average value of $f(t)$, it can be found by a more direct integration:

$$\int_{t_0}^{t_0+T} f(t)\, dt = \int_{t_0}^{t_0+T} \left[ a_0 + \sum_{k=1}^{\infty} (a_k \cos k\omega_0 t + b_k \sin k\omega_0 t) \right] dt = a_0 T$$

where, again, all of the sinusoids are integrated over integer numbers of periods. Therefore,

$$a_0 = \frac{1}{T} \int_{t_0}^{t_0+T} f(t)\, dt$$

Because the formulas for computing the Fourier coefficients are of such central importance, they are repeated here for easy reference:

$$a_0 = \frac{1}{T} \int_{t_0}^{t_0+T} f(t)\, dt \tag{12.7a}$$

$$a_n = \frac{2}{T} \int_{t_0}^{t_0+T} \cos n\omega_0 t\, f(t)\, dt \tag{12.7b}$$

$$b_n = \frac{2}{T} \int_{t_0}^{t_0+T} \sin n\omega_0 t\, f(t)\, dt \tag{12.7c}$$

The integrations of Equations 12.7a, 12.7b, and 12.7c can be performed over any complete period of the function $f(t)$. Usually, a period centered on or beginning at $t = 0$ is chosen. An interesting property of the Fourier coefficients is that they can all be computed independently of one another. The interpretation of Equations 12.7a, 12.7b, and 12.7c is self-evident, but performing the required calculations can be a time-consuming and tedious job. An appreciation of this chore will come only from looking at some examples. There is not enough space here to include every step in the calculations. On your own, you should work through missing steps as you read the examples.

**Example 12.1**   *Fourier Series of a Square Wave*

Let's compute the Fourier coefficients of the square wave function shown in Figure 12.1 and repeated in Figure 12.3.

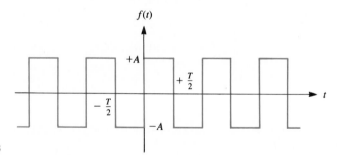

**FIGURE 12.3**

In the period between $-T/2 < t < +T/2$, the function can be written as

$$f(t) = \begin{cases} -A, & -\dfrac{T}{2} < t < 0 \\[2mm] +A, & 0 < t < \dfrac{T}{2} \end{cases}$$

*(continues)*

**Example 12.1**  *Continued*

By inspection, we can see that the average value of $f(t)$ is zero. This fact is easily verified:

$$a_0 = \frac{1}{T} \int_{-T/2}^{T/2} f(t)\, dt = \frac{1}{T} \left[ \int_{-T/2}^{0} (-A)\, dt + \int_{0}^{T/2} A\, dt \right]$$

$$= \frac{1}{T} \left[ -\frac{AT}{2} + \frac{AT}{2} \right] = 0$$

Next, we compute $a_n$, the coefficients of the cosine terms:

$$a_n = \frac{2}{T} \int_{-T/2}^{T/2} \cos n\omega_0 t\, f(t)\, dt$$

$$= \frac{2}{T} \left[ \int_{-T/2}^{0} (-A) \cos n\omega_0 t\, dt + \int_{0}^{T/2} A \cos n\omega_0 t\, dt \right]$$

$$= \frac{2A}{Tn\omega_0} \left[ -\sin 0 + \sin\left(-n\omega_0 \frac{T}{2}\right) + \sin\left(n\omega_0 \frac{T}{2}\right) - \sin 0 \right]$$

Using the fact that $\sin(-x) = -\sin x$, we conclude that

$$a_n = 0, \qquad \text{for all } n$$

Finally, we calculate the $b_n$'s:

$$b_n = \frac{2}{T} \left[ \int_{-T/2}^{0} (-A) \sin n\omega_0 t\, dt + \int_{0}^{T/2} A \sin n\omega_0 t\, dt \right]$$

$$= \frac{2A}{Tn\omega_0} \left[ \cos 0 - \cos\left(\frac{n\omega_0 T}{2}\right) - \cos\left(-\frac{n\omega_0 T}{2}\right) + \cos 0 \right]$$

When we make the substitution $\omega_0 = 2\pi/T$, this equation can be evaluated as follows:

$$b_n = \frac{2A}{n\pi} [1 - \cos n\pi]$$

While the expression that has been found for $b_n$ is correct, it is customary to put it in a somewhat different form. We note that

$$\cos n\pi = \begin{cases} +1, & n \text{ even} \\ -1, & n \text{ odd} \end{cases}$$

from which it can be concluded that

$$b_n = \begin{cases} 0, & n \text{ even} \\ \dfrac{4A}{n\pi}, & n \text{ odd} \end{cases}$$

Now that we have calculated all of the coefficients, we can write the Fourier series for $f(t)$:

$$f(t) = \frac{4A}{\pi} \sin \omega_0 t + \frac{4A}{3\pi} \sin 3\omega_0 t + \frac{4A}{5\pi} \sin 5\omega_0 t + \frac{4A}{7\pi} \sin 7\omega_0 t + \cdots$$

**Example 12.2**    *Fourier Series of a Triangle Wave*

Represent the triangular waveform of Figure 12.4 as a Fourier series. Our first step is to find an analytical expression for $f(t)$ over a complete period.

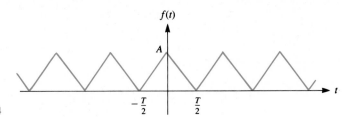

**FIGURE 12.4**

The function can be written as

$$f(t) = \begin{cases} \left(\dfrac{2A}{T}\right)\left(t + \dfrac{T}{2}\right), & -\dfrac{T}{2} < t < 0 \\ \left(\dfrac{2A}{T}\right)\left(-t + \dfrac{T}{2}\right), & 0 < t < +\dfrac{T}{2} \end{cases}$$

The average value of the signal is

$$a_0 = \frac{A}{2}$$

You can verify this by using Equation 12.7a.

*(continues)*

**Example 12.2**   *Continued*

Next, $a_n$ is found:

$$a_n = \frac{2}{T} \int_{-T/2}^{T/2} \cos n\omega_0 t \; f(t) \; dt$$

$$= \frac{2}{T} \int_{-T/2}^{0} \left(\frac{2A}{T}\right)\left(t + \frac{T}{2}\right) \cos n\omega_0 t \; dt$$

$$+ \frac{2}{T} \int_{0}^{T/2} \left(\frac{2A}{T}\right)\left(-t + \frac{T}{2}\right) \cos n\omega_0 t \; dt$$

Because a cosine integrated over an integer number of periods is zero, this equation can be simplified as follows:

$$a_n = \frac{4A}{T^2} \left( \int_{-T/2}^{0} t \cos n\omega_0 t \; dt - \int_{0}^{T/2} t \cos n\omega_0 t \; dt \right)$$

We integrate by parts defining

$$u = t \qquad \text{and} \qquad dv = \cos n\omega_0 t \; dt$$

This leads to

$$du = dt \qquad \text{and} \qquad v = \left(\frac{1}{n\omega_0}\right) \sin n\omega_0 t$$

which results in

$$a_n = \frac{4A}{T^2} \left( \frac{t}{n\omega_0} \sin n\omega_0 t \Big|_{-T/2}^{0} - \int_{-T/2}^{0} \frac{1}{n\omega_0} \sin n\omega_0 t \; dt \right)$$

$$- \frac{4A}{T^2} \left( \frac{t}{n\omega_0} \sin n\omega_0 t \Big|_{0}^{T/2} - \int_{0}^{T/2} \frac{1}{n\omega_0} \sin n\omega_0 t \; dt \right)$$

Laborious but straightforward calculations simplify this equation to

$$a_n = \begin{cases} \dfrac{4A}{(n\pi)^2}, & n \text{ odd} \\[2ex] 0, & n \text{ even} \end{cases}$$

Finally, you should verify that $b_n = 0$ for this function. The Fourier series can now be written:

$$f(t) = \frac{A}{2} + \frac{4A}{\pi^2} \cos \omega_0 t + \frac{4A}{(3\pi)^2} \cos 3\omega_0 t + \frac{4A}{(5\pi)^2} \cos 5\omega_0 t + \cdots$$

## 12.4 THE USE OF SYMMETRY IN FOURIER SERIES CALCULATIONS

You probably noticed that the Fourier series found in Example 12.1 consisted only of sine waves and that in Example 12.2 had only cosine terms. Also, both functions contained only odd-numbered harmonics. You may have wondered if some method exists for knowing without calculation whether certain components will be missing from the Fourier series of a function. In fact, there is such a method. It is based on symmetries that might be present in the periodic function $f(t)$. Recognizing such symmetries can greatly reduce the number of calculations needed.

### Even and Odd Symmetry

You may remember from earlier studies of mathematics the two most widely recognized symmetries that a function can have. A function $f(t)$ displays **even symmetry** if

$$f(-t) = f(t) \tag{12.8a}$$

A function has **odd symmetry** if

$$f(-t) = -f(t) \tag{12.8b}$$

Plots of various even functions are shown in Figure 12.5. Two additional examples are $\cos \omega t$ and $t^2 + 2$. Some odd functions are shown in Figure 12.6. $A \sin \omega t$ and $-10t^3$ are two more examples that have odd symmetry.

Not all functions have even or odd symmetry. Nevertheless, it is possible to write an arbitrary function $f(t)$ as the sum of an even and an odd part. That is,

$$f(t) = f_e(t) + f_o(t) \tag{12.9}$$

where $f_e(t)$ is the even component and $f_o(t)$ is the odd component. These even and odd parts can be found from $f(t)$ by using Equations 12.8 and 12.9. In Problem 12.6 at the end of the chapter, you are asked to show that

$$f_e(t) = \frac{1}{2} [f(t) + f(-t)] \tag{12.10a}$$

$$f_o(t) = \frac{1}{2} [f(t) - f(-t)] \tag{12.10b}$$

Figure 12.7 shows an example of a periodic function and the successive steps needed to graphically find its even and odd parts.

FIGURE 12.5    *Even Functions*

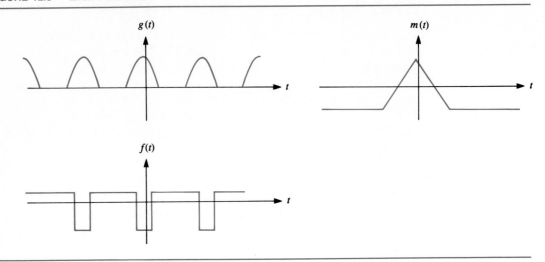

FIGURE 12.6    *Odd Functions*

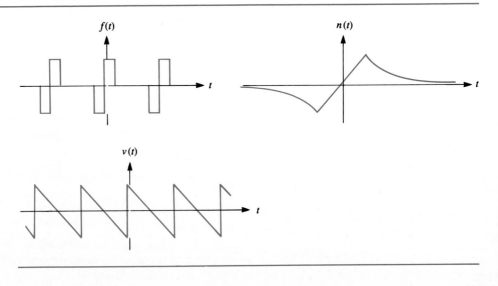

FIGURE 12.7    *The Even and Odd Parts of a Periodic Function*

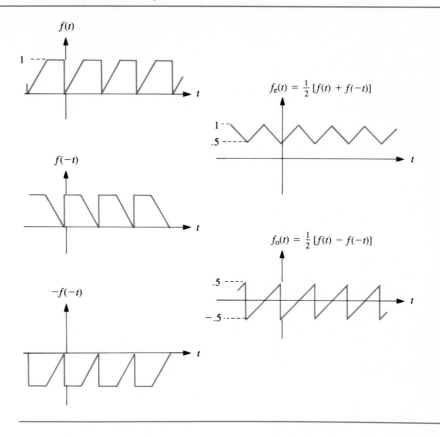

Several additional properties of symmetry must be introduced before we can apply it to the task of computing Fourier coefficients. The first three of these properties are summarized as follows:

1. The product of two even functions is an even function.
2. The product of two odd functions is an even function.
3. The product of an even function and an odd function is an odd function.

You are asked to prove these properties in Problem 12.8 at the end of the chapter.

Another important property of symmetrical functions is seen when they are integrated over limits that are symmetrical about the origin. Figure 12.8A shows that integrating an odd function over symmetrical limits results in an answer of zero because of equal positive and negative areas under the curve. As shown in Figure 12.8B, even functions will not, in general, integrate to zero in this way. Their symmetry, however, does allow for a simplification. Because the area under

FIGURE 12.8   *Integrating Odd and Even Functions over Symmetrical Limits*

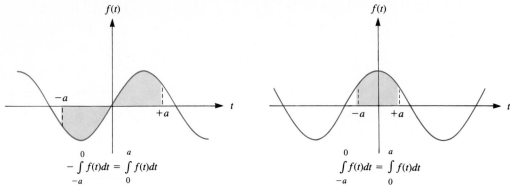

A. Odd function results in area of zero          B. Even function does not

the curve of an even function from $-a$ to 0 is the same as that from 0 to $+a$, the total area is exactly twice that from 0 to $+a$. These observations can be expressed mathematically as follows:

$$\int_{-a}^{+a} f(t)\, dt = 2 \int_{0}^{+a} f(t)\, dt \qquad \text{(even function)} \tag{12.11a}$$

$$\int_{-a}^{+a} f(t)\, dt = 0 \qquad \text{(odd function)} \tag{12.11b}$$

Let's apply these properties to the task of finding the Fourier coefficients of a periodic function. We rewrite Equations 12.7b and 12.7c using symmetrical limits of integration:

$$a_n = \frac{2}{T} \int_{-T/2}^{+T/2} \cos n\omega_0 t \, f(t) \, dt \tag{12.12a}$$

$$b_n = \frac{2}{T} \int_{-T/2}^{+T/2} \sin n\omega_0 t \, f(t) \, dt \tag{12.12b}$$

Consider what happens when $f(t)$ is even. Then, Equation 12.12a is the integration of an even function over symmetrical limits. In general, the result is nonzero. However, $b_n$ is found by integrating an odd function over symmetrical limits. The result will always be zero. We conclude that *the Fourier series of an even function contains only cosine terms*. The formulas for finding $a_0$ and $a_n$ can be simplified by using Equation 12.11a. We can summarize our observations for even functions as follows:

*Even Function*

$$a_0 = \frac{2}{T} \int_0^{T/2} f(t)\, dt \qquad (12.13a)$$

$$a_n = \frac{4}{T} \int_0^{T/2} \cos n\omega_0 t\, f(t)\, dt \qquad (12.13b)$$

$$b_n = 0 \qquad (12.13c)$$

If $f(t)$ is odd, just the opposite happens. *The Fourier series of an odd function contains only sine terms*. All of the $a_n$ coefficients are zero. We can summarize the results for odd functions as follows:

*Odd Function*

$$a_0 = 0 \qquad (12.14a)$$

$$a_n = 0 \qquad (12.14b)$$

$$b_n = \frac{4}{T} \int_0^{T/2} \sin n\omega_0 t\, f(t)\, dt \qquad (12.14c)$$

---

**Example 12.3**  *Fourier Series of a Rectified Sine Wave*

Let's find the Fourier series for the full-wave rectified sine wave, $f(t) = |A \sin(\pi/T)t|$, shown in Figure 12.9.

By inspection, we see that $f(t)$ is an even function. Thus, $b_n = 0$ for all $n$, and no sine terms will appear in the series.

We use Equations 12.13a and 12.13b to find the coefficients. First, $f(t)$ must be analytically described:

$$f(t) = A \sin \frac{\pi}{T} t, \qquad 0 < t < \frac{T}{2}$$

It is left to you to show that the average value of $f(t)$ is $a_0 = 2A/\pi$.

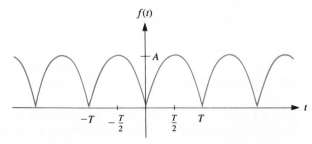

**FIGURE 12.9**

(*continues*)

**Example 12.3**   *Continued*

Next, $a_n$ is found:

$$a_n = \frac{4A}{T} \int_0^{T/2} \cos n \frac{2\pi}{T} t \sin \frac{\pi}{T} t \, dt$$

We use a trigonometric identity for the product of a sine and a cosine:

$$a_n = \frac{2A}{T} \int_0^{T/2} \left[ \sin(1 - 2n) \frac{\pi}{T} t + \sin(1 + 2n) \frac{\pi}{T} t \right] dt$$

$$= \frac{2A}{T} \left[ -\frac{\cos(1 - 2n)(\pi/T)t}{(1 - 2n)(\pi/T)} - \frac{\cos(1 + 2n)(\pi/T)t}{(1 + 2n)(\pi/T)} \right] \Bigg|_0^{T/2}$$

$$= \frac{4A}{\pi(1 - 4n^2)}$$

The Fourier series can now be compactly written as

$$f(t) = \frac{2A}{\pi} + \sum_{n=1}^{\infty} \frac{4A}{\pi} \frac{1}{1 - 4n^2} \cos n\omega_0 t$$

---

## Half-Wave Symmetry

There is an additional type of symmetry that is helpful in Fourier series calculations. A function displays **half-wave symmetry** when

$$f(t) = -f\left( t \pm \frac{T}{2} \right) \tag{12.15}$$

Unlike even and odd symmetry, the property of half-wave symmetry applies only to periodic functions. Some examples are shown in Figure 12.10. Figure 12.10A can be used to visually interpret the meaning of Equation 12.15. First, a half-cycle of a periodic function is identified. Then, it is shifted a half-period $(T/2)$ in either direction. If the shifted segment can be inverted about the horizontal axis to exactly match the function, the function has half-wave symmetry. It is stated without proof that all functions with half-wave symmetry possess only odd harmonics. Recognizing half-wave symmetry is an important check on the correctness of calculations used in finding the Fourier coefficients.

Half-wave symmetry leads to simplifications in the formulas for calculating Fourier coefficients similar to those in Equations 12.13 and 12.14. These simplifications will be developed in Problem 12.12 at the end of the chapter.

FIGURE 12.10   *Functions with Half-Wave Symmetry*

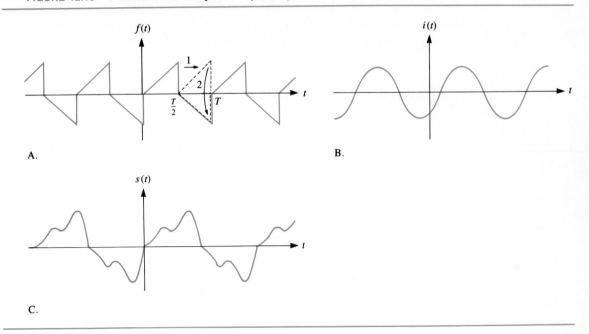

A.

B.

C.

## Hints for Using Symmetry

The recognition of symmetry in functions is not always as easy as it would seem from the discussion so far. For instance, the triangular waveform of Example 12.2 contains only odd harmonics, and, yet, if we try to apply Equation 12.15, the function fails the test for half-wave symmetry. One of the tricks to making use of symmetry is to understand that the harmonics of a function are independent of its DC value. Shifting a function up or down by changing its DC value affects only $a_0$ and none of the other coefficients. If a waveform does not seem to have any symmetries, you should imagine removing its DC component and checking the function again. This approach will sometimes uncover odd or half-wave symmetries that would otherwise be overlooked.

In a similar way, a function can be shifted to the left or right along the time axis without affecting the amplitudes of any of the harmonics (the coefficients $A_n$ of Equation 12.3). Functions can sometimes be adjusted in this way to create even or odd symmetries. This approach eliminates the need to calculate half of the coefficients. The answer must then be shifted back an equal amount in order to arrive at the correct Fourier series. Some of the problems at the end of the chapter will give you practice in these methods.

## 12.5   THE FOURIER SERIES IN EXPONENTIAL FORM

Another equivalent form for the Fourier series of Equation 12.2 will be shown in this section. We begin by noting that Euler's theorem can be used to find sines and cosines as sums of complex exponentials:

$$\cos n\omega_0 t = \frac{1}{2}\left(e^{jn\omega_0 t} + e^{-jn\omega_0 t}\right) \tag{12.16a}$$

$$\sin n\omega_0 t = \frac{1}{j2}\left(e^{jn\omega_0 t} - e^{-jn\omega_0 t}\right) \tag{12.16b}$$

These expressions can be substituted into Equation 12.2 to arrive at, after some reorganization,

$$f(t) = a_0 + \sum_{n=1}^{\infty}\left[\left(\frac{a_n - jb_n}{2}\right)e^{jn\omega_0 t} + \left(\frac{a_n + jb_n}{2}\right)e^{-jn\omega_0 t}\right] \tag{12.17}$$

We can use the term with the positive exponent to define a new coefficient:

$$c_n = \frac{a_n - jb_n}{2}, \qquad n = 1, 2, 3, \ldots \tag{12.18}$$

We now substitute into Equation 12.18 the known formulas for $a_n$ and $b_n$. The integrations will be performed over the interval $-T/2 < t < +T/2$:

$$c_n = \frac{1}{T}\int_{-T/2}^{T/2} f(t)(\cos n\omega_0 t - j\sin n\omega_0 t)\, dt$$

$$= \frac{1}{T}\int_{-T/2}^{T/2} f(t)e^{-jn\omega_0 t}\, dt, \qquad n = 1, 2, 3, \ldots \tag{12.19}$$

Next, we turn our attention to the coefficient of the term in Equation 12.17 that has the negative exponent. This coefficient is $(a_n + jb_n)/2$. It is the complex conjugate of $c_n$ and, thus, of the integral of Equation 12.19. That is,

$$c_n^* = \frac{a_n + jb_n}{2} = \frac{1}{T}\int_{-T/2}^{T/2} f(t)e^{jn\omega_0 t}\, dt, \qquad n = 1, 2, 3, \ldots \tag{12.20}$$

If we compare Equations 12.19 and 12.20, we see that $c_n^*$ can be found by replacing $n$ with $-n$ in Equation 12.19:

$$c_{-n} = c_n^* = \frac{a_n + jb_n}{2}, \qquad n = 1, 2, 3, \ldots \tag{12.21}$$

With the additional definition that $c_0 = a_0$ (let $n = 0$ in Equation 12.19), we can rewrite Equation 12.17 as follows:

$$f(t) = c_0 + \sum_{n=1}^{\infty} c_n e^{jn\omega_0 t} + \sum_{n=1}^{\infty} c_{-n} e^{-jn\omega_0 t}$$

Now, negate the variable of summation in the last term:

$$f(t) = c_0 + \sum_{n=1}^{\infty} c_n e^{jn\omega_0 t} + \sum_{n=-1}^{-\infty} c_n e^{jn\omega_0 t}$$

The first term on the right-hand side of the equation corresponds to $n = 0$; the middle term is a summation over all positive $n$; and the last term a summation over all negative $n$. These three terms, then, can be combined into one grand summation:

$$f(t) = \sum_{n=-\infty}^{\infty} c_n e^{jn\omega_0 t} \qquad (12.22)$$

where

$$c_n = \frac{1}{T} \int_{-T/2}^{T/2} f(t) e^{-jn\omega_0 t}\, dt, \qquad -\infty < n < \infty \qquad (12.23)$$

Equation 12.22 is the complex exponential form of the Fourier series. Its coefficients can be found through Equation 12.23. This form has the advantage over the trigonometric series in that all of the coefficients $c_n$ can be found with just one formula. Note that in contrast to the trigonometric form of the Fourier series, Equation 12.22 is summed over $n$ from $-\infty$ to $+\infty$. Thus, both methods use the same number of coefficients (two) to describe each harmonic and the same number to describe the complete function. However, the symmetry properties of Section 12.4 cannot be used here. All of the quantities of the exponential series are complex numbers, although some coefficients may be purely real or purely imaginary. Complex functions do not display simple symmetries of the types that we have described.

---

**Example 12.4**   *Fourier Series of a Square Wave (Exponential Form)*

Let's find the exponential-form Fourier series of the square wave function of Figure 12.11. It is the same function as previously analyzed in Example 12.1.

*(continues)*

**Example 12.4**  *Continued*

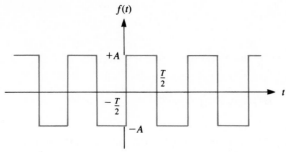

**FIGURE 12.11**

The coefficients are found as

$$c_n = \frac{1}{T}\int_{-T/2}^{+T/2} f(t)e^{-jn\omega_0 t}\, dt$$

$$= \frac{1}{T}\left[\int_{-T/2}^{0} -Ae^{-jn\omega_0 t}\, dt + \int_{0}^{+T/2} Ae^{-jn\omega_0 t}\, dt\right]$$

$$= \frac{1}{T}\left[\frac{A}{jn\omega_0}\, e^{-jn\omega_0 t}\Big|_{-T/2}^{0} + \frac{A}{-jn\omega_0}\, e^{-jn\omega_0 t}\Big|_{0}^{+T/2}\right]$$

$$= \frac{1}{T}\left[\frac{A}{jn\omega_0} - \frac{A}{jn\omega_0}\, e^{jn\omega_0 T/2} - \frac{A}{jn\omega_0}\, e^{-jn\omega_0 T/2} + \frac{A}{jn\omega_0}\right]$$

$$= \frac{A}{jn\omega_0 T}\left[2 - (e^{jn\omega_0 T/2} + e^{-jn\omega_0 T/2})\right]$$

$$= \frac{A}{jn\omega_0 T}\left[2 - 2\cos\left(\frac{n\omega_0 T}{2}\right)\right]$$

After making the substitution $\omega_0 = 2\pi/T$ and performing some manipulations, we arrive at

$$c_n = \begin{cases} 0, & n \text{ even} \\[2mm] -j\,\dfrac{2A}{n\pi}, & n \text{ odd} \end{cases}$$

As was said earlier, the DC constant can also be found from Equation 12.23. Sometimes, this calculation results in a division by zero. This difficulty can be avoided either by setting $n$ to zero before the integration or by making use of l'Hôpital's rule after the integration. In this case, inspection tells us that $c_n = 0$, which is in agreement with the result arrived at above. We can now write the complex exponential form of the Fourier series:

$$f(t) = \sum_{n \text{ odd}} \left( -j \frac{2A}{n\pi} \right) e^{jn\omega_0 t}$$

---

The results of Examples 12.1 and 12.4 should, of course, be consistent. We can check this by comparing $c_n$ with the earlier values of $a_n$ and $b_n$. We do so by separating

$$c_n = \frac{a_n - jb_n}{2}, \qquad n > 0$$

into

$$a_n = 2Re[c_n], \qquad n > 0 \tag{12.24a}$$

and

$$b_n = -2Im[c_n], \qquad n > 0 \tag{12.24b}$$

The demonstration that, indeed, the Fourier series of the two examples are consistent with each other is left for your investigation.

## 12.6  FREQUENCY SPECTRA

The **spectrum** of a signal is the distribution of amplitudes and phases of its components as functions of frequency. If you encountered the term before, it was probably with respect to the spectral content of light sources such as stars. Such spectra are often displayed as vertical lines along a frequency axis to indicate which frequency components are present in the source and, just as important, which are not.

A signal's Fourier coefficients are an analytical description of its spectrum. Very often, they are plotted as functions of frequency. The plots can be based either on the coefficients $a_0$, $A_n$, and $\phi_n$ of Equation 12.3 or on the coefficients $c_n$ of Equation 12.22. These equations are repeated here, with the second in an expanded form:

$$f(t) = a_0 + \sum_{n=1}^{\infty} A_n \cos(n\omega_0 t + \phi_n)$$

$$f(t) = \sum_{n=-\infty}^{+\infty} c_n e^{jn\omega_0 t}$$

$$= \cdots + c_{-2} e^{-j2\omega_0 t} + c_{-1} e^{-j\omega_0 t} + c_0 + c_1 e^{j\omega_0 t} + c_2 e^{j2\omega_0 t} + \cdots$$

Before we look at some examples, let's see how the coefficients $c_n$ relate to the magnitudes and phases of the harmonics. Note that while $A_n$ and $\phi_n$ are real numbers, $c_n$ is, in general, complex. Assume a polar form of

$$c_n = |c_n|e^{j\theta_n} \quad \text{and} \quad c_{-n} = |c_n|e^{-j\theta_n}$$

Now, consider one of the conjugate pairs that forms a harmonic in the exponential series:

$$c_{-n}e^{-jn\omega_0 t} + c_n e^{+jn\omega_0 t} = |c_n|e^{-j\theta_n}e^{-jn\omega_0 t} + |c_n|e^{j\theta_n}e^{jn\omega_0 t}$$

$$= |c_n|[e^{-j(n\omega_0 t+\theta_n)} + e^{j(n\omega_0 t+\theta_n)}]$$

$$= 2|c_n|\cos(n\omega_0 t + \theta_n)$$

When this last expression is compared with Equation 12.3, we see that $A_n = 2|c_n|$. That is, $|c_n|$ is half the magnitude of the $n$th harmonic. We see also that the angle of $c_n$ ($n > 0$) is equal to the phase angle of the $n$th harmonic. We use this last fact to rewrite $c_n$ to conform to our earlier notation for phase angle:

$$c_n = |c_n|e^{j\phi_n} \tag{12.25}$$

We plot the spectra of periodic signals in either of two ways: (1) The phase and magnitude of the coefficients $c_n$ are plotted for all $n$, or (2) $A_n$ and $\phi_n$ are plotted for $n > 0$ (along with $a_0$). The plots can be shown as functions either of frequency or of the harmonic number $n$. Figure 12.12 shows examples of each of these two approaches. Figure 12.12A shows the **magnitude spectrum** and **phase spectrum** for the square wave of Example 12.1. These plots are based on the coefficients of Equation 12.3. Figure 12.12B shows the alternate method of presenting magnitude and phase spectra for the same function. These plots are based on the coefficients that were found in Example 12.4. In both cases, the spectral components are plotted as functions of the harmonic number $n$. If the fundamental frequency ($\omega_0 = 2\pi/T$) is known, then the scales of the abscissas can be converted to frequency.

Plots like those in Figure 12.12 are referred to as **line spectra** or **discrete spectra.** Part A is a one-sided line spectrum, while part B is a two-sided line spectrum. Periodic functions have only certain discrete frequency components (the harmonics). The graphical presentation of their spectra is in the form of a number of vertical lines whose lengths correspond to the values of amplitudes or phase shifts.

Engineers and scientists are often more interested in the frequency distribution of the energy content of a signal rather than in the amplitudes of its harmonic components. For a periodic signal, however, it makes more sense to talk of the power of the signal rather than of its energy. The energy that a periodic signal can deliver is limited only by the length of time that the signal is active; the average

FIGURE 12.12    *Magnitude and Phase Spectra of Square Wave of Figure 12.11*

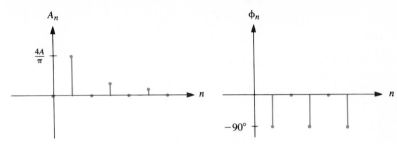

A.  Spectrum based on Equation 12.3 and found in Example 12.1

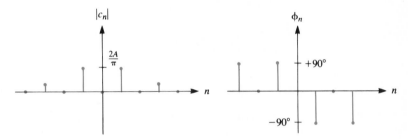

B.  Spectrum based on Equation 12.22 and found in Example 12.4

power or rate at which energy is delivered is a constant property of a periodic signal. This power is proportional to the square of the signal's magnitude function. Plots of such magnitude-squared functions are known as **power spectra.** Clearly, the power spectra for periodic signals, like their magnitude spectra, have nonzero components only at certain frequencies. More will be said about energy content when nonperiodic signals are introduced in the next section.

The spectral plots just described are very much like the frequency response curves of Chapter 11. In both cases, the magnitude and phase of a complex quantity are plotted as functions of frequency. There is, however, a significant difference in the two situations. Frequency response curves describe properties of systems; spectra describe the nature of signals. Taken together, these two methods can be used to determine the spectrum of a system's output signal in response to an input of known frequency content. This is shown by the simple example in Figure 12.13. Note that although the spectrum of the response is easily found in this way, it is, in general, not easy to find the response as a simple closed-form function of time.

Before we turn to the last topic of this chapter, let's look at one more example of finding and plotting the frequency content of a signal.

**FIGURE 12.13**  *Finding Spectrum of a Circuit's Response*

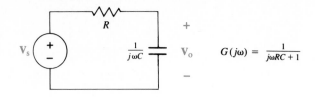

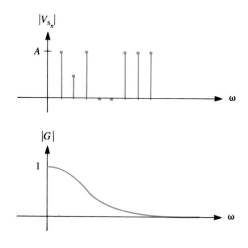

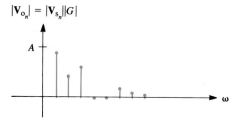

**Example 12.5**  *Fourier Series of a Rectangular Pulse Train (Exponential Form)*

Consider the rectangular pulse train of Figure 12.14. The width of an individual pulse is $\tau$, and the period is $T$.

We find the Fourier series coefficients of Equation 12.23 as follows:

$$c_n = \frac{1}{T} \int_{-T/2}^{T/2} f(t) e^{-jn\omega_0 t}\, dt = \frac{1}{T} \int_{-\tau/2}^{\tau/2} A e^{-jn\omega_0 t}\, dt$$

$$= \frac{A}{jn\omega_0 T} [e^{jn\omega_0 \tau/2} - e^{-jn\omega_0 \tau/2}] = \frac{A\tau}{T} \frac{\sin(n\omega_0 \tau/2)}{n\omega_0 \tau/2} \tag{12.26}$$

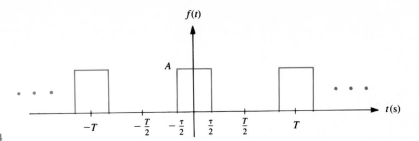

FIGURE 12.14

The variable $n\omega_0$ is meaningful only for integer values of $n$. However, the values of $c_n$ are more easily plotted if, for the moment, we treat $n\omega_0$ as if it were a continuous frequency. The dashed line in Figure 12.15 is a plot of Equation 12.26 when $n\omega_0$ is treated as a continuous variable. This dashed line is the **envelope** of the spectrum. Its value is meaningful only when $n$ is an integer—that is, where the spectral lines are shown in the figure. The single plot of the figure accounts for both magnitude and phase information because, in this particular example, all of the $c_n$ are purely real numbers. Where the spectrum is positive, its angle is $0°$; where it is negative, its angle is $\pm 180°$. The spectrum has been plotted as a function of the frequency $f$ for convenience of presentation.

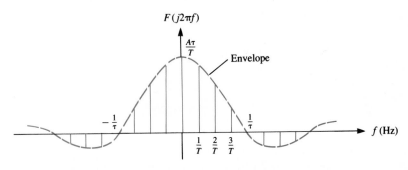

FIGURE 12.15

The function that describes the envelope of the spectrum of Example 12.5 is extremely important in Fourier analysis. It appears so often in the spectra of signals that it is given its own name, sinc (pronounced "sink") and defined as follows:

$$\text{sinc } \pi x = \frac{\sin \pi x}{\pi x} \tag{12.27}$$

**FIGURE 12.16    *Sinc Function***

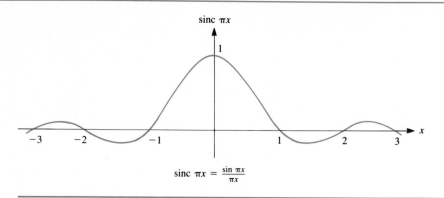

$$\text{sinc } \pi x = \frac{\sin \pi x}{\pi x}$$

Sometimes, you will see the sinc function defined without $\pi$. That is,

$$\text{sinc } x = \frac{\sin x}{x}$$

The $\pi$ is included as a normalization factor. The sinc function is shown plotted in Figure 12.16. Note that it has a value of 1 at the origin. This fact is often a point of confusion when it is first encountered. However, when you remember that $\sin \pi x \approx \pi x$ for small $x$, you can see that

$$\lim_{x \to 0} \text{sinc } \pi x \approx \frac{\pi x}{\pi x} = 1$$

## 12.7  THE FOURIER TRANSFORM

Periodic functions are very useful in electrical engineering, but they in no way account for all of the signals that are encountered in practical applications. The Fourier transform provides us with a method of computing the frequency content of some signals that are aperiodic (not periodic). The Fourier transform can also be applied to periodic functions in some circumstances, but we will not do so here.

We begin our development of the Fourier transform by considering more carefully the function and spectrum of Example 12.5. The discussion here is entirely heuristic but will greatly aid us in interpreting the results of the more mathematical development that will follow. The rectangular pulse train of Figure 12.14 along with its already computed magnitude spectrum (Figure 12.15) is repeated in Figure 12.17. The important parameters of the function are its period ($T$) and the width of an individual rectangular pulse ($\tau$). Both affect the amplitude of the spectrum. The basic shape of the envelope, however, depends only on $\tau$. The envelope has zero crossings at $f = 1/\tau$, $2/\tau$, $3/\tau$, and so on.

**FIGURE 12.17**  *Spectrum of Rectangular Pulse Train*

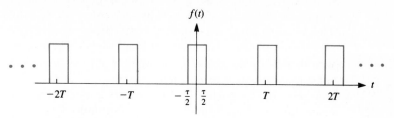

A. Rectangular pulse train

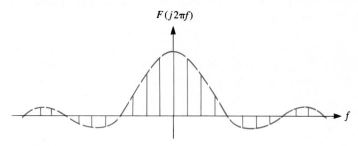

B. Spectrum

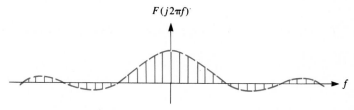

C. Spectrum as $T$ doubles

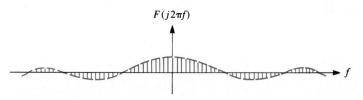

D. Spectrum as $T$ doubles again

However, the locations of individual harmonics depend only on the period of the time function. Harmonics exist at $f = 1/T$ and its integer multiples. In addition, there is a frequency component at DC ($f = 0$ Hz).

Let's see what happens if $\tau$ is kept constant and $T$ is allowed to increase, eventually approaching infinity. Clearly, $f(t)$ will become an aperiodic function consisting of a single rectangular pulse centered on the origin. Figures 12.17C and

12.17D show the spectrum after $T$ has been doubled and then doubled again from its original value. The general shape of the envelope is unchanged, but its amplitude decreases. Also seen to decrease is the spacing between the harmonics. That is, over any range of frequency there are more harmonics, but each is of smaller amplitude than in the original signal. These two effects interact in such a way that the sum of the amplitudes of the harmonics over any frequency range remains approximately constant. Finally, as $T$ approaches infinity and the function $f(t)$ becomes aperiodic, all frequencies are infinitesimally represented. We conclude that aperiodic functions have continuous spectra. Thus, aperiodic signals can contain, in general, some energy at all frequencies, which is in contrast to periodic time functions that have energy (power) only at those frequencies that are harmonically related to the period.

Now, let's proceed with a semi-rigorous proof of the Fourier transform. It is most conveniently done in terms of the radian frequency $\omega$. As with the preceding heuristic discussion, we begin with a periodic function and allow its period to approach infinity. The fundamental frequency of a periodic function is $\omega_0 = 2\pi/T$. This same value is the spacing between adjacent harmonics of the spectrum as a function of $\omega$:

$$\Delta\omega = \frac{2\pi}{T}$$

In the limit as $T$ approaches infinity, $\Delta\omega$ becomes the differential

$$d\omega = \lim_{T\to\infty}\left(\frac{2\pi}{T}\right)$$

We apply the definition of $\Delta\omega$ to Equation 12.23, the formula for finding the Fourier coefficients, $c_n$:

$$c_n = \frac{1}{T}\int_{-T/2}^{T/2} f(t)e^{-jn\omega_0 t}\,dt = \frac{\Delta\omega}{2\pi}\int_{-T/2}^{T/2} f(t)e^{-jn\omega_0 t}\,dt$$

$$2\pi\,\frac{c_n}{\Delta\omega} = \int_{-T/2}^{T/2} f(t)e^{-jn\omega_0 t}\,dt \qquad\qquad (12.28)$$

The ratio on the left-hand side of Equation 12.28 does not change as $T$ approaches infinity because both the numerator, $c_n$, and the denominator, $\Delta\omega$, are inversely proportional to $T$. Individual frequency components become vanishingly small, but they also become infinitesimally close to one another in such a way that the total "strength" or energy of the spectral components within a narrow range of frequency remains constant. The ratio $c_n/\Delta\omega$, then, is seen to be a **spectral density** (per radian). A spectral density is a measure of the distribution of the strength of a signal as a function of frequency. We use this complex quantity to define the function:

$$F(jn\omega_0) = 2\pi \frac{c_n}{\Delta\omega} = \int_{-T/2}^{T/2} f(t)e^{-jn\omega_0 t}\, dt$$

The final step is to note that, as $T$ approaches infinity, the limits of integration approach $\pm\infty$ and $n\omega_0$ becomes a continuous frequency variable. Thus,

$$F(j\omega) = \int_{-\infty}^{\infty} f(t)e^{-j\omega t}\, dt \tag{12.29}$$

Equation 12.29 defines the **Fourier transform** of the aperiodic function $f(t)$. It describes the spectral density of the function $f(t)$. For electrical variables, spectral densities have dimensions of volts per unit frequency or amperes per unit frequency. A similar relationship, the **inverse Fourier transform,** can be used to find $f(t)$ if $F(j\omega)$ is known. It is defined as follows:

$$f(t) = \frac{1}{2\pi} \int_{-\infty}^{\infty} F(j\omega)e^{j\omega t}\, d\omega \tag{12.30}$$

Equation 12.30 is derived by starting with the summation of Equation 12.22 and letting $T$ approach infinity.

The Fourier transform and its inverse are mathematical operations. They are sometimes indicated with an operational notation, as follows:

$$F(j\omega) = \mathscr{F}\{f(t)\} \qquad \text{(direct)}$$
$$f(t) = \mathscr{F}^{-1}\{F(j\omega)\} \qquad \text{(inverse)}$$

The Fourier transform and its inverse can be proven in rigorous terms mathematically. Fourier transforms exist for all time functions satisfying the Dirichlet conditions discussed earlier with the one modification that the function be absolutely integrable in the sense that

$$\int_{-\infty}^{\infty} |f(t)|\, dt < \infty$$

These conditions are sufficient but not necessary. Fourier transforms exist for some functions that do not satisfy the Dirichlet conditions. Some of these functions will be considered in the next section.

Let's apply this new tool to the aperiodic function that began our heuristic development of the Fourier transform.

**Example 12.6**   *Fourier Transform of a Rectangular Pulse*

Find the Fourier transform of the single rectangular pulse shown in Figure 12.18A.

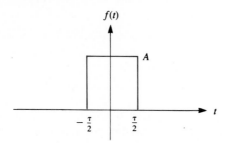

A. Single rectangular pulse

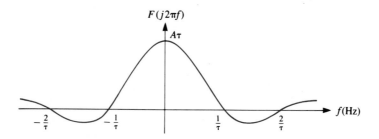

**FIGURE 12.18**   B. Fourier transform

Note that $f(t) = A$ for $-\tau/2 < t < \tau/2$ and is zero otherwise:

$$F(j\omega) = \int_{-\infty}^{\infty} f(t)e^{-j\omega t}\, dt = \int_{-\tau/2}^{\tau/2} Ae^{-j\omega t}\, dt = \frac{A}{-j\omega}\left(e^{-j\omega\tau/2} - e^{j\omega\tau/2}\right)$$

$$= A\tau\left[\frac{\sin(\omega\tau/2)}{\omega\tau/2}\right]$$

Because this transform is purely real, $F(j2\pi f)$ rather than its magnitude is shown plotted in Figure 12.18B. The phase of $F(j2\pi f)$ is $0°$ when $F(j2\pi f)$ is positive and $\pm180°$ when $F(j2\pi f)$ is negative.

If we compare the Fourier transform of Example 12.6 with the Fourier series of Example 12.5, we see that the Fourier transform of a single rectangular pulse is the same as the envelope of the spectrum of a periodic function made of repeated rectangular pulses. This observation is very significant and is not limited to this specific example. If a periodic function is created by an infinite repetition of identical and equally spaced pulses of arbitrary shape, then the envelope of its spectrum is the same as the Fourier transform of an individual pulse. The locations of the harmonics of the Fourier series depend only on the rate of repetition of the periods. The relationship between the hypothetical spectrum of a signal $p(t)$ and that of its periodic extension $f(t)$ is shown in Figure 12.19.

Always remember that, unlike the discrete spectrum of a periodic function, the magnitude of the Fourier transform at any frequency is not equal to the amplitude of a component at that frequency. If such were the case, the signal would have infinite energy because there would be an infinite number of such components. Instead, the height of the transform's magnitude function is its spectral density. The area under the curve over a frequency range is proportional to the strength of the signal in that range. More correctly, we should say that the area under the square of the magnitude function over any frequency range is proportional to the signal's total energy in that frequency range. This relationship is shown in Figure 12.20.

**FIGURE 12.19**  *Spectra of a Pulse and Its Periodic Extension*

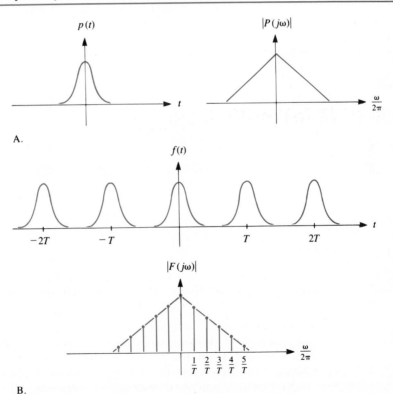

FIGURE 12.20   *Relationship between Magnitude and Power Spectra*

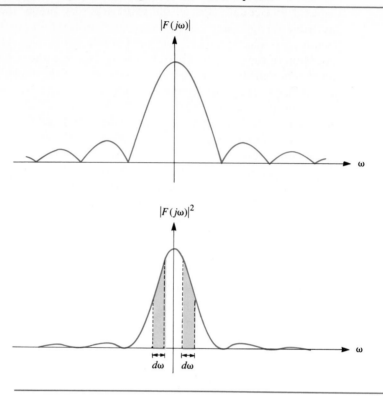

## 12.8   FOURIER TRANSFORM PAIRS AND OPERATIONS

A time function and its Fourier transform (if it exists) form a unique pair:

$$f(t) \leftrightarrow F(j\omega)$$

If either of the pair is known, the other automatically follows.* So that we do not

---

* To say that if $f(t)$ is known, its transform "automatically follows" makes the process sound more straightforward than it might possibly be. Application of the Fourier transform or its inverse sometimes results in indeterminate forms that cannot be evaluated. In this case, indirect methods must be used to find the transform.

have to use the integral transform relationships every time a transform is needed, tables of transform pairs and operational relationships have been developed. In many cases, these tables can be used to find Fourier transforms and their inverses without resorting to Equations 12.29 and 12.30. In this section, we develop short tables of transform pairs and transform operations. Table 12.1 gives the transforms of a number of basic time functions. Table 12.2 shows the effects that certain mathematical operations in the time domain have on Fourier transforms.

**TABLE 12.1** *Some Basic Fourier Transform Pairs*

| $f(t)$ | $F(j\omega)$ |
|---|---|
| 1. $u(t)$ | $\pi\delta(\omega) + \dfrac{1}{j\omega}$ |
| 2. $\delta(t)$ | $1$ |
| 3. $1$ | $2\pi\delta(\omega)$ |
| 4. $e^{-at}u(t)$ | $\dfrac{1}{j\omega + a}$ |
| 5. $e^{j\omega_0 t}$ | $2\pi\delta(\omega - \omega_0)$ |
| 6. $\cos \omega_0 t$ | $\pi[\delta(\omega + \omega_0) + \delta(\omega - \omega_0)]$ |
| 7. $\sin \omega_0 t$ | $j\pi[\delta(\omega + \omega_0) - \delta(\omega - \omega_0)]$ |
| 8. $\cos \omega_0 t\, u(t)$ | $\dfrac{j\omega}{\omega_0^2 - \omega^2} + \dfrac{\pi}{2}[\delta(\omega - \omega_0) + \delta(\omega + \omega_0)]$ |
| 9. $\sin \omega_0 t\, u(t)$ | $\dfrac{\omega_0}{\omega_0^2 - \omega^2} + \dfrac{\pi}{j2}[\delta(\omega - \omega_0) - \delta(\omega + \omega_0)]$ |
| 10. $e^{-at}\cos \omega_0 t\, u(t)$ | $\dfrac{j\omega + a}{(j\omega + a)^2 + \omega_0^2}$ |
| 11. $e^{-at}\sin \omega_0 t\, u(t)$ | $\dfrac{\omega_0}{(j\omega + a)^2 + \omega_0^2}$ |
| 12. $u\left(t + \dfrac{T}{2}\right) - u\left(t - \dfrac{T}{2}\right)$ | $T\,\dfrac{\sin(\omega T/2)}{\omega T/2}$ |

**TABLE 12.2**    *Fourier Transform Operational Pairs*

| | Property | $f(t)$ | $F(j\omega)$ |
|---|---|---|---|
| 1. | Linearity | $af_1(t) + bf_2(t)$ | $aF_1(j\omega) + bF_2(j\omega)$ |
| 2. | Differentiation | $\dfrac{d^n f}{dt^n}$ | $(j\omega)^n F(j\omega)$ |
| 3. | Integration | $\displaystyle\int_{-\infty}^{t} f(x)\, dx$ | $\dfrac{1}{j\omega} F(j\omega) + \pi F(0)\delta(\omega)$ |
| 4. | Time shifting | $f(t - t_0)$ | $e^{-j\omega t_0} F(j\omega)$ |
| 5. | Frequency shifting | $e^{j\omega_0 t} f(t)$ | $F[j(\omega - \omega_0)]$ |
| 6. | Time scaling | $f(at)$ | $\dfrac{1}{a} F\!\left(j\dfrac{\omega}{a}\right), \quad a > 0$ |
| 7. | Multiplication by $t$ | $tf(t)$ | $j\dfrac{dF(j\omega)}{d\omega}$ |
| 8. | Convolution | $\left.\begin{array}{l} \displaystyle\int_{-\infty}^{\infty} h(\lambda)x(t-\lambda)\, d\lambda \\[2em] \displaystyle\int_{-\infty}^{\infty} h(t-\lambda)x(\lambda)\, d\lambda \end{array}\right\}$ | $H(j\omega)X(j\omega)$ |
| 9. | Multiplication | $f_1(t)f_2(t)$ | $\dfrac{1}{2\pi}\displaystyle\int_{-\infty}^{\infty} F_1(j\lambda)F_2[j(\omega - \lambda)]\, d\lambda$ |
| 10. | Modulation | $f(t)\cos\omega_c t$ | $\dfrac{1}{2}[F(j(\omega - \omega_c)) + F(j(\omega + \omega_c))]$ |

## Simple Exponential

Let's begin by finding the Fourier transform of a decaying exponential that begins at $t = 0$:

$$f(t) = e^{-at}u(t)$$

We use Equation 12.29 to find its transform:

$$F(j\omega) = \int_{-\infty}^{\infty} e^{-at}u(t)e^{-j\omega t}\, dt = \int_{0^+}^{\infty} e^{-at}e^{-j\omega t}\, dt$$

$$= \frac{1}{-(a + j\omega)}\, e^{-(a+j\omega)t}\,\bigg|_{0^+}^{\infty}$$

Evaluated at the upper limit, the complex exponential becomes

$$\lim_{t \to \infty} e^{-at} (\cos \omega t - j \sin \omega t) = 0$$

We are left with

$$F(j\omega) = \frac{1}{a + j\omega}$$

We have thus established our first Fourier transform pair:

$$e^{-at}u(t) \leftrightarrow \frac{1}{a + j\omega} \tag{12.31}$$

This relationship is valid only for $a > 0$. An increasing exponential has no Fourier transform because it is not absolutely integrable.

## Unit Impulse

Next, let's consider the unit impulse function. Although a perfect impulse cannot be generated in the physical world, it is of immense theoretical importance. For generality, we consider an impulse at the arbitrary location $t_0$. Its transform is

$$F(j\omega) = \int_{-\infty}^{\infty} \delta(t - t_0)e^{-j\omega t} \, dt$$

The integral can be evaluated by applying the sifting property of the impulse function, which states that

$$\int_{-\infty}^{\infty} \delta(t - t_0)f(t) \, dt = f(t_0)$$

This relationship was proven in Chapter 9. Applied to the present situation, it leads directly to the transform pair

$$\delta(t - t_0) \leftrightarrow e^{-j\omega t_0} \tag{12.32}$$

For the special case of $t_0 = 0$, Equation 12.32 becomes

$$\delta(t) \leftrightarrow 1$$

From either of the two preceding transform expressions, we can see a very important property of the impulse function. The magnitude of the impulse function's spectrum is a constant. That is, *all* frequencies are equally represented in the ideal impulse! Therefore, the impulse can be thought of as the most versatile of forcing functions; it simultaneously stimulates the system with all frequencies. However, no physical device could possibly generate a signal with energy spread equally over all frequencies. Therefore, we reluctantly conclude that one of the impulse function's most interesting properties is the very reason that it is impossible to generate one. Nonetheless, the impulse can be reasonably approximated by a very narrow pulse if the pulse is used to excite a system whose time constants are large compared to the duration of the pulse.

### Complex Exponential

At this point, let's do something that may seem a little backward. Let's ask ourselves the question, is there a time function whose *transform* is the impulse? That is, can we find $f(t)$ such that

$$F(j\omega) = \delta(\omega - \omega_0)$$

If so, then $f(t)$ can be found through Equation 12.30:

$$f(t) = \frac{1}{2\pi} \int_{-\infty}^{\infty} \delta(\omega - \omega_0) e^{j\omega t} \, d\omega$$

We can again use the sifting property of the impulse to find the transform pair:

$$e^{j\omega_0 t} \leftrightarrow 2\pi\delta(\omega - \omega_0) \tag{12.33}$$

This result is in perfect agreement with our work in Chapter 10. There we said that the exponential $e^{j\omega_0 t}$ is characterized by the single complex frequency $j\omega_0$. Fourier analysis neatly arrives at the same result. The spectrum of $e^{j\omega_0 t}$ is confined exclusively to the frequency $+\omega_0$.

Comparing the transform pairs of Equations 12.32 and 12.33 provides a good example of one of the central principles of Fourier analysis. We have shown that an impulse in the time domain transforms into a complex exponential in the frequency domain. The reverse is also true. Transform pairs are always complementary in this way except for a scale factor. Reflection upon Equations 12.29 and 12.30 indicates that this should be so. The two equations are identical in all important respects.

### Constant

A special case of a complex exponential in the time domain occurs when $\omega_0 = 0$. We are left with a constant whose transform can be written as

$$1 \leftrightarrow 2\pi\delta(\omega) \tag{12.34}$$

Again, Fourier analysis nicely supports our intuition and earlier experience. The spectrum of a constant is seen to be confined to $\omega = 0$ (DC).

The complex exponential, in itself, is not of much practical use. However, we can use its known transform to compute the transforms of the sine and cosine. This is left for you to do in Problem 12.29 at the end of the chapter.

## Modulation

Finally, let's demonstrate two operational transforms. Unless you have been introduced to the basic principles of communications theory, the usefulness of the first might seem a bit puzzling. We wish to show the effect that multiplying a time function by a sinusoid will have on the function's spectrum. That is,

$$\mathcal{F}\{f(t)\cos\omega_c t\} = ?$$

The product $f(t)\cos\omega_c t$ is a form of the process of amplitude modulation. Modulation is the mathematical basis of radio transmission. The frequency $\omega_c$ is the carrier frequency. It is the frequency (expressed as $f_c = \omega_c/2\pi$ Hz) to which you tune your AM radio. The process of frequency modulation (FM) transmission is similar but somewhat more involved mathematically. A derivation of the effects of modulation as defined here follows:

$$\mathcal{F}\{f(t)\cos\omega_c t\} = \frac{1}{2}\int_{-\infty}^{\infty} f(t)(e^{j\omega_c t} + e^{-j\omega_c t})e^{-j\omega t}\,dt$$

$$= \frac{1}{2}\int_{-\infty}^{\infty} f(t)e^{-j(\omega-\omega_c)t}\,dt + \frac{1}{2}\int_{-\infty}^{\infty} f(t)e^{-j(\omega+\omega_c)t}\,dt$$

$$= \frac{1}{2}\{F[j(\omega-\omega_c)] + F[j(\omega+\omega_c)]\} \qquad (12.35)$$

In a similar way, it can be shown that

$$\mathcal{F}\{f(t)\sin\omega_c t\} = \frac{1}{2j}\{F[j(\omega-\omega_c)] - F[j(\omega+\omega_c)]\} \qquad (12.36)$$

This result is very interesting and important. Notice that the *shape* of the original spectrum is not changed. It is merely shifted along the frequency axis to be centered at both $\omega = +\omega_c$ and $\omega = -\omega_c$. The original signal can, therefore, be recaptured by reversing the frequency shifting process and appropriately filtering the result.

## Differentiation

To conclude this section, let's see what effect differentiation in the time domain has on a spectrum. Equation 12.30 tells us that

$$f(t) = \frac{1}{2\pi} \int_{-\infty}^{\infty} F(j\omega)e^{j\omega t}\, d\omega$$

$$\frac{df(t)}{dt} = \frac{1}{2\pi} \frac{d}{dt} \int_{-\infty}^{\infty} F(j\omega)e^{j\omega t}\, d\omega = \frac{1}{2\pi} \int_{-\infty}^{\infty} [j\omega F(j\omega)]e^{j\omega t}\, d\omega$$

Thus,

$$\frac{df(t)}{dt} \leftrightarrow j\omega F(j\omega) \tag{12.37}$$

In words, differentiation in the time domain corresponds to multiplication by $j\omega$ in the frequency domain. This property can be applied to higher-order derivatives and allows us to solve differential equations with Fourier transform techniques. We arrived at similar results in developing the use of phasors and of the Laplace transform. In the former case, differentiation corresponds to multiplication by $j\omega$. In the latter, it corresponds to multiplication by $s$. Keep in mind that all of the techniques for going from the time domain to the frequency domain are just different aspects of one general method.

## 12.9  THE USE OF FOURIER TRANSFORMS IN CIRCUIT ANALYSIS

Because the Fourier transform is a linear operation, it can be used to solve linear electrical circuit problems. Of course, the signals present in the circuit must be Fourier transformable. One possible use of the method is to write the necessary differential equations and then apply the transform operation to them. A more direct application of Fourier analysis is to transform the circuit itself. All of the signals in the circuit are treated as Fourier variables. The passive elements become impedances based on the differential and integral operations of Table 12.2. That is, for the inductor,

$$v(t) = L\,\frac{di(t)}{dt} \qquad \rightarrow \qquad V(j\omega) = j\omega L I(j\omega) \tag{12.38}$$

and for the capacitor,

$$i(t) = C\,\frac{dv(t)}{dt} \qquad \rightarrow \qquad I(j\omega) = j\omega C V(j\omega) \tag{12.39}$$

These impedance relationships are similar to those used with the phasor notation of Chapter 8 with the important difference that here $\omega$ is treated as a variable of the transformation and is not evaluated at the frequency of the forcing function.

**Example 12.7** *Circuit Analysis Problem*

Use the Fourier transform to find the current in the circuit of Figure 12.21A. Its transformed version is shown in Figure 12.21B.

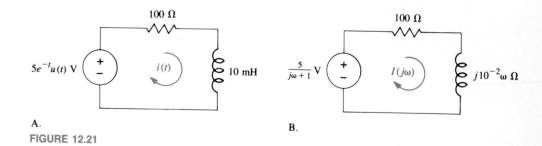

A.

B.

**FIGURE 12.21**

First, write a KVL equation:

$$\frac{5}{j\omega + 1} = 100I(j\omega) + j10^{-2}\omega I(j\omega)$$

$$I(j\omega) = \frac{500}{(j\omega + 1)(j\omega + 10{,}000)}$$

To find $i(t)$, put the expression in a form that has a recognizable inverse transform. Using partial fraction expansion techniques (see Section 9.5) yields

$$I(j\omega) = \frac{500/9999}{j\omega + 1} - \frac{500/9999}{j\omega + 10{,}000}$$

Now, consult Table 12.1 and write the inverse Fourier transform as follows:

$$i(t) = \frac{500}{9999}(e^{-t} - e^{-10{,}000t})u(t) \text{ A}$$

## 12.10 COMPARING THE FOURIER AND LAPLACE TRANSFORMS

You have probably come to the realization that the Fourier transform is very much like the Laplace transform. Important similarities exist between the two transform operations, but there are some significant differences.

Let's begin with the similarities. Both transforms are used to find frequency-domain representations of time functions. Thus, we can use either transform to solve circuit problems in the frequency domain. The process is as follows. First, a time-domain representation of the problem (either a circuit model or a set of appropriate equations) is found. Then, the transform is applied to either the circuit model or its describing equations to take the problem into the frequency domain. This step in the process has the advantage of transforming differential equations into algebraic equations, although complex variables must be used. Next, these equations are manipulated to find the transformed version of the solution. Finally, the frequency-domain answer is reverse transformed back to the time domain. For both the Fourier and the Laplace transforms, this inverse operation is usually done by referring to tables of transform pairs rather than by applying the integral relationships.

Although the procedures in using the two transforms are identical, some fundamental differences in their applications should be understood. The Laplace transform that we have defined in Chapter 9 is one-sided; it is most usefully applied to time functions that are zero for $t < 0$. The Fourier transform can be applied to functions defined for all time, sinusoids being one such case. Engineers consider the Laplace transform, however, to be applicable to a wider class of functions. The real part of the exponential contained within the Laplace transform integral forces the transform to converge for many one-sided functions that are not absolutely integrable. A good example is the ramp function, which is Laplace transformable but which has no Fourier transform.

As we saw in many of the examples in Chapter 9, the Laplace transform easily handles problems involving circuits not initially at rest. The Fourier transform can also be used in this way, but the mathematical manipulations can be very messy because impulse functions are needed to account for the initial conditions. It is dangerous to make sweeping statements, but the mathematical manipulations needed in Fourier analysis are usually more intricate than they are with the Laplace transform.

The Laplace transform has its greatest advantage in the analysis of transient problems involving initial conditions. It also has the attribute of describing important properties of the systems themselves. The system functions arrived at through Laplace analysis contain the natural frequencies and zeros of the system. When evaluated at $s = j\omega$, the system function gives the frequency response characteristics.

The Fourier transform more obviously gives us information about signals themselves than does the Laplace transform. The frequency spectrum of a time signal is directly related to its Fourier transform. Although the Fourier transform can be used to analyze nearly any circuit problem that the Laplace transform can, it is usually the better choice only for problems in the steady state or when the nature of the signals themselves is of prime interest.

If we had approached the Laplace and Fourier transforms more rigorously in the mathematical sense, we would have seen that the Fourier transform can be treated as a special case of the Laplace transform. Based on the presentation in this book, this is true in a somewhat limited sense. Because the one-sided Laplace transform cannot be used for functions that are defined for negative time, let's restrict our attention to the two transformations applied to a function $f(t)$, $t > 0$.

The resulting two transforms, $F(s)$ and $F(j\omega)$, may be related by

$$F(j\omega) = F(s)\big|_{s=j\omega} \tag{12.40}$$

This relationship is true only if $f(t)$ strictly meets the Dirichlet conditions. Thus, from Table 12.1 and Table 9.1, we see that Equation 12.40 applies to simple exponentials and to exponentially damped sinusoids, but not to the unit step function $u(t)$ because the unit step function does not satisfy the convergence criterion. Such mathematical details are left to later studies. We see from Equation 12.40 that although the Laplace transform is related to the entire $s$ plane, the Fourier transform is restricted to the $j\omega$ axis. We see also that it is no accident that similar notation has been used to represent the Fourier and Laplace transforms!

We have not considered system functions with respect to the Fourier transform, but they exist just as they do for Laplace transforms or for phasors. For a single-input/single-output system, a system function is the ratio of the Fourier transform of the output to the Fourier transform of the input function. It is left to you to demonstrate that this version of the system function is related to that of the Laplace transform by

$$H(j\omega) = H(s)\big|_{s=j\omega} \tag{12.41}$$

Note that Equation 12.41 is only meaningful for stable systems. Otherwise, use of the system function would lead to responses that increase without bound and for which Fourier transforms do not exist.

## 12.11  SUMMARY

Fourier analysis is a branch of mathematics used to describe functions in terms of their frequency content. This chapter introduced its two main areas. The Fourier series is an infinite series representation of a periodic function based on its harmonics. As such, only those frequencies that are harmonically related to the periodic function are present in its Fourier series. There are two distinct, equivalent forms for the Fourier series. The first has a basis set of real-valued trigonometric functions. Its main advantage lies in the intuitive ease with which it can be interpreted. Fourier series can also be written as infinite sums of complex exponentials. This form leads to a more compact notation for the derivation of the Fourier coefficients. This exponential series also has the advantage of leading more directly to the Fourier transform.

Although very useful in the study of periodic signals, Fourier series calculations are seldom applied to the analysis of electrical circuits. The reason for this is that, while Fourier series calculations let us easily determine the frequency content of a periodic function of time, no corresponding inverse operation is available. From what we already know, the response of a linear system to a periodic input can be found, but it usually must be written as an infinite summation of sinusoids rather than in a more useful closed-form representation.

The Fourier transform was used to find the spectra of aperiodic functions. In contrast to periodic functions, these spectra are continuous. A time function and its Fourier transform form a unique pair. Therefore, the Fourier transform can be conveniently applied to the analysis of electrical circuits. The existence of the inverse operation allows us to arrive at time-domain solutions of closed form. The inverse operation is usually done by referring to tables of transform pairs and operations.

The Fourier transform, thus, joins the collection of mathematical methods that can be applied to circuit analysis problems. There are times when it is the best choice—for example, when it is desired to know the spectra of signals within a circuit. There are other situations, however, when the Fourier transform is not the best choice. For problems involving transient response and initial energy storage, Laplace transform or time-domain methods are preferable to Fourier analysis. It is to the great benefit of the engineer not only to understand how to apply the many analytical tools at her or his disposal, but also to understand how to best select among them for particular problems.

Fourier analysis and transform theory, in general, are among the more sophisticated mathematical topics of engineering and physics. They are also among the most useful. Because the presentation here has been introductory, many of the details of Fourier analysis have, of necessity, been omitted.

## ■ PROBLEMS

**12.1** Prove that the basis set of constants and harmonically related sines and cosines is orthogonal.

**12.2** Verify Equation 12.7c.

**12.3** For Example 12.2, show by direct calculation that

    **a.** $a_0 = A/2$ (Use Equation 12.7a.)

    **b.** The coefficients $b_n = 0$ (Use Equation 12.7c.)

**12.4** Find trigonometric-form Fourier series for each of the periodic functions shown in Figure P12.4. Do not make use of symmetry. Instead, find every coefficient by direct calculation. Use tables of integrals where necessary.

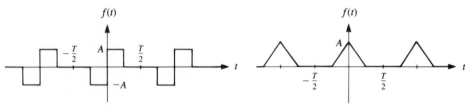

**FIGURE P12.4**   A.                                                      B.

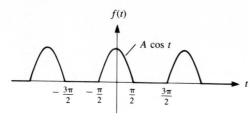

C.

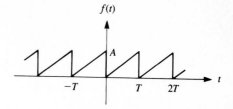

D.

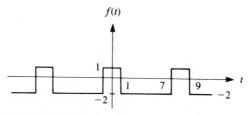

E.

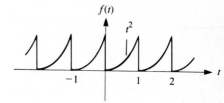

F.

**FIGURE P12.4** (*continued*)

12.5 The two functions shown in Figure P12.5 are similar to those in Examples 12.1 and 12.2. Use the results of those examples to write the Fourier series for each in the form of Equation 12.3.

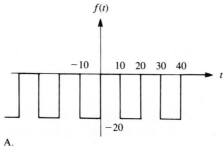

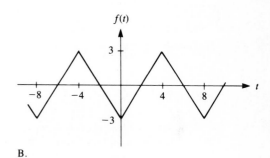

**FIGURE P12.5**   A.

B.

12.6 Prove that the even and odd parts of a function can be found according to Equations 12.10a and 12.10b.

12.7 Sketch the even and odd parts of the functions shown in Figure P12.7.

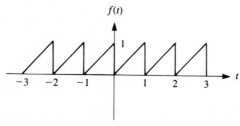

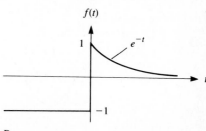

**FIGURE P12.7**   A.

B.

(*continues*)

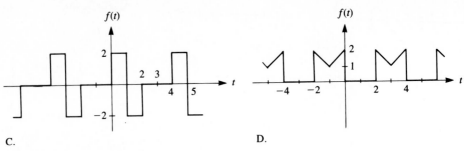

C.                                                D.

**FIGURE P12.7**   *(continued)*

**12.8 a.** Prove that the product of two even functions is even.

**b.** Prove that the product of two odd functions is even.

**c.** Prove that the product of an even and an odd function is odd.

**12.9** Use symmetry as an aid in finding the trigonometric-form Fourier series for the functions shown in Figure P12.9.

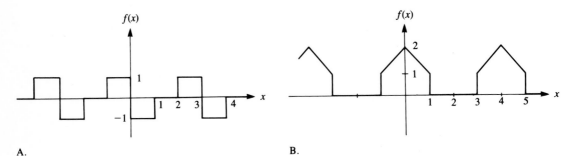

A.                                                B.

**FIGURE P12.9**

**12.10** To investigate how distortion of a sinusoid can affect its frequency content, find the trigonometric form of the Fourier series for the following functions, which are also shown in Figure P12.10. Invoke symmetry, where appropriate.

**a.**   The sinusoid $\sin \pi t$

**b.**   The same sinusoid clipped in amplitude at $1/\sqrt{2}$

**c.**   A distorted sinusoid approximated by $.9 \sin \pi t$ for $-1 < t < 0$ and by $\sin \pi t$ for $0 < t < 1$

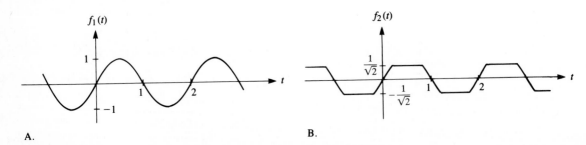

A.                                                B.

**FIGURE P12.10**

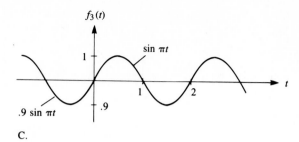

C.

FIGURE P12.10 (continued)

12.11 The periodic functions in Figure P12.11 can be made either even or odd with an appropriate shift along the time axis. Use this trick to simplify the calculations of the coefficients of Equation 12.2. Then, shift the result an equal but opposite amount in time to arrive at the correct answer. Final results should be presented as in Equation 12.3. *Caution:* Each harmonic is shifted an equal amount of time, but by a different phase!

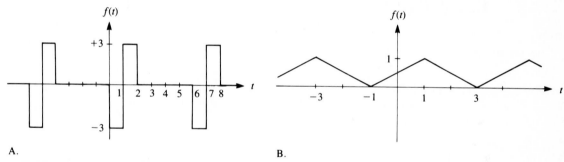

A.

B.

FIGURE P12.11

12.12 Prove that, for a function of half-wave symmetry,

$$a_n = \begin{cases} \dfrac{4}{T} \displaystyle\int_0^{T/2} f(t) \cos n\omega_0 t \, dt, & n \text{ odd} \\ \\ 0, & n \text{ even} \end{cases}$$

and that similar expressions exist for $b_n$. *Hint:* Center the integrations on the origin and make a change of variables, $t = \lambda - (T/2)$, for negative time.

12.13 Use the results of Problem 12.12 to find the Fourier coefficients, $a_n$ and $b_n$, for the functions shown in Figure P12.13.

12.14 Prove that the basis set of complex exponentials is orthogonal.

12.15 Find the complex exponential form of the Fourier series of the functions of Problem 12.4 by making use of the relationship $c_n = (a_n - jb_n)/2$, $n > 0$.

12.16 Find the Fourier coefficients $c_n$ for the periodic functions shown in Figure P12.16.

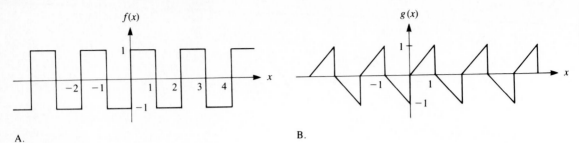

A.                                                              B.

**FIGURE P12.13**

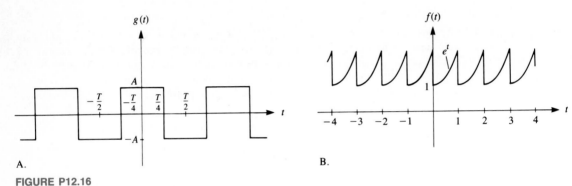

A.                                                              B.

**FIGURE P12.16**

**12.17** Find the complex exponential form of the Fourier series for the infinite impulse train shown in Figure P12.17. This function is significant in the theory of digital signals.

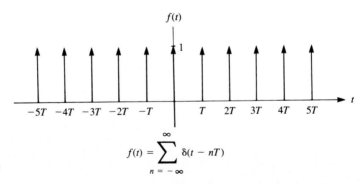

$$f(t) = \sum_{n=-\infty}^{\infty} \delta(t - nT)$$

**FIGURE P12.17**

**12.18** Sketch both the one-sided and two-sided spectra for the function in Example 12.2.

**12.19** Plot both the one-sided and two-sided line spectra for the following functions:

a. $f(t) = 4 + \sum_{n=1}^{\infty} \left[ \frac{4}{n} \cos 1000nt - \frac{5}{n+3} \sin 1000nt \right]$

b. $g(t) = \sum_{n=-\infty}^{\infty} \left( \frac{4}{n} - j\frac{10}{n^2} \right) e^{-j10nt}, \quad n \neq 0$

**12.20** Plot the frequency spectra for the functions of the following problems:

    **a.** Problem 12.4 (both one-sided and two-sided)

    **b.** Problem 12.5 (one-sided)

    **c.** Problem 12.6 (two-sided)

    **d.** Problem 12.10 (one-sided)

    **e.** Problem 12.11 (one-sided and two-sided)

    **f.** Problem 12.13 (two-sided)

    **g.** Problem 12.15 (one-sided and two-sided)

    **h.** Problem 12.16 (two-sided)

**12.21** Sketch the function sinc $a\pi x$ for $a = 1/4,\ 1/2,\ 1,\ 2,$ and 4.

**12.22** For the rectangular pulse train of Example 12.5, plot the spectrum (Equation 12.26) for

    **a.** $T = 1$ and $\tau = 1/2,\ 1/4,$ and $1/8$

    **b.** $\tau = 1$ and $T = 2,\ 4,$ and 8

**12.23** Write Fourier series for the functions whose spectra are shown in Figure P12.23. In each case, write both forms of the series—that is, those based on Equation 12.3 and on Equation 12.22.

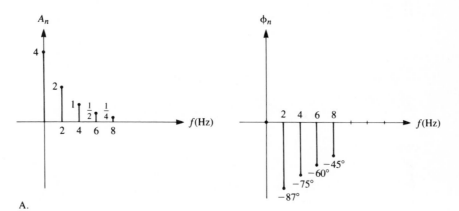

A.

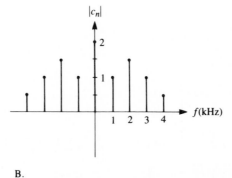
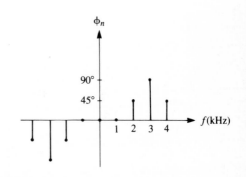

**FIGURE P12.23**　B.

**12.24** A circuit with a periodic input is shown in Figure P12.24.

a. Calculate the trigonometric-form Fourier series of the input.

b. Determine the frequency response characteristics of the circuit.

c. Use parts a and b to plot the one-sided spectrum of the response.

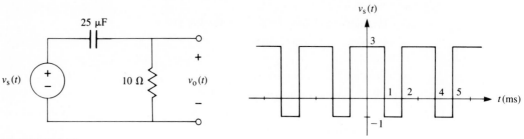

FIGURE P12.24

**12.25** Find the Fourier transform of the triangular pulse shown in Figure P12.25. Hint: Use Euler's relationship to expand $e^{-j\omega t}$ and make use of symmetry. Compare the result to the Fourier series of Example 12.2.

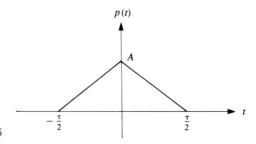

FIGURE P12.25

**12.26** a. Find the Fourier transform of the pulse shown in Figure P12.26.

b. Sketch its spectrum.

c. Sketch the spectra of periodic signals made up of the signal being repeated every 2, 4, and 16 seconds.

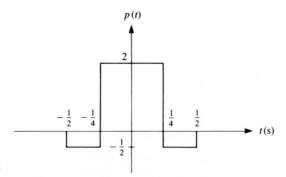

FIGURE P12.26

12.27 Refer to Table 12.2 and prove the following theorems:
    **a.** Linearity (operational pair 1)
    **b.** Time scaling (operational pair 6)
    **c.** Time shifting (operational pair 4)

12.28 Use the following steps to prove the Fourier transform of the unit step function.
    **a.** Find the transform of the function shown in Figure P12.28.
    **b.** Plot $f(t)$ and write $F(j\omega)$ for $a \to 0$.
    **c.** $u(t) = f(t) + 1/2$. Use this fact and the linearity of the Fourier transform to find $\mathscr{F}\{u(t)\}$.

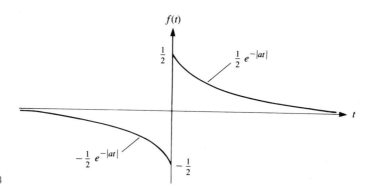

**FIGURE P12.28**

12.29 Use linearity and the known transform of the complex exponential to find the transforms of $\cos \omega t$ and $\sin \omega t$.

12.30 Prove Equation 12.36.

12.31 $f(t)$ is a rectangular pulse that is equal to 1 for $-.5\,\text{ms} < t < .5\,\text{ms}$ and is equal to 0 otherwise.
    **a.** Sketch its spectrum.
    **b.** Sketch the spectrum of $f(t) \cos 200{,}000\pi t$.
    **c.** Sketch the spectrum of $f(t) \cos 20{,}000\pi t$.
    **d.** Sketch the spectrum of $f(t) \cos 2000\pi t$.
    **e.** Sketch the spectrum of $f(t) \cos 200\pi t$.

12.32 Find the Fourier transform of an infinite impulse train. Start with the Fourier series version of the time function that was found in Problem 12.17.

12.33 A function and its transform are shown in Figure P12.33. It is sampled (made into a digital signal) by multiplication with an infinite impulse train of 1 ms spacing and amplitude 1.
    **a.** Sketch the sampled signal.
    **b.** Sketch also the spectrum of the sampled signal by making use of operational transform pair 9 in Table 12.2, which states that multiplication in the time domain corresponds to convolution in frequency. The spectrum of the impulse train was found in Problem 12.32.

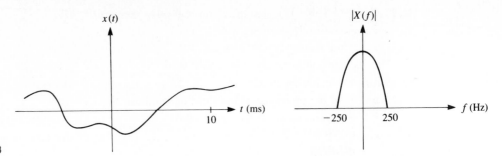

**FIGURE P12.33**

**12.34** If the signal from Problem 12.33 is put through the ideal low-pass filter whose characteristics are shown in Figure P12.34, what will be the resultant spectrum? What do you conclude?

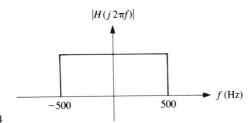

**FIGURE P12.34**

**12.35** Solve the circuit problem of Figure P12.35 by using Fourier transform techniques.

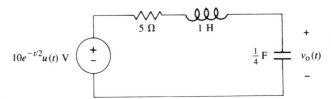

**FIGURE P12.35**

**12.36** Sketch, as functions of frequency, the effect that a pure delay in time will have on the spectrum of an arbitrary time function.

**12.37** A system is described by the following differential equation:

$$b_n \frac{d^n y}{dt^n} + b_{n-1} \frac{d^{n-1} y}{dt^{n-1}} + \cdots + b_0 y = a_m \frac{d^m x}{dt^m} + \cdots + a_0 x$$

Assume all initial conditions to be zero.

a.  Apply the Laplace transform operation to find the system function $H(s) = Y(s)/X(s)$.

b.  Apply the Fourier transform to find the system function $H(j\omega) = Y(j\omega)/X(j\omega)$.

c.  Compare the answers to parts a and b.

**12.38** Show that Equation 12.36 is consistent with the Fourier transform of $\sin \omega t$ by letting $f(t) = 1$.

12.39 From Table 12.1, sketch the magnitude spectra of the following:

    **a.**   $u(t)$

    **b.**   $e^{-at}u(t)$

    **c.**   $\cos \omega_0 t$

    **d.**   $\cos \omega_0 t \, u(t)$

# Power and Energy

## 13.1  INTRODUCTION

Electrical engineering traces its beginnings as a profession to the latter decades of the nineteenth century when electrical power was first generated and distributed commercially. At that time, a debate existed as to whether electrical power should be distributed in DC (direct current) or AC (alternating current) form. Each form has its advantages and disadvantages. AC distribution won out primarily because of the ease with which AC power can be transformed from one voltage level to another. Today, most electrical energy in the world is generated and distributed in the form of sinusoidally varying voltages and currents. Although it is not the primary purpose of this chapter to explain the practical aspects of AC power systems, it is of interest to briefly discuss their basic operating parameters and to examine the reasoning behind the selection of the parameter values commonly used.

We know that power is calculated as the product of voltage and current. If a given amount of power is to be transmitted over a long distance, it is desirable that it be done at high voltage and low current levels to minimize the resistance losses, which are proportional to the current squared. Long-distance transmission lines operate at up to 765 kV and higher in order to take advantage of the increased efficiencies at these high voltage levels. Consumers of electrical power, however, are not equipped to operate at such high voltages. Most people are aware of the fact that in North America, residential customers receive electrical power at a voltage level of 120 V AC. Low voltages are chosen primarily for reasons of safety. Between the generator and the user, the voltage must be stepped down in amplitude, perhaps in several stages, through the use of transformers. There is no physical principle that requires the distribution level of 120 V AC. This level was simply agreed upon within the industry in order to avoid the confusion that would result if varying standards were established in different regions. In Europe, the standard of 240 V was chosen. A traveler who wishes to take a small electrical appliance across the Atlantic Ocean must also take a transformer to accommodate the change in voltage.

Besides voltage, the other main parameter of an AC power system is the frequency at which it operates. The selection of this parameter value, like that of voltage, is somewhat arbitrary. Certain physical considerations, however, do limit the useful range of frequencies that can be chosen. These considerations have to do mainly with the sizes of reactances in power systems. These reactances are, of course, frequency dependent. In North America, 60 Hz (377 rad/s) has been selected as the standard frequency for power systems; in Europe, the operating frequency is 50 Hz. In some parts of Asia, frequencies as low as 25 Hz have been used in the past. At the opposite extreme, AC power aboard aircraft operates at 400 Hz. Many competing factors are involved in the selection of an operating frequency. For instance, transformers must be made physically very large in order to operate effectively at low frequencies. However, their internal losses go up with increasing frequencies. The properties of some types of motors also limit the range of useful frequencies because their rotational speeds increase with frequency. The frequencies used in various parts of the world are compromises between these and other considerations.

AC power is of central importance to electrical engineers and is, thus, the main topic of this chapter. However, we will begin with a general discussion of the power associated with any periodic signal, not just those that are sinusoidal. Then, we will investigate in detail the power delivered by sinusoidal signals to resistive, capacitive, inductive, and more general loads. One of our findings will be that some general impedance loads can, on average, absorb only a small amount of power even though their supplied voltage and current levels may be very high. A method for correcting this situation will be examined. Finally, the principles and notation of three-phase power will be introduced.

## 13.2  AVERAGE POWER AND rms SIGNAL VALUE

Consider Figure 13.1, which shows a source supplying power to a load. The instantaneous power absorbed by the load is $p(t) = v(t)i(t)$. The power varies as a function of time in a way that depends on the properties of both the source and the load.

The expression $p(t)$ is a complete description of the instantaneous power absorbed by the load. Sometimes, not all of the information contained in $p(t)$ is needed. For example, in the study of AC power systems many calculations use the average rate at which energy is delivered as a central variable. Average power is easily calculated when $p(t)$ is periodic. The power $p(t)$ is periodic whenever $v(t)$ and $i(t)$ are periodic.

Consider a resistor with a periodic voltage across it. The current through the resistor will just be a scaled version of the voltage, differing only in amplitude. The periods of voltage and current will be the same. As we will see in the next section, the period of the power is not necessarily equal to that of the voltage and current. Assume the power's period to be $T$. Then, the average power absorbed by the resistor can be written in any one of several ways:

**FIGURE 13.1**  *Source Supplying Electrical Power to a General Load*

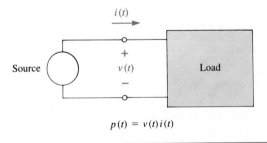

$$p(t) = v(t)i(t)$$

$$P = \frac{1}{T} \int_{t_0}^{t_0 + T} p(t)\, dt \qquad (13.1a)$$

$$P = \frac{1}{T} \int_{t_0}^{t_0 + T} Ri^2(t)\, dt \qquad (13.1b)$$

$$P = \frac{1}{T} \int_{t_0}^{t_0 + T} \frac{v^2(t)}{R}\, dt \qquad (13.1c)$$

In this chapter, the symbol $P$ will be used to denote the average rate at which energy is absorbed. The integrations are performed over one complete period of the power where $t_0$ is any arbitrary starting point.

The definition of the **effective value** of a periodic voltage or current comes from considering the average power that the signal delivers to a resistor. This value can be calculated with Equations 13.1b or 13.1c. A periodic signal is said to be effectively equal to the DC signal that delivers the same average power to the resistor. In the case of current, we can equate the average power from the periodic signal with that supplied by the corresponding DC current, called here $I_{\text{eff}}$:

$$P = \frac{1}{T} \int_{t_0}^{t_0 + T} Ri^2(t)\, dt = R(I_{\text{eff}})^2$$

which leads to the following relationship:

$$I_{\text{eff}} = \sqrt{\frac{1}{T} \int_{t_0}^{t_0 + T} i^2(t)\, dt} \qquad (13.2a)$$

The effective value of a periodic voltage is found in exactly the same way:

$$V_{\text{eff}} = \sqrt{\frac{1}{T} \int_{t_0}^{t_0 + T} v^2(t)\, dt} \qquad (13.2b)$$

The effective value of a periodic signal is usually referred to as its **root mean square** (rms) value because it is found by taking the square *root* of the *mean* (average) value of the *square* of the periodic signal. The discussion that follows will use only the term *root mean square* value.

Both voltages and currents can be expressed in terms of their rms values. The same is true of some nonelectrical physical quantities. Therefore, the definition is expanded to include any periodic function $f(t)$:

$$F_{rms} = \sqrt{\frac{1}{T} \int_{t_0}^{t_0+T} f^2(t)\, dt}$$

(13.2c)

When you refer to other electrical engineering books, remember that the effective and root mean square values of electrical signals are identical.

**Example 13.1**    *Calculating Average and rms Values*

Figure 13.2 shows the form of a periodic current. Calculate its average and rms values.

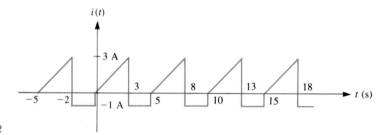

**FIGURE 13.2**

From the figure, note that

$$i(t) = \begin{cases} t\, \text{A}, & 0 < t < 3 \\ -1\, \text{A}, & 3 < t < 5 \end{cases}$$

and that $T = 5$ is the fundamental period of the signal. Then, calculate

$$I_{ave} = \frac{1}{5} \int_0^5 i(t)\, dt = \frac{1}{5}\left[ \int_0^3 t\, dt + \int_3^5 (-1)\, dt \right] = \frac{1}{2}\, \text{A}$$

The rms value of the current is found by

$$I_{rms} = \sqrt{\frac{1}{5} \int_0^5 i^2(t)\, dt} = \sqrt{\frac{1}{5}\left[ \int_0^3 t^2\, dt + \int_3^5 (-1)^2\, dt \right]} = \sqrt{\frac{11}{5}}\, \text{A}$$

If $i(t)$ existed in a 10 Ω resistor, the average power absorbed by the resistor would be

$$P = R(I_{rms})^2 = 22\ \text{W}$$

The peak instantaneous power absorbed will occur at those times when the current is maximum, which happens here when $t = 3, 8, 13$, and so on. At these times,

$$p(t)_{max} = 10(3)^2 = 90 \text{ W}$$

Although the rms value of a signal is related to its effective ability to deliver power, the rms value should never be confused with the power itself. Any rms signal value is proportional to the square root of power. Thus, the rms value of a current signal is a current and has the units of amperes. The rms value of a voltage has the units of volts.

Most of this chapter is concerned with AC, or sinusoidal, power. The rms value of a sinusoidal signal is easily found, as the next example shows.

**Example 13.2**      *The rms Value of a Sinusoid*

Assume a sinusoidal voltage of arbitrary amplitude and frequency:

$$v(t) = V_m \sin \omega t \text{ V}$$

The period is $T = 2\pi/\omega$.
Then,

$$V_{rms} = \sqrt{\frac{1}{2\pi/\omega} \int_0^{2\pi/\omega} (V_m \sin \omega t)^2 \, dt} = \sqrt{\frac{\omega}{2\pi} \int_0^{2\pi/\omega} \left(\frac{V_m^2}{2}\right)(1 - \cos 2\omega t) \, dt}$$

$$= \frac{V_m}{\sqrt{2}} \text{ V}$$

Example 13.2 demonstrates a relationship that is very useful in AC power calculations—namely, a sinusoid of amplitude $V_m$ has an rms value of

$$V_{rms} = \frac{V_m}{\sqrt{2}} = .707V_m \tag{13.3}$$

Notice that the rms value of a sinusoid depends only on its amplitude and is independent of the frequency $\omega$. The relationship of Equation 13.3 is used so often that you should commit it to memory. In so doing, do not forget that it applies only to sinusoidal signals. The rms values of other periodic waveforms must be calculated from the defining formula of Equation 13.2c.

An interesting result that will be verified in Problem 13.5 at the end of the chapter relates to periodic functions that are composed of the sum of sinusoids of different frequencies. For instance, consider the current

$$i(t) = I_{DC} + I_1 \cos \omega_1 t + \cdots + I_N \cos \omega_N t \qquad (13.4a)$$

The DC component of Equation 13.4a can be thought of as a sinusoid with a frequency of zero. The current $i(t)$ has an rms value that is simply related to the rms values of its components:

$$I_{rms} = \sqrt{(I_{DC})^2 + (I_{1_{rms}})^2 + \cdots + (I_{N_{rms}})^2} \qquad (13.4b)$$

The average power delivered by such a signal to a resistor would be

$$P = (I_{rms})^2 R = [(I_{DC})^2 + (I_{1_{rms}})^2 + \cdots + (I_{N_{rms}})^2]R$$
$$= P_{DC} + P_1 + P_2 + \cdots + P_N \qquad (13.4c)$$

Stated in words, the total average power delivered by a signal that is a sum of sinusoidal components is equal to the sum of the average powers that would be delivered by the individual components acting separately. This result is true only for signals that are the sum of *sinusoidal* components with different frequencies. Average power, in general, does not follow this rule of superposition.

---

**Example 13.3**   *The rms Value of a Mixed-Frequency Signal*

What average power is delivered to a 47 $\Omega$ resistor by the current $i(t) = [2 - 4 \cos(100t - 45°) + .5 \sin 2000t]$ A?

The rms value of a sinusoid is not changed by a shift of phase. Therefore, from Equation 13.4b,

$$I_{rms} = \sqrt{2^2 + \frac{4^2}{2} + \frac{.5^2}{2}} = \sqrt{12.125} \text{ A rms}$$

and the average power delivered is

$$P = R(I_{rms})^2 = (47)(12.125) = 569.875 \text{ W}$$

---

## 13.3   AC POWER IN BASIC CIRCUIT ELEMENTS

The rest of this chapter is limited to a discussion of AC power. Let's begin by considering power delivered to the three basic passive circuit elements: resistors, capacitors, and inductors.

Figure 13.3 shows a resistor with a sinusoidal current in it. It is assumed that

FIGURE 13.3    *Power and Net Energy Absorbed by a Sinusoidally Excited Resistor*

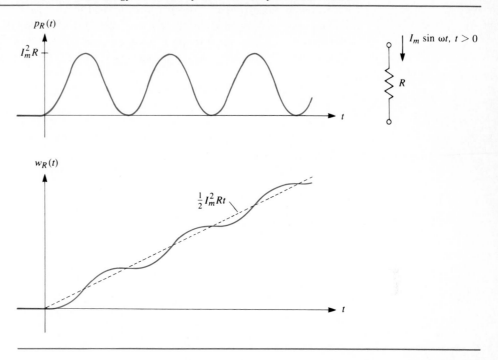

the current and voltage of the resistor are zero for $t < 0$ so that the resistor has absorbed no energy prior to $t = 0$. The instantaneous power absorbed by the resistor is

$$p_R(t) = R(I_m \sin \omega t)^2 = \frac{RI_m^2}{2} - \frac{RI_m^2}{2} \cos 2\omega t \text{ W}, \qquad t > 0 \qquad \textbf{(13.5a)}$$

Let's also calculate $w_R(t)$, the total or net energy absorbed by the resistor for all time up until time $t$. Because $w_R(0) = 0$, the calculation becomes

$$w_R(t) = \int_0^t p_R(t)\,dt = \int_0^t \frac{RI_m^2}{2}(1 - \cos 2\omega t)\,dt$$

$$= \frac{1}{2}RI_m^2\left(t - \frac{\sin 2\omega t}{2\omega}\right) \text{J}, \qquad t > 0 \qquad \textbf{(13.5b)}$$

Both $w_R(t)$ and $p_R(t)$ are shown plotted in Figure 13.3. Several important observations can be made:

1. The instantaneous power, $p(t)$, has a positive average value that equals

$$P = \frac{RI_m^2}{2} = R\left(\frac{I_m}{\sqrt{2}}\right)^2 = R(I_{\text{rms}})^2 = \frac{(V_{\text{rms}})^2}{R}$$

2. The power is periodic. Its time-varying component is sinusoidal with a frequency *twice* that of the current and voltage.
3. The power absorbed by a resistor is always positive except at those instants of time when the current (and voltage) is zero. It follows that $w_R(t)$ never decreases. That is, electrical energy absorbed by a resistor cannot be recaptured in electrical form. It is transformed into heat. Thus, the power absorbed by a resistor is **dissipated.**

Figure 13.4 shows a sinusoidally excited capacitor along with plots of its absorbed power and energy. Because the voltage across the capacitor is assumed to be $V_m \sin \omega t$, it follows that its current is

$$i(t) = C \frac{d}{dt} (V_m \sin \omega t) = \omega C V_m \cos \omega t$$

The power absorbed, then, is

$$p_C(t) = (\omega C V_m \cos \omega t)(V_m \sin \omega t) = \frac{1}{2} \omega C V_m^2 \sin 2\omega t \text{ W} \qquad (13.6a)$$

**FIGURE 13.4**     *Power and Net Energy Absorbed by a Sinusoidally Excited Capacitor*

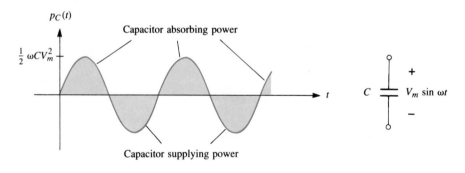

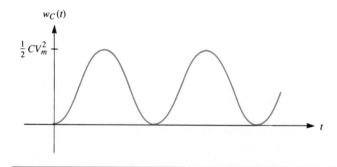

As with the resistor, we assume that no *net* energy has been absorbed by the capacitor up until $t = 0$. That is, $w_C(0) = 0$. Then, we can calculate the net energy absorbed by the capacitor between 0 and time $t$:

$$w_C(t) = \int_0^t \frac{1}{2} \omega C V_m^2 \sin 2\omega t \, dt = \frac{1}{4} C V_m^2 (1 - \cos 2\omega t) \text{ J} \qquad \text{(13.6b)}$$

When the AC power behavior of capacitors is compared with that of resistors, important similarities as well as differences are evident:

1. The AC power absorbed by a capacitor is sinusoidal with a frequency *twice* that of the voltage and current.
2. The average value of power absorbed by a capacitor under AC excitation is zero.
3. The *net* energy absorbed by a sinusoidally excited capacitor periodically returns to zero.

Although the second property in the preceding list follows from the first, it is listed separately because it so clearly shows an important difference between resistors and capacitors. Energy absorbed by resistors is not recoverable in electrical form. It is dissipated as heat. In contrast, the electrical energy absorbed by an ideal capacitor is only temporarily stored in the capacitor's electrical field until such time as the capacitor is discharged. When $p_C(t) > 0$, the capacitor is absorbing energy. When $p_C(t) < 0$, the capacitor is discharging or supplying energy to the circuit to which it is connected. Under AC excitation, energy that is stored in the capacitor during one quarter-cycle of the voltage (or current) will be given back to the circuit during the next quarter-cycle. Remembering from Chapter 5 that the net energy stored in a capacitor is $w_C(t) = (1/2)C v_C^2(t)$, we see that $w_C(t)$ is zero whenever $v_C(t)$ is zero.

The AC power behavior of inductors is very much like that of capacitors. The power absorbed varies sinusoidally with time, and no net energy is absorbed on average. The mathematical details will be worked out in Problem 13.13 at the end of the chapter.

## 13.4  AC POWER DELIVERED TO A GENERAL LOAD

Now, let's consider how AC power is delivered to a general load that consists of both resistive and reactive elements. The situation is shown in Figure 13.5. The AC source symbol used in the figure was introduced in Chapter 1. We could use either the voltage or the current as a reference and give it a phase angle of 0° without loss of generality. For completeness, however, let's begin by assigning arbitrary phase angles to both. The relationship between the phase angles is $\alpha - \beta = \theta$, where $\theta$ is the angle of the impedance $Z$.

Calculation of the power absorbed by the load is straightforward. Assume

**FIGURE 13.5**   *AC Power Delivered to a General Impedance Load*

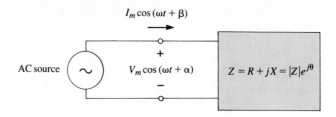

the following:

$$v(t) = V_m \cos(\omega t + \alpha)\ \text{V} \tag{13.7a}$$

$$i(t) = I_m \cos(\omega t + \beta)\ \text{A} \tag{13.7b}$$

Then,

$$p(t) = V_m \cos(\omega t + \alpha) \cdot I_m \cos(\omega t + \beta)$$

$$= \frac{V_m I_m}{2} \cos(\alpha - \beta) + \frac{V_m I_m}{2} \cos(2\omega t + \alpha + \beta)$$

$$= \frac{V_m I_m}{2} \cos\theta + \frac{V_m I_m}{2} \cos(2\omega t + 2\alpha - \theta)\ \text{W} \tag{13.7c}$$

As seen in Equation 13.7c, the instantaneous power absorbed by the load consists of a constant or average part and a sinusoidally varying portion that has a frequency twice that of the voltage and current. Figure 13.6 shows this power plotted when $\alpha = 0°$ and $\theta = +45°$. This situation lies between the purely resistive and purely reactive cases already considered. As with the purely resistive example, $p(t)$ has a nonzero average value. And, as with the purely reactive case,

**FIGURE 13.6**   *AC Power Absorbed by an Impedance of $1\underline{/45°}\ \Omega$*

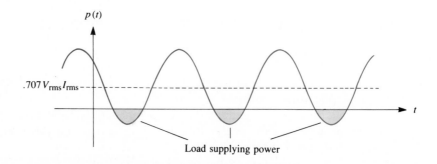

$p(t)$ is sometimes positive (energy flowing into the load) and sometimes negative (energy flowing out of the load).

Because the average part of $p(t)$ represents electrical energy irretrievably lost to the load, it must be associated with the resistive part of $Z$. The time-varying portion is due to both the resistive and reactive parts of the load. These points are seen more easily after an additional manipulation of Equation 13.7c:

$$p(t) = \frac{V_m I_m}{2} \cos \theta + \frac{V_m I_m}{2} \cos \theta \cdot \cos (2\omega t + 2\alpha)$$

$$+ \frac{V_m I_m}{2} \sin \theta \cdot \sin (2\omega t + 2\alpha) \text{ W} \tag{13.7d}$$

In order to simplify this expression for power, it can be presented assuming the phase angle for voltage is zero; that is, $\alpha = 0°$. This assumption does not change any important properties of the result; it merely has the effect of shifting plots of $v(t)$, $i(t)$, and $p(t)$ to the left or right along the time axis. Under this assumption, Equations 13.7a, 13.7b, and 13.7d become

$$v(t) = V_m \cos \omega t \text{ V} \tag{13.8a}$$

$$i(t) = I_m \cos (\omega t - \theta) \text{ A} \tag{13.8b}$$

$$p(t) = \frac{V_m I_m}{2} \cos \theta + \frac{V_m I_m}{2} \cos \theta \cdot \cos 2\omega t$$

$$+ \frac{V_m I_m}{2} \sin \theta \cdot \sin 2\omega t \text{ W} \tag{13.8c}$$

The first of the three terms making up $p(t)$ in Equations 13.7d or 13.8c is, again, the average power absorbed by the load. We define $P$ as follows:

$$P = \frac{V_m I_m}{2} \cos \theta \tag{13.9a}$$

$$= \left(\frac{V_m}{\sqrt{2}}\right)\left(\frac{I_m}{\sqrt{2}}\right) \cos \theta$$

$$= V_{\text{rms}} I_{\text{rms}} \cos \theta \tag{13.9b}$$

Several other useful forms for $P$ can be demonstrated:

$$P = (I_{\text{rms}})^2 |Z| \cos \theta \tag{13.9c}$$

$$= \frac{(V_{\text{rms}})^2}{|Z|} \cos \theta \tag{13.9d}$$

Keep in mind that Equation 13.9 in its various forms is valid only in the sinusoidal steady state. The average power $P$ is maximum for a purely resistive load since $\theta = 0°$. It is minimum (equal to zero, actually) for any purely reactive load because then $\theta = \pm 90°$ for which $\cos \theta = 0$.

The second term of Equation 13.8c accounts for the sinusoidal variation in the absorption of power by the resistive portion of the load. Note that its amplitude is equal to $P$. When $\theta = \pm 90°$ (purely reactive load), Equation 13.8c is reduced to just its last term. This term accounts for power oscillating into and out of the load because of its reactive elements. We take its amplitude and define $Q$ as follows:

$$Q = \frac{V_m I_m}{2} \sin \theta \tag{13.10a}$$

$$= V_{rms} I_{rms} \sin \theta \tag{13.10b}$$

$$= (I_{rms})^2 |Z| \sin \theta \tag{13.10c}$$

$$= \frac{(V_{rms})^2}{|Z|} \sin \theta \tag{13.10d}$$

The properties of $Q$ are complementary to those of $P$. $Q$ is zero for a purely resistive load and has a maximum value of $V_{rms} I_{rms}$ for reactive loads either capacitive or inductive. $Q$ is positive for inductive loads and negative for capacitive loads. $P$ is always positive (or zero) for any passive load.

With the definitions of the terms $P$ and $Q$, we can write Equation 13.8c in a more compact form:

$$p(t) = P + P \cos 2\omega t + Q \sin 2\omega t \text{ W} \tag{13.11}$$

All of the three terms of Equation 13.11 are absorbed powers of the load and, as such, share the same dimensions. Because of their individual meanings, however, engineers apply two different units to them. The first two terms describe the dissipation of power by the resistive portion of the load. They are given the units of **watts** (W) and together are called the resistive or **real power** absorbed by the impedance $Z$. The third term does not describe a net absorption of power but, as has been said earlier, represents the oscillation of power into and out of the load due to the presence of reactive elements. It is thus called **reactive power** and is given the units of **volt-ampere reactive** (VAR).

The discussion to this point may give the impression that $P$ is due just to the resistive elements of a load and that $Q$ depends only on the reactance. This is not true. For instance, if the load $Z = R + jX$ is supplied by a voltage $V_m \sin \omega t$, then the magnitude of the resulting current will equal $V_m/|Z|$. That is, the current will depend on both $R$ and $X$. Thus, the average power is a function of all the elements that are present, not just those that are resistive. Similarly, $Q$ depends on both the resistive and reactive elements of the load.

Real power represents energy that is transformed from electrical to nonelectrical form by the load. As has already been pointed out, resistors transform electrical energy into heat. Other loads, such as motors, are able to turn absorbed real power into mechanical work. Reactive power is never transformed into nonelectrical form. Instead, it is temporarily stored in electromagnetic fields and then returned to the source network as voltages and currents. Real power is what allows us to make practical use of electricity. Reactive power is an often undesirable side effect of energy transfer.

---

**Example 13.4**    *Complete Power Calculation for a Complex Load*

Assume that a load with an impedance of $Z = 75 - j25 \ \Omega$ is supplied by a voltage of $169.7 \cos 377t$ V. Calculate the instantaneous power absorption of the load.

The rms value of the voltage is $169.7/\sqrt{2} = 120$ V rms. The rms value of the current can be found from

$$I_{rms} = \frac{V_{rms}}{|Z|} = \frac{120}{79.06} = 1.52 \ \text{A rms}$$

The angle of the impedance is

$$\theta = \tan^{-1}\left(\frac{-25}{75}\right) = -18.43°$$

We can now write the terms

$$P = V_{rms}I_{rms} \cos \theta = (120)(1.52) \cos(-18.43°) = 173.04 \ \text{W}$$

and

$$Q = V_{rms}I_{rms} \sin \theta = (120)(1.52) \sin(-18.43°) = -57.66 \ \text{VAR}$$

The power absorbed is

$$p(t) = P + P \cos 2\omega t + Q \sin 2\omega t$$
$$= 173.04 + 173.04 \cos 754t - 57.66 \sin 754t$$
$$= 173.04 + 182.4 \cos(754t + 18.43°) \ \text{W}$$

---

## 13.5 COMPLEX POWER

In Chapter 8, it was useful to represent sinusoidal signals as complex valued *phasors*. If phasor notation is used for the voltage and current of our present problem, it leads to the definition of complex power. Let's begin by writing the appropriate phasor forms for the general voltage and current of Equations 13.7a and 13.7b:

$$\mathbf{V} = V_{rms} e^{j\alpha} \text{ V} \tag{13.12a}$$

$$\mathbf{I} = I_{rms} e^{j\beta} \text{ A} \tag{13.12b}$$

As the two phasors of Equation 13.12 show, it is customary in AC power calculations to use the rms values of signals rather than their amplitudes because they lead to a more compact notation. To emphasize the use of rms amplitudes, units are often expressed in the forms V AC or $I_{rms}$. Otherwise, you should assume that amplitudes are not rms values.

The product of a voltage and a current is power. Let's determine the product of voltage and the complex conjugate of the current and define the result as the **complex power S:**

$$\begin{aligned} \mathbf{S} = \mathbf{V}\mathbf{I}^* &= V_{rms} e^{j\alpha} I_{rms} e^{-j\beta} = V_{rms} I_{rms} e^{j(\alpha - \beta)} \\ &= V_{rms} I_{rms} e^{j\theta} = V_{rms} I_{rms} \cos\theta + j V_{rms} I_{rms} \sin\theta \text{ VA} \end{aligned} \tag{13.13}$$

Because the complex power **S** is the product of a phasor voltage and a phasor current, it is given the units of **volt-ampere** (VA). Note that

$$Re[\mathbf{S}] = V_{rms} I_{rms} \cos\theta = P \text{ (W)}$$

and

$$Im[\mathbf{S}] = V_{rms} I_{rms} \sin\theta = Q \text{ (VAR)}$$

S, P, and Q all have dimensions of power. Their units of volt-amperes, watts, and volt-amperes reactive are somewhat arbitrarily defined to distinguish between the three powers.

Notice that the real part of S equals P, which has already been shown to be the average power absorbed by the load. The imaginary part of the complex power **S** is equal to Q, which is the amplitude of the oscillations of power into and out of the load due to the presence of reactance.

We can now see that the use of the term *real power* has basis both in the physical process of power transfer and in the mathematics used to describe it. To complete the picture, some books refer to Q as the *imaginary power*. However, because this term is not widely used in the power industry, it is avoided here.

The association of $P$ with the resistive elements of the load and $Q$ with its reactive elements can be directly seen through the following calculation:

$$\mathbf{S} = \mathbf{VI}^* = (\mathbf{ZI})\mathbf{I}^* = Z|\mathbf{I}|^2 \tag{13.14a}$$

$$= R|\mathbf{I}|^2 + jX|\mathbf{I}|^2 \tag{13.14b}$$

$$= P + jQ \tag{13.14c}$$

Thus, the average power $P$ is due entirely to power dissipated by the resistance of the load, and the reactive power $Q$ is the result of power "absorbed" by its reactive element. Again, note that although the average power $P$ is entirely accounted for by resistive losses, its magnitude depends on the reactive elements also.

---

**Example 13.5**    *Calculations Using Complex Power* **S**

If $v(t) = 10 \cos 377t$ V  and  $Z = 5 + j10\,\Omega$, find the average and reactive powers absorbed.

The calculations are as follows:

$$\mathbf{V} = 7.07e^{j0°} \text{ V rms}$$

$$Z = 11.18e^{j63.4°}\,\Omega$$

$$\mathbf{I} = \frac{\mathbf{V}}{Z} = .632e^{-j63.4°} \text{ A rms}$$

$$\mathbf{S} = \mathbf{VI}^* = 4.47e^{j63.4°} = 2 + j4 \text{ VA}$$

$$P = 2 \text{ W}$$

$$Q = 4 \text{ VAR}$$

---

## 13.6  IMPEDANCE AND POWER TRIANGLES

The impedance, $Z$, of a load and the complex power, $\mathbf{S}$, which is delivered to it, are complex numbers that can be represented graphically on the complex number plane. If the real part, imaginary part, and magnitude of a complex number are drawn, the result is in the form of a triangle. Figure 13.7A, for instance, shows an **impedance triangle.** Assume that the impedance carries a current, $\mathbf{I}$. Further assume, for now, that the current is of zero phase so that $\mathbf{I} = I$, a real number. Then, the voltage across $Z$ is $\mathbf{V} = Z\mathbf{I} = RI + jXI$. The corresponding **voltage triangle** is shown in Figure 13.7B. Note that the impedance and voltage triangles are geometrically similar. The **power triangle,** shown in Figure 13.7C, is found

FIGURE 13.7    *Impedance, Voltage, and Power Triangles*

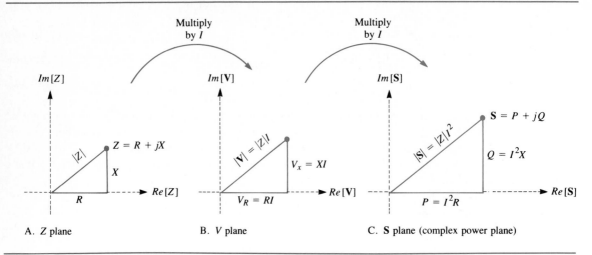

A. *Z* plane              B. *V* plane              C. **S** plane (complex power plane)

from an additional multiplication by $I$:

$$\mathbf{S} = \mathbf{V}\mathbf{I}^* = \mathbf{V}\mathbf{I} = RI^2 + jXI^2$$

The power triangle is also similar to the impedance triangle.

Power and impedance triangles are used widely in the practice of power engineering. As with most methods of graphical presentation, the triangles are used not so much as tools of calculation but instead because they are helpful in understanding the principles involved in power calculations. They also provide a strong visual check as to the reasonableness of the results of such calculations. Voltage triangles are not commonly encountered in practice. The concept has been included here because it helps to explain the relationship between the other two triangles.

As should be clear from Figure 13.7, the impedance triangle is a function of the load alone. The power triangle depends on both the load and the way in which it is excited. Although our development assumed a phase angle of zero for the current, the conclusions are entirely general. The minor differences caused by a nonzero phase will be dealt with in one of the problems at the end of the chapter.

The base of the power triangle has a length equal to $P$, which was earlier assigned the units of watts. The length of the vertical side is equal to $Q$ and has the units of volt-amperes reactive. The hypotenuse, then, is

$$|\mathbf{S}| = \sqrt{P^2 + Q^2} = \sqrt{(V_{rms}I_{rms}\cos\theta)^2 + (V_{rms}I_{rms}\sin\theta)^2} = V_{rms}I_{rms}$$

$$(13.15)$$

$V_{\text{rms}}I_{\text{rms}}$ is the **apparent power** delivered to the load. If an ammeter is used to measure the rms value of the current into the load and a voltmeter is used to measure the rms voltage, then the product of the two would apparently suggest an average power of $V_{\text{rms}}I_{\text{rms}}$. This interpretation is wrong, of course; it does not take into account the phase angle between the voltage and current and can be correct only in the case of a purely resistive load. Apparent power has units of volt-amperes, the same as for **S** itself. The volt-ampere unit is very important in the power industry. Contrary to what you might think, heavy-duty industrial equipment, such as transformers and motors, is often rated according to kVA rather than kW capacities.

The measurement of power can be of critical importance in the operation of a safe and efficient power supply system. Apparent power is measured indirectly by measuring separately the voltage and current and then taking the product of their rms values. An instrument called a **wattmeter** is required to directly measure the average real power that is dissipated by a system's loads. Similarly, a VAR meter is used to directly measure the reactive power absorbed by a load. This measurement is just as important as knowing the real power because the reactive flow of energy into and out of a load must be accounted for in sizing a power distribution system. Wattmeters and VAR meters are placed on all of the major loads of power systems to monitor the system's overall behavior.

The ratio of the average and apparent powers absorbed by a load is known as the load's **power factor** (PF). For AC operation, this becomes

$$\text{PF} = \frac{V_{\text{rms}}I_{\text{rms}}\cos\theta}{V_{\text{rms}}I_{\text{rms}}} = \cos\theta \tag{13.16}$$

As you can see from Equation 13.16, the power factor for a passive load in an AC system must be between 0 and +1. It is quite possible for different loads to have identical power factors. This situation is most easily seen when two loads whose impedances are complex conjugates of one another are considered. For instance, $Z_1 = 150 + j90\ \Omega$ and $Z_2 = 150 - j90\ \Omega$ both have power factors of $\cos[\tan^{-1}(90/150)]$, or .857, because the cosine is an even function. To distinguish between complex conjugate loads with the same numerical power factors, it is common to state whether the load's *current* is **leading** or **lagging** in phase relative to the voltage. Loads with inductive reactance are of the lagging type, while loads with capacitive reactances are of the leading type. Thus, a load of $Z_3 = 45 + j45\ \Omega$ has a power factor of .707, lagging; $Z_4 = 100 - j200\ \Omega$ is described as having a power factor of .447, leading. Most practical loads are of the lagging type, their inductance coming from the windings of motors and transformers. Power factors are important considerations in the design and operation of commercial power systems. More will be said about them in the next section.

Because power and impedance triangles are understood to represent complex quantities, the axes of the complex number plane can be omitted in any

**FIGURE 13.8** *Impedance and Power Triangle Definitions*

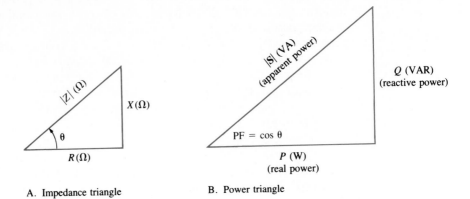

A. Impedance triangle

B. Power triangle

drawing. We already know that the impedance and power triangles are similar in that they share the same set of three angles. The angle $\theta$ is often shown in the lower left-hand corner of an impedance triangle. In a power triangle, the power factor is usually written in the same corner. These details are shown in the general example of Figure 13.8.

**Example 13.6**    *Calculating a System's Power Triangle*

The load shown in Figure 13.9A is powered by a 60 Hz, 120 V AC source. Find the average and apparent powers and the power factor.

Because it will not take much additional work, develop the impedance and power triangles of the system in order to have a complete description. The reactance of the load is found by

$$X_L = \omega L = (377)(.3) = 113.1\ \Omega$$

so that the impedance is

$$Z = 100 + j113.1\ \Omega$$

Next, the complex power is found by

$$\mathbf{S} = \mathbf{V}\mathbf{I}^* = \mathbf{V}\left(\frac{\mathbf{V}^*}{Z^*}\right) = \frac{|\mathbf{V}|^2}{Z^*} = \frac{120^2}{100 - j113.1} = 63.18 + j71.46\ \text{VA}$$

The power and impedance triangles are shown in Figures 13.9B and 13.9C.

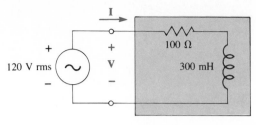

A.

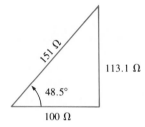

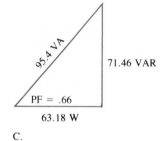

**FIGURE 13.9**    B.                                    C.

Complete the problem by calculating $\theta$ and the power factor:

$$\theta = \angle(100 + j113.1) = +48.5°$$
$$PF = \cos\theta = \cos 48.5° = .66, \text{ lagging}$$

In the shorthand notation of power engineers, loads in AC systems are often specified by giving their power factors and average power absorption. You have probably noticed that many small electrical appliances are labeled according to power absorption. For example, a light bulb might be rated for 75 W and a hair dryer for 1200 W. Inspection of Figure 13.8 should convince you that if a load's average power absorption is known along with its power factor (including a description of the load as leading or lagging), then the entire power triangle can be determined. If, in addition, the value of the source's voltage or current is known, the impedance of the load can be found.

**Example 13.7**    *Finding Impedance and Power Triangles*

The load shown in Figure 13.10A is described as having a power factor of .6, leading, and is rated at 1000 W when supplied by a 240 V AC source operating at 50 Hz. Find the impedance of the load.

First, note that since the load is leading, $\theta$ and $Q$ are negative and the

(*continues*)

**Example 13.7**   *Continued*

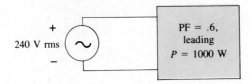

A.

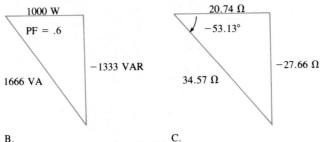

FIGURE 13.10    B.                                        C.

reactive part of the impedance is capacitive. $\theta$ and $Q$ are found by

$$\theta = -\cos^{-1}(\text{PF}) = -\cos^{-1}(.6) = -53.13°$$

$$\frac{Q}{P} = \tan(-53.13°) = -1.333$$

$$Q = -1333 \text{ VAR}$$

The power triangle is drawn as in Figure 13.10B, and the apparent power is calculated:

$$V_{\text{rms}}I_{\text{rms}} = \sqrt{P^2 + Q^2} = 1666 \text{ VA}$$

from which it follows that

$$I_{\text{rms}} = \frac{1666}{240} = 6.94 \text{ A rms}$$

Finally, use of Equation 13.14a leads to

$$Z = \frac{S}{|I|^2} = \frac{1000 - j1333}{48.21} = 20.74 - j27.65 \ \Omega$$

If the load is assumed to be in the form of a series *RC* combination, its resistance is $20.74\ \Omega$ and its capacitance is

$$C = \frac{1}{\omega X_C} = \frac{1}{2\pi(50)(27.65)} = 115.1\ \mu\text{F}$$

---

The steps followed in the solution of Example 13.7 are not unique. Other approaches could have arrived at the same result. Regardless of the analytical method used, however, drawing the impedance and power triangles is a useful aid in following a problem through.

## 13.7 CORRECTING FOR LOW POWER FACTORS

Power companies charge their customers for the real power ($P$) that they consume. More correctly, customers pay for the total net energy that they use. Thus, power rates are generally given in terms of price per kilowatt-hour (kWh) of electrical energy. The rates are set according to the cost that the company incurs in supplying the power. As we will see, these costs depend on the power factors of the loads and may differ from one customer to another. Loads with good power factors (PF close to 1) can be more efficiently served than can loads with lower power factors. Thus, it is in the power company's (and its customers') best interest that major loads on the system have power factors as close to 1 as possible. To encourage this, power companies charge higher rates to their large customers if they have poor power factors. This consideration is not important for residential and small commercial customers because their power factors are virtually 1.

Let's see exactly why power companies set rates, in part, based on the power factors of their customers. We reason as follows: Assume that a power company has two loads, Customer A and Customer B. Both customers perform the same amount of work electrically and, so, consume energy at the same average rate. However, Customer B's PF is much lower than that of Customer A. One consequence of the low PF is that the apparent power delivered to Customer B is greater than the apparent power to Customer A. It is usual that power systems supply energy to their customers at certain set voltage levels. If the supplied voltages are equal for the two customers, it follows that the transmission lines supplying power to Customer B carry larger currents than do those supplying Customer A. This situation is undesirable because resistive losses along a transmission line are proportional to the square of the current. As a result, the power system will be less efficient in supplying the customer with the lower PF. The power company could try to reduce its resistance losses by installing larger conductors, but this solution would increase its equipment costs. Also, transformers, generators, and other pieces of equipment in the system would have to be sized larger in order to accommodate the higher currents, which would further add to the cost. Therefore, to be fair to all of its customers, the power company

**FIGURE 13.11**   *Improving Power Factor of a Load*

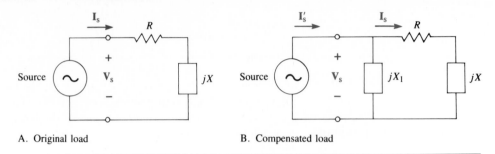

A. Original load                                          B. Compensated load

sets higher rates for those customers with low power factors in order to offset the higher costs incurred in supplying them. These higher rates encourage major customers to increase their power factors and along with them the overall efficiency of the power supply system.

If an electrical load has been designed as efficiently as possible for a high PF but the power factor is still not good enough, compensating elements external to the load itself can be added to improve the situation. Refer to Figure 13.11A. Power is supplied to a load at a specified voltage level. The load is given as a power factor (assumed to be low) and a certain average power absorption. Figure 13.11B shows the same load with a purely reactive impedance placed in parallel. The presence of this added reactance does not change the average power absorbed by the actual load because it still has the same voltage across it. In addition, the added reactance does not itself absorb any net energy. The effect of the compensating reactance, if properly chosen, is to increase the power factor seen by the source supplying the load. Stated without proof, the value of reactance that should be added is

$$X_1 = \frac{R^2 + X^2}{kR \tan(\cos^{-1} PF) - X} \tag{13.17}$$

where   $Z = R + jX$ = impedance of the uncompensated load

PF = desired value of the *corrected* power factor

$k = \pm 1$

If $k = +1$, the lagging/leading property of the load will not be affected; if $k = -1$, the property will be reversed. In practical power systems, the compensating reactance $X_1$ is almost always capacitive.

---

**Example 13.8**   *Improving a Load's Power Factor*

Consider the load shown in Figure 13.12A. Its power factor is $\cos 39.8°$, or .768. To add an element in parallel in order to increase the power factor to

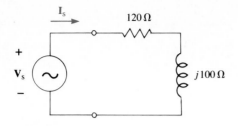
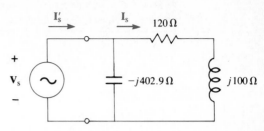

A. Original load

B. Compensated load

**FIGURE 13.12**

.95, we choose

$$X_1 = \frac{120^2 + 100^2}{120 \tan (\cos^{-1} .95) - 100} = -402.9$$

In a 60 Hz system, $X_1$ corresponds to a capacitor of

$$C = 6.58 \ \mu F$$

It is left to Problem 13.28 at the end of the chapter to demonstrate that the system supplies less current to the compensated load (Figure 13.12B) than it does to the uncompensated.

Other methods can be used to compensate for the poor power factors of loads and the unnecessary system losses to which they lead. One theoretically possible approach is to add a reactive element in series with the actual load. The calculations are more straightforward than for parallel compensation, but the approach is impractical because it changes $P$ for the load unless the supplied voltage is also adjusted.

## 13.8   IMPEDANCE MATCHING

This section will extend a result from Chapter 4 to see how to match the impedance of a load to that of a source in order to maximize the average power transfer to the load. Impedance matching is justified here with a heuristic argument; the analytical details of a more formal proof are left to Problem 13.32 at the end of the chapter.

Chapter 4 demonstrated that a source with an internal resistance of $R_{eq}$ will transfer the greatest power to a load resistor $R_L$ if

$$R_L = R_{eq} \tag{13.18}$$

In this chapter, we are considering the more general case of complex impedances. Figure 13.13 shows a source with a general internal impedance and a general

FIGURE 13.13 *Impedance Matching*

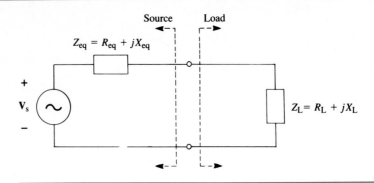

impedance load. All of the average power absorbed by the load is due to its resistance. For a fixed value $R_L$, the power absorbed will be maximum if the current, $I_s$, is made maximum. Choosing $X_L = -X_{eq}$ minimizes the total impedance and maximizes the current to the load. Thus, we have a purely resistive case to which we can apply the result of Chapter 4. Maximum power transfer occurs when

$$Z_L = R_{eq} - jX_{eq}$$

or

$$Z_L = Z_{eq}^* \qquad (13.19)$$

---

**Example 13.9** *Choosing a Load for Maximum Power Transfer*

Figure 13.14A shows an AC source with a complex internal impedance and a load that is absorbing energy. Maximize the average power transfer by selecting the optimal load impedance. The frequency-domain representation of the source circuit is given in Figure 13.14B.

Begin by calculating the internal impedance of the source:

$$Z_{eq} = 100 + 50\|(-j50) = 125 - j25 \ \Omega$$

As shown in Figure 13.14C, the desired load impedance will be

$$Z_L = Z_{eq}^* = 125 + j25 \ \Omega$$

This value corresponds most simply to a $125 \ \Omega$ resistor in series with an inductor. At 60 Hz, the inductor has a value of 66.3 mH.

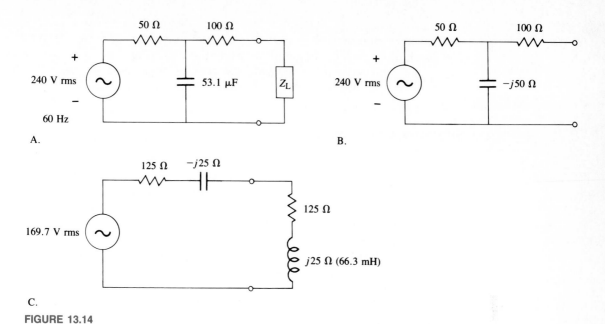

A.

B.

C.

**FIGURE 13.14**

Next, find the magnitude of the Thévenin or open circuit AC voltage of the source circuit:

$$V_{oc_{rms}} = \left| \frac{-j50}{50 - j50} \right| 240 = 169.7 \text{ V rms}$$

The current supplied to the load is then found by

$$I_{rms} = \frac{V_{oc_{rms}}}{|Z_{eq} + Z_L|} = .68 \text{ A rms}$$

from which the average power delivered can be calculated as

$$P = R_L (I_{rms})^2 = 57.6 \text{ W}$$

Although impedance matching and power factor correction might seem, at first glance, to be related to one another, they are actually very different techniques with very different goals. Power factor correction is used in power systems to minimize losses within the distribution network and thus improve the efficiency of the overall system. Impedance matching, on the other hand, is concerned only with maximizing power to the load. A difficulty with impedance matching is that it leads to an efficiency ($P_{load}/P_{source}$) of only 50%. That is, only 50% of the generated power is actually delivered to the load. The rest is dissipated in the supply system itself. This is obviously undesirable for power systems where efficiency of operation is of prime importance. However, impedance matching is used in communications systems where we are more willing to suffer a poor efficiency in order to deliver the maximum possible power to the load.

## 13.9   THREE-PHASE POWER SYSTEMS

Each of the examples considered thus far in this chapter has suggested the presence of only one source in a power system. Such systems would be entirely adequate and the discussion of power and energy here nearly complete if it were not for one very important type of load: three-phase AC induction motors. They require more elaborate power supply systems than do other types of electrical motors, but they also have several distinct advantages. Primary among them is the fact that they absorb electrical energy at a constant rate. Also, the resulting mechanical torque is nearly constant. The intention here is not to explain the theory of induction motors except to say that their operation is based on the presence of rotating electromagnetic fields. Such fields are most commonly created by the controlled interaction of several AC voltages of equal amplitude and frequency, but different phases. The details of the mechanisms are left to a course in energy conversion.

Power systems that have several AC sources whose voltages are identical except for phase are called **polyphase** power systems. Each separate voltage is called a **phase** of the system. Usually, each of the phase voltages is referenced to a common ground point, or **neutral.** Induction motors can be designed for operation in systems of any number of phases. For instance, a two-phase (sometimes abbreviated as $2\phi$) power system has two separately conducted voltages with a 90° phase shift between them. A three-phase ($3\phi$) system has three voltages, each with the same amplitude and frequency, but shifted in phase 120° from each of the other two. From some points of view, three-phase systems are the most economical of polyphase systems to implement. They supply most users' needs and still keep equipment costs low. When, in some specialized industrial applications, more phases are needed, they are created at the point of use by manipulating the incoming three phases. The aluminum industry, for instance, uses up to 48 phases in its smelting process!

Figure 13.15A shows one possible configuration for the three related sources of a three-phase power system. The common reference point is labeled "n." The combined three-phase power source can be connected to the outside world at points a, b, and c and also at the neutral point (n). By convention, the **phase voltages** are named $V_{an}$, $V_{bn}$, and $V_{cn}$. These voltages are shown as functions of time in Figure 13.15B and as phasors in Figure 13.15C. Phasor diagrams are the customary method for displaying the voltage and current variables of a power system.

The three phases of a power system are not treated randomly but instead have a particular **phase sequence** or order. When a rotating machine is to be

FIGURE 13.15    *Voltages of a Three-Phase Power System*

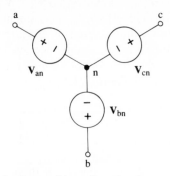

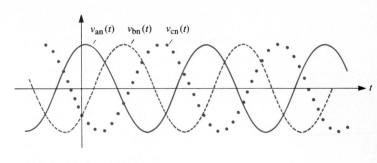

A. One possible configuration for
   the three sources

B. Phase voltages as functions of time

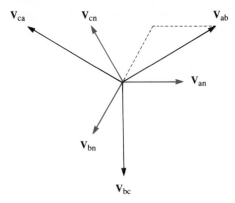

C. Phasor diagram of phase and line voltages

powered, it is important to know the sequence of the phases because it affects the direction of rotation. Consider Figure 13.15B. If you were to simultaneously observe the three phases as functions of time, you would see, in order, a peak of $v_{an}$ followed by a peak of $v_{bn}$ followed, in turn, by a peak of $v_{cn}$. This pattern would then repeat itself and continue on without end. Thus, we would say that the system has a phase sequence of a–b–c.

A three-phase transmission line generally has four conductors: one conductor for each of the three phases and one for the neutral. Long-distance transmission lines, however, usually have conductors present in multiples of three because the earth itself is used as the neutral conductor. This arrangement is possible primarily because in properly operating three-phase systems the neutral carries virtually no current. Systems designed in this way are well grounded at all critical locations.

The voltage between any line and the system neutral has already been defined as a phase voltage. These are the voltages $V_{an}$, $V_{bn}$, and $V_{cn}$ shown in

Figure 13.15C. The voltage between any two of the lines is a line-to-line voltage or simply a **line voltage.** $V_{ab} = V_{an} - V_{bn}$ is a line voltage, as are $V_{bc} = V_{bn} - V_{cn}$ and $V_{ca} = V_{cn} - V_{an}$. These voltages are also shown in Figure 13.15C. The magnitudes of line and phase voltages are always given in rms values. Often, in casual usage, the magnitudes themselves are referred to as the line and phase voltages of the system. It is left to Problem 13.34 at the end of the chapter to demonstrate that for three-phase power systems of the form shown in Figure 13.15,

$$|V_{ab}| = \sqrt{3}\,|V_{an}| \tag{13.20}$$

There are two standard ways in which the loads and sources of three-phase power systems are modeled in circuit diagrams. They reflect the different ways in which the three individual elements of a source or a load can be connected in a three-phase system. One of these configurations has already been introduced (Figure 13.15), and Figure 13.16 gives a simple rationale for its use.

**FIGURE 13.16**    *Development of Wye Form for Load of a Three-Phase Power System*

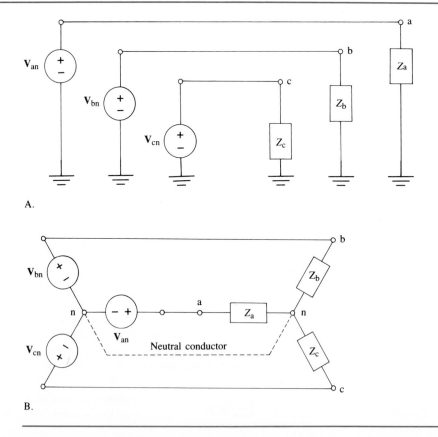

Notice, from Figure 13.16A, that each phase appears as an independently operating system and is drawn separately with its source and load. Once it is realized that each of these "separate" phases is referenced to the same ground point, they can be rearranged as shown in Figure 13.16B. When they are drawn in this way, both the load and the source are said to be modeled in **wye** form because each is in the shape of the letter "Y."

A three-phase wye load is said to be **balanced** if the impedances of each of its branches are identical. We have already assumed that the source is balanced by saying that its three voltages are identical in magnitude and equally spaced in phase. It is not too hard to prove (Problem 13.33) that the neutral conductor of a balanced three-phase system will carry no current. That is, $\mathbf{I}_a + \mathbf{I}_b + \mathbf{I}_c = 0$. If such is the case, the neutral conductor could theoretically be removed without altering the system's behavior. In large power systems, great care is taken to ensure that the loads of long-distance transmission lines are balanced, at least in the statistical sense. Even if the loads became slightly unbalanced, the current in the neutral would be small.

When a load is supplied by all three phases of a power system and when

**FIGURE 13.17**  *Standard Forms for Three-Phase Power System Loads and Sources*

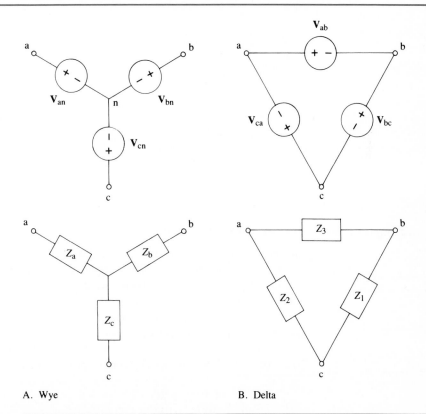

A. Wye                B. Delta

each phase is referenced directly to ground, the wye form of circuit model, as shown in Figure 13.17A, is used because it best reflects the physical configuration of the connections. Not all loads are arranged in this way. In some connections, the phases are referenced not to a common ground but to each other. The result is a triangular or **delta** ($\Delta$) form, such as is shown in Figure 13.17B.

Although the physical arrangement of a load or source in a three-phase system may be best modeled with a wye or a delta form in order to reflect the actual connections, it is mathematically quite possible to switch back and forth between the two. An equivalent delta form always exists for any wye-arranged load or source and vice versa. Chapter 4 first introduced this wye–delta transformation, which is also known as the tee–pi transformation. Chapter 4 showed its applicability only to resistive networks, but it is also valid here with more general impedances. The impedances for wye and delta models are defined in Figure 13.17. The rules for moving back and forth between wye and delta models for loads are given in Equations 13.21 and 13.22:

$\Delta$ to Y:  $\quad Z_a = \dfrac{Z_2 Z_3}{Z_1 + Z_2 + Z_3}$ $\hspace{2cm}$ (13.21a)

$\quad\quad\quad Z_b = \dfrac{Z_1 Z_3}{Z_1 + Z_2 + Z_3}$ $\hspace{2cm}$ (13.21b)

$\quad\quad\quad Z_c = \dfrac{Z_1 Z_2}{Z_1 + Z_2 + Z_3}$ $\hspace{2cm}$ (13.21c)

Y to $\Delta$:  $\quad Z_1 = \dfrac{Z_a Z_b + Z_b Z_c + Z_c Z_a}{Z_a}$ $\hspace{1.5cm}$ (13.22a)

$\quad\quad\quad Z_2 = \dfrac{Z_a Z_b + Z_b Z_c + Z_c Z_a}{Z_b}$ $\hspace{1.5cm}$ (13.22b)

$\quad\quad\quad Z_3 = \dfrac{Z_a Z_b + Z_b Z_c + Z_c Z_a}{Z_c}$ $\hspace{1.5cm}$ (13.22c)

---

**Example 13.10**   *A Wye–Delta Transformation*

Find the equivalent delta form of the three-phase wye load shown in Figure 13.18A. Then, verify the equivalence of the two forms.

Use of Equation 13.22 shows that the delta form, shown in Figure 13.18B, is completely symmetrical, with each leg of the load equal to 300 $\Omega$. To verify the equivalence of the two forms, we calculate the average power absorbed by each. Because the load is balanced, its central node in the wye form is at ground and each of the 100 $\Omega$ resistors has the phase voltage of 120 V AC across it. The total average power absorbed is

$$P = \frac{120^2}{100} + \frac{120^2}{100} + \frac{120^2}{100} = 432 \text{ W}$$

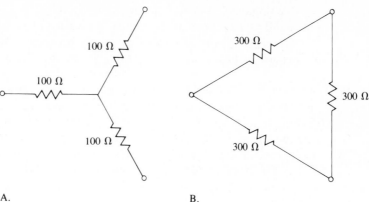

**FIGURE 13.18**    A.                                    B.

Each of the resistors of the delta form has the line voltage placed across it. From Equation 13.20, we know this voltage to be $120\sqrt{3}$ V AC. The total average power absorbed is

$$P = \frac{(120\sqrt{3})^2}{300} + \frac{(120\sqrt{3})^2}{300} + \frac{(120\sqrt{3})^2}{300} = 432 \text{ W}$$

Our faith in equivalency has once again been justified!

Although the wye and delta models are mathematically equivalent, several distinctions exist between the two in actual practice. One is that a wye-connected system is firmly grounded, while the delta form generally is not. Another distinction between the two is that the wye gives direct access to the phase voltages of the system; with delta connections, the larger line voltages are easily available. In some three-phase applications, the wye connection is preferable; in others, the delta form has decided advantages. We will leave a discussion of these differences to a course on power systems.

Analyzing a system that contains three-phase sources and loads can sometimes be a bit involved, but it is simply a special-case application of the use of the steady-state phasors and impedances introduced in Chapter 8. Some of the problems at the end of this chapter will give you practice in this analysis.

## 13.10  SUMMARY

This chapter introduced some of the fundamental concepts of AC power in electrical systems. Although the term *AC power* is limited in application to systems with sinusoidal voltages and currents, several of the principles discussed apply to other signal types. For instance, the effective or rms value of any periodic electrical signal is a concise measure of its ability to deliver power (on average) to a resistor.

A single AC source delivers power to a load in a way that depends on the phase difference between the voltage and current at the load's terminals. If the two are in phase, the load must be purely resistive, and power absorption by the load is always positive. In such cases, we say that the power is dissipated because it is not directly recoverable in electrical form. If the voltage and current are out of phase by 90°, the load is purely reactive, the average power absorbed is zero, and energy oscillates into and out of the load. The AC power absorbed by a general impedance $Z = R + jX$ is the sum of two parts. The average power absorbed can be accounted for entirely by the power dissipated in the load's resistive elements. The time-varying portion of the power has a frequency twice that of the exciting source and is due to the properties of both the resistance and the reactance of the load. In this chapter, we saw how these ideas are conveniently contained in the complex power $\mathbf{S}$, a mathematical formulation based on the phasor forms of voltage and current.

The methods developed in the first five sections of this chapter are commonly used in all areas of electrical engineering in which power is important. Sections 13.6, 13.7, and 13.9, however, introduced techniques used almost exclusively in the engineering of commercial power distribution systems. Power factors were defined, and the transmission inefficiencies caused by loads with low power factors were discussed. Impedance and power triangles were introduced as convenient methods of displaying results graphically and of keeping track of calculations as they are done.

Loads with poor power factors are costly to serve because of the large currents that they cause to exist in power systems. Especially in an era of energy conservation, power companies encourage their major customers to have power factors as close to 1 as possible. This is usually done by adding reactive elements in parallel with the load itself. When the reactance is properly chosen, the power factor is increased without changing the average power absorbed by the load.

Finally, three-phase power systems were introduced. Three-phase systems enjoy several advantages over single-phase systems. One is that they can be used to conveniently supply AC induction motors. The load and sources of three-phase systems are modeled in either wye or delta forms. These forms are mathematically equivalent, but in particular applications one or the other of these forms is likely to be preferable. Power engineers must be well versed in both.

## ■ PROBLEMS

13.1 For each of the periodic signals in Figure P13.1, determine the average and rms values of the signal and the average and peak powers delivered by the signal to a 10 Ω resistor.

13.2 **a.** Find the average and rms values of the half-wave rectified sinusoid of Figure P13.2A.

**b.** Find the average and rms values of the full-wave rectified sinusoid of Figure P13.2B.

13.3 Show that the rms value of a sinusoid is independent of its phase.

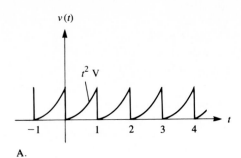

A.

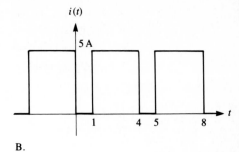

B.

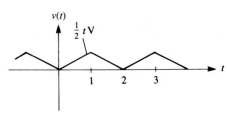

FIGURE P13.1    C.

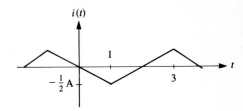

D.

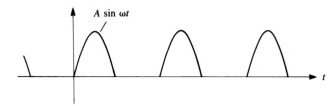

A.

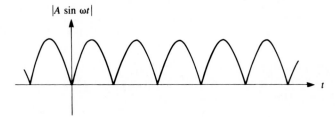

FIGURE P13.2    B.

13.4 In addition to their ohmic values, resistors are rated according to their ability to dissipate power. How much average power should resistors of values $10\,\Omega$, $100\,\Omega$, and $1000\,\Omega$ be able to handle when they are placed in either a standard 120 V AC system or in a system with a source of $20\sin 100t$ V?

13.5 Derive Equation 13.4b.

13.6 Find the rms values of the following functions:

    **a.**  $x(t) = 5 - \sin 10t + \cos(100t + 45°)$

    **b.**  $y(t) = 2 \cos 30t + 5 \sin 50t - 10 \cos 50t$

    **c.**  $g(t) = -10 + 7 \sin 1000t - 3 \sin 60t + 4 \cos(60t + 120°)$

**13.7** Verify Equation 13.6b by making direct use of $w_C(t) = (1/2)Cv_C^2(t)$.

**13.8** For each of the three passive elements in Figure P13.8, sketch the voltage across, current in, and power absorbed by the element. The three functions for each element should be plotted on one set of axes with time ranging from 0–10 s. For each element, indicate the regions of time when the net energy absorbed is increasing and when it is decreasing.

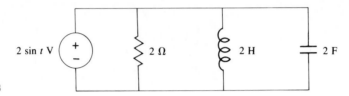

**FIGURE P13.8**

**13.9** Plot $v(t)$, $i(t)$, and $p(t)$ on the same axes for a load of $.5 + j1.3\ \Omega$ supplied from a source of $4 \sin 2\pi t$ V.

**13.10** Show that Equation 13.8c is consistent with Equation 13.5a for purely resistive loads. *Caution:* Equation 13.8c assumes a cosine form for voltage, while Equation 13.5a assumes a sine form.

**13.11** Show that Equation 13.8c reduces to Equation 13.6a for purely capacitive loads. Note the caution in Problem 13.10.

**13.12** Show that, under sinusoidal excitation, Ohm's law relates the rms values of voltage and current. That is, $V_{rms} = RI_{rms}$.

**13.13** Assume that an inductor of value $L$ has had no current prior to $t = 0$. At that time, the current becomes $I_m \sin \omega t$. Find expressions for the power absorbed $p_L(t)$ and the net energy absorbed $w_L(t)$. Sketch the results.

**13.14** In deriving some of the relationships in Section 13.4, it was assumed that $v(t) = V_m \cos \omega t$ V and $i(t) = I_m \cos(\omega t - \theta)$ A. Develop an expression for power equivalent to Equation 13.8c under the conditions that $v(t) = V_m \cos(\omega t + \theta)$ V and $i(t) = I_m \cos \omega t$ A.

**13.15** Find $p(t)$, the power absorbed by loads, under each of the following voltage and current conditions. For each case, identify $P$ and $Q$ and state, if you can, whether the load contains active elements.

    **a.**  $v(t) = 10 \cos(50t + 60°)$ V

        $i(t) = .5 \cos 50t$ A

    **b.**  $v(t) = 70 \cos(377t - 40°)$ V

        $i(t) = 3 \cos 377t$ A

    **c.**  $v(t) = 20 \cos(1000t + 110°)$ V

        $i(t) = .01 \cos 1000t$ A

    **d.**  $v(t) = 250 \cos(100t + 225°)$ V

        $i(t) = 2 \cos 100t$ A

Do the results suggest any general observations?

**13.16** Find $p(t)$, the power absorbed by the load shown in Figure P13.16. What is the average power absorption?

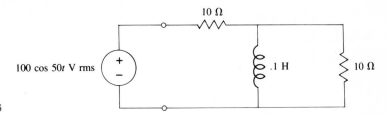

**FIGURE P13.16**

**13.17** For the circuit of Problem 13.16, show by direct calculation that the total average power absorbed is the sum of the average powers absorbed by each of the two resistors.

**13.18** Show that Equations 13.9c and 13.9d follow from Equation 13.9a.

**13.19 a.** Complex power is conserved. That is, the complex power **S** supplied to a load is equal to the total sum of complex powers absorbed by individual elements. Show that this statement is true for the special case of a sinusoidally excited load consisting of a series connection of elements.

   **b.** Show that since **S** is conserved, $P$ and $Q$ must be individually conserved.

**13.20** Determine the complex power for each of the following conditions. Which conditions can you say, for certain, involve active elements?

   **a.** $v(t) = 5 \cos (50t + 70°)$ V, $i(t) = 2 \cos (50t + 20°)$ A

   **b.** $v(t) = 100 \cos (1000t + 80°)$ V, $i(t) = 5 \cos (1000t - 30°)$ A

   **c.** $v(t) = 20 \cos (t - 135°)$ V, $i(t) = .25 \cos t$ A

   **d.** $v(t) = -10 \cos (10t + 75°)$ V, $i(t) = 1.5 \cos (10t - 45°)$ A

   In which quadrants of $\theta$ are $P$ and $Q$ positive and negative?

**13.21 a.** For $C = 200$ μF, show that the average power absorbed by the load in Figure P13.21 is accounted for entirely by the two resistors.

   **b.** Do the same for $C = 50$ μF.

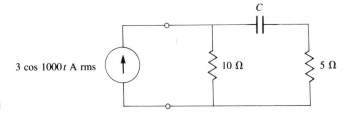

**FIGURE P13.21**

Use phasors and complex frequency notation in your work. What conclusions do you draw from comparing the results of the two cases?

**13.22** Repeat the development of impedance, voltage, and power triangles given in Section 13.6, but assume a nonzero phase angle for current.

**13.23** Draw the impedance and power triangles for each of the four separate situations shown in Figure P13.23.

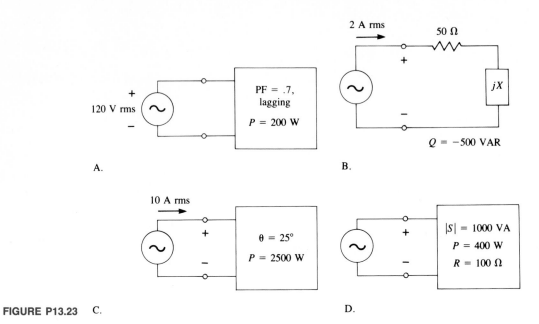

A.                                                                      B.

FIGURE P13.23   C.                                                      D.

**13.24** Partially labeled power and impedance triangles for two different loads are shown in Figures P13.24A and P13.24B. Find the rms value of current for part A of the figure and voltage for part B.

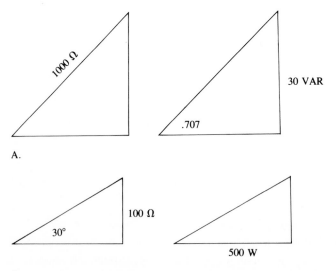

A.

**FIGURE P13.24**   B.

**13.25** A load consisting of a resistor and an inductor along with an AC voltage source is shown in Figure P13.25. If the inductor is replaced by a capacitor, what value of capacitor would cause the same power factor for the load?

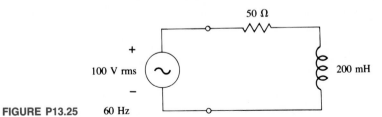

**FIGURE P13.25**     60 Hz

**13.26 a.** What is the power factor of the load shown in Figure P13.26? Assume that it operates at $f = 60$ Hz.

**b.** What value of inductor or capacitor should be added in parallel to the load to bring its power factor to .9 without changing the lead/lag status?

**c.** How much current should the added element be capable of carrying?

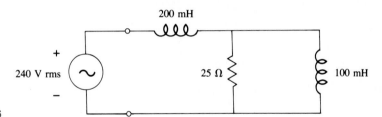

**FIGURE P13.26**

**13.27** Prove Equation 13.17 for $k = +1$. *Hint:* First determine the input impedance of the compensated load, and then find its power factor.

**13.28** Show that in Example 13.8 the load corrected for a higher power factor draws less current from the source than the original load does. Assume $V_s = 200$ V rms.

**13.29** Calculate the rms value of the current supplied by the source in Figure P13.29 under the following conditions:

**a.** When it is connected to the load shown

**b.** When an element has been added in parallel to bring the power factor to .98

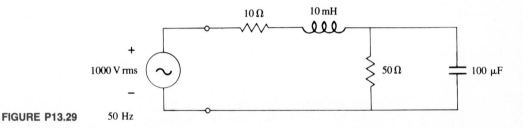

**FIGURE P13.29**     50 Hz

**13.30** Find the input impedance of the load of Figure P13.30 after it has been corrected to PF = 1. Explain the result.

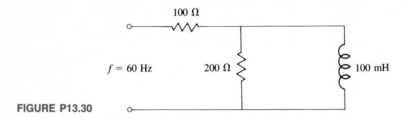

**FIGURE P13.30**

**13.31** To investigate a possible alternate way of correcting for a low power factor, a reactive element will be added in series, as indicated in Figure P13.31.

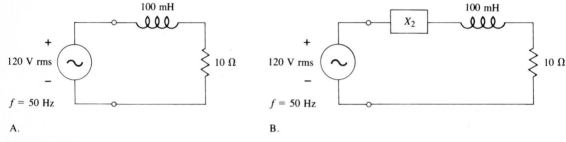

**FIGURE P13.31**

    **a.** Compute the current supplied to and average power absorbed by the load circuit in Figure P13.31A. Find its power factor.

    **b.** What reactance should be added in series, as shown in Figure P13.31B, to bring the power factor to .95?

    **c.** Compute the current supplied to and average power absorbed by this "corrected" circuit. Compare these values to those of the original circuit and comment on the differences. Is this a good method for power factor correction?

**13.32** Derive Equation 13.19 by finding the power delivered to the load and, by taking derivatives, maximizing it with respect to $X_L$ and $R_L$.

**13.33** Refer to Figure P13.33. Show that, for a balanced load, the neutral conductor of a three-phase system carries no current and can theoretically be omitted.

**13.34** Prove Equation 13.20.

**13.35** Most residences are serviced by three-wire, single-phase power supplies. These power supplies consist of a neutral (ground) conductor, a conductor carrying one of the phase voltages of the more general three-phase system, and a third conductor carrying the negative of that phase voltage. Consider an electric heater modeled as a 10 Ω resistor. It can be wired with the neutral at one end and the phase voltage at the other or by the phase voltage and its negative at opposite ends. If the phase voltage is 60 Hz, 120 V AC, what power is delivered by the two different wiring approaches?

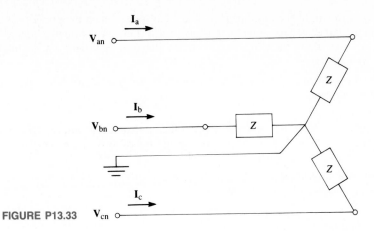

FIGURE P13.33

**13.36** What will be the maximum possible average power delivered to a load by the source shown along with its internal impedance in Figure P13.36?

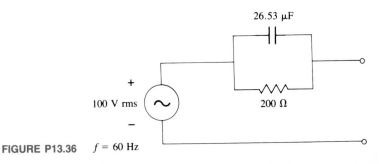

FIGURE P13.36    $f = 60$ Hz

**13.37** The circuit shown in Figure P13.37 is an example of an unbalanced three-phase load. It can be used as a simple phase sequence meter to determine the order in which a power system's phases occur. The "meter" consists of a capacitor and two light bulbs that can be modeled as resistors. One end of each element is tied to a central point, and the other end

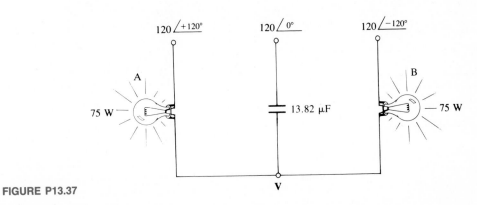

FIGURE P13.37

is connected to one of the three phases. The meter operates on the principle that the light bulb with the larger voltage magnitude across it will shine more brightly. In the circuit, will Bulb A or Bulb B be the brighter? A phasor diagram may help you to interpret your results. Assume $f = 60$ Hz.

**13.38** Consider the equivalent wye and delta loads of Example 13.10. Show that the current supplied to the two loads is identical—that is, that the current in each of the supply lines is the same in both cases.

**13.39** A three-phase load consists of three $500\,\Omega$ resistors that can be arranged in either wye or delta connections. Which draws the most current and absorbs the most power on average when it is supplied by a 120 V AC three-phase system?

**13.40** Draw a phasor diagram showing all of the line and phase voltages and the phase currents for the load shown in Figure P13.40.

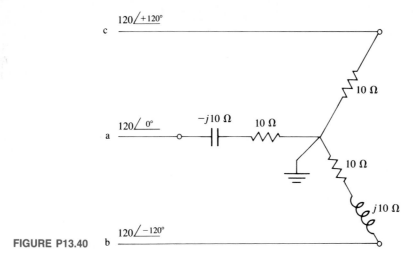

**FIGURE P13.40**

**13.41** An unbalanced wye load and its three-phase supply is shown in Figure P13.41. What average power is absorbed by the load? *Note:* Central node is not grounded.

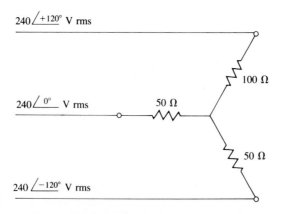

**FIGURE P13.41**

# Two-Port Networks

## 14.1  INTRODUCTION

In Chapter 4, we studied one-port electrical circuits and saw that it is often useful to find equivalent forms for them. Known as Thévenin and Norton equivalent circuits, these equivalents can be used to model complicated circuits with just two elements: one source and one impedance. A one-port circuit and its Thévenin and Norton forms are equivalent because all three have exactly the same external $V–I$ characteristics. Therefore, no electrical measurements made at the port can distinguish between the three circuits. As a result, a one-port circuit contained within a larger network can be replaced with either of its equivalent forms and the behavior of the system will not change. The advantage of using the Thévenin or Norton models in analysis is that they can considerably simplify subsequent calculations.

Many practical electrical circuits have two ports instead of just one. Any network with a single input and a single output can be thought of as a two-port device. Some practical networks have more than two ports, and several important theorems are based on networks with arbitrarily large numbers of ports. We will consider just the two-port case, however. Commonly encountered two-port devices include amplifiers, filters, transformers and transistors.

Large systems are sometimes formed by combining several one- and two-port circuits. The block diagram of the cascaded circuits in Figure 14.1, for example, represents one channel of a stereo system. Circuit A is the input device, perhaps a record turntable. Circuit B is an amplifier and C is the audio speaker. The output of a turntable is an electrical signal, but it has no such electrical input. A turntable does have a mechanical input in the form of a vibrating stylus or needle, but we can ignore this nonelectrical input by considering the turntable and the record it plays to be a complete system. Speakers have electrical signal inputs but no electrical outputs. Both turntables and speakers can be modeled as one-port circuits. Amplifiers, which have both inputs and outputs, are modeled as two-port circuits.

The analysis of systems like the one shown in Figure 14.1 can be made much easier if each of the individual components is replaced with an equivalent, simpler model. Chapter 4 showed how this replacement is done for one-port circuits. This chapter introduces methods used for modeling two-port devices. Some of the assumptions used will be slightly different than with one-ports, but the basic principle will be the same. A two-port circuit and any of its equivalent models

FIGURE 14.1  *One Channel of a Stereo System*

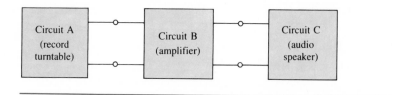

have identical external characteristics, and replacing one with another in a system will not change the behavior of the system. As with Thévenin and Norton circuits, the advantage of these equivalent models is that, once they are known, their use considerably simplifies any additional analysis.

## 14.2  BASIC PRINCIPLES OF ONE- AND TWO-PORT CIRCUITS

The mathematical modeling of two-port circuits is based on their external $V$–$I$ characteristics. Before we analyze these circuits, however, let's review some basic principles. In the terminology of electrical circuits, a **port** is a pair of leads or nodes through which electrical signals can enter or leave a circuit. Often, the port does not exist in a well-defined physical sense with two nicely paired wires emerging from a circuit. Instead, the analyst recognizes that any node pair in a circuit can be treated as a port for purposes of modeling. A voltage and current can be defined at any port. Current entering the circuit through one of the leads of a port will be equal to the current leaving by the other lead.

The basics of one-port circuit modeling are reviewed in Figure 14.2. There are two variables at the port of such an electrical circuit. When the circuit is linear, it is a simple matter to consider either of the variables to be independent and the other dependent. These alternate assumptions lead to the following equations:

$$V = V_{oc} - Z_{eq}I \tag{14.1a}$$

$$I = I_{sc} - Y_{eq}V \tag{14.1b}$$

FIGURE 14.2    *One-Port Circuit Modeling*

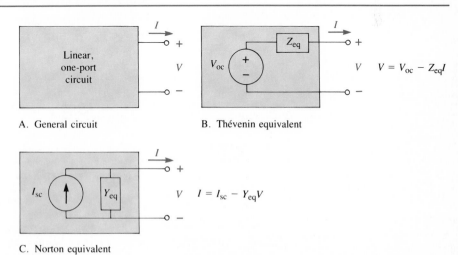

A.  General circuit

B.  Thévenin equivalent

C.  Norton equivalent

$V_{oc}$ is the open circuit value of $V$, and $I_{sc}$ is the current that exists when a short circuit is placed across the port. $Z_{eq}$ is the equivalent impedance seen looking into the port, and $Y_{eq} = 1/Z_{eq}$. In Equation 14.1a, the current $I$ is treated as the independent variable; the voltage $V$ is dependent. In Equation 14.1b, the opposite treatment is used. Although they contain somewhat different terms or parameters, the two equations are entirely equivalent. The one that is chosen is largely a matter of convenience and will depend on how the model fits into a larger problem.

If a circuit has only two external leads, there is no problem in identifying the port. For devices of three, four, or more leads, however, the pairings of leads into ports are not necessarily unique, and the ports can be identified only by knowing how the device is to be used. The bipolar junction transistor shown in Figure 14.3A has three terminals. They are its base (B), collector (C), and emitter (E) leads. In many circuit applications, one of the three leads is used as a common voltage reference for each of the other two leads. Thus, we can think of the transistor as a two-port device. Two commonly used transistor configurations are shown in Figures 14.3B and 14.3C. In each case, one of the leads is shared in common by the two ports.

The identification of ports for a circuit of several leads is usually no problem, especially if you can see how the circuit is used. An amplifier, for instance, is designed so that two of its four terminals must be paired to be the input port; the other two obviously form the output.

A circuit with two ports has four external electrical variables: a voltage and a current at the input port and a voltage and a current at the output. Such a

FIGURE 14.3   *Bipolar Junction Transistor*

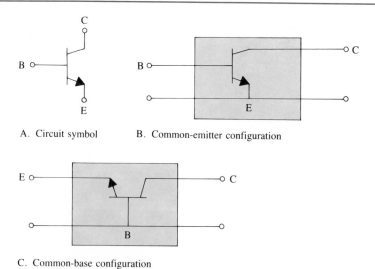

A. Circuit symbol          B. Common-emitter configuration

C. Common-base configuration

FIGURE 14.4    *Current and Voltage Conventions for a Two-Port Circuit*

circuit is shown in general terms in Figure 14.4. It is not always necessary or particularly useful to think of one of the ports as the input and the other as the output. Instead, they can simply be labeled as port 1 and port 2. The usual convention with multiport circuits is to define the current at a port to be entering the terminal given the positive polarity. At each port, the current entering the positive lead is equal to that exiting the negative lead. Before you apply the techniques of this chapter to any particular circuit problem, be sure to note whether or not the assumed directions for current in the problem agree with this stated convention. If they do not, you may either redefine the current directions or modify the techniques of this chapter to fit the requirements of the problem. Uppercase symbols are used in Figure 14.4 because the variables may be expressed as Fourier or Laplace transforms or they may be phasors.

Any two of the four external variables of a two-port circuit may be treated as independent; equations can be written that show how the remaining two variables depend on them. There are six possible ways in which the four variables can be grouped into independent and dependent pairs. Just as any one-port circuit can be modeled in either a Norton or a Thévenin equivalent form, normally any of the six possible variable groupings can be used as a basis for modeling a two-port circuit.* The approach that is chosen in any particular application will depend on how the two-port circuit fits into a larger problem.

## 14.3    *z* PARAMETERS

Figure 14.5 shows a linear, two-port circuit. It is assumed that the network contains no *independent* sources of energy and that it is initially at rest (no stored energy). In addition, any controlled sources within the circuit cannot depend on variables that are outside of the circuit. Let's begin by considering the currents at the ports to be independent variables, which is emphasized by the presence of the

---

* Some special situations exist in which one or several of these six modeling approaches may not be well defined for a given circuit. It is assumed in this chapter that this is not a problem.

FIGURE 14.5   *Two-Port Circuit with Independent External Currents*

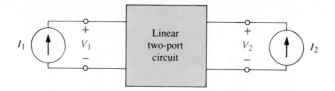

current sources $I_1$ and $I_2$. The argument that follows does not actually require that the current sources be present, but they are helpful for purposes of visualization.

Superposition tells us that the voltage $V_1$ of the circuit in Figure 14.5 is the sum of two parts, one due to $I_1$ and the other to $I_2$. The same can be said of the voltage $V_2$. Because the network is linear, these effects can be mathematically described by the following two algebraic equations:

$$V_1 = z_{11}I_1 + z_{12}I_2 \qquad (14.2a)$$

$$V_2 = z_{21}I_1 + z_{12}I_2 \qquad (14.2b)$$

The terms $z_{11}$, $z_{12}$, $z_{21}$, and $z_{22}$ are known as the **z parameters** of the two-port network. As we will see, these parameters can be calculated analytically, but they can also be found through measurement in the laboratory. Finding the parameters of a circuit by laboratory measurement is an example of the process of system identification. With the results of the measurements, the system's response behavior to any excitation can be predicted even though the internal makeup of the circuit might be totally unknown. One of the measurements needed to find the z parameters of a circuit can be understood by considering Equation 14.2a. It tells us that if $I_2$ is constrained to be zero by leaving port 2 open-circuited, then

$$z_{11} = \left.\frac{V_1}{I_1}\right|_{I_2=0}$$

The other parameters can be found in similar ways so that the entire set can be written as follows:

$$z_{11} = \left.\frac{V_1}{I_1}\right|_{I_2=0} \qquad z_{12} = \left.\frac{V_1}{I_2}\right|_{I_1=0}$$

$$z_{21} = \left.\frac{V_2}{I_1}\right|_{I_2=0} \qquad z_{22} = \left.\frac{V_2}{I_2}\right|_{I_1=0} \qquad (14.3)$$

Now you can see why the symbol $z$ was chosen to represent these parameters. Each of the $z$ parameters is a system function with the dimensions of impedance; $Z$ has long been the general symbol for impedance in electrical engineering. From Equation 14.3, note that these system functions are valid only under the restrictions that one port or the other is open-circuited. Thus, $z$ parameters are also known as **open circuit impedance parameters.** Knowing all four of the $z$ parameters of a two-port network allows you to completely model the network mathematically.

There are several ways to calculate two-port parameters. Two methods are demonstrated in the following example.

**Example 14.1** *Finding z Parameters*

Find the $z$ parameters of the circuit shown in Figure 14.6A.

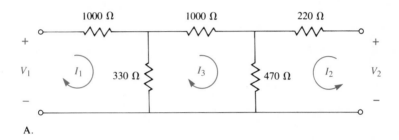

A.

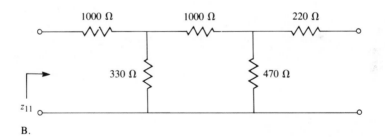

B.

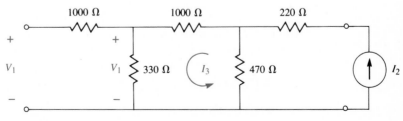

**FIGURE 14.6** C.

*(continues)*

**Example 14.1**   *Continued*

The circuit contains more variables than just those at the two ports. By using mesh analysis, a complete set of equations can be written:

$$V_1 = 1330I_1 - 330I_3$$
$$0 = -330I_1 + 470I_2 + 1800I_3$$
$$V_2 = 690I_2 + 470I_3$$

Notice that the usual convention of having all mesh currents in a clockwise direction has been violated here in order to have the mesh current $I_2$ agree with the assigned direction of the current at port 2. The three equations contain the four external port variables and one additional current. $I_3$ can be eliminated to yield

$$V_1 = 1269.5I_1 + 86.2I_2$$
$$V_2 = 86.2I_1 + 567.3I_2$$

The four $z$ parameters are easily identified by comparing the last two equations with Equation 14.2:

$$z_{11} = 1269.5 \ \Omega \qquad z_{12} = 86.2 \ \Omega$$
$$z_{21} = 86.2 \ \Omega \qquad z_{22} = 567.3 \ \Omega$$

Another way to find the parameters is by using the formulas of Equation 14.3. We calculate $z_{11}$ and $z_{12}$ here and leave it to you to find the others. Finding each in this way requires that a current be set equal to zero. These modifications are shown in Figures 14.6B and 14.6C.

The impedance seen looking into port 1 when port 2 is open-circuited is $z_{11}$. The easiest way to find it is to make the necessary parallel and series combinations of resistors:

$$z_{11} = 1000 + 330 \| (1000 + 470) = 1269.5 \ \Omega$$

Because there can be no current in the 220 $\Omega$ resistor when port 2 is open-circuited, it has no effect on this calculation. Finding $z_{21}$ is a little more complicated. Port 1 should be open-circuited. An applied current $I_2$ will result in a voltage $V_1$, and their ratio is $z_{12}$, as is shown in Figure 14.6C. One possible way to calculate $z_{12}$ is through the following steps:

1. Write a KVL equation around the central mesh:

$$1800I_3 - 470I_2 = 0$$
$$I_3 = \frac{470I_2}{1800}$$

**2.** Calculate $V_1$ as a function of $I_2$:

$$V_1 = 330 I_3 = \frac{(330)(470)I_2}{1800}$$

**3.** Find the ratio $V_1/I_2$:

$$z_{12} = \frac{V_1}{I_2} = \frac{(330)(470)}{1800} = 86.2 \ \Omega$$

So, we see that the two methods agree, just as they should.

---

Because the circuit of Example 14.1 is purely resistive, its $z$ parameters are real numbers. Circuits containing inductors and capacitors require the use of impedances, and the parameters will be functions of the complex frequency $s$ or of the sinusoidal frequency $j\omega$. We will see examples of frequency-dependent parameters later.

## 14.4  CIRCUIT MODELS OF TWO-PORT NETWORKS

Equations 14.2a and 14.2b mathematically describe the external behavior of two-port networks based on their $z$ parameters. Before additional parameter sets are introduced, let's see how this mathematical description (and the corresponding network) can be modeled by simple electrical circuits. For convenience, the $z$ parameter based equations that describe the behavior of a two-port circuit are repeated here:

$$V_1 = z_{11}I_1 + z_{12}I_2 \tag{14.4a}$$

$$V_2 = z_{21}I_1 + z_{22}I_2 \tag{14.4b}$$

All of the terms of Equations 14.4a and 14.4b have the dimensions of voltage. They can thus be treated as a set of two mesh equations that relate the four external variables of the two-port circuit. Each equation contains three terms: the external voltage at a port, a voltage due to the current at that same port, and another voltage due to the current at the opposite port. The most straightforward way in which to model these equations is shown in Figure 14.7A. From the figure, we see that two-port circuits, no matter how complicated, can be equivalently modeled with just four elements: two impedances and two dependent sources. Shown as a dashed line is a short circuit connection between points x and x′ that should be included if $V_1$ and $V_2$ share a common reference point. It is easy to prove that it carries no current.

There are other, less obvious, circuit models that satisfy the external $V$–$I$ characteristics of Equation 14.4. They can be found by manipulating the equa-

**FIGURE 14.7**  *Two z Parameter Models for a Two-Port Circuit*

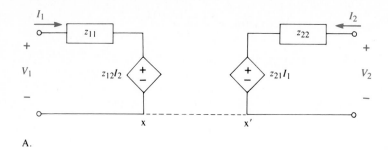

A.

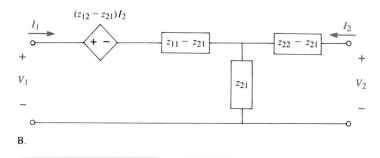

B.

tions into other forms. The result of one possible manipulation is

$$V_1 - (z_{12} - z_{21})I_2 = (z_{11} - z_{21})I_1 + z_{21}(I_1 + I_2) \tag{14.5a}$$

$$V_2 = z_{21}(I_2 + I_1) + (z_{22} - z_{21})I_2 \tag{14.5b}$$

You should verify that Equations 14.4 (a and b) and 14.5 (a and b) are consistent with one another. Equations 14.5a and 14.5b can, again, be interpreted as mesh equations because all of their individual terms are voltages. The corresponding circuit model is shown in Figure 14.7B. Like the model of Figure 14.7A, it has four circuit elements, although three are impedances and only one is a dependent source. Note, however, that for this model, $V_1$ and $V_2$ must share a common reference point.

## 14.5  ADDITIONAL PARAMETER SETS

The open circuit $z$ parameters are just one of six possible ways in which two-port circuits can be modeled. Figure 14.8 defines the six sets and gives the most direct circuit models for four of them. These six possible representations, when they exist, are completely equivalent to one another, but some of them are more useful than others for certain applications. This section gives examples of the calculation

**FIGURE 14.8** *Defining the Six Sets of Two-Port Parameters*

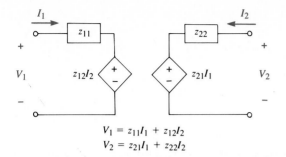

$$V_1 = z_{11}I_1 + z_{12}I_2$$
$$V_2 = z_{21}I_1 + z_{22}I_2$$

A. Open circuit impedance parameters

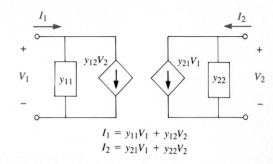

$$I_1 = y_{11}V_1 + y_{12}V_2$$
$$I_2 = y_{21}V_1 + y_{22}V_2$$

B. Short circuit admittance parameters

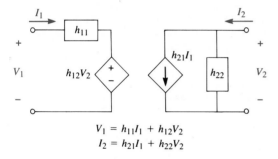

$$V_1 = h_{11}I_1 + h_{12}V_2$$
$$I_2 = h_{21}I_1 + h_{22}V_2$$

C. Hybrid parameters

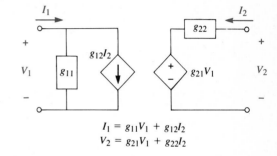

$$I_1 = g_{11}V_1 + g_{12}I_2$$
$$V_2 = g_{21}V_1 + g_{22}I_2$$

D. Inverse hybrid parameters

$$V_1 = a_{11}V_2 - a_{12}I_2$$
$$I_1 = a_{21}V_2 - a_{22}I_2$$

E. Transmission parameters

$$V_2 = b_{11}V_1 - b_{12}I_1$$
$$I_2 = b_{21}V_1 - b_{22}I_1$$

F. Inverse transmission parameters

of some of these parameter sets. Some of the advantages of the different sets will be considered later.

## *y* Parameters

The **y parameters** are known as **short circuit admittance parameters.** (Refer to Figure 14.8B.)

**Example 14.2** *Finding y Parameters*

Compute the *y* parameters for the circuit in Figure 14.9. Find each parameter separately by direct computation.

*(continues)*

**Example 14.2**  *Continued*

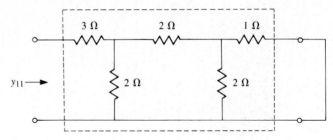

A.

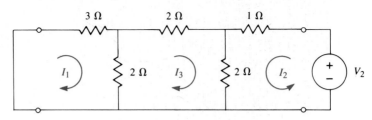

B.

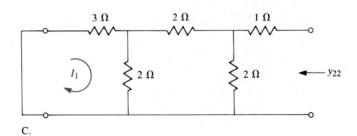

C.

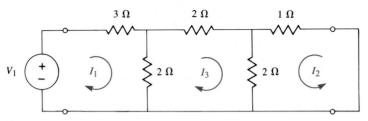

**FIGURE 14.9**  D.

**1.**  $y_{11} = \dfrac{I_1}{V_1}\bigg|_{V_2=0}$

This parameter is the input admittance at port 1 with port 2 short-circuited. Therefore, if we refer to Figure 14.9A,

$$\frac{1}{y_{11}} = 3 + 2\|(2 + 2\|1) = \frac{29}{7} \ \Omega$$

$$y_{11} = \frac{7}{29} \ S$$

2. $y_{12} = \dfrac{I_1}{V_2}\bigg|_{V_1=0}$

Using the variables defined in Figure 14.9B, we can write a complete set of equations for the circuit. Note that port 1 has been shorted.

$$5I_1 - 2I_3 = 0$$

$$3I_2 + 2I_3 = V_2$$

$$-2I_1 + 2I_2 + 6I_3 = 0$$

We then solve for $I_1$ in terms of $V_2$:

$$I_1 = -\frac{2}{29} \ V_2$$

so

$$y_{12} = \frac{I_1}{V_2} = -\frac{2}{29} \ S$$

3. $y_{22} = \dfrac{I_2}{V_2}\bigg|_{V_1=0}$

This parameter is the input admittance at port 2 with port 1 short-circuited. Therefore, if we refer to Figure 14.9C,

$$\frac{1}{y_{22}} = 1 + 2\|(2 + 2\|3) = \frac{29}{13} \ \Omega$$

$$y_{22} = \frac{13}{29} \ S$$

4. $y_{21} = \dfrac{I_2}{V_1}\bigg|_{V_2=0}$

Referring to Figure 14.9D for the case of port 2 being short-circuited, we can write the following set of equations:

$$5I_1 - 2I_3 = V_1$$

$$3I_2 + 2I_3 = 0$$

$$-2I_1 + 2I_2 + 6I_3 = 0$$

(continues)

**Example 14.2**  *Continued*

We then solve for $I_2$ in terms of $V_1$:

$$y_{21} = \frac{I_2}{V_1} = -\frac{2}{29}\,S$$

## *h* Parameters

The *h* **parameters** are known as **hybrid parameters.** The most direct two-port circuit model based on *h* parameters (see Figure 14.8C) has an input side like that of the *z*-parameter model and an output side like that of the *y*-parameter model. Therefore, this model is a combination, or hybrid, of the other two. The parameter $h_{11}$ has the dimensions of impedance, $h_{12}$ is a voltage gain, $h_{21}$ is a current gain, and $h_{22}$ is an admittance. This parameter set is particularly important to understand because the specifications for bipolar junction transistors are most often given in terms of their *h* parameters.

**Example 14.3**  *Finding h Parameters*

Find the *h* parameters for the two-port circuit shown in Figure 14.10. Do so by writing a complete set of node equations that use the variables defined in the figure.

The equations are as follows:

$$I_1 = \left(\frac{1}{220} + \frac{1}{150}\right) V_1 - \frac{1}{220} V_3 - \frac{1}{150} V_2 + .001V_x$$

$$0 = -\frac{1}{220} V_1 + \left(\frac{1}{220} + \frac{1}{100} + \frac{1}{330}\right) V_3 - \frac{1}{330} V_2$$

$$I_2 = -\frac{1}{150} V_1 - \frac{1}{330} V_3 + \left(\frac{1}{330} + \frac{1}{150}\right) V_2 - .001V_x$$

$$V_x = V_3 - V_2$$

For the *h*-parameter model, we treat $I_1$ and $V_2$ as the independent variables. We group them to one side, simplify the equations, and, at the same time, eliminate $V_x$:

$$.01121V_1 - .00355V_3 = I_1 + .00767V_2$$

$$-.00455V_1 + .01758V_3 = .00303V_2$$

$$I_2 + .00667V_1 + .00403V_3 = .00107V_2$$

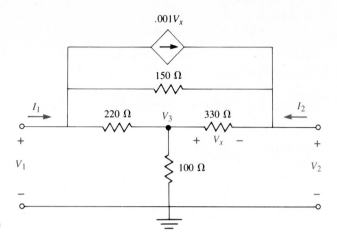

**FIGURE 14.10**

We are not interested in $V_3$, but we use this set of three equations for finding $V_1$ and $I_2$ in terms of $I_1$ and $V_2$:

$$I_2 = \frac{\begin{vmatrix} I_1 + .00767V_2 & .01121 & -.00355 \\ .00303V_2 & -.00455 & .01758 \\ .00107V_2 & .00667 & .00403 \end{vmatrix}}{\begin{vmatrix} 0 & .01121 & -.00355 \\ 0 & -.00455 & .01758 \\ 1 & .00667 & .00403 \end{vmatrix}}$$

$$= -.749I_1 - .0058V_2$$

Similarly,

$$V_1 = 97.17I_1 + .805V_2$$

The $h$ parameters can be seen to be $h_{11} = 97.17\,\Omega$, $h_{12} = .805$, $h_{21} = -.749$, and $h_{22} = -.0058$. It is common for the $h$ parameters of a circuit to vary from one another by several orders of magnitude.

## Transforming from One Parameter Set to Another

Once one parameter set for a circuit is known, it can be used as the starting point in calculating any of the other five sets. You just have to rearrange the two equations properly in terms of the desired independent and dependent variables.

**Example 14.4**    *Transforming from z Parameters to g Parameters*

It is easy to verify that the $z$-parameter equations for the circuit shown in Figure 14.11 are

$$V_1 = (s + 1)I_1 + sI_2$$
$$V_2 = sI_1 + (s + 2)I_2$$

Determine the system's $g$ parameters.

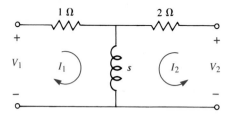

**FIGURE 14.11**

From Figure 14.8, we see that $V_1$ and $I_2$ should be treated as independent variables. They are put to one side of the two equations:

$$(s + 1)I_1 = V_1 - sI_2$$
$$V_2 - sI_1 = (s + 2)I_2$$

Solving for $I_1$ and $V_2$ leads to

$$I_1 = \frac{1}{s + 1} V_1 - \frac{s}{s + 1} I_2$$

$$V_2 = \frac{s}{s + 1} V_1 + \frac{3s + 2}{s + 1} I_2$$

Comparing these equations to the equations of Figure 14.8D leads to the conclusion that

$$g_{11} = \frac{1}{s + 1}, \qquad g_{12} = -\frac{s}{s + 1}$$

$$g_{21} = \frac{s}{s + 1}, \qquad g_{22} = \frac{3s + 2}{s + 1}$$

Table 14.1 gives formulas for going from any one of the six parameter sets directly to any of the other five. Verification of some of these relationships is

**TABLE 14.1** *Conversion Table for Two-Port Parameters*

| | [z] Impedance | [y] Admittance | [h] Hybrid | [g] Inverse Hybrid | [a] Transmission | [b] Inverse Transmission |
|---|---|---|---|---|---|---|
| **[z]** | $\begin{matrix} z_{11} & z_{12} \\ z_{21} & z_{22} \end{matrix}$ | $\begin{matrix} \dfrac{y_{22}}{\Delta_y} & -\dfrac{y_{12}}{\Delta_y} \\ -\dfrac{y_{21}}{\Delta_y} & \dfrac{y_{11}}{\Delta_y} \end{matrix}$ | $\begin{matrix} \dfrac{\Delta_h}{h_{22}} & \dfrac{h_{12}}{h_{22}} \\ -\dfrac{h_{21}}{h_{22}} & \dfrac{1}{h_{22}} \end{matrix}$ | $\begin{matrix} \dfrac{1}{g_{11}} & -\dfrac{g_{12}}{g_{11}} \\ \dfrac{g_{21}}{g_{11}} & \dfrac{\Delta_g}{g_{11}} \end{matrix}$ | $\begin{matrix} \dfrac{a_{11}}{a_{21}} & \dfrac{\Delta_a}{a_{21}} \\ \dfrac{1}{a_{21}} & \dfrac{a_{22}}{a_{21}} \end{matrix}$ | $\begin{matrix} \dfrac{b_{22}}{b_{21}} & \dfrac{1}{b_{21}} \\ \dfrac{\Delta_b}{b_{21}} & \dfrac{b_{11}}{b_{21}} \end{matrix}$ |
| **[y]** | $\begin{matrix} \dfrac{z_{22}}{\Delta_z} & -\dfrac{z_{12}}{\Delta_z} \\ -\dfrac{z_{21}}{\Delta_z} & \dfrac{z_{11}}{\Delta_z} \end{matrix}$ | $\begin{matrix} y_{11} & y_{12} \\ y_{21} & y_{22} \end{matrix}$ | $\begin{matrix} \dfrac{1}{h_{11}} & -\dfrac{h_{12}}{h_{11}} \\ \dfrac{h_{21}}{h_{11}} & \dfrac{\Delta_h}{h_{11}} \end{matrix}$ | $\begin{matrix} \dfrac{\Delta_g}{g_{22}} & \dfrac{g_{12}}{g_{22}} \\ -\dfrac{g_{21}}{g_{22}} & \dfrac{1}{g_{22}} \end{matrix}$ | $\begin{matrix} \dfrac{a_{22}}{a_{12}} & -\dfrac{\Delta_a}{a_{12}} \\ -\dfrac{1}{a_{12}} & \dfrac{a_{11}}{a_{12}} \end{matrix}$ | $\begin{matrix} \dfrac{b_{11}}{b_{12}} & -\dfrac{1}{b_{12}} \\ \dfrac{\Delta_b}{b_{12}} & \dfrac{b_{22}}{b_{12}} \end{matrix}$ |
| **[h]** | $\begin{matrix} \dfrac{\Delta_z}{z_{22}} & \dfrac{z_{12}}{z_{22}} \\ -\dfrac{z_{21}}{z_{22}} & \dfrac{1}{z_{22}} \end{matrix}$ | $\begin{matrix} \dfrac{1}{y_{11}} & -\dfrac{y_{12}}{y_{11}} \\ \dfrac{y_{21}}{y_{11}} & \dfrac{\Delta_y}{y_{11}} \end{matrix}$ | $\begin{matrix} h_{11} & h_{12} \\ h_{21} & h_{22} \end{matrix}$ | $\begin{matrix} \dfrac{g_{22}}{\Delta_g} & -\dfrac{g_{12}}{\Delta_g} \\ -\dfrac{g_{21}}{\Delta_g} & \dfrac{g_{11}}{\Delta_g} \end{matrix}$ | $\begin{matrix} \dfrac{a_{12}}{a_{22}} & \dfrac{\Delta_a}{a_{22}} \\ -\dfrac{1}{a_{22}} & \dfrac{a_{21}}{a_{22}} \end{matrix}$ | $\begin{matrix} \dfrac{b_{12}}{b_{11}} & \dfrac{1}{b_{11}} \\ -\dfrac{\Delta_b}{b_{11}} & \dfrac{b_{21}}{b_{11}} \end{matrix}$ |
| **[g]** | $\begin{matrix} \dfrac{1}{z_{11}} & -\dfrac{z_{12}}{z_{11}} \\ \dfrac{z_{21}}{z_{11}} & \dfrac{\Delta_z}{z_{11}} \end{matrix}$ | $\begin{matrix} \dfrac{\Delta_y}{y_{22}} & \dfrac{y_{12}}{y_{22}} \\ -\dfrac{y_{21}}{y_{22}} & \dfrac{1}{y_{22}} \end{matrix}$ | $\begin{matrix} \dfrac{h_{22}}{\Delta_h} & -\dfrac{h_{12}}{\Delta_h} \\ -\dfrac{h_{21}}{\Delta_h} & \dfrac{h_{11}}{\Delta_h} \end{matrix}$ | $\begin{matrix} g_{11} & g_{12} \\ g_{21} & g_{22} \end{matrix}$ | $\begin{matrix} \dfrac{a_{21}}{a_{11}} & -\dfrac{\Delta_a}{a_{11}} \\ \dfrac{1}{a_{11}} & \dfrac{a_{12}}{a_{11}} \end{matrix}$ | $\begin{matrix} \dfrac{b_{21}}{b_{22}} & -\dfrac{1}{b_{22}} \\ \dfrac{\Delta_b}{b_{22}} & \dfrac{b_{12}}{b_{22}} \end{matrix}$ |
| **[a]** | $\begin{matrix} \dfrac{z_{11}}{z_{21}} & \dfrac{\Delta_z}{z_{21}} \\ \dfrac{1}{z_{21}} & \dfrac{z_{22}}{z_{21}} \end{matrix}$ | $\begin{matrix} -\dfrac{y_{22}}{y_{21}} & -\dfrac{1}{y_{21}} \\ -\dfrac{\Delta_y}{y_{21}} & -\dfrac{y_{11}}{y_{21}} \end{matrix}$ | $\begin{matrix} -\dfrac{\Delta_h}{h_{21}} & -\dfrac{h_{11}}{h_{21}} \\ -\dfrac{h_{22}}{h_{21}} & -\dfrac{1}{h_{21}} \end{matrix}$ | $\begin{matrix} \dfrac{1}{g_{21}} & \dfrac{g_{22}}{g_{21}} \\ \dfrac{g_{11}}{g_{21}} & \dfrac{\Delta_g}{g_{21}} \end{matrix}$ | $\begin{matrix} a_{11} & a_{12} \\ a_{21} & a_{22} \end{matrix}$ | $\begin{matrix} \dfrac{b_{22}}{\Delta_b} & \dfrac{b_{12}}{\Delta_b} \\ \dfrac{b_{21}}{\Delta_b} & \dfrac{b_{11}}{\Delta_b} \end{matrix}$ |
| **[b]** | $\begin{matrix} \dfrac{z_{22}}{z_{12}} & \dfrac{\Delta_z}{z_{12}} \\ \dfrac{1}{z_{12}} & \dfrac{z_{11}}{z_{12}} \end{matrix}$ | $\begin{matrix} -\dfrac{y_{11}}{y_{12}} & -\dfrac{1}{y_{12}} \\ -\dfrac{\Delta_y}{y_{12}} & -\dfrac{y_{22}}{y_{12}} \end{matrix}$ | $\begin{matrix} \dfrac{1}{h_{12}} & \dfrac{h_{11}}{h_{12}} \\ \dfrac{h_{22}}{h_{12}} & \dfrac{\Delta_h}{h_{12}} \end{matrix}$ | $\begin{matrix} -\dfrac{\Delta_g}{g_{12}} & -\dfrac{g_{22}}{g_{12}} \\ -\dfrac{g_{11}}{g_{12}} & -\dfrac{1}{g_{12}} \end{matrix}$ | $\begin{matrix} \dfrac{a_{22}}{\Delta_a} & \dfrac{a_{12}}{\Delta_a} \\ \dfrac{a_{21}}{\Delta_a} & \dfrac{a_{11}}{\Delta_a} \end{matrix}$ | $\begin{matrix} b_{11} & b_{12} \\ b_{21} & b_{22} \end{matrix}$ |

$\Delta_x = x_{11}x_{22} - x_{12}x_{21}$

handled in Problem 14.17 at the end of the chapter. Notice that in a few rare cases, it is possible to encounter the problem of division by zero. In these cases, the parameter set you are trying to find is not properly defined. Remember that in this chapter, however, all of the parameters sets for a circuit are assumed to be well behaved!

## 14.6 RECIPROCITY

You may have noticed that when the $z$ parameters of the circuit in Example 14.1 were calculated, $z_{12}$ and $z_{21}$ were found to be equal. In Example 14.2, $y_{12}$ was equal to $y_{21}$. These outcomes were not by mere coincidence. This property is always true of passive two-port circuits that contain only resistors, capacitors, and inductors. Circuits for which $z_{12} = z_{21}$ are said to be **reciprocal.** Looking at the relationships of Equation 14.3 leads to the conclusion that for reciprocal circuits,

$$z_{12} = z_{21}$$

$$\left.\frac{V_1}{I_2}\right|_{I_1=0} = \left.\frac{V_2}{I_1}\right|_{I_2=0} \tag{14.6}$$

In words, we say that the circuit's two transimpedances are equal. A current applied to one port will generate a voltage at the other, and the voltage will be the same regardless of which port is the input and which is the output. It is also true of reciprocal circuits that their transadmittance functions are equal. That is, if a voltage is applied to one port and a current measured at the opposite (shorted) port, the results will be the same in both directions. That is,

$$y_{12} = y_{21}$$

$$\left.\frac{I_1}{V_2}\right|_{V_1=0} = \left.\frac{I_2}{V_1}\right|_{V_2=0} \tag{14.7}$$

While the validity of Equation 14.6 will be demonstrated in this section, that of Equation 14.7 is dealt with in Problem 14.8 at the end of the chapter. The properties of reciprocal circuits are summarized in Figure 14.12.

What kinds of circuits are reciprocal? The **reciprocity theorem** states that any electrical network consisting entirely of bilateral elements (defined in the next paragraph) is reciprocal. If a current source placed in branch A generates a voltage across the open-circuited branch B, then the same current source placed in branch B will produce an equal voltage at the open-circuited branch A.

A two-terminal element is **bilateral** if its behavior is the same regardless of

**FIGURE 14.12**    *Conditions of Reciprocity*

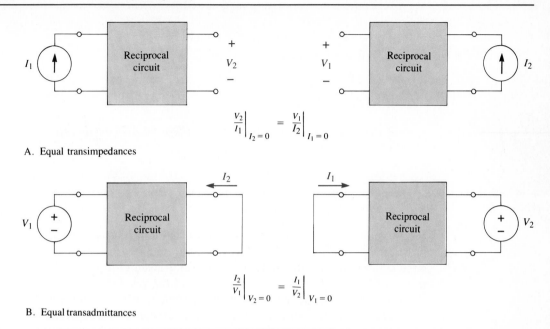

A. Equal transimpedances

B. Equal transadmittances

the direction of the current. Bilateral elements are also sometimes defined by saying that their leads are interchangeable. This definition applies to resistors, inductors, and capacitors but eliminates sources of any kind. Not all passive elements are bilateral. The behavior of a diode, for instance, changes if its leads are reversed.

The reciprocity theorem is easily proven when you realize that the impedance notation used throughout this chapter applies only to bilateral elements. If you study Figure 14.7B, you will see that if $z_{12} = z_{21}$ (the condition of reciprocity), then the dependent source drops out and we are left with a totally passive circuit. Any passive two-port circuit that consists entirely of resistors, capacitors, and inductors must be reciprocal. Figure 14.13 shows a specific example of a reciprocal circuit. You should verify that the ammeter reads the same value of current in each location.

Although a reciprocal circuit's transimpedance and transadmittance functions are unchanged when the input and output ports are reversed, the same cannot be said about voltage or current transfer between the ports unless the additional restriction is made that $z_{11} = z_{22}$. You are asked to show this in Problem 14.9 at the end of the chapter. If you look at Figure 14.7B again, you will see that this additional condition is met by symmetrical circuits. Symmetrical circuits look just the same when they are viewed from either port.

Reciprocity can be used as a check on calculations for passive circuits. If the

**FIGURE 14.13**  *A Reciprocal Circuit*

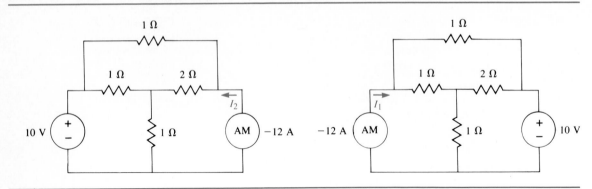

results show that $z_{12} \neq z_{21}$, then we know that a mistake has been made. A more practical application of reciprocity comes from the area of antennas rather than circuits. Antennas can be used either to transmit or to receive signals. If a transmitter is used to excite an antenna and the resulting electromagnetic field is measured in space, the pattern will be the same as when a source is moved around the antenna and the strength of the received signal is measured. Therefore, an antenna can be tested as either a transmitter or a receiver, and the results will apply to the opposite case.

## 14.7  ANALYSIS OF SYSTEMS CONTAINING TWO-PORT CIRCUITS

As was noted in Section 14.1, the use of two-port parameters in a problem can considerably simplify the calculations needed to analyze a system. No matter how complicated a two-port circuit is, its external characteristics can be reduced to two algebraic equations in four variables. As Figure 14.1 suggests, two-port circuits are often used in a simple combination with a one-port source circuit and a load. The diagram is repeated in Figure 14.14 with some additional detail. When the circuits are connected, $V_1 = V_3$ and $I_1 = -I_3$. Also, $V_2 = V_4$ and $I_2 = -I_4$. The

**FIGURE 14.14**  *Interconnection of One- and Two-Port Circuits*

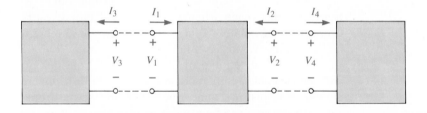

two-port circuit, by itself, does not set the values of $V_1$, $V_2$, $I_1$, and $I_2$, but it does establish certain *constraints* among them in the form of two algebraic equations. Similarly, the source circuit's presence places constraints on $V_1$ and $I_1$, and the load does so for $V_2$ and $I_2$.

Viewed in this way, the overall system has four central variables and four algebraic equations to describe the relationship among them. These equations are generally quite easy to solve. Even in a situation where one of the devices, such as the load, is nonlinear, it is a straightforward matter to program a computer to generate the solution.

---

**Example 14.5**  *Using Two-Port Parameters in Analysis*

Figure 14.15 shows a circuit that includes a commonly used *h*-parameter model of a bipolar junction transistor under AC operation. The parameter values given are typical of actual transistors.

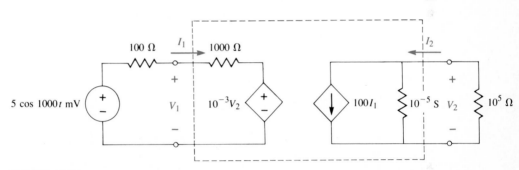

**FIGURE 14.15**

Find the voltage delivered to the load. To do so, you must determine the four constraining *V–I* equations.

From the transistor,

$$V_1 = 1000I_1 + 10^{-3}V_2$$

$$I_2 = 100I_1 + 10^{-5}V_2$$

From the source circuit,

$$.005 \cos 1000t - 100I_1 = V_1$$

And, from the load (note the polarity!),

$$V_2 = -10^5 I_2$$

*(continues)*

**Example 14.5**   *Continued*

This set of four equations can be solved to find

$$V_2 = 6.4 \cos 1000t \text{ V}$$

Notice that this circuit acts like an amplifier with a voltage gain greater than 1000!

---

**Example 14.6**   *Finding a Transfer Function of a Two-Port Circuit*

For something a little different, find the system function $V_2/V_s$ in the circuit of Figure 14.16. The two-port circuit is given only in terms of its *a* parameters:

$$a_{11} = \frac{s+2}{2} \qquad a_{12} = 1$$

$$a_{21} = \frac{s^2 + 4s + 2}{2} \qquad a_{22} = \frac{s+2}{2}$$

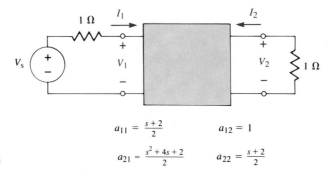

**FIGURE 14.16**

The equations for the two-port circuit are written as

$$V_1 = \left(\frac{s+2}{2}\right) V_2 - I_2$$

$$I_1 = \left(\frac{s^2 + 4s + 2}{2}\right) V_2 - \left(\frac{s+2}{2}\right) I_2$$

From the load, we know that

$$V_2 = -I_2$$

And, from the source,

$$V_1 = V_s - I_1$$

We have a set of four equations and five variables. They can be reduced to one equation containing $V_s$ and $V_2$:

$$\frac{V_2}{V_s} = \frac{2}{s^2 + 6s + 8}$$

---

## 14.8  COUPLED INDUCTORS

Coupled inductors and transformers were first introduced in Chapter 5. We return to them in this chapter because we are better able to describe and understand their behaviors with the analytical tools that we now possess. Let's begin by reviewing some basic facts about inductors.

An inductor's voltage is proportional to the time rate of change of its current. That is,

$$v(t) = L\,\frac{di}{dt}$$

This equation is correct as far as it goes. However, it is possible for the current in an inductor to affect not only its own voltage but also the voltages of nearby inductors. This behavior is in contrast to ideal resistors, which always behave independently of such outside effects. Classically, capacitors have also been modeled as operating independently of outside effects. This behavior, however, can no longer be thought of as entirely true. Modern integrated circuits are being fabricated with very densely packed capacitors operating at very high speeds. Under these circumstances, it has been found that the electrical fields of different capacitors can affect one another. This topic must still be considered as very specialized, however, and is a digression from the present discussion.

Inductors whose behaviors are linked are said to be magnetically coupled. Before the behavior of **coupled inductors** can be explained, let's briefly discuss the physical properties of single inductors operating in isolation of outside effects.

An inductor is a circuit element that is based on the interaction between a current in a conductor and the magnetic field that it generates. Physically, an inductor is made of a wire coiled so that it links maximally with the surrounding magnetic field. The magnetic field, although strongest within the coil, extends into the space surrounding the inductor. Figure 14.17 shows how a second inductor, placed near the first, can also link with the magnetic field.

Imagine a current in $L_1$. If it is time-varying, it will generate a time-varying magnetic field about the inductor. This, according to Faraday's law, causes a

**FIGURE 14.17**   *Magnetically Coupled Inductors*

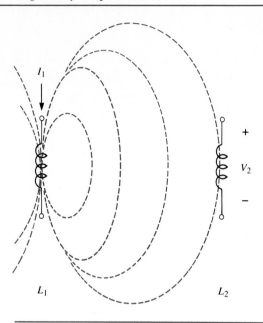

$L_1$ $L_2$

voltage to exist across $L_1$. This phenomenon of **self-inductance** is embodied in Equation 14.7. The parameter $L$ is known as the *coefficient of self-inductance*. Now, a second inductor, $L_2$, placed near the first, will also link with the magnetic field so that a voltage will be induced across $L_2$. In a complementary way, a current in $L_2$ will induce a voltage across $L_1$. This phenomenon is known as **mutual inductance.** The total voltage across an inductor is the algebraic sum of the self-induced and mutually induced voltages.

The magnitude of the interactive effect between two nearby inductors is the same in both directions. The *coefficient of mutual inductance* is given the symbol $M$. The value of $M$ is related to those of $L_1$ and $L_2$ and to the geometry between the two inductors. It can be demonstrated that $M$ has an upper bound of

$$M \leq \sqrt{L_1 L_2} \tag{14.8}$$

The *V–I* relationship of an isolated inductor has a plus or minus sign attached to it according to the passive sign convention. The algebraic sign given to mutual inductance can also be either plus or minus depending on the relative directions in which the two inductors are coiled and the assumed directions of currents and polarities of voltages for the inductors. The **dot convention** is a circuit notation that tells us which sign to apply to $M$. As is seen in Figure 14.18, each inductor of a coupled pair has a dot assigned to one of its terminals. The details will not be pursued here, but suffice it to say that the dot locations for any pair of

**FIGURE 14.18** *Dot Convention Applied to Two Sets of Coupled Inductors*

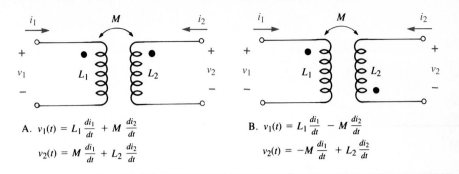

A. $v_1(t) = L_1 \dfrac{di_1}{dt} + M \dfrac{di_2}{dt}$

$v_2(t) = M \dfrac{di_1}{dt} + L_2 \dfrac{di_2}{dt}$

B. $v_1(t) = L_1 \dfrac{di_1}{dt} - M \dfrac{di_2}{dt}$

$v_2(t) = -M \dfrac{di_1}{dt} + L_2 \dfrac{di_2}{dt}$

coupled inductors can be established through applications of the right-hand rule and Lenz's law from basic physics.[*]

The dot convention states that current entering the dotted terminal of one of a pair of coupled inductors will cause a voltage across the other inductor that is algebraically positive at its dotted terminal. Independent of this, the current and self-induced voltage of either inductor obey the passive sign convention. Two possible combinations of dotted terminals and current directions are shown in Figure 14.18. Also shown are the corresponding KVL equations. Other possibilities are left to Problem 14.24 at the end of the chapter.

The terminal equations for coupled inductors can be written either as differential equations or in impedance notation. For the circuit of Figure 14.18A, these equations are

$$v_1 = L_1 \frac{di_1}{dt} + M \frac{di_2}{dt} \tag{14.9a}$$

$$v_2 = M \frac{di_1}{dt} + L_2 \frac{di_2}{dt} \tag{14.9b}$$

and

$$V_1 = sL_1 I_1 + sMI_2 \tag{14.10a}$$

$$V_2 = sMI_1 + sL_2 I_2 \tag{14.10b}$$

From Figure 14.18, you can clearly see that a pair of coupled inductors is a two-port circuit. Equation 14.10 can be used to identify its $z$ parameters.

---

[*] For an explanation, see, for instance, M.E. Van Valkenberg, *Network Analysis*, 3d ed., Prentice-Hall, 1974.

**Example 14.7** *Analysis of a Coupled-Inductor Circuit*

Figure 14.19 shows a simple circuit containing coupled inductors. It is shown in Laplace variable terms for finding the system's step response. We can write KVL equations around each of the two meshes:

$$\frac{1}{s} - I_1 = 2sI_1 + sI_2$$

$$-2I_2 = sI_1 + 3sI_2$$

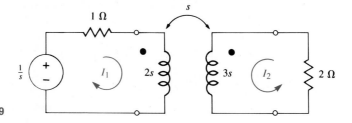

**FIGURE 14.19**

From these equations, we can identify the $z$ parameters of the coupled inductor: $z_{11} = 2s$, $z_{12} = z_{21} = s$, and $z_{22} = 3s$. The equations can be rearranged and solved for $I_2$:

$$(2s + 1)I_1 + sI_2 = \frac{1}{s}$$

$$sI_1 + (3s + 2)I_2 = 0$$

$$I_2 = \frac{\begin{vmatrix} 2s+1 & 1/s \\ s & 0 \end{vmatrix}}{\begin{vmatrix} 2s+1 & s \\ s & 3s+2 \end{vmatrix}} = \frac{-1}{5s^2 + 7s + 2} = \frac{1/3}{s+1} - \frac{1/3}{(s+2/5)}$$

Taking the inverse Laplace transform leads to

$$i_2(t) = \left[\frac{1}{3} e^{-t} - \frac{1}{3} e^{-2/5\,t}\right] u(t) \text{ A}$$

and

$$v_2(t) = \left[\frac{2}{3} e^{-2/5\,t} - \frac{2}{3} e^{-t}\right] u(t) \text{ V}$$

We can use the answer to Example 14.7 to make some important observations about the response behavior of coupled inductors and the circuits that contain them. The first observation has to do with the excitation in this particular example. It is a step function, and, like all step functions, it acts like a DC signal as $t$ approaches infinity. Because no constant (or step) terms appear in the response variables, we can conclude that DC excitations cannot pass across a coupled inductor. The second observation that we can make is that the response variables have two natural frequencies: $s = -2/5$ and $s = -1$. This is just what you would expect of a circuit containing two energy storage elements.

## 14.9  IDEAL TRANSFORMERS

Mutual inductance is sometimes an undesirable effect in circuits when more than one inductive element is present. Various design steps can be taken to minimize such stray effects. Transformers, on the other hand, are the result of a design approach that desires just the opposite result.

**Transformers** are coupled inductors designed to maximize both the mutual inductance between the two windings and the impedances of the individual windings themselves. Physically, these conditions can be approximated by winding two coils of many turns concentrically around an iron core. Mathematically, this leads to the assumptions that

$$M = \sqrt{L_1 L_2} \qquad \text{(14.11a)}$$

$$L_1 \to \infty \qquad \text{and} \qquad L_2 \to \infty \qquad \text{(14.11b)}$$

Equation 14.11a tells us that the coefficient of mutual coupling for an ideal transformer should be at its maximum limit. Equation 14.11b says that the self-inductances and their corresponding impedances are very large. The word *large* is a relative term. Here, it means that the impedances of the individual windings of the transformer should be much greater than any external impedances to which they may be connected. In addition to Equation 14.11, it is assumed that the coils and their shared core are lossless. These are the basic assumptions for ideal transformers. Real transformers can approach but never quite equal them. More will be said about real transformers later.

Very simple relationships exist between the input and output variables of ideal transformers. They were introduced but not proven in Chapter 5. We can derive them most easily by assuming sinusoidal excitation. This assumption does not mean that our results will apply only to sinusoidal functions, however. Figure 14.20 shows an ideal transformer excited by a sinusoidal source and terminated by a resistive load. The transformer is shown within the shaded area. The symbol for a transformer sometimes differs from that of a coupled inductor by the inclusion of vertical lines that suggest its iron core, which is usually laminated. The input

**FIGURE 14.20**  *Ideal Transformer Circuit*

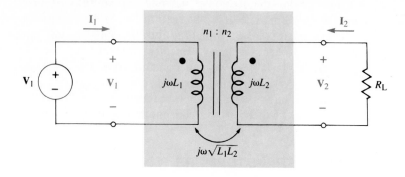

side is called the **primary winding,** and the output side is called the **secondary winding.** For our present purposes, the coefficient of mutual inductance is included in the diagram, but it is customary instead to indicate the numbers of turns of the two windings, which are labeled $n_1$ and $n_2$.

KVL equations can be written around the two magnetically coupled meshes of the circuit:

$$\mathbf{V}_1 = j\omega L_1 \mathbf{I}_1 + j\omega\sqrt{L_1 L_2}\mathbf{I}_2 \tag{14.12a}$$

$$\mathbf{V}_2 = j\omega\sqrt{L_1 L_2}\mathbf{I}_1 + j\omega L_2 \mathbf{I}_2 = -\mathbf{I}_2 R_L \tag{14.12b}$$

In writing these equations, the assumed value of $M = \sqrt{L_1 L_2}$ has been used. From Equation 14.12b, we find that

$$\frac{\mathbf{I}_2}{\mathbf{I}_1} = \frac{-j\omega\sqrt{L_1 L_2}}{R_L + j\omega L_2} \approx -\sqrt{\frac{L_1}{L_2}}$$

The latter expression results from the assumption that $L_1$ and $L_2$ are very large and, thus, that $\omega L_2$ is much larger than $R_L$. You may recall from physics that $L$ for an inductor is proportional to the square of the number of turns in its coil. If $L_1$ has $n_1$ turns and $L_2$ has $n_2$, we can conclude that

$$\boxed{\frac{\mathbf{I}_2}{\mathbf{I}_1} = -\frac{n_1}{n_2}} \tag{14.13}$$

There is a similar relationship for the input and output voltages of a transformer, and it can be found by an indirect method. Combining Equations

14.12a and 14.12b gives

$$V_1 = j\omega L_1 I_1 - \frac{(j\omega\sqrt{L_1 L_2})^2}{j\omega L_2 + R_L} I_1$$

which can be rearranged as

$$\frac{V_1}{I_1} = \frac{j\omega R_L L_1}{R_L + j\omega L_2} \approx R_L \frac{L_1}{L_2}$$

The last relationship comes from again using the assumption that $L_2$ is very large. Next, we use the fact that the ratio of $L_1/L_2$ for a transformer is equal to the ratio of the squares of the number of turns in the windings:

$$\frac{V_1}{I_1} = R_L \frac{n_1^2}{n_2^2} \qquad (14.14)$$

Because $R_L = -V_2/I_2$, Equation 14.14 can be rewritten as

$$\frac{V_1}{I_1} = -\frac{V_2}{I_2} \cdot \frac{n_1^2}{n_2^2}$$

$$\frac{V_1}{V_2} = -\frac{I_1}{I_2} \cdot \frac{n_1^2}{n_2^2} = \frac{n_2}{n_1} \cdot \frac{n_1^2}{n_2^2}$$

$$\frac{V_2}{V_1} = \frac{n_2}{n_1} \qquad (14.15)$$

Equations 14.13, 14.14, and 14.15 describe the behavior of an ideal transformer. They all depend on $n_2/n_1$, the **turns ratio** for the transformer. If $n_2/n_1 > 1$, the current in the secondary will be smaller than the primary's current, and the voltage across the secondary will be larger than the primary voltage. If $n_2/n_1 < 1$, then the opposite is true. An ideal transformer is a passive, resistanceless device. Therefore, no power is lost to or supplied by it. You are asked to prove this in Problem 14.30 at the end of the chapter.

Equation 14.14 tells us that a transformer terminated on its secondary side by a resistor $R_L$ will have an equivalent resistance seen looking into the primary coil that is $R_L(n_1/n_2)^2$. Transformers are sometimes used for matching the impedances of loads to those of sources for purposes of maximum power transfer. Figure 14.21 shows a simple example of a transformer used for impedance matching.

FIGURE 14.21    *Impedance Matching with a Transformer*

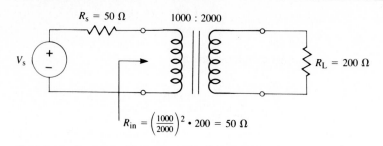

The presence of an ideal transformer in a circuit will change somewhat our analytical approach, as can be seen in the following example.

**Example 14.8**    *Analysis of an Ideal Transformer Circuit*

What is the average power delivered to the load in the circuit of Figure 14.22? Because the circuit is operating in the sinusoidal steady state, the circuit is also shown in phasor form. The turns ratio is indicated as $n_2/n_1 = 1/2$.

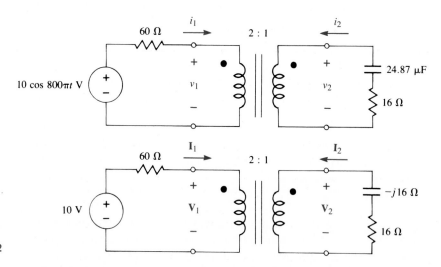

**FIGURE 14.22**

The input and output voltages of the transformer must, at first, be given variable names. Around the primary circuit, we can write

$$10 = 60\mathbf{I}_1 + \mathbf{V}_1$$

and, around the secondary,

$$(16 - j16)\mathbf{I}_2 + \mathbf{V}_2 = 0$$

We have four unknowns but only two KVL equations. The transformer provides two additional relationships in the form of Equations 14.12 and 14.14:

$$\mathbf{V}_1 = \left(\frac{n_1}{n_2}\right)\mathbf{V}_2 = 2\mathbf{V}_2$$

$$\mathbf{I}_1 = -\left(\frac{n_2}{n_1}\right)\mathbf{I}_2 = \frac{-\mathbf{I}_2}{2}$$

These four equations can be solved for $\mathbf{I}_2$:

$$\mathbf{I}_2 = .1433e^{-j152.7°}\ \text{A}$$

From Chapter 13, we know that the average power absorbed by the load is due entirely to the resistor and is equal to

$$P_L = R(I_{2_{\text{rms}}})^2 = 16\left(\frac{.1433}{\sqrt{2}}\right)^2 = 164.3\ \text{mW}$$

---

Because the input–output relationships of Equations 14.13 and 14.15 do not depend on frequency, we might be tempted to conclude that transformers are able to pass any waveform without distortion. That is, if the input side of the transformer is excited by a step function or an exponential, then the response should be exactly the same except for a change in amplitude and, perhaps, a delay in time. This might be true of our ideal transformer model, but it is certainly not the case with real transformers. The reason is simple: Actual physical transformers do not meet the assumptions needed in order to arrive at the mathematical model of the ideal transformer. For instance, the primary and secondary coils of a real transformer are never perfectly coupled. More seriously, the coils of physical transformers cannot have infinite impedances. No matter how many turns are placed on a winding, its impedance will go to zero at DC. In addition, there are secondary modeling effects, such as the capacitance between adjacent turns of a coil, that we have not even considered. Despite these difficulties, actual transformers can be designed to behave nearly ideally over fairly wide ranges of frequency and load. The performance of any transformer will deteriorate at low frequencies because the reactive impedance of the windings will become too small. At very high frequencies, the internal losses of the transformer increase.

As has just been said, an ideal transformer does not change the shape of an applied signal although the amplitude may be scaled. Real transformers do not

**FIGURE 14.23**   *Measured Step Response of an Actual Transformer Circuit*

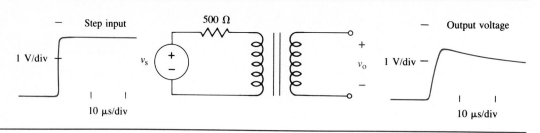

live up to this perfect standard. Figure 14.23 shows the measured step input and response waveforms of an actual transformer. Note the distortion present both at the leading edge of the response and in the steady state as $t$ approaches infinity.

Transformers are probably the least complicated and most widely used of two-port circuit elements. Nearly every solid-state electronic device that is powered from an AC source contains a transformer in its power supply circuit. You should not think that transformers are less important than other circuit elements just because they have not been discussed in detail until this last chapter. Transformers are essential to the areas of electronics and power distribution. The most detailed discussion of them was deferred until now because they are obviously two-port circuits and because we now have the analytical tools at our disposal to fully appreciate them.

## 14.10   COMPARING THE USEFULNESS OF THE PARAMETER SETS

Most two-port circuits can be described by any of the six two-port parameter sets. Because the descriptions are entirely equivalent, it makes no theoretical difference which set is chosen. However, some of the sets have very definite advantages over others in certain applications. These advantages are most easily seen when we consider the different ways that two-port circuits can be interconnected to form larger systems.

Figure 14.24 shows three ways in which two two-port circuits can be connected together. They are the parallel, series, and cascade connections. In each case, the larger system is itself a two-port network. It should, therefore, be possible to describe the overall system with two-port parameters.

When a number of two-port circuits are connected in parallel, the $y$ parameters for the combination are equal to the sum of the individual $y$ parameters. When the circuits are in series, their individual $z$ parameters can be added to find the $z$ parameters of the total system. Finally, when two-port circuits are cascaded, the transmission or $a$ parameters of the string can be found through the matrix multiplication of the transmission parameters of the individual sections. These statements can be expressed in mathematical terms by defining $[x_{jk}]$ to be the $2 \times 2$ matrix of a parameter set and $[x_{jk}^A]$ and $[x_{jk}^B]$ to be the parameter

**FIGURE 14.24** *Three Connections of Two Two-Port Circuits*

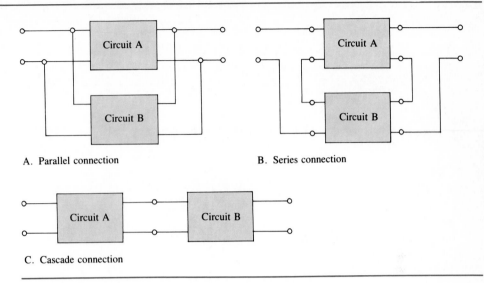

A. Parallel connection

B. Series connection

C. Cascade connection

sets for a circuit $A$ and circuit $B$, respectively. Then, we write the following relationships:

Two-ports in parallel:   $[y_{jk}] = [y_{jk}^A] + [y_{jk}^B]$ (14.16a)

Two-ports in series:   $[z_{jk}] = [z_{jk}^A] + [z_{jk}^B]$ (14.16b)

Two-ports in cascade:   $[a_{jk}] = [a_{jk}^A][a_{jk}^B]$* (14.16c)

You are asked to prove these relationships in Problem 14.35 at the end of the chapter. You can see that certain two-port configurations are much more easily modeled with some parameter sets than with others.

Figure 14.24A shows that when two circuits are in parallel, their port voltages are equal and the input currents of the total system are the sums of the individual input currents. When the circuits are in series, their input currents are the same, but their voltages add. The interconnections are not always so simple. For instance, it is often useful in the design of voltage amplifiers to construct a **feedback circuit** that senses the amplifier's output voltage and uses it in some way to affect or modify the input voltage. As Figure 14.25 shows, a series connection is then required on the input side and a parallel connection on the output, so that a choice of hybrid parameters might be useful.

---

* This form assumes that the current at port 2 is treated as leaving rather than entering the positive terminal. That is, $-I_2$ rather than $I_2$ is used in the equations.

FIGURE 14.25   *Voltage Amplifier Circuit A with Feedback Circuit f*

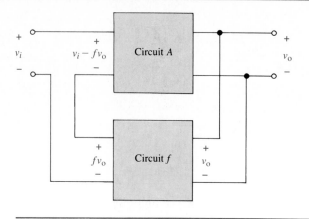

There are no simple rules for selecting the "best" parameter set for a particular problem. There are only guidelines. For instance, voltage amplifiers are often modeled with $y$ parameters because the models for this set include voltage-controlled sources. However, one of the simplest amplifying devices of all, the bipolar junction transistor, is almost always specified in terms of its $h$ parameters. It is much easier to measure a transistor's $h$ parameters in the laboratory than it is to measure any of the other parameter sets.

If you have not seen $h$-parameter data for a transistor already, you probably will in the future. An $h$-parameter model for a transistor is shown in Figure 14.26. The subscripts used are somewhat different here than those introduced earlier in this chapter and, so, require some explanation. Instead of $h_{11}$, $h_{12}$, $h_{21}$, and $h_{22}$, transistors are usually specified with $h_{ie}$, $h_{re}$, $h_{fe}$, and $h_{oe}$. This notation assumes that the transistor is to be used in the common-emitter configuration (the letter e of each subscript stands for *emitter*). These four terms are defined as follows:

FIGURE 14.26   *Commonly Used Model for a Common-Emitter Transistor*

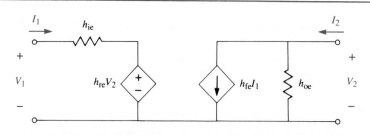

$$h_{ie} = \left.\frac{V_1}{I_1}\right|_{V_2=0} \quad \text{(transistor's } \textit{input} \text{ impedance)}$$

$$h_{re} = \left.\frac{V_1}{V_2}\right|_{I_1=0} \quad (\textit{reverse} \text{ voltage ratio)}$$

$$h_{fe} = \left.\frac{I_2}{I_1}\right|_{V_2=0} \quad (\textit{forward} \text{ current gain)}$$

$$h_{oe} = \left.\frac{I_2}{V_2}\right|_{I_1=0} \quad (\textit{output} \text{ admittance)}$$

We have only begun to suggest the usefulness of the specific parameter sets. Choosing a set to work with depends on the configuration of the system, the devices contained within the system, and the types of manipulations that you will be doing.

## 14.11  SUMMARY

The parameter sets introduced in this chapter are concise ways of modeling the behavior of circuits with two ports. They are based entirely on the relationships between the four external variables of the circuit. Each of the six possible parameter sets is defined in terms of two algebraic equations relating the four external variables of the circuit. The parameters are the coefficients of these equations. Most generally, they are functions of the complex frequency variable $s$.

The $y$ and $z$ parameters are closely related to the principle of reciprocity in two-port circuits. A circuit is reciprocal if its transadmittance functions are equal or if its transimpedance functions are equal. Mathematically, that is, $y_{12} = y_{21}$ and $z_{12} = z_{21}$. Passive linear circuits with no controlled sources are always reciprocal.

Using two-port parameters to describe complicated networks has the advantage of simplifying any subsequent calculations relating to them. This is particularly true when certain system configurations such as series, parallel, and cascade connections are encountered. Two-port parameters can also be very useful in describing the behavior of physical devices such as transistors, which otherwise may not be so simply modeled.

Transformers are special cases of coupled inductors. They are among the simplest and most widely used of two-port circuit devices. They have the very interesting property of being able to connect two sections of a circuit magnetically rather than through a shared common branch. Although the coupling is not via conductors capable of carrying current, it is nonetheless a very real connection. The energy contained within a magnetic field exists just as surely as a resistor or a wire!

Ideal transformers will pass any signal except DC without distortion. Only a change in amplitude will occur. Real transformers come close to this ideal under most intended uses, but the limitations should be kept in mind.

## ■ PROBLEMS

**14.1** Consider the cascaded string of one- and two-port circuits shown in Figure P14.1. Show that the current entering one lead of a port is equal to the current leaving the other lead of the same port. Use cut-sets.

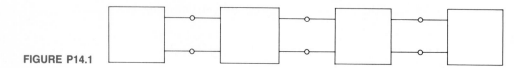

**FIGURE P14.1**

**14.2** By direct computation (Equation 14.3), find $z_{21}$ and $z_{22}$ for the circuit of Example 14.1.

**14.3** The z parameters are often calculated by node analysis. The port currents are thought of as being supplied by independent sources as in Figure 14.5. Use this approach to find the z parameters for the circuit of Example 14.1.

**14.4** Show that the z parameters of the circuit of Figure 14.11 are as given in Example 14.4.

**14.5** Find the z parameters for each of the circuits shown in Figure P14.5. For part A of the figure, use node analysis; for part B, use the formulas of Equation 14.3.

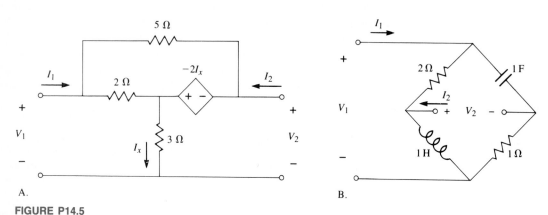

A.                                                         B.

**FIGURE P14.5**

**14.6** Draw two different circuit models for a two-port circuit described by the following equations:

$$V_1 = 100I_1 + 20I_2$$
$$V_2 = 40I_1 + 200I_2$$

**14.7** Draw two different equivalent circuit models for each of the two-port circuits of Problem 14.5.

**14.8** Prove that the two transadmittance functions of a circuit are equal if it is reciprocal. That is, if $z_{12} = z_{21}$, then $y_{12} = y_{21}$.

**14.9** Use the circuit model of Figure 14.7B to show that when a circuit is reciprocal and $z_{11} = z_{22}$, the circuit is symmetrical.

**14.10 a.** Show that for a two-port network, the open circuit voltage transfer functions $V_2/V_1$ and $V_1/V_2$ are equal if the network is reciprocal and if $z_{11} = z_{22}$.

**b.** Show that under the same conditions, the short circuit current transfer functions $I_2/I_1$ and $I_1/I_2$ are also equal.

**14.11** Find the $h$ parameters of the circuits in the following examples and problems:

**a.** Example 14.2

**b.** Example 14.4

**c.** Problem 14.5, part A of the figure

**d.** Problem 14.5, part B of the figure

**14.12** Write the formulas by which the $g$ parameters of a two-port network can be directly measured (or calculated).

**14.13** Find the dimensions of each of the following:

**a.** Inverse hybrid parameters

**b.** Transmission parameters

**c.** Inverse transmission parameters

**14.14** Some measurements are taken on a two-port circuit:

**1.** When a current of $-.5$ A is applied to port 1, $V_2 = -2$ V and $V_1 = -1.5$ V.

**2.** $I_2$ is set to 3 A, which results in $V_1 = (1/2)I_2$ and $V_2 = 30$ V.

Draw a $z$-parameter model for the circuit.

**14.15** Two experiments are performed on a circuit:

**1.** With port 1 short-circuited, $I_1 = 2I_2$ and $V_2 = .5I_2$.

**2.** With port 2 open-circuited, $I_1 = -5V_1$ and $V_2 = 10V_1$.

Find the circuit's short circuit admittance parameters.

**14.16** A transistor circuit is shown in Figure P14.16. Use the AC model given in Chapter 1. Assume $r_x = 500\ \Omega$, $\beta = 100$, and $r_o = 10\ k\Omega$. Find the $y$ parameters of the part of the overall circuit that is contained within the dashed box.

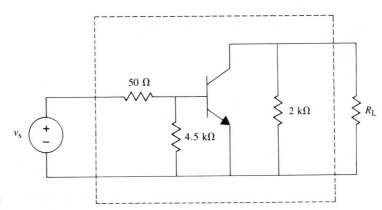

**FIGURE P14.16**

**14.17** Verify the following entries in Table 14.1:

**a.** $z$ parameters as functions of $y$ parameters

    **b.** *a* parameters as functions of *h* parameters

    **c.** *g* parameters as functions of *b* parameters

**14.18 a.** Use direct methods for computing separately each of the inverse hybrid parameters for the circuit of Example 14.2.

    **b.** Use the appropriate entry in Table 14.1 to verify your answers when compared with the results of Example 14.2.

**14.19 a.** Find the inverse transmission parameters for the two-port circuit shown within the dashed box in Figure P14.19A.

    **b.** Find the Thévenin equivalent circuit of everything to the left of the load element.

    **c.** Use graphical methods for finding the voltage and current delivered to the load when it is (1) a $10\,\Omega$ resistor and (2) a nonlinear element with the $V$–$I$ characteristics shown in Figure P14.19B.

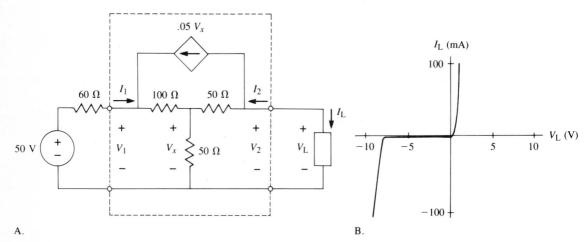

A.                                                                          B.

**FIGURE P14.19**

**14.20** The two-port circuit shown within the larger network of Figure P14.20 has *g* parameters of $g_{11} = 1\,\text{S}$, $g_{12} = -4$, $g_{21} = 3$, and $g_{22} = 10\,\Omega$. What voltage will be delivered to the following loads:

    **a.** $1\,\Omega$

    **b.** $10\,\Omega$

    **c.** $100\,\Omega$

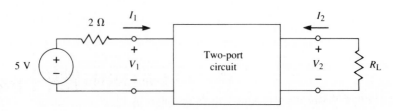

**FIGURE P14.20**

**14.21** A model that can be used for a bipolar junction transistor operated at high frequencies is shown in Figure P14.21. Find its $y$ parameters as functions of $\omega$.

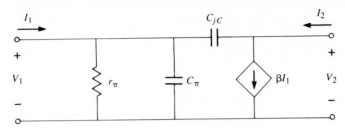

**FIGURE P14.21**

**14.22** Find the $a$ parameters of the circuit shown in Figure P14.22. Is this circuit equivalent to the two-port circuit of Example 14.6? If not, complete the steps of that example to find $V_2/V_s$ when the two-port circuit of Figure P14.22 is substituted.

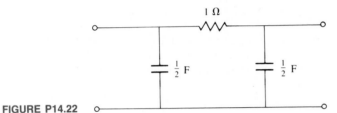

**FIGURE P14.22**

**14.23** Derive a model similar to that of Figure 14.7B but based on $y$ parameters. Use it to prove that, for a passive circuit, $y_{12} = y_{21}$.

**14.24** Write equations relating $v_1$, $i_1$, $v_2$, and $i_2$ for the coupled inductors shown in Figure P14.24. Caution: Take note of the assigned relationship between $v_2$ and $i_2$ in part B.

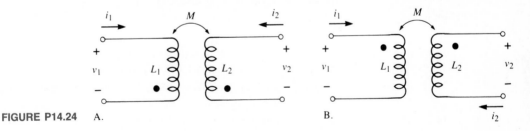

**FIGURE P14.24**   A.                                                           B.

**14.25** Write a set of time-domain mesh equations for the circuit shown in Figure P14.25.

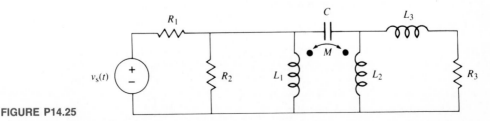

**FIGURE P14.25**

14.26 **a.** Find, for the frequency of excitation $\omega = 4\ \text{rad/s}$, the steady-state $z$ parameters for the part of the circuit shown within the dashed lines in Figure P14.26.

**b.** Use the results of part a as a starting point for finding $i_s(t)$.

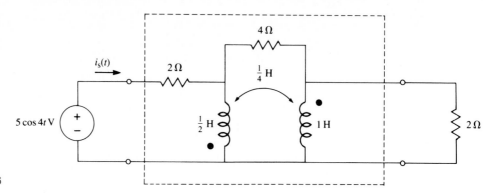

**FIGURE P14.26**

14.27 A real (nonideal) transformer is part of the circuit shown in Figure P14.27. Use Laplace transforms to find the circuit's output voltage in response to a unit step input. Sketch the result and compare it to the response if the transformer was made ideal and had the same turns ratio.

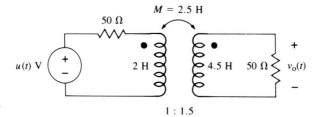

**FIGURE P14.27**

14.28 Reconsider the circuit of Problem 14.27 with both real and ideal transformers. Find the response of each to input sinusoids of $\cos t$, $\cos 10t$, and $\cos 100t$.

14.29 Use Equation 14.10 to derive an equivalent model for a pair of coupled inductors. The model should consist entirely of inductors, none of which is coupled to any other.

14.30 Show that an ideal transformer is lossless. That is, all of the power supplied to a transformer with a resistive load is delivered to that load.

14.31 For the transformer circuit shown in Figure P14.31, select a turns ratio that will cause a maximum average power transfer to the load. What will that power be?

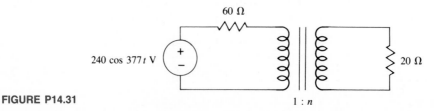

**FIGURE P14.31**

**14.32** Repeat Example 14.8 by first finding the equivalent impedance seen looking into the transformer. (Equation 14.14 is true for general loads, not just those that are resistive.)

**14.33** A transformer with losses is modeled in the circuit shown in Figure P14.33. Determine the efficiency of the transformer $(P_{out}/P_{in})$ when it is operating in the larger circuit.

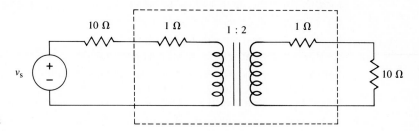

**FIGURE P14.33**

**14.34** Usually, when a transformer is described, the turns ratio rather than the absolute number of turns is given. For example, for $n_1 = 3000$ and $n_2 = 1500$, a turns ratio of 2:1 is given. With this in mind, would there be any difference in behavior if a real transformer had $n_1 = 20$ and $n_2 = 200$ as opposed to $n_1 = 200$ and $n_2 = 2000$?

**14.35** Verify Equations 14.16a, 14.16b, and 14.16c.

**14.36** Find the input impedance of the terminated two-port circuit shown in Figure P14.36 in terms of the following:

   **a.** Its $z$ parameters and $Z_L$

   **b.** Its $y$ parameters and $Z_L$

   **c.** Its $g$ parameters and $Z_L$

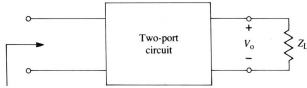

**FIGURE P14.36**   $Z_{in}$

**14.37** Find the voltage transfer function $V_o/V_s$ for the circuit shown in Figure P14.37 in terms of the following:

   **a.** Its $z$ parameters and external element values

   **b.** Its $b$ parameters and external element values

   **c.** Its $h$ parameters and external element values

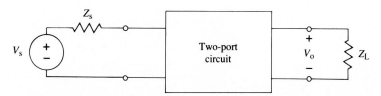

**FIGURE P14.37**

**14.38** Look at the network of Figure 14.25. Assume that circuit $A$ is modeled in terms of $h$ parameters and that circuit $f$ is modeled by $g$ parameters. Find the $h$ parameters of the overall system. (Assume that the input side of circuit $f$ is connected to the output side of circuit $A$ and vice versa.)

**14.39** Figure 14.25 displays one type of feedback: Voltage is sensed on the output of circuit $A$ and used to affect the input voltage to circuit $A$. Investigate three other basic arrangements:

**a.** Output voltage sensed, input current modified

**b.** Output current sensed, input voltage modified

**c.** Output current sensed, input current modified

For each arrangement, draw a block diagram and identify the most natural parameter set to use for circuit $A$ and the most natural to use for circuit $f$. Assume that the input side of circuit $f$ is connected to the output of circuit $A$ and vice versa.

**14.40** If two circuits are equivalent, their parameters will be equal. Use this fact to verify the following:

**a.** Delta–wye transformation of Equation 13.21 (use $z$ parameters)

**b.** Wye–delta transformation of Equation 13.22 (use $y$ parameters)

# Matrices and Vectors

## A.1  INTRODUCTION

A **matrix** is a rectangular array of numbers arranged in horizontal *rows* and vertical *columns*. Several examples of matrices are as follows:

$$\mathbf{A} = \begin{bmatrix} 1 & 3 & 0 \\ -3 & 2 & -1 \\ 4 & 6 & -5 \end{bmatrix}$$

$$\mathbf{B} = \begin{bmatrix} 1 & 2 & 3 & 4 & 5 \\ 6 & 7 & 8 & 9 & 10 \\ 11 & 12 & 13 & 14 & 15 \\ 16 & 17 & 18 & 19 & 20 \end{bmatrix}$$

$$\mathbf{c} = \begin{bmatrix} 1 & 3 & -2 & 0 & 4 \end{bmatrix}$$

$$\mathbf{d} = \begin{bmatrix} x_1 \\ x_2 \\ x_3 \\ x_4 \end{bmatrix}$$

$$y = 20$$

The **dimension** of a matrix is an ordered pair of numbers $n \times m$, where $n$ is the number of rows and $m$ is the number of columns. For example, the dimension of matrix $\mathbf{A}$ is

$$\dim \mathbf{A} = 3 \times 3$$

A particular entry in matrix $\mathbf{A}$ is written as $A(i, j)$ or $a_{ij}$, where $i$ is the row index and $j$ is the column index. For example,

$$A(1, 1) = a_{11} = 1$$
$$A(1, 3) = a_{13} = 0$$
$$A(3, 2) = a_{32} = 6$$

The other matrices given at the beginning of this appendix have the following dimensions:

$$\dim \mathbf{B} = 4 \times 5$$
$$\dim \mathbf{c} = 1 \times 5$$
$$\dim \mathbf{d} = 4 \times 1$$
$$\dim \mathbf{y} = 1 \times 1$$

If $n = m$, then the matrix is a *square* matrix. Matrix $\mathbf{A}$ is a square matrix. The entries $A(1, 1)$, $A(2, 2)$, and $A(3, 3)$ are the *diagonal* elements of the square matrix. The other entries are the *off-diagonal* elements.

If either $n = 1$ or $m = 1$, then the matrix is commonly termed a **vector** and is usually represented with a lowercase letter. Matrix $\mathbf{c}$ is a *row* vector, while $\mathbf{d}$ is a *column* vector. We usually drop the row index for row vectors and the column index for column vectors. For example, we write entries in $\mathbf{c}$ and $\mathbf{d}$ as

$$c(1) = 1 \qquad c(3) = -2$$
$$d(1) = x_1 \qquad d(3) = x_3$$

We know from a problem's definition whether a given vector is a row or a column vector.

The variable $\mathbf{y}$ shown in the example matrices is a **scalar** and can be considered to be a $1 \times 1$ matrix.

Two matrices are equal if they have the same row and column dimensions and if each row and column contains the same entries. That is

$$\mathbf{A} = \mathbf{B} \qquad \text{if } A(i, j) = B(i, j) \text{ for all } i \text{ and } j \tag{A1.1}$$

A matrix of special interest is the **identity matrix I**. The identity matrix is a square matrix with diagonal elements equal to 1 and off-diagonal elements equal to 0. The $4 \times 4$ identity matrix is

$$\mathbf{I}_4 = \begin{bmatrix} 1 & 0 & 0 & 0 \\ 0 & 1 & 0 & 0 \\ 0 & 0 & 1 & 0 \\ 0 & 0 & 0 & 1 \end{bmatrix}$$

where the subscript indicates the dimension of the identity matrix.

The **transpose** of a matrix is formed by interchanging its rows and columns. That is,

$$\mathbf{B} = \mathbf{A}^T \text{ if } B(i, j) = A(j, i)$$

## A.2  MATRIX ADDITION AND SUBTRACTION

For two matrices to be added or subtracted, they must have the same row and column dimensions. Therefore, none of the matrices shown in the preceding section may be added or subtracted.

Adding two matrices of equal dimension results in a third matrix with the same dimension. That is,

$$\mathbf{A}^{n \times m} + \mathbf{B}^{n \times m} = \mathbf{C}^{n \times m} \tag{A1.2}$$

where

$$C(i, j) = A(i, j) + B(i, j) \tag{A1.3}$$

The superscript $n \times m$ indicates the row and column dimensions. An example of matrix addition is

$$\begin{bmatrix} 1 & 0 & -2 \\ -3 & 2 & 5 \end{bmatrix} + \begin{bmatrix} 2 & -3 & 5 \\ 3 & -6 & 2 \end{bmatrix} = \begin{bmatrix} (1+2) & (0-3) & (-2+5) \\ (-3+3) & (2-6) & (5+2) \end{bmatrix}$$

$$= \begin{bmatrix} 3 & -3 & 3 \\ 0 & -4 & 7 \end{bmatrix}$$

Matrix subtraction proceeds in a similar fashion:

$$\mathbf{A}^{n \times m} - \mathbf{B}^{n \times m} = \mathbf{C}^{n \times m} \tag{A1.4}$$

where

$$C(i, j) = A(i, j) - B(i, j) \tag{A1.5}$$

For example,

$$\begin{bmatrix} 3 & -2 \\ 0 & 4 \\ -1 & -3 \end{bmatrix} - \begin{bmatrix} 1 & 4 \\ -3 & 2 \\ 5 & -3 \end{bmatrix} = \begin{bmatrix} (3-1) & (-2-4) \\ (0+3) & (4-2) \\ (-1-5) & (-3+3) \end{bmatrix} = \begin{bmatrix} 2 & -6 \\ 3 & 2 \\ -6 & 0 \end{bmatrix}$$

## A.3  MATRIX MULTIPLICATION

Matrix multiplication is an ordered operation. That is, **AB** does not, in general, equal **BA.** Matrix multiplication is possible only if the column dimension of the first matrix is equal to the row dimension of the second matrix. The matrix product has the row dimension of the first matrix and the column dimension of the second matrix. That is,

$$\mathbf{A}^{n \times m} \cdot \mathbf{B}^{m \times p} = \mathbf{C}^{n \times p} \tag{A1.6}$$

As an example, we first define two vectors:

$$\mathbf{x}_1 = \begin{bmatrix} 5 & -1 & 2 & .5 \end{bmatrix}$$

and

$$\mathbf{x}_2 = \begin{bmatrix} 3 \\ 4 \\ -2 \\ 10 \end{bmatrix}$$

We now take the product of $\mathbf{x}_1$ and $\mathbf{x}_2$:

$$\mathbf{x}_3 = \mathbf{x}_1 \cdot \mathbf{x}_2 = \begin{bmatrix} 5 & -1 & 2 & .5 \end{bmatrix} \cdot \begin{bmatrix} 3 \\ 4 \\ -2 \\ 10 \end{bmatrix}$$

where $\mathbf{x}_1$ has dimension $1 \times 4$ and $\mathbf{x}_2$ has dimension $4 \times 1$. Thus, $\mathbf{x}_3$ has dimension $1 \times 1$ and is a scalar. This product of a row vector times a column vector is known as an **inner product.** An inner product is found as follows:

$$\mathbf{x}_3 = x_1(1) \cdot x_2(1) + x_1(2) \cdot x_2(2) + x_1(3) \cdot x_2(3) + x_1(4) \cdot x_2(4)$$

so

$$\mathbf{x}_3 = (5 \cdot 3) + (-1 \cdot 4) + (2 \cdot -2) + (.5 \cdot 10) = 12$$

In general terms, if $\mathbf{x}_1$ has dimension $1 \times n$ and $\mathbf{x}_2$ has dimension $n \times 1$, then the inner product of $\mathbf{x}_1 \cdot \mathbf{x}_2$ is given by

$$\mathbf{x}_3 = \mathbf{x}_1 \cdot \mathbf{x}_2 = \sum_{i=1}^{n} \mathbf{x}_1(i) \cdot \mathbf{x}_2(i) \tag{A1.7}$$

where $\mathbf{x}_1$ is a row vector and $\mathbf{x}_2$ is a column vector, and $\mathbf{x}_3$ is a scalar.

Matrix multiplication is performed as a sequence of inner products. For example,

$$\begin{bmatrix} 2 & 0 & 4 & 8 \\ 1 & -1 & 2 & .5 \\ 3 & 1 & 6 & 5 \end{bmatrix} \cdot \begin{bmatrix} -1 & 5 & 2 & 1 & 3 \\ 3 & 1 & 4 & 0 & 2 \\ 5 & 2 & 1 & -2 & 4 \\ -2 & 4 & 8 & 6 & 10 \end{bmatrix} = \mathbf{C}$$

The product $\mathbf{C}$ is a $3 \times 5$ matrix and has 15 entries. We find these entries one at a time with the following rule:

$$C(i, j) = \text{Inner product of row } i \text{ of first matrix} \\ \text{and column } j \text{ of second matrix} \tag{A1.8}$$

Some entries of **C** for this example are

$$C(1, 1) = \begin{bmatrix} 2 & 0 & 4 & 8 \end{bmatrix} \cdot \begin{bmatrix} -1 \\ 3 \\ 5 \\ -2 \end{bmatrix} = 2$$

$$C(1, 2) = \begin{bmatrix} 2 & 0 & 4 & 8 \end{bmatrix} \cdot \begin{bmatrix} 5 \\ 1 \\ 2 \\ 4 \end{bmatrix} = 50$$

$$C(2, 1) = \begin{bmatrix} 1 & -1 & 2 & .5 \end{bmatrix} \cdot \begin{bmatrix} -1 \\ 3 \\ 5 \\ -2 \end{bmatrix} = 5$$

$$C(3, 5) = \begin{bmatrix} 3 & 1 & 6 & 5 \end{bmatrix} \cdot \begin{bmatrix} 3 \\ 2 \\ 4 \\ 10 \end{bmatrix} = 85$$

The complete answer is

$$\mathbf{C} = \begin{bmatrix} 2 & 50 & 72 & 42 & 102 \\ 5 & 10 & 4 & 0 & 16 \\ 20 & 48 & 56 & 11 & 85 \end{bmatrix}$$

## Special Cases

An important matrix multiplication result is that the product of a square matrix and the appropriately dimensioned identity matrix is equal to the original square matrix. That is,

$$\mathbf{A}^{n \times n} \cdot \mathbf{I}_n = \mathbf{A}^{n \times n} \qquad \text{or} \qquad \mathbf{I}_n \cdot \mathbf{A}^{n \times n} = \mathbf{A}^{n \times n} \tag{A1.9}$$

Another special case is the multiplication of a scalar times a matrix. In this case, each entry in the matrix is multiplied by the scalar. Scalar multiplication of a matrix does commute:

$$a \cdot \mathbf{A} = \mathbf{A} \cdot a = \mathbf{B} \tag{A1.10}$$

where $a$ is a scalar and $B(i, j) = a \cdot A(i, j)$.

## Simultaneous Equations

It is a simple matter to represent a set of simultaneous equations in matrix form. For example,

$$10v_1 - 5v_2 = 1$$

$$-5v_1 + 4v_2 - 2v_3 = 0$$

$$-2v_2 + 5v_3 = -4$$

becomes

$$\begin{bmatrix} 10 & -5 & 0 \\ -5 & 4 & -2 \\ 0 & -2 & 5 \end{bmatrix} \cdot \begin{bmatrix} v_1 \\ v_2 \\ v_3 \end{bmatrix} = \begin{bmatrix} 1 \\ 0 \\ -4 \end{bmatrix}$$

The coefficients of the output voltage variables become entries in the square matrix. The input and output variables are represented as column vectors. You should verify that this matrix multiplication results in the given set of simultaneous equations.

## A.4 MATRIX INVERSES AND DETERMINANTS

Matrices cannot be divided in the usual sense of the word. However, the matrix equivalent to scalar division can be accomplished by first defining the inverse of a matrix. If a matrix $\mathbf{A}$ has an inverse $\mathbf{A}^{-1}$, then we can write

$$\mathbf{A}^{-1} \cdot \mathbf{A} = \mathbf{I} \quad \text{or} \quad \mathbf{A} \cdot \mathbf{A}^{-1} = \mathbf{I} \tag{A1.11}$$

Only square matrices can have inverses. Also, certain conditions must be met even for a square matrix to have an inverse. These conditions are discussed later.
Using Equation A1.11, we can solve the following matrix equation:

$$\mathbf{A} \cdot \mathbf{B} = \mathbf{C}$$

If $\mathbf{A}$ is a square matrix and its inverse exists, we can multiply both sides of this equation by $\mathbf{A}^{-1}$ to obtain

$$(\mathbf{A}^{-1} \cdot \mathbf{A}) \cdot \mathbf{B} = \mathbf{A}^{-1} \cdot \mathbf{C}$$

Because matrix multiplication is ordered, $\mathbf{A}^{-1}$ must appear on the left in both sides of the equation. The equation becomes

$$\mathbf{I} \cdot \mathbf{B} = \mathbf{A}^{-1} \cdot \mathbf{C}$$

so

$$\mathbf{B} = \mathbf{A}^{-1} \cdot \mathbf{C} \tag{A1.12}$$

If **B** is a square invertible matrix, we can find **A** as follows:

$$\mathbf{A} \cdot \mathbf{B} = \mathbf{C}$$

$$\mathbf{A} \cdot (\mathbf{B} \cdot \mathbf{B}^{-1}) = \mathbf{C} \cdot \mathbf{B}^{-1}$$

$$\mathbf{A} \cdot \mathbf{I} = \mathbf{C} \cdot \mathbf{B}^{-1}$$

so

$$\mathbf{A} = \mathbf{C} \cdot \mathbf{B}^{-1} \tag{A1.13}$$

The matrix equivalent of scalar division is multiplication by the matrix inverse. Finding the inverse, however, can be time consuming. The availability of calculator and computer programs alleviates this problem. Let's examine the inverse technique for a $2 \times 2$ matrix and then an alternative approach that is a bit easier to work with when we are using pen and paper.

We want to find the vector **x** in the following equation:

$$\underbrace{\begin{bmatrix} 3 & -1 \\ -2 & 4 \end{bmatrix}}_{\mathbf{A}} \cdot \begin{bmatrix} x_1 \\ x_2 \end{bmatrix} = \begin{bmatrix} 4 \\ 10 \end{bmatrix}$$

We have to find the inverse of the $2 \times 2$ matrix **A**. We can do this without any special techniques by solving

$$\begin{bmatrix} 3 & -1 \\ -2 & 4 \end{bmatrix} \cdot \underbrace{\begin{bmatrix} b_{11} & b_{12} \\ b_{21} & b_{22} \end{bmatrix}}_{\mathbf{B}} = \begin{bmatrix} 1 & 0 \\ 0 & 1 \end{bmatrix}$$

The matrix **B** is the inverse we are looking for—that is, $\mathbf{A}^{-1} = \mathbf{B}$. We now perform the multiplication indicated:

$$\begin{bmatrix} (3b_{11} - b_{21}) & (3b_{12} - b_{22}) \\ (-2b_{11} + 4b_{21}) & (-2b_{12} + 4b_{22}) \end{bmatrix} = \begin{bmatrix} 1 & 0 \\ 0 & 1 \end{bmatrix}$$

For these two matrices to be equal, we must equate their entries:

$$3b_{11} - b_{21} = 1$$

$$3b_{12} - b_{22} = 0$$

$$-2b_{11} + 4b_{21} = 0$$

$$-2b_{12} + 4b_{22} = 1$$

Solving this set of equations leads to

$$b_{11} = \frac{4}{(3 \cdot 4) - (-1 \cdot -2)} = \frac{4}{10} \qquad \text{(A1.14a)}$$

$$b_{12} = \frac{1}{(3 \cdot 4) - (-1 \cdot -2)} = \frac{1}{10} \qquad \text{(A1.14b)}$$

$$b_{21} = \frac{2}{(3 \cdot 4) - (-1 \cdot -2)} = \frac{2}{10} \qquad \text{(A1.14c)}$$

$$b_{22} = \frac{3}{(3 \cdot 4) - (-1 \cdot -2)} = \frac{3}{10} \qquad \text{(A1.14d)}$$

So we have

$$\mathbf{B} = \mathbf{A}^{-1} = \begin{bmatrix} .4 & .1 \\ .2 & .3 \end{bmatrix}$$

and

$$\mathbf{x} = \mathbf{A}^{-1} \cdot \begin{bmatrix} 4 \\ 10 \end{bmatrix} = \begin{bmatrix} .4 & .1 \\ .2 & .3 \end{bmatrix} \cdot \begin{bmatrix} 4 \\ 10 \end{bmatrix} = \begin{bmatrix} 2.6 \\ 3.8 \end{bmatrix}$$

The answer is easily checked by substituting for **x** in the original matrix equation.

Returning to Equations A1.14a–d, we see that each element in the inverse is divided by the same factor. This factor is the **determinant** of **A** and is denoted either as det **A** or $|\mathbf{A}|$. The determinant of a $2 \times 2$ matrix is given by

$$\det \mathbf{A} = |\mathbf{A}| = (a_{11} \cdot a_{22}) - (a_{12} \cdot a_{21}) \qquad \text{(A1.15)}$$

The inverse of **A** does *not* exist if det **A** = 0.

The inverse of a $2 \times 2$ matrix is very simple and is given by

$$\mathbf{A}^{-1} = \frac{1}{\det \mathbf{A}} \cdot \begin{bmatrix} a_{22} & -a_{12} \\ -a_{21} & a_{11} \end{bmatrix} \qquad \text{(A1.16)}$$

where the matrix on the right is known as the **adjoint matrix** of **A** ($\mathbf{A}^{+}$). An example of a $2 \times 2$ matrix inverse is

$$\mathbf{A} = \begin{bmatrix} 10 & -5 \\ -20 & 20 \end{bmatrix}$$

so

$$\mathbf{A}^{-1} = \frac{1}{100} \cdot \begin{bmatrix} 20 & 5 \\ 20 & 10 \end{bmatrix} = \begin{bmatrix} .2 & .05 \\ .2 & .1 \end{bmatrix}$$

As a check,

$$\mathbf{A} \cdot \mathbf{A}^{-1} = \begin{bmatrix} 10 & -5 \\ -20 & 20 \end{bmatrix} \cdot \begin{bmatrix} .2 & .05 \\ .2 & .1 \end{bmatrix} = \begin{bmatrix} 1 & 0 \\ 0 & 1 \end{bmatrix} = \mathbf{I}$$

The inverse of any square invertible matrix is equal to the adjoint matrix divided by the matrix determinant. That is,

$$\mathbf{A}^{-1} = \frac{\mathbf{A}^+}{\det \mathbf{A}}$$

## Determinants and Cofactors

A determinant of a square matrix $\mathbf{A}$ is a scalar quantity that is dependent on the elements $a_{ij}$ of $\mathbf{A}$. The determinant of a $2 \times 2$ matrix was given in Equation A1.16. The determinant of a $3 \times 3$ is also easily found. For example, if

$$\mathbf{A} = \begin{bmatrix} 2 & 4 & 6 \\ 1 & 3 & 5 \\ 8 & 4 & 2 \end{bmatrix}$$

We write the matrix as given and repeat the first two columns as shown:

$$\begin{bmatrix} 2 & 4 & 6 \\ 1 & 3 & 5 \\ 8 & 2 & 4 \end{bmatrix} \begin{matrix} 2 & 4 \\ 1 & 3 \\ 8 & 2 \end{matrix}$$

Now, we draw diagonal lines as shown, multiply the three numbers along each diagonal, and add the results:

$$\begin{bmatrix} 2 & 4 & 6 \\ 1 & 3 & 5 \\ 8 & 2 & 4 \end{bmatrix} \begin{matrix} 2 & 4 \\ 1 & 3 \\ 8 & 2 \end{matrix} = (2 \cdot 3 \cdot 4) + (4 \cdot 5 \cdot 8) + (6 \cdot 1 \cdot 2) = 196$$

Next, we draw diagonal lines in the opposite direction and proceed as before:

$$\begin{bmatrix} 2 & 4 & 6 \\ 1 & 3 & 5 \\ 8 & 2 & 4 \end{bmatrix} \begin{matrix} 2 & 4 \\ 1 & 3 \\ 8 & 2 \end{matrix} = (6 \cdot 3 \cdot 8) + (2 \cdot 5 \cdot 2) + (4 \cdot 1 \cdot 4) = 180$$

Finally, we subtract the second result from the first:

$$\det \mathbf{A} = 196 - 180 = 16$$

The general expression for a matrix determinant involves the use of the cofactors of the matrix.

The **minors** of the matrix $\mathbf{A}$, $M_{ij}$, are found by deleting from $\mathbf{A}$ the $i$th row and the $j$th column and finding the determinant of the sub-matrix that results. The cofactor $C_{ij}$ is found as follows:

$$C_{ij} = (-1)^{i+j} \cdot M_{ij}$$

That is, if the row and column indices sum to an even number, the cofactor equals the minor. If the row and column indices sum to an odd number, the cofactor is the negative of the minor. The next example shows how to use the cofactors to find the determinant of a $3 \times 3$ matrix.

Given the matrix,

$$\mathbf{A} = \begin{bmatrix} 2 & 4 & 6 \\ 1 & 3 & 5 \\ 8 & 2 & 4 \end{bmatrix}$$

we find det $\mathbf{A}$ by multiplying each entry in any *one* row or column by its corresponding cofactor and summing the three products. For example, we can use the entries in the first row:

$$\det \mathbf{A} = (2 \cdot C_{11}) + (4 \cdot C_{12}) + (6 \cdot C_{13})$$

We find $C_{11}$ by deleting the first row and first column of $\mathbf{A}$ and finding the determinant of the remainder:

$$C_{11} = M_{11} = \det \begin{bmatrix} 3 & 5 \\ 2 & 4 \end{bmatrix} = 2$$

We find $C_{12}$ by deleting the first row and second column of $\mathbf{A}$:

$$C_{12} = -M_{12} = -\det \begin{bmatrix} 1 & 5 \\ 8 & 4 \end{bmatrix} = 36$$

We find $C_{13}$ by deleting the first row and third column of $\mathbf{A}$:

$$C_{13} = M_{13} = \det \begin{bmatrix} 1 & 3 \\ 8 & 2 \end{bmatrix} = -22$$

The determinant of $\mathbf{A}$ is, therefore,

$$\det \mathbf{A} = (2 \cdot 2) + (4 \cdot 36) + (6 \cdot -22) = 16$$

We can also find the determinant of $\mathbf{A}$ by using any other row or column. Let's use the third column; then,

$$\det \mathbf{A} = (6 \cdot C_{13}) + (5 \cdot C_{23}) + (4 \cdot C_{33})$$

where

$$C_{13} = M_{13} = (2 \cdot 1) - (3 \cdot 8) = -22$$
$$C_{23} = -M_{23} = -(2 \cdot 2) - (4 \cdot 8) = 28$$
$$C_{33} = M_{33} = (2 \cdot 3) - (4 \cdot 1) = 2$$

and

$$\det \mathbf{A} = (6 \cdot -22) + (5 \cdot 28) + (4 \cdot 2) = 16$$

The determinant of a $3 \times 3$ matrix is found by calculating three $2 \times 2$ determinants. The determinant of a $4 \times 4$ matrix is found from four $3 \times 3$ determinants, each of which requires finding three $2 \times 2$ determinants. Computational effort greatly increases with increased matrix order.

In general, the determinant of an $n \times n$ can be found by expanding along the $i$th row

$$\det \mathbf{A} = \sum_{j=1}^{n} a_{ij} C_{ij} \tag{A1.17a}$$

or by expanding along the $j$th column

$$\det \mathbf{A} = \sum_{i=1}^{n} a_{ij} C_{ij} \tag{A1.17b}$$

## Adjoint Matrix

The adjoint matrix of A is formed by transposing the matrix of the cofactors of A. That is,

$$\mathbf{A}^{+}(i, j) = C_{ji} \tag{A1.18}$$

## A.5  CRAMER'S RULE

Cramer's rule provides an alternative technique for solving a matrix–vector equation:

$$\begin{bmatrix} a_{11} & a_{12} & a_{13} \\ a_{21} & a_{22} & a_{23} \\ a_{31} & a_{32} & a_{33} \end{bmatrix} \cdot \begin{bmatrix} x_1 \\ x_2 \\ x_3 \end{bmatrix} = \begin{bmatrix} y_1 \\ y_2 \\ y_3 \end{bmatrix} \tag{A1.19}$$

The formal solution of this matrix equation is

$$\mathbf{x} = \mathbf{A}^{-1} \cdot \mathbf{y}$$

However, finding the inverse can be very time consuming. Cramer's rule allows us to find all three entries of $\mathbf{x}$ by finding four $3 \times 3$ determinants. We first find $\det \mathbf{A}$. The *first* entry, $x_1$, is found by replacing the *first* column of $\mathbf{A}$ with the vector $\mathbf{y}$, finding the determinant of the new matrix, and, finally, dividing this determinant by $\det \mathbf{A}$. The *second* entry, $x_2$, is found by replacing the *second* column of $\mathbf{A}$ with $\mathbf{y}$, finding the determinant, and dividing by $\det \mathbf{A}$. The *third* entry, $x_3$, is found by replacing the *third* column of $\mathbf{A}$ with $\mathbf{y}$ and proceeding as before. That is,

$$x_1 = \frac{\begin{vmatrix} y_1 & a_{12} & a_{13} \\ y_2 & a_{22} & a_{23} \\ y_3 & a_{32} & a_{33} \end{vmatrix}}{\det \mathbf{A}}$$

$$x_2 = \frac{\begin{vmatrix} a_{11} & y_1 & a_{13} \\ a_{21} & y_2 & a_{23} \\ a_{31} & y_3 & a_{33} \end{vmatrix}}{\det \mathbf{A}}$$

$$x_3 = \frac{\begin{vmatrix} a_{11} & a_{12} & y_1 \\ a_{21} & a_{22} & y_2 \\ a_{31} & a_{32} & y_3 \end{vmatrix}}{\det \mathbf{A}}$$

Again note that if $\det \mathbf{A} = 0$, we cannot use this technique to find $x_1$, $x_2$, and $x_3$. As an example, we solve the typical node voltage equation:

$$\underbrace{\begin{bmatrix} 10 & -5 & 0 \\ -5 & 4 & -2 \\ 0 & -2 & 5 \end{bmatrix}}_{\mathbf{A}} \cdot \begin{bmatrix} v_1 \\ v_2 \\ v_3 \end{bmatrix} = \begin{bmatrix} 1 \\ 0 \\ -4 \end{bmatrix}$$

$$\det \mathbf{A} = [(10 \cdot 4 \cdot 5) + (-5 \cdot -2 \cdot 0) + (0 \cdot -5 \cdot -2)]$$
$$- [(0 \cdot 4 \cdot 0) + (10 \cdot -2 \cdot -2) + (-5 \cdot -5 \cdot 5)]$$
$$= 35$$

$$v_1 = \frac{\begin{vmatrix} 1 & -5 & 0 \\ 0 & 4 & -2 \\ -4 & -2 & 5 \end{vmatrix}}{35} = -\frac{24}{35} \text{ V}$$

$$v_2 = \frac{\begin{vmatrix} 10 & 1 & 0 \\ -5 & 0 & -2 \\ 0 & -4 & 5 \end{vmatrix}}{35} = -\frac{55}{35} \text{ V}$$

$$v_3 = \frac{\begin{vmatrix} 10 & -5 & 1 \\ -5 & 4 & 0 \\ 0 & -2 & -4 \end{vmatrix}}{35} = -\frac{50}{35} \text{ V}$$

Note that if we are just interested in one node voltage (e.g., $v_1$), we do not need to solve for the other variables.

# Fundamental Identities

## B.1 ANGLES AND RADIANS

$$360° = 2\pi \text{ rad}$$

$$180° = \pi \text{ rad}$$

$$90° = \frac{\pi}{2} \text{ rad}$$

$$45° = \frac{\pi}{4} \text{ rad}$$

$$57.3° \approx 1 \text{ rad}$$

## B.2 SIMPLE PHASE SHIFT RELATIONSHIPS*

$$\sin (x \pm 180°) = -\sin x$$

$$\cos (x \pm 180°) = -\cos x$$

$$\cos (x - 90°) = \sin x$$

$$\cos (x + 90°) = -\sin x$$

$$\sin (x + 90°) = \cos x$$

$$\sin (x - 90°) = -\cos x$$

## B.3 TRIGONOMETRIC IDENTITIES

$$\sin (A \pm B) = \sin A \cos B \pm \cos A \sin B$$

$$\cos (A \pm B) = \cos A \cos B \mp \sin A \sin B$$

$$\cos A \cos B = \frac{1}{2} [\cos (A + B) + \cos (A - B)]$$

$$\sin A \sin B = \frac{1}{2} [\cos (A - B) - \cos (A + B)]$$

$$\sin A \cos B = \frac{1}{2} [\sin (A + B) + \sin (A - B)]$$

$$\sin A \pm \sin B = 2 \sin \frac{1}{2} (A \mp B) \cos \frac{1}{2} (A \mp B)$$

$$\cos A + \cos B = 2 \cos \frac{1}{2} (A + B) \cos \frac{1}{2} (A - B)$$

---

* Engineers often use degrees rather than the more correct radians to express phase shifts.

$$\cos A - \cos B = 2 \sin \frac{1}{2}(A+B) \sin \frac{1}{2}(B-A)$$

$$\sin 2A = 2 \sin A \cos A$$

$$\cos 2A = 2 \cos^2 A - 1 = 1 - 2 \sin^2 A = \cos^2 A - \sin^2 A$$

$$\sin \frac{1}{2} A = \sqrt{\frac{1}{2}(1 - \cos A)}$$

$$\cos \frac{1}{2} A = \sqrt{\frac{1}{2}(1 + \cos A)}$$

$$\sin^2 A = \frac{1}{2}(1 - \cos 2A)$$

$$\cos^2 A = \frac{1}{2}(1 + \cos 2A)$$

$$\sin x = \frac{e^{jx} - e^{-jx}}{2j}$$

$$\cos x = \frac{e^{jx} + e^{-jx}}{2}$$

$$e^{jx} = \cos x + j \sin x$$

$$A \cos x + B \sin x = \sqrt{A^2 + B^2} \cos\left(x - \tan^{-1}\frac{B}{A}\right)$$

$$= \sqrt{A^2 + B^2} \cos\left(x + \tan^{-1}\frac{-B}{A}\right)$$

## B.4   INTEGRALS

$$\int \sin ax \, dx = \frac{1}{a}\cos ax$$

$$\int \cos ax \, dx = \frac{1}{a}\sin ax$$

$$\int \sin^2 ax \, dx = \frac{x}{2} - \frac{\sin 2ax}{4a}$$

$$\int x \sin ax \, dx = \frac{1}{a^2}(\sin ax - ax \cos ax)$$

$$\int x^2 \sin ax \, dx = \frac{1}{a^3}(2ax \sin ax + 2 \cos ax - a^2x^2 \cos ax)$$

$$\int \cos^2 ax \, dx = \frac{x}{2} + \frac{\sin 2ax}{4a}$$

$$\int x \cos ax \, dx = \frac{1}{a^2} (\cos ax + ax \sin ax)$$

$$\int x^2 \cos ax = \frac{1}{a^3} (2ax \cos ax - 2 \sin ax + a^2x^2 \sin ax)$$

$$\int \sin ax \sin bx \, dx = \frac{\sin (a-b)x}{2(a-b)} - \frac{\sin (a+b)x}{2(a+b)}, \qquad a^2 \neq b^2$$

$$\int \sin ax \cos bx \, dx = -\left[ \frac{\cos (a-b)x}{2(a-b)} + \frac{\cos (a+b)x}{2(a+b)} \right], \qquad a^2 \neq b^2$$

$$\int \cos ax \cos bx \, dx = \frac{\sin (a-b)x}{2(a-b)} + \frac{\sin (a+b)x}{2(a+b)}, \qquad a^2 \neq b^2$$

$$\int e^{ax} \, dx = \frac{1}{a} e^{ax}$$

$$\int x e^{ax} \, dx = \frac{e^{ax}}{a^2} (ax - 1)$$

$$\int x^2 e^{ax} \, dx = \frac{e^{ax}}{a^3} (a^2x^2 - 2ax + 2)$$

$$\int e^{ax} \sin bx \, dx = \frac{e^{ax}}{a^2 + b^2} (a \sin bx - b \cos bx)$$

$$\int e^{ax} \cos bx \, dx = \frac{e^{ax}}{a^2 + b^2} (a \cos bx + b \sin bx)$$

# Complex Numbers

## C.1 INTRODUCTION

Knowledge of the concept and use of complex numbers is an assumed prerequisite for anyone reading this book. As such, this appendix is intended to be only a brief outline of the subject that you can use to review the fundamentals. If this review is not understandable to you, consult a mathematics textbook that covers the topic.

Complex numbers can be written in either of two equivalent forms, both of which are shown graphically in Figure C.1. Such graphical representations are known as **Argand diagrams.**

FIGURE C.1 *Graphical Representations of Complex Numbers*

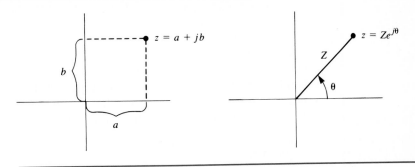

## C.2 RECTANGULAR NOTATION (CARTESIAN COORDINATES)

$$z = a + jb \tag{C.1}$$

Electrical engineers define $j = \sqrt{-1}$. The real part of $z$ is $Re[z] = a$, and its imaginary part is $Im[z] = b$, also a real number. If $z_1 = a + jb$ and $z_2 = c + jd$, then $z_1 = z_2$ if and only if $a = c$ and $b = d$.

## C.3 EXPONENTIAL NOTATION (POLAR COORDINATES)

$$z = Ze^{j\theta} \tag{C.2}$$

The magnitude of $z$ is $|z| = Z$, and its angle or argument is $\arg[z] = \theta$. Two complex numbers are equal if and only if their magnitudes are equal and their arguments are equal.

## C.4 CONVERSION BETWEEN RECTANGULAR AND EXPONENTIAL NOTATION

Euler's identity states that

$$e^{j\theta} = \cos\theta + j\sin\theta \qquad (C.3)$$

This equation can be applied to Equation C.2 to see that

$$z = Z\cos\theta + jZ\sin\theta \qquad (C.4)$$

Comparing Equation C.3 to Equation C.1 leads to the conclusion that

$$a = Re[z] = Z\cos\theta \qquad (C.5)$$

and

$$b = Im[z] = Z\sin\theta \qquad (C.6)$$

Equations C.5 and C.6 are rules for converting from exponential to rectangular notation. To develop rules for rectangular to exponential conversion, Equations C.5 and C.6 can be squared and then added to find that

$$Z = \sqrt{a^2 + b^2} \qquad (C.7)$$

Next, consider the ratio of Equation C.5 to Equation C.4 to see

$$\theta = \tan^{-1}\frac{b}{a} \qquad (C.8)$$

## C.5 SUMMATION OF COMPLEX NUMBERS

Adding and subtracting complex numbers is most easily done when they are in rectangular form. If $z_1 = a + jb$ and $z_2 = c + jd$, then

$$z_1 + z_2 = (a + c) + j(b + d) \qquad (C.9)$$

or

$$z_1 - z_2 = (a - c) + j(b - d) \qquad (C.10)$$

## C.6 MULTIPLICATION OF COMPLEX NUMBERS

Multiplying complex numbers is most easily done when they are in exponential form. If $z_1 = Z_1 e^{j\theta_1}$ and $z_2 = Z_2 e^{j\theta_2}$, then

$$z_1 z_2 = Z_1 Z_2 e^{j(\theta_1 + \theta_2)} \qquad (C.11)$$

That is, the product of complex numbers has a magnitude that is the product of the individual magnitudes and an angle that is the sum of the angles. Division takes a similar form:

$$\frac{z_1}{z_2} = \frac{Z_1}{Z_2}\, e^{j(\theta_1 - \theta_2)} \tag{C.12}$$

When done in rectangular notation, these operations become

$$z_1 z_2 = (a + jb)(c + jd)$$
$$= (ac - bd) + j(ad + bc) \tag{C.13}$$

and

$$\frac{z_1}{z_2} = \frac{a + jb}{c + jd}$$
$$= \frac{ac + bd}{c^2 + d^2} + j\, \frac{bc - ad}{c^2 + d^2} \tag{C.14}$$

## C.7   COMPLEX CONJUGATES

The roots of quadratic equations often exist in the form

$$z = a + jb \qquad \text{and} \qquad z^* = a - jb \tag{C.15}$$

The numbers $z$ and $z^*$ are said to be complex conjugates of each other. They have the interesting property that when they are multiplied together, the result is a real number that is the square of the magnitude of the complex number:

$$zz^* = (a + jb)(a - jb) = a^2 + b^2 = Z^2 \tag{C.16}$$

## C.8   POWERS AND ROOTS OF COMPLEX NUMBERS

Raising a complex number to a power or taking its root is most easily done when the number is in exponential form:

$$z^n = (Ze^{j\theta})^n = Z^n e^{jn\theta} \tag{C.17}$$

Raising any number to the power $n$ yields one unique result. In contrast, the $m$th root of a number has $m$ distinct values:

$$\sqrt[m]{z} = \sqrt[m]{Ze^{j\theta}} = \sqrt[m]{Ze^{j(\theta + k2\pi)}} = \sqrt[m]{Z}\, e^{j(\theta + k2\pi)/m} \tag{C.18}$$

The variable $k$ is an integer. The expression of Equation C.18 will repeat itself after $k$ is increased from 0 to $m - 1$.

# Answers to
# Selected Problems

## Chapter 1

1.2   a.  $-500 \sin 50t \ \mu A$

1.3   30.83 C

1.5   a.  $P_{N_1} = -50$ mW, delivering;
           $P_{N_2} = 50$ mW, absorbing

1.6   a.  $50t$ mJ

1.9   a.  4 A      b.  .375 $\Omega$

1.13  a.  14.4 $\Omega$      b.  167 cm

1.18  $-10 \sin 100t$ V

1.20  35 A

## Chapter 2

2.1   $I_1 = -3$ A, $I_2 = 5$ A, $I_3 = -3$ A

2.3   $V_1 = 3$ V, $V_2 = -5$ V, $V_3 = -1$ V

2.5   $I_1 = 3.5$ A, $I_2 = 6$ A

2.9   $I_1 = 3.2$ A, $I_2 = 3.4$ A, $I_3 = 1.8$ A

2.16  3.55 k$\Omega$

2.17  .4 S

2.22  100 $\Omega$, 200 $\Omega$, 300 $\Omega$, 600 $\Omega$

2.26  24 V

2.28  466.25 V

2.32  2.43 V

2.35  $.38 \cos 10t$ V

2.37  a.  $-302.67 \cos 100t$ mV

2.39  a.  8.5 V      b.  $-.44 \cos 1000t$ V

## Chapter 3

3.1   a.  $V_a = 3.33$ V, $V_b = 8$ V
       b.  $I_1 = 3.33$ mA, $I_2 = -2.33$ mA, $I_3 = 2.67$ mA

3.3   a.  $V_a = 13.33$ V, $V_b = 18$ V
       b.  $I_1 = 3.33$ mA, $I_2 = -2.33$ mA, $I_3 = 2.67$ mA

3.5   $V_a = 246$ V, $V_b = 246$ V, $V_c = 414$ V

3.9   a.  $V_a = 4.02$ V, $V_b = 4$ V, $V_c = 3$ V, $V_d = 5$ V,
           $V_e = 0$ V (datum), $V_f = 2.08$ V

3.11  $v_o \approx -10v_{in}$

3.14  $v_o = -79.6 \cos 10t$ mV

3.16  $v_{R_1} = 5.87 \sin t$ V, $v_{R_2} = 5.1 \sin t$ V

3.19  $I_1 = 3.64$ mA, $I_2 = .19$ mA, $I_3 = 3.45$ mA,
       $I_4 = 3$ mA, $I_5 = -6.64$ mA, $I_6 = 6.45$ mA

3.23  $i_2 = .06 \cos t - .36e^{-5t} - 1.21$ mA
       $i_4 = .06 \cos t - .36e^{-5t} + .79$ mA
       $i_6 = .11 \cos t + .36e^{-5t} - .12$ mA
       $i_8 = .05 \cos t - .28e^{-5t} + .09$ mA

3.26  $-.92v_{in}$

3.28  $i_y = .04$ A, $i_{R_1} = .096$ A, $v_x = .88$ V

## Chapter 4

4.2   $R_{Th} = 873 \ \Omega$, $V_{Th} = .28$ V

4.4   a.  $R_{Th} = 3.2$ k$\Omega$, $V_{Th} = 3.4$ V

4.7   a.  $R_{Th} = 8.875 \ \Omega$, $V_{Th} = -25.67$ V

4.9   $R_{eq} = -24 \ \Omega$

4.12  $R_{out} = R_{Th} = 1.41$ M$\Omega$,
       $V_{Th} = -1.14 \cos 100t$ V

4.14  $R_N = .87$ k$\Omega$, $I_N = .32$ mA

4.16  $R_N = 3.2$ k$\Omega$, $I_N = 1.06$ mA

4.18  $R_N = 1414$ k$\Omega$, $I_N = -.8 \cos 100t \ \mu A$

4.21  a.  3.25 k$\Omega$      b.  11.26 mW      c.  12.1 V
       d.  3.72 mA

4.23  .83 $\Omega$

4.26  $V_{oc} = 12.1$ V, $I_{sc} = 3.72$ mA

4.28  2.4 k$\Omega$

4.34  $v_o = -11v_{in}$

## Chapter 5

5.1   a.  $(-5 \sin 100t + 100 \cos 100t)e^{-5t}$ A
       b.  $10 \cos (50t + 30°) - 500t \sin (50t + 30°)$ A

5.3   a.  $2.1 \times 10^{-28}$ J      b.  6.05 mJ

5.5   a.  2.32 $\mu F$      b.  $i_s(t) = 2.32 \cos t \ \mu A$,
                                 $v_o(t) = .03 \sin t$ V

5.10  a.  1.5 J      b.  1.33 J

5.12  a.  $(-105 + 50t)e^{-10t}$ V, $t > 0$
       d.  $-2\pi \sin 2\pi t$ V, $t > 0$

5.14  a.  $4.19 \times 10^{-44}$ J      b.  .09 J      c.  1.25 mJ

5.16  a.  .77 mH      b.  $v_1(t) = -.77 \sin t \ \mu V$,
                          $i_o(t) = .03 \cos t$ mA

5.21  a.  $R_{in} = 50$ k$\Omega$, $v_o(t) = 33 \cos 1000t$ mV

5.22  b.  $v_o = -RC \dfrac{dv_{in}}{dt}$

## Chapter 6

6.1  a. $5e^{-10t}$, $t>0$    b. 1.84    c. .0693 s

6.3  b. $(1+10t)e^{-10t}$, $t>0$

6.4  b. .17 S

6.6  a. $v_0(t) = e^{-t/.267}$ V, $i_1(t) = .25e^{-t/.267}$ mA

6.7  a. $i(t) = 2e^{-40,000t}$ mA,
$v_1(t) = -.4e^{-40,000t}$ V

6.10  $R = 4\,\Omega$, $C = 50\,\mu$F

6.13  $L = .5$ H, $C = 2\,\mu$F, $v_C(0) = 8$ V,
$i_L(0) = -16.02$ mA

6.16  a. $v_C(t) = -1.57e^{-112t} + 3.57e^{-893t}$ V, $t>0$
$i_L(t) = .14e^{-112t} - .04e^{-893t}$ A, $t>0$

6.20  a. $2.34e^{-t} - .44e^{-10t} + .11e^{-20t}$, $t>0$

6.22  a. $-2.84e^{-2t} + .71e^{-8t} + 3.13$, $t>0$

6.24  $i_L = 11.67$ mA, $v_1 = 6.67$ V

6.31  $\dfrac{d^2y_2}{dt^2} + \dfrac{60}{9}\dfrac{dy_2}{dt} + \dfrac{50}{9}y_2 = \dfrac{10}{9}$

6.34  a. $v_C(t) = 8(1 - e^{-166.67t})$ V, $t>0$

6.36  a. .25 mA    b. $.25e^{-3.75t}$ mA, $t>0$

6.38  a. $i_R(0) = 2$ mA, $\dfrac{di_R(0)}{dt} = -.15$ A/s

## Chapter 7

7.2  a. $2e^{-t/.2}$ V, $t>0$    c. $-2e^{-t/.2} + 4$ V, $t>0$

7.4  $v_C(t) = -13.33e^{-t/2.5} + 20$ V, $t>0$
$i_C(t) = 5.33e^{-t/2.5}$ mA, $t>0$
$i_R(t) = 1.33e^{-t/2.5} - 2$ mA, $t>0$

7.6  a. $i_L(t) = 2e^{-t/15\times10^{-6}}$ mA, $t>0$
c. $i_L(t) = -8e^{-t/15\times10^{-6}} + 10$ mA, $t>0$

7.8  $i_L(t) = 9.5e^{-t/10^{-5}} + .5$ mA, $t>0$
$v_L(t) = -9.5e^{-t/10^{-5}}$ V, $t>0$
$v_R(t) = -9.5e^{-t/10^{-5}} - 2$ V, $t>0$

7.10  $v_R(t) = -1.41e^{-29.4t} + 8$ V, $t>0$

7.12  $v_R(t) = 5 - 3e^{-2t} + 3e^{-4998t}$ V, $t>0$

7.14  $v_C(t) = 12.5e^{-.1t} - (1.25\times10^{-6})e^{-10^6t}$ V, $t>0$
$i_L(t) = -1.25e^{-.1t} + 1.25e^{-10^6t}$ mA, $t>0$
$i_R(t) = -.69e^{-.1t} + .69e^{-10^6t}$ mA, $t>0$

7.16  $i_L(t) = 71 + e^{-.018t}(179 \cos .707t$
$+ 4.6 \sin .707t)$ mA, $t>0$
$v_C(t) = -.127e^{-.018t} \sin .707t$ V, $t>0$
$v_R(t) = .714 - .036e^{-.018t} \sin .707t$ V, $t>0$

7.20  $v_C(0^+) = 12.5$ V, $i_L(0^+) = 0$ A,
$i_R(0^+) = 0$ A
$dv_C(0^+)/dt = 0$ V/s, $di_L(0^+)/dt = -1250$ A/s,
$di_R(0^+)/dt = 693.75$ A/s

7.22  $v_C(0^+) = 0$ V, $i_L(0^+) = 3.33$ mA,
$i_R(0^+) = 1.12$ mA
$dv_C(0^+)/dt = -3.33$ V/s, $di_L(0^+)/dt = 0$ A/s,
$di_R(0^+)/dt = -3.73$ mA/s

7.24  $i_L(t) = 5.33e^{-.041t} - 5.33e^{-.979t} + 5$ A, $t>0$
$v_C(t) = 51.12e^{-.041t} - 1.12e^{-.979t} + 50$ V, $t>0$

7.28  $v_C(t) = 100 - 98.33e^{-t/.113}$ V, $t>0$

7.30  $v_C(t) = 4.12e^{-126.28t} - .05e^{-9898.72t}$
$- .07e^{-500t}$ V
$i_L(t) = -4.17e^{-126.28t} + 5.11e^{-9898.72t}$
$+ .07e^{-500t}$ mA

7.32  a. $i_L(t) = (1 - e^{-(t-3\times10^{-6})/.5\times10^{-6}})$
$\times u[t - (3\times10^{-6})]$ mA

7.36  $i_L(t) = 1.67 + 1.67e^{-t/16.67\times10^{-6}}$ mA,
$0 < t < .05$ ms
$i_L(t) = 1.75e^{-(t-.05\times10^{-3})/.15\times10^{-3}}$ mA, $.05$ ms $< t$

## Chapter 8

8.2  a. $10 \sin\left(\dfrac{\pi}{1.3}\cdot10^6t - 96.9°\right)$

b. $5 \cos\left(\dfrac{\pi}{3}\cdot t - 60°\right)$

8.4  a. $y$ lags $x$ by 155°.    c. $y$ leads $x$ by 160°.

8.6  $v_0(t) = 10.73 \cos(100t - 153.4°)$ V

8.12  $i_L(t) = 1.12 \cos(5000t + 161°)$ A

8.20  $\mathbf{V}_C$ is largest for $0 < \omega < 1$.
$\mathbf{V}_R$ is largest for $1 < \omega < 10$.
$\mathbf{V}_L$ is largest for $\omega > 10$.

8.26  $\mathbf{V}_0 = 1.98e^{j172.8°}\mathbf{V}_i$,    $f = 10$ Hz
$\mathbf{V}_0 = 0.159e^{j94.5°}\mathbf{V}_i$,    $f = 1000$ Hz

8.30  $v_0(t) = 12.52 \cos(10t - 116.6°)$ V

8.32  $v(t) = 10.6 \cos(100t + 122°)$ V

**8.36** $I_{sc} = 1.26e^{-j71.6°}$ A

$Z_{eq} = 10 - j20$ Ω

# Chapter 9

**9.2** a. $A\dfrac{s}{s^2 + \omega^2}$    b. $A\dfrac{s\cos\theta + \omega\sin\theta}{s^2 + \omega^2}$

c. $Ae^{-(\theta/\omega)s}\dfrac{s}{s^2 + \omega^2}$

**9.10** a. $V(s) = \dfrac{4}{s(s+2)} - \dfrac{5}{s+2}$

d. $V(s) = \dfrac{-5}{(s+1)(s^2 + 2s + 101)}$

**9.11** a. $I(s) = \dfrac{3s}{s^2 + 6s + 8}$

d. $I(s) = 3\dfrac{s+3}{s^2 + 4s + 4}$

**9.12** a. $I_o(s) = \dfrac{-4s^2}{(s^2 + 1)(s + 2)} + \dfrac{4}{s+2}$

**9.13** b. $f(t) = [2t + 1 - e^{-2t}]u(t)$

c. $f(t) = \left[1 - \dfrac{5}{3}e^{-t} + \dfrac{5}{3}e^{-4t}\right]u(t)$

**9.18** a. $i(t) = [-3e^{-2t} + 6e^{-4t}]u(t)$

c. $i(t) = \left[-\dfrac{1}{3}te^{-t} - \dfrac{2}{9}e^{-t} + \dfrac{20}{9}e^{-4t}\right]u(t)$

**9.22** $F(s) = \dfrac{2}{s+1} + \dfrac{-5}{(s+2)^3} + \dfrac{-2}{(s+2)^2} + \dfrac{-2}{s+2}$

**9.28** a. $p(t) = 0.5r(t) - 0.5r(t-4) - 2u(t-4)$

$P(s) = \dfrac{1}{2s^2} - \dfrac{e^{-4s}}{2s^2} - \dfrac{2e^{-4s}}{s}$

$W(s) = \dfrac{P(s)}{1 - e^{-4s}}$

**9.32** $i(t) = \dfrac{5}{9}[e^{-2t} - e^{-20t}]u(t)$ A

**9.34** $i_o(t) = [-1.5e^{-t}\cos t + 2.5e^{-t}\sin t - e^{-2t}]$
$\times u(t)$ A

**9.38** $v_C(0^-) = -50$ V and $i_L(0^-) = 6$ A

**9.40** $\dfrac{V_C}{V_i} = \dfrac{1/sC}{R + sL + (1/sC)}$

**9.42** $\dfrac{V_o}{V_i} = \dfrac{s^2}{s^2 + s(1/R_2C_1 + 1/R_2C_2) + 1/R_1R_2C_1C_2}$

**9.43** a. $y(t) = \begin{cases} 0, & t < 0 \\ t^2/2, & 0 < t < 4 \\ 4t - 8, & t > 4 \end{cases}$

**9.44** Step response: $v_o(t) = [25e^{-2t} - 25e^{-3t}]u(t)$ V

# Chapter 10

**10.2** b. $x(t) = 2e^{-693t}$; $\omega = 0$, $\sigma = -693$

d. $x(t) = .5e^{.22t}\sin\dfrac{\pi}{2}t$; $\omega = \dfrac{\pi}{2}$, $\sigma = .22$

**10.5** a. $x(t) = Ke^t\cos(2t + \theta)$

c. $x(t) = A + Be^{-t}$

**10.7** a. $y(t) = 1.41e^{-t}\cos(t - 135°)$

c. $x(t) = 1.84e^t\cos(3t - 100.6°)$

**10.8** $i_R(t) = .895e^{-t}\cos(1000t - 116.58°)$ A

**10.11** c. $y(t) = .3e^{-t}\cos(100t - 130°)$

**10.13** $i_2(t) = .115e^{-t}\cos(2t + 165.3°)$ A

**10.16** $v_o(t) = -5e^{-t} + .47\cos(t + 45°)$ V

**10.18** $\dfrac{V_o}{V_s} = -\dfrac{R_2}{R_1}\dfrac{sR_1C_1 + 1}{sR_2C_2 + 1}$

**10.20** $\dfrac{V_2}{V_1} = \dfrac{1}{s^2R_1R_2C_1C_2 + s(R_1 + R_2)C_1 + 1}$

**10.24** b. $G(-1 + j) = 5.55e^{-j36.1°}$

c. $A(-1 + j) = 6.71e^{+j1.51°}$

**10.26** b. $s = 0$ and $s = -1000$

c. $s = -.5 + j\sqrt{3}/2$ and $s = -.5 - j\sqrt{3}/2$

**10.28** $v_o(t) = 5e^{-2t} - 5e^{-3t}$ V, $t > 0$

**10.32** $\dfrac{V_o}{V_i} = -\dfrac{s^2 - 16}{(s+2)(s+8)}$

**10.34** b. $v_o(t) = 100e^{-990t} - 100e^{-1000t}$ V

c. $v_o(t) = 1000te^{-1000t}$ V

**10.39** Stable for $K < 2$; overdamped for $1 < K < 2$; critically damped for $K = 1$; underdamped for $K < 1$

**10.41** $\dfrac{I_o}{I_s} = \dfrac{s}{s^2 + (1 + K)s + 1}$

## Chapter 11

11.1  b.  $v_o(t) = 68.6 \cos (2t - 30.96°)$ V

    h.  $v_o(t) = 4.62e^{2t}$ V

11.2  a.  $y(t) = 5 \cos (500t + 53.1°)$

    b.  $y(t) = 1.93 \cos (1000t - 68.8°)$

11.3  a.  $v(t) = 4.5 \sin (10t - 105°)$ V

11.9  $C = 1$ μF, $R = 31.8$ Ω (not unique)

11.12  a.  $\omega_0 = 1$ rad/s, $\omega_1 = .618$ rad/s,

       $\omega_2 = 1.618$ rad/s

    b.  $\omega_c = .786$ rad/s

    c.  $\omega_c = 1.272$ rad/s

11.23  Approximate cutoffs at

    $\omega = \omega_n (1 \pm 1/2Q)$ rad/s

11.26  d.  $R = 100$ Ω for $Q = 10$

11.33  $H(s) = \dfrac{s + .1}{(s + 1)(s + 10)}$

11.36  $\dfrac{V_o}{V_s} = -\dfrac{90.9s^2}{(s + 27.3)(s + 110)}$

## Chapter 12

12.4  a.  $a_0 = a_n = 0$,  $b_n = \dfrac{2A}{n\pi} \left(1 - \cos n \dfrac{\pi}{2}\right)$

    f.  $a_0 = 1/3$,  $a_n = 1/(n\pi)^2$,  $b_n = -1/n\pi$

12.9  a.  $f(x)$ is an odd function. Therefore,

    $a_0 = a_n = 0$.

    $b_n = \begin{cases} 0, & n \text{ divisible by 3} \\ -3/n\pi, & \text{otherwise} \end{cases}$

12.11  b.  $f(t) = \dfrac{1}{2} - \sum_{n \text{ odd}} \dfrac{4}{(n\pi)^2} \cos (n\omega_0 t + n\omega_0)$

12.13  b.  Odd harmonics only

    $a_n = -4/(n\pi)^2$,  $b_n = 2/n\pi$

12.16  b.  $c_n = \dfrac{e(\cos 2n\pi) - 1}{1 - j2n\pi}$

12.17  $f(t) = \dfrac{1}{T} \sum_{n=-\infty}^{+\infty} e^{jn\omega_0 t}$

12.23  b.  $f(t) = 2 + 2 \cos 2000\pi t$

      $+ 3 \cos (4000\pi t + 45°)$

      $+ 2 \cos (6000\pi t + 90°)$

      $+ \cos (8000\pi t + 45°)$

12.26  a.  $P(j\omega) = \dfrac{5}{4} \left(\text{sinc } \dfrac{\omega}{4}\right) - \dfrac{1}{2} \left(\text{sinc } \dfrac{\omega}{2}\right)$

12.28  a.  $F(j\omega) = \dfrac{-j\omega}{a^2 + \omega^2}$

12.35  $v_o(t) = \left[\dfrac{80}{21} e^{-4t} - \dfrac{80}{3} e^{-t} + \dfrac{160}{7} e^{-t/2}\right] u(t)$

## Chapter 13

13.2  b.  $X_{\text{ave}} = \dfrac{2A}{\pi}$,  $X_{\text{rms}} = \dfrac{A}{\sqrt{2}}$

13.6  a.  $X_{\text{rms}} = 5.1$  b.  $Y_{\text{rms}} = 8.03$

13.14  $p(t) = P + P \cos 2\omega t - Q \sin 2\omega t$ W

13.15  a.  $p(t) = 1.25 + 1.25 \cos 100t$

      $- 2.16 \sin 100t$ W

    c.  $p(t) = - .034 - .034 \cos 2000t$

      $- .094 \sin 2000t$ W

13.16  $p(t) = 750 + 750 \cos 100t$

      $+ 250 \sin 100t$ W

13.20  b.  $S = -85.5 + j234.9$ W

    d.  $S = 3.75 - j6.5$ W

13.23  a.  $Z = 35.28 + j35.99$ Ω,

      $S = 200 + j204$ W

    d.  $Z = 100 + j229.1$ Ω,

      $S = 400 + j916.5$ W

13.24  a.  $I_{\text{rms}} = .21$ A

13.26  a.  PF = .196   b.  $C = 26.5$ μF

13.28  With compensation, $I_s$ decreases from 1.41 to 1.056 A rms.

13.36  $P_{\text{max}} = 62.5$ W

## Chapter 14

14.5  a.  $V_1 = 5I_1 + \dfrac{25}{7} I_2$,  $V_2 = 5I_1 + 5I_2$

14.11  a.  $V_1 = 4.14I_1 + .29V_2$,  $I_2 = -.29I_1 + .43V_2$

    d.  $V_1 = \dfrac{s^2 + (3/2)s + 1}{(s + 1)(s + 1/2)} I_1$

      $- \dfrac{s/2}{(s + 1)(s + 1/2)} V_2$

$$I_2 = \frac{s/2}{(s+1)(s+1/2)} I_1$$

$$+ \frac{s^2 + 3s + 1}{2(s+1)(s+1/2)} V_2$$

14.14 $V_1 = 3I_1 + \dfrac{1}{2} I_2, \quad V_2 = 4I_1 + 10I_2$

14.16 $I_1 = \dfrac{1}{500} V_1, \quad I_2 = \dfrac{90}{500} V_1 + \dfrac{1}{1667} V_2$

14.20 a. $V_2 = \dfrac{5}{19}$ V    b. $V_2 = \dfrac{25}{14}$ V

14.22 $V_1 = \left(\dfrac{s+2}{2}\right)V_2 - I_2$

$$I_1 = \left[\frac{s(s+4)}{4}\right]V_2 - \left[\frac{s+2}{2}\right]I_2$$

14.26 a. $\mathbf{V}_1 = (2.45 + j2.9)\mathbf{I}_1 + (-.75 - j2.5)\mathbf{I}_2$
  $\mathbf{V}_2 = (-.75 - j2.5)\mathbf{I}_1 + (1.25 + j6.5)\mathbf{I}_2$
  b. $i_s(t) = 1.636 \cos(4t - 40.08°)$ A

14.28 For $v_s(t) = \cos t$ V:
  $v_o(t) = .0496 \cos(t + 82.6°)$ V (real transformer)
  $v_o(t) = .37 \cos t$ V (ideal)

14.33 $\dfrac{P_{out}}{P_{in}} = .67$

14.36 c. $Z_{in} = \dfrac{1}{g_{11} - \dfrac{g_{12}g_{21}}{g_{22} + Z_L}}$

14.38 $V_{in} = (h_{11} + g_{22})I_{in} + (h_{12} + g_{21})V_o$
  $I_o = (h_{21} + g_{12})I_{in} + (h_{22} + g_{11})V_o$

# Index